Signals and Systems

Signals and Systems

Theory and Practical Explorations with Python

Fatoş Tunay Yarman Vural
Middle East Technical University
Ankara, Turkey

Emre Akbaş
Middle East Technical University
Ankara, Turkey

WILEY

This edition first published 2025

Registered Offices
John Wiley & Sons, Inc., 111 River Street, Hoboken, NJ 07030, USA
John Wiley & Sons Ltd, New Era House, 8 Oldlands Way, Bognor Regis, West Sussex, PO22 9NQ, UK

For details of our global editorial offices, customer services, and more information about Wiley products visit us at www.wiley.com.

Wiley also publishes its books in a variety of electronic formats and by print-on-demand. Some content that appears in standard print versions of this book may not be available in other formats.

Library of Congress Cataloging-in-Publication Data:

Names: Yarman Vural, Fatoş Tunay, author. | Akbaş, Emre, author.
Title: Signals and systems : theory and practical explorations with Python / Fatoş Tunay Yarman Vural, Emre Akbaş.
Description: Hoboken, NJ : Wiley, 2025. | Includes bibliographical references and index.
Identifiers: LCCN 2024017990 (print) | LCCN 2024017991 (ebook) | ISBN 9781394215751 (hardback) | ISBN 9781394215768 (adobe pdf) | ISBN 9781394215775 (epub)
Subjects: LCSH: Signal theory (Telecommunication) | System analysis. | Python (Computer program language)
Classification: LCC TK5102.5 .Y37 2025 (print) | LCC TK5102.5 (ebook) | DDC 621.382/23 – dc23/eng/20240515
LC record available at https://lccn.loc.gov/2024017990
LC ebook record available at https://lccn.loc.gov/2024017991

Cover Design: Wiley
Cover Image: © MR.Cole_Photographer/Getty Images

Set in 9.5/12.5pt STIXTwoText by Straive, Chennai, India
Printed and bound by CPI Group (UK) Ltd, Croydon, CR0 4YY

C9781394215751_030125

The manufacturer's authorized representative according to the EU General Product Safety Regulation is Wiley-VCH GmbH, Boschstr. 12, 69469 Weinheim, Germany, e-mail: Product_Safety@wiley.com.

Dedicated to those,

who endeavor with love and wisdom

to better the world!

Contents

About the Authors

Fatoş Tunay Yarman Vural received her B.S. and M.S. degrees from the Technical University of Istanbul and Bogazici University, Turkey. She received her Ph.D. degree from Princeton University in 1981. She was an Assistant Professor at Drexel University and a research fellow at the Massachusetts Institute of Technology in the United States between 1985 and 1987. She joined the Department of Computer Engineering, Middle East Technical University as a faculty member in 1992. In 1996, she became the Department Chairperson until 2000. Between 2000 and 2008, she served as the Assistant President responsible for research and industrial relationships at Middle East Technical University. Dr. Yarman Vural organized dozens of workshops, national and international meetings, and conferences during her academic career.

Her research areas include neuroinformatics, machine learning, pattern recognition, computer vision, and image processing, where she has published more than 200 papers. She accomplished about 100 national and international research and industrial projects. She founded research centers at Middle East Technical University related to modeling, simulation, ethics, science and society, and e-government. She has founded various programs in Turkey for faculty development, research project development, science and society relations, distance education, and postdoctoral studies. She was the Turkish representative for research in NATO between 2004 and 2007. Dr. Yarman Vural is a member of IEEE Computer Society. She is also a member of Turkish Informatics Association, Turkish Intelligence Association, and Turkish Informatics Foundation.

Emre Akbaş is an Associate Professor at the Department of Computer Engineering, Middle East Technical University (METU). Prior to joining METU, he was a postdoctoral research associate at the Department of Psychological and Brain Sciences, University of California, Santa Barbara. He received his PhD degree from the Department of Electrical and Computer Engineering, University of Illinois at Urbana-Champaign in 2011. His BS and MS degrees are both from the Department of Computer Engineering, METU. Dr. Akbas's research received the Beckman Institute's Cognitive Science AI Award, "METU thesis of the year" award (three times), the Parlar Foundation Research Incentive Award, and the Young Scientist Award of Science Academy, Turkey. His research interests are in computer vision and deep learning with a focus on object detection and human pose estimation.

Preface

Through the course of civilization, humankind has been driven by a great curiosity to grasp the relationships between the signals observed in nature and the systems that orchestrate them. From the rhythmic motion of celestial bodies to the intricate processes of life, we humans develop tools and methodologies that both illuminate and empower us in our interactions within the universe of which we are an integral part.

This book presents the fundamental principles and methodologies for modeling natural phenomena and creating human-made systems by blending theory and practical exploration. The chapters provide insights into systems and their observable signals, going beyond mere abstract formulations. The readers are invited to appreciate studying rigorous mathematical concepts such as symmetry while they walk around the beautiful ceramic tile decorations of Alhambra Palace and to be inspired by the impossible tiles of Roger Penrose. They enjoy the musical harmony while they grasp the harmonically related complex exponential functions. They find themselves in the infinite dimensions of Hilbert spaces while they study the abstract concepts of Fourier series and transforms. They appreciate the aesthetics of formalism of intricate systems, which can be represented as an ensemble of simple systems interconnected by signals, revealing the inherent wholeness and implicate order.

This approach is reflected in the following topics covered in this book:

- Chapter 1 introduces the holistic approach of the book to modeling natural and human-made systems and their manifestations as signals. Mathematical foundations of signals and systems, together with their relationships, are provided.
- Chapter 2 explains the Lego functions, namely trigonometric functions, exponential functions, unit steps, and unit impulse functions. Properties of these basic functions, such as symmetry, harmony, continuity, and discreteness, are investigated. In the rest of the chapters, we assemble the Lego functions to construct more elaborate signals, which are observed both in nature and in human-made machinery.
- Chapter 3 explores the crucial properties of continuous time and discrete time systems, such as linearity, time invariance, stability, invertibility, and memory. These properties enable us to simplify the design and analysis of elaborate systems. Specifically, we model and implement systems with well-defined methodologies, when a system is linear and time-invariant (LTI). We also define some building blocks of LTI systems, such as adders, integrators, differentiators, delay operators, and scalar multipliers.
- Chapter 4 defines a unique function, called the impulse response, which represents linear time-invariant systems. We describe an essential operation, called convolution, which relates the input–output pair of an LTI system through the impulse response.

- Chapter 5 models continuous time and discrete time LTI systems by their dynamic nature using differential and difference equations. We study the solution to the differential and difference equations to investigate the relative rate of change between the input and output signals of an LTI system.
- Chapters 6 and 7 introduce continuous time and discrete time Fourier series representations in Hilbert spaces, respectively. A periodic function is represented as a vector in this infinite-dimensional space, where the spectral coefficients correspond to the coordinates of the function.
- Chapters 8 and 9 introduce continuous- and discrete time Fourier transforms for an aperiodic function, which satisfy a set of conditions, called Dirichlet conditions. We also extend the Fourier transform to the Laplace transform for continuous functions and to the z-transform for discrete functions, respectively.
- Chapter 10 explains how signals are reshaped by linear time-invariant filters in time and frequency domains, covering low-pass, high-pass, band-pass, and band-reject filters.
- Chapters 11 and 12 explain the pioneering sampling theorem of Claude Shannon for continuous time and discrete time signals and systems, which bridges the continuous world to the discrete world, opening the door to the digital era.

The chapters of the book are crafted by a rigorous formalism with a wide range of practical exercises and problems.

The theoretical content of the book is enriched by Python code snippets, providing practical implementation of key concepts. For example, elementary signal operations for time scale, time reverse, and translation are implemented, where the students plug and play a wide range of parameters to observe the operations on the functions. Also, students can run the Python codes for convolution, Fourier transforms and series, sampling, and interpolation for a wide range of functions with different sets of parameters. The companion website of the book enables the students to interact with systems and signals by changing the parameters of functions, providing a rich educational experience.

Some of the concepts behind the classic formulations are explained and interpreted with their historical progresses. For example, the origin and meaning of Euler's number, the rationale behind complex numbers, the impact of complex exponentials as a building block of many complicated functions, the meaning of convolution operations, the concept of harmony, and the impact of the sampling theorem on the development of digital era are explained in a simple nontechnical language along with a mathematical formalism.

The book is designed as a one-semester (14 weeks with 3 lecture hours per week) undergraduate course material for science and engineering students, specifically computer science and engineering, electrical and electronics engineering, physics, informatics, and applied mathematics. Each chapter can be covered in 3 lecture hours per week, excluding Chapters 8 and 9, which can be split into two weeks each. For shorter semesters, the instructors can skip the mathematical preliminaries of Chapter 1. Laplace and z-transforms of Chapters 8 and 9 and Chapter 10 in full can also be omitted.

The students are expected to gain the basic knowledge and skills, which will be very useful for their future education and research in their undergraduate and graduate studies in the field of communication, computer vision, machine learning, signal, and image and video processing, among others.

Bridging the gap between theory and practice, the book also serves to the professionals, working in the above disciplines and a wide range of multidisciplinary fields, including bioinformatics, robotics, neuroscience, remote sensing, aeronautics, seismology, biomedical engineering, process control, astrophysics, cosmology, energy, and mechatronics.

Ironically, we owe this book to COVID-19 pandemic, where we ceased the class lectures and switched to online education. At that time, we urgently needed some course materials for our students to conduct the online education. We rapidly converted our handwritten notes, accumulated over years of experience, into digital files, adding a few animations and videos along the way. On the way, we crafted numerous examples and figures with the invaluable assistance of our students and colleagues. After five years of dedicated effort, we completed this book, which is now available in both PDF format with a hardcover option and for e-book readers. Additionally, the book is complemented by a companion website offering interactive materials, where readers can visualize functions, view videos, and execute Python programs. This is a brief history of our journey on the production of this book.

Now, we invite you to join us on this journey of exploring and designing systems and their expressions through signals, with the interplay of theory, practice, aesthetics, and curiosity.

Ankara
May 2024

Fatoş Tunay Yarman Vural
Emre Akbaş

Acknowledgments

Our journey in crafting this book was a dynamic process, commencing with our handwritten notes, which we diligently transcribed into Markdown and then converted into HTML pages. Subsequently, we transitioned to interactive Jupyter notebook files before ultimately transforming the content into LaTeX format alongside the creation of a companion website. This back-and-forth iterative process involved continuous refinement, with both new content additions and existing content revisions. We would not be able to overcome all the challenges without the assistance of our students. We extend our sincere gratitude to the students who contributed to the creation of this textbook. Their dedication to excellence is reflected in the high-quality figures, meticulous typesetting, and adept problem-solving skills they exhibited throughout the project. This book would not have been realized in a short amount of time without their support.

We would like to thank our students Anıl Eren Göçer, Baran Yancı, Can Ufuk Ertenli, Kıvanç Tezören, Robin Koç, and Zafer Bora Yılmazer for their general assistance with technical issues, typesetting, and creation of figures. Their dedication and enthusiastic participation were invaluable.

We have a long list of students to thank for Matplotlib figures and TIKZ drawings: Emin Sak, Hakan Gürsoy, Meriç Buğra Haliloğlu, Kenan Kartal, Deniz Eren Yılmaz, Arda Çavuşoğlu, Adil Ahmadli, Buket Zeren, Görkem Yiğit, Çağla B. Çam, and Tuğba Tümer. We thank Furkan Genç and Tahira Kazimi for creating the initial typesets of LaTeX tables.

Special acknowledgment is due to Güneş Sucu, Can Ufuk Ertenli, and Çağlar Seylan for their assistance in the initial transcribing of handwritten notes to Markdown format, and to Ulaş Aydın, Alpay Özkan, Ali Doğan, and Ömer Köse for their invaluable aid in resolving typesetting challenges across various Markdown and LaTeX setups. We thank Oğuz Gödelek and Yavuz Durmazkeser for their assistance with Python programming problems.

Our students' support extended to reviewing the content for typos, grammatical errors, and mathematical errors, for which we thank Gökçen Gökçeoğlu, Fırat Ağış, Elif Sena Kuru, Kıvanç Filizci, Defne Ekin, and Syed Ebad Hyder. We would like to extend our gratitude to all of the students who played a part in creating this book. While we have endeavored to acknowledge each of them by name, it is possible that we may have inadvertently overlooked some. We wish to emphasize that any omissions were entirely unintentional, and we express our sincere gratitude for the support and assistance of all who have contributed.

We are grateful to Göktürk Üçoluk for the icons that we use to denote interactive content (video or code). He designed these icons and generously donated them to our book.

We appreciate the supportive environment fostered by our colleagues in the Department of Computer Engineering at Middle East Technical University, particularly Sinan Kalkan and Göktürk Üçoluk, who offered invaluable guidance and shared their experiences in book writing.

We also acknowledge the invaluable assistance provided by Becky Cowan, Sandra Grayson, Sundaramoorthy Balasubramani, and Nandhini Karuppiah from Wiley, whose expertise and guidance were essential in shaping the content.

Additionally, we thank our PhD advisors, Bradley William Dickinson and Narendra Ahuja, for their mentorship and influence on our development as scientists and authors.

Finally, we extend our heartfelt gratitude to our beloved family members, Huseyin Vural, Eren Karaca Akbaş, Derviş Can Vural, Ekin Akbaş, Tolga Yarman, Ayşe Yarman Öztekin, Binboğa Sıddık Yarman, and Faruk Ağa Yarman. Their unwavering support and understanding throughout the book-writing journey have been invaluable.

This textbook stands as a testament to the collective efforts and contributions of all those involved.

Fatoş Tunay Yarman Vural
Emre Akbaş

About the Companion Website

This book is accompanied by a companion website:

www.wiley.com/go/signalsandsystems

This website offers interactive materials, where readers can visualize functions, view videos, and execute Python programs.

1

Introduction to Systems and Signals

"There is nothing more practical than a good theory!"

Vladimir Vapnik

This book is about the mathematical representation of systems and signals.

Let us start by describing the meaning of the words **systems** and **signals**.

The origin of the word **systems** dates back to 15th century, when it was used as a Latin word **systema**, which means the entire universe. Since then, this very wide meaning has narrowed to a set of connected items or devices that operate together. In the context of this book, **a system can be defined as a unified collection of interrelated and interdependent parts**. And in many cases, it is more than the summation of its parts.

The aforementioned definition is quite flexible and may cover both natural or human-made systems. It can be as large as a planet, a star, or a galaxy, or as small as a single cell, a molecule, or a microchip.

In this book, we shall use the **systems approach** to model, analyze and investigate the natural systems, and design and implement human-made systems.

Motivating Question: What is the systems approach?

The systems approach is a holistic paradigm to mathematically represent a system. Holism is the philosophy, which accepts a system as a whole, not only as a collection of its parts. It is the opposite of the reductionist paradigm, which assumes that a complex system can be represented by its simpler components. For example, in a reductionist paradigm, a puzzle can be represented by the collection of its pieces, which come in a box. However, when we turn the box of puzzle over a table, we see all the pieces, but we cannot perceive the theme of it. On the other hand, in the systems approach, we need to do the puzzle and look at the ordered puzzle to see that it consists of a picture (Figure 1.1).

In order to model a system using the holistic paradigm, we not only represent the attributes of its multiple components, but also formulate their inter-relationships, considering the objective of the entire system. This approach implicitly models the synergy created by a system.

WATCH

Learn more about the systems approach @ https://384book.net/v0101

Signals and Systems: Theory and Practical Explorations with Python,
First Edition. Fatoş Tunay Yarman Vural and Emre Akbaş.

Companion website: www.wiley.com/go/signalsandsystems

Figure 1.1 *Waterfall* by M.C. Escher.[1] The puzzle on the left consists of the pieces of the entire lithograph, but have no meaning. In order to observe the falling water of the watermill, we need to solve the puzzle. Source: The M.C. Escher Company B.V/https://mcescher.com/gallery/impossible-constructions/#iLightbox[galleryimage1]/5/last accessed March 09, 2024.

The origin of the word **signal** is even older than that of systems, dating back to 13th. century. It comes from the Latin word **signale**, which means **anything that serves to indicate or communicate information**.

When we observe a signal, we assume that there is a source system, which generates the signal. Thus, signals can be considered as partial information about the systems. In most cases, systems can be modeled and represented by a collection of subsystems. The interrelations among the subsystems of a system can be modeled by the received input signal(s) and the generated output signal(s), i.e., signals, of each subsystem.

In summary, the response of a system to a specific set of input signals provides information about the properties of systems. Signals describe the interrelations among the parts of a system. Loosely speaking, signals are the measurements of our varying observations about a system and/or its parts.

1.1 Example Applications

Models for representing signals and systems are widely used in electrical engineering and computer science for **filter design, control, communications, computer vision, machine learning, speech, image**, and **video processing.** The formalism of signals and systems is also used in a wide range of multidisciplinary areas, including **bioinformatics, robotics, neuroscience, remote sensing, aeronautics, seismology, biomedical engineering, chemical process control, energy and mechatronics, astronomy**, and **cosmology**.

Let us give some examples, where the methodologies of the systems approach are intensively utilized, in the modeling, design, and implementation stages of natural and human-made systems. Most of these models are generated by using the signals measured at the input and/or output of the systems.

---

1 https://mcescher.com/gallery/impossible-constructions/#iLightbox[gallery_image_1]/5.

Figure 1.2 Digital terrain model of a transportation area obtained from LIDAR scanning. Source: black_mts/Adobe Stock.

1.1.1 Three-Dimensional World Models by LIDAR Signals

Light detection and ranging (LIDAR) signals are generated by a source that emits laser beams. These signals bounce off the surrounding objects and return to a sensor. Systems approach is, then, used to create a three-dimensional representation of the physical environment by measuring the elapsed time for each laser pulse to return to the sensor (Figure 1.2).

WATCH Learn more about the LIDAR example @ https://384book.net/v0102

1.1.2 Modeling the Brain Networks from the Brain Signals

Functional magnetic resonance imaging (fMRI) technique records the brain signals, which indirectly measure the activities in the anatomical regions. It is possible to model and analyze the cognitive processes, such as vision, speech, and memory of the human brain from the fMRI signals.

Representing brain activities by networks is crucial to understand various cognitive states. It is possible to extract brain networks from the fMRI recorded while the subjects perform a predefined cognitive task. Figure 1.3 shows two brain networks for planning and execution phases, while the subject solves a complex problem. The suggested computational network model can successfully

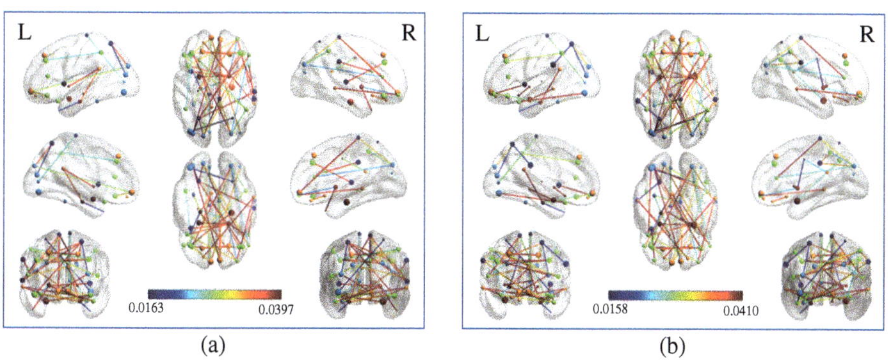

Figure 1.3 Visualizing anatomical regions during both the planning (a) and execution (b) phases while the selected subject solves a complex problem.[2] Source: With permission of IEEE.

2 Yarman Vural and Değirmendereli (2020).

Figure 1.4 Example of building detection using remote sensing applications.[3] Source: With permission of IEEE.

discriminate the planning and execution phases of complex problem-solving process with more than 90% accuracy, when the estimated dynamic networks, extracted from the fMRI data, are classified by a machine learning algorithm.

WATCH An example: speech synthesis from neural decoding of spoken sentences @ https://384book.net/v0103

1.1.3 Detecting the Buildings from the Remote-Sensed Satellite Images

Remote sensing images are recorded by measuring the signals of several electromagnetic waveforms reflected from the earth's surface. These signals are used to extract various information, such as measuring environmental pollution or climate change, the growth rate of cities or green areas, etc.

One important application of remote sensing is to detect the buildings in municipalities. For this purpose, a multidimensional signal measured from the earth's surface is modeled to filter the buildings in the remotely sensed data, as shown in Figure 1.4.

1.1.4 Noise Reduction in Old Records

Due to the technological limitations of their time, the old gramophone recordings are mostly noisy. These recordings can be cleaned by using methodologies of signal processing. Additive noise is partially eliminated by estimating a mathematical model for the noise and subtracting it from the corrupted signal. An example of noise reduction can be found in the companion website of the book.

WATCH Noise reduction on "O'sole mio" @ https://384book.net/v0104

1.2 Relationship Between Signals and Systems

The aforementioned brief descriptions and examples of signals and systems show that there is a remarkable relationship between the signals and the underlying system, which generates the signal. Philosophically, one may consider the signals as the manifestation of systems. We, humans,

3 Senaras et al. (2013).

can perceive the physical world through these manifestations. **Heraclitus of Ephesus** summarizes this view by his famous saying:

τα παντα ῥεῖ (ta panta rhei),

which translates to English as:

All flows!

Almost 2500 years ago, Heraclitus claimed that everything changes. Since then, as we study the nature, we discover some invariant laws, which lie behind the changes. Although we can only perceive the world of flux, these invariant laws govern our changing observations. In other words, we can only perceive variances, generated by the invariant laws, which govern the natural systems. Our aim is to find these invariant laws, manifested through our varying observations.

To analyze and understand a natural system or design and implement a human-made system, we need rigorous mathematical representations of signals, which correspond to our varying observations. Based on these observations, we can model the invariant rules of a system, which administers a set of prescribed tasks.

Motivating Question: How can we analyze and understand the laws that govern the natural systems? How can we design a human-made system to achieve a specific goal?

The answers to these questions require mathematical representation of systems and/or their subsystems. To follow the holistic approach, we also need mathematical representation of signals, which describe interrelations among the subsystems and the interaction between a system with its environment.

1.3 Mathematical Representation of Signals and Systems

There are many ways to formally represent signals and systems. In the context of this book,

- signals are represented by **functions**,
- systems are represented by **equations and/or algorithms**.

Loosely speaking, a system receives an input signal, represented by a function $x(t)$, and generates an output signal, represented by a function $y(t)$, for this particular input. The relationship between the input and output signals provides us with system equation or algorithm (Figure 1.5).

Throughout this book, we shall study the signals by representing and manipulating them with well-known mathematical objects, namely functions. We shall, also, study the systems by establishing the relationship between the input and output signals using a class of equations, namely the differential and integral equations. We pay special attention to linear systems, not only because

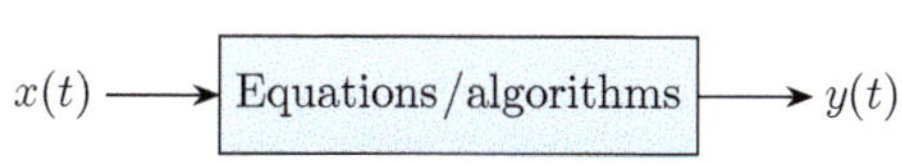

Figure 1.5 Schematic representation of a system by a box, which consists of an equation or an algorithm. A system receives an input signal, $x(t)$ and generates an output signal, $y(t)$. The equation or algorithm relates the input signal to the output signal.

of their mathematical tractability, but also because they open the door to analyze and design a wide range of nonlinear systems.

In the rest of this chapter, we shall provide a brief overview of functions to represent and manipulate signals. We shall, also, study a very interesting property of functions, called **symmetry**.

1.3.1 Signals Represented by Functions

Let us start by recalling the definition of a function. A function associates the elements in a domain set and the elements in a range set. Formally, a function, x, maps the values in the domain to values in the range,

$$x : D \to R, \tag{1.1}$$

where D is the domain set and R is the range set. Therefore, a function is represented by a triplet,

$$(D, R, x), \tag{1.2}$$

where x is a set of ordered pairs, $(d \in D, r \in R)$.

1.3.2 Types of Signals

The elements of the domain and range sets define the type of the signals. The domain and range sets may consist of multi-dimensional vectors, with real-, complex-, or integer-valued entries. In other words, the domain of the function, x, can be an m-dimensional vector with entries, defined over the set of real numbers, integer numbers, or complex numbers,

$$d \in \mathbb{R}^m \quad \text{or} \quad d \in \mathbb{I}^m \quad \text{or} \quad d \in \mathbb{C}^m, \tag{1.3}$$

respectively. Note that, the sets, $\mathbb{R}^m$ and $\mathbb{C}^m$ form a vector space over the field of real numbers and complex numbers, respectively. However, the set of integers, $\mathbb{I}^m$ is not closed under scalar multiplication, thus it is not a vector space.

Similarly, the range of a function can be defined as an n-dimensional vector, with entries defined over the set of real numbers, integer numbers, and complex numbers,

$$r \in \mathbb{R}^n \quad \text{or} \quad r \in \mathbb{I}^n \quad \text{or} \quad r \in \mathbb{C}^n, \tag{1.4}$$

respectively.

When the dimension of the domain, $m > 1$, the function, x, is called a **multivariate function**. For $m = 1$, the function, x, is called a **univariate function.**

In this book, we focus on univariate functions, where the domain variable is either real or integer-valued **scalar time measures**. When the domain variable is real, we indicate the time measure by $t \in \mathbb{R}$ and the corresponding function by $x(t)$. When the domain variable is integer-valued, we indicate the time measure by $n \in \mathbb{I}$ and the corresponding function by $x[n]$. The elements of the range can also be real, complex, or integer-valued. Depending on the elements of the domain and range, we define the following types of signals.

1) **Continuous Time Signals**: Continuous time signals are represented by continuous functions, where both the domain, t and the range, $x(t)$, consists of real numbers, i.e.,

$$t \in \mathbb{R}, \quad x(t) \in \mathbb{R}. \tag{1.5}$$

As an example, a continuous time sinusoidal signal is represented by the following function:

$$x(t) = A\sin(\omega_0 t), \tag{1.6}$$

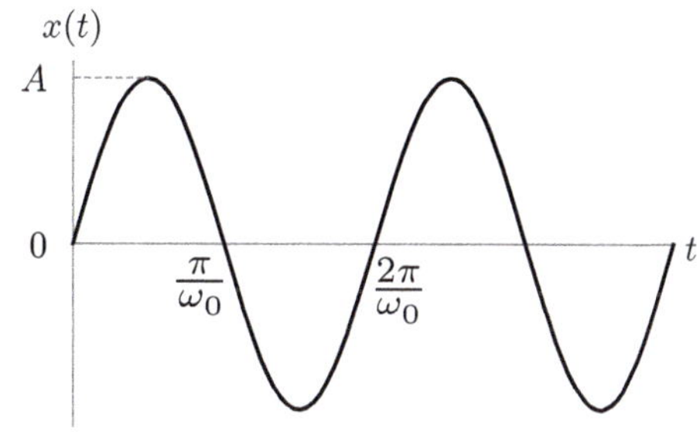

Figure 1.6 Plot of the continuous time function, $x(t) = A\sin(\omega_0 t)$.

where the time $t \in \mathbb{R}$ is the domain and $x(t) \in \mathbb{R}$ is the range. The parameter A is called the **amplitude** of the sinusoidal function (Figure 1.6).

Note that the range of the sinusoidal function is **bounded** by its amplitude, i.e., $-A < x(t) < A$, where the amplitude, $A \in \mathbb{R}$, is a finite number. The signal, represented by the sinusoidal function of Equation (1.6) repeats itself at every time instance, $T = 2\pi/\omega_0$, where T is called period.

2) **Discrete Time Signals**: Discrete time signals are represented by discrete functions, where the domain variable, n, is an integer number and range, $x[n]$, is a real number, i.e.,

$$n \in \mathbb{I}, \quad x[n] \in \mathbb{R}. \tag{1.7}$$

As an example, a discrete time sinusoidal signal can be represented by the following function:

$$x[n] = A\sin(\Omega_0 n). \tag{1.8}$$

The discrete time function of Equation (1.8), repeats itself at every integer time period, $N = 2\pi/\Omega_0$. An example is illustrated in Figure 1.7 for $N = 25$. Note that the range of a discrete time function consists of the vector space of real numbers, $x[n] \in \mathbb{R}$. Thus, all the values in the range are well defined. On the other hand, the domain, $n \in \mathbb{I}$, does not form a vector space. Since the values between two integers are **not** defined, we cannot perform scalar multiplication operations in the domain. This fact requires special attention, when we deal with discrete time signals. For example, when we multiply the time variable, n of a function, $x[n]$ by a rational number, the resulting time instances may not yield integer values. In such cases, the range of the function is not defined. As a result, if the period N is not integer-valued, the integer multiples of the period, kN of the domain of a sinusoidal function becomes undefined. Thus, the function does not satisfy the periodicity property.

Discrete time signals can be inherently discrete or they can be obtained by quantizing the domain of a continuous time signal. The signal of Figure 1.7 can be obtained from its continuous counterpart, $x(t) = \sin(\frac{2\pi}{25}t)$ by quantizing the domain of t. On the other hand, inherently

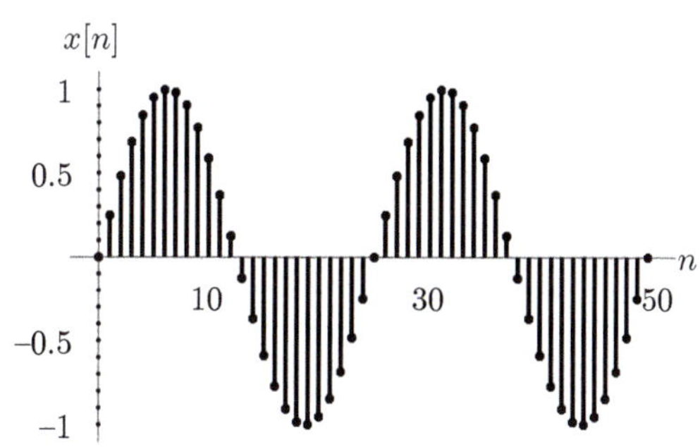

Figure 1.7 Plot of the discrete time sinusoidal function, $x[n] = \sin(\frac{2\pi}{25}n)$. Note that the period is $N = 25$.

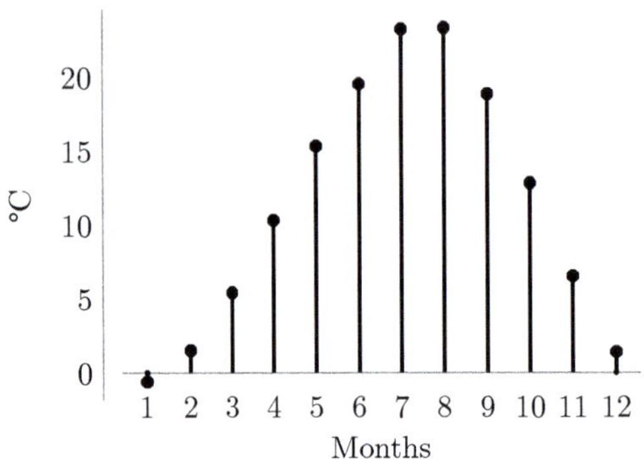

Figure 1.8 Average monthly temperature in Ankara Source: Adopted from https://en.climate-data.org.

discrete signals can be obtained by recording the data at specific time instances. Popular examples include meteorological data (temperature, humidity, etc.) recorded on a daily basis and economic or social indicators (growth rate, population, disease distributions, etc.) of countries on a yearly basis. For example, the signal showing average monthly temperature values in Figure 1.8 is inherently discrete.

3) **Digital Signals**: Digital signals are represented by range-quantized discrete functions. Thus, both the domain and range of the digital signals consist of integer numbers, i.e., $n \in \mathbb{Z}$ and $x[n] \in \mathbb{Z}$, i.e.,

$$d \in \mathbb{Z}, \quad r \in \mathbb{Z}. \tag{1.9}$$

Digital signals can be obtained by quantizing the domain and range of a continuous time or the range of the discrete time signal. On the other hand, some signals are inherently digital. As an example, we can measure the existence and nonexistence of an event on a timely basis by a binary function, such as daily records of rain and no rain. As another example, Figure 1.9 shows a digital signal obtained by quantizing the domain and the range of a continuous time signal into eight quantization levels, in the interval $[0, 7]$.

There is a bridge, which relates the continuous time signals to discrete time signals and/or digital signals through the famous **Sampling Theorem of Claude Shannon**. This bridge will be established in the last two chapters of this book.

The modern information and communication technology is based on digital signals and systems. Even if a signal is inherently continuous, it is digitized prior to processing by a digital computing device. After the process is completed, the digital signal may be converted to its continuous time counterpart, if necessary.

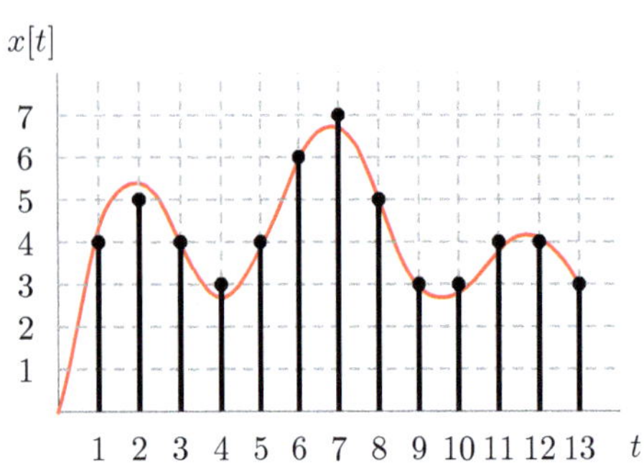

Figure 1.9 The digital signal (black) is obtained by quantizing the range of a continuous time signal (red) into 8 quantization levels and that of the domain into 13 levels.

INTERACTIVE

Explore different types of signals @ https://384book.net/i0101

1.3.3 Energy of a Signal

An important characteristic of signals is the concept of energy. In physics, energy is a quantitative property, which is transferred to an object to perform a work. Although the energy of a signal is somewhat related to the energy of a physical system, there is no one-to-one correspondence between the energy in physics and the following definitions. However, it is customary to use the term energy in many fields, including signal processing.

The energy of a signal provides us with useful information about the mathematical tractability and realizability of the signals and systems, during the design and implementation phases.

Definition 1.1 In continuous time, the **energy of a signal** in the interval of $[t_1, t_2]$ is defined as follows:

$$E_{[t_1,t_2]}(x(t)) = \int_{t_1}^{t_2} |x(t)|^2 dt. \tag{1.10}$$

In discrete time, the energy of the signal in the interval of $[n_1, n_2]$ is defined as follows:

$$E_{[n_1,n_2]}(x[n]) = \sum_{n=n_1}^{n_2} |x[n]|^2. \tag{1.11}$$

Definition 1.2 **Total energy of a signal** for continuous time and discrete time signals are defined as:

$$E(x(t)) = \int_{-\infty}^{\infty} |x(t)|^2 dt \tag{1.12}$$

and

$$E(x[n]) = \sum_{-\infty}^{\infty} |x[n]|^2, \tag{1.13}$$

respectively.

The definitions of Equations (1.12) and (1.13) reveal that the energy of a signal is the area under the squared magnitude of the corresponding function. The amount of this area gives us important information about the characteristics of a signal. If the area is large, then we suppose that the signal consists of very large amplitudes. In some cases, the area may approach to infinity, which makes the signal mathematically intractable and physically unrealizable in real life applications.

Definition 1.3 A signal $x(\cdot)$ is called an **energy signal** if its total energy is finite, i.e., $E(x) < \infty$.

Energy signals are absolutely summable for discrete time, and absolutely integrable, for continuous time signals. This property brings a bound to the signals and enables us to apply many tractable mathematical operations on the functions, which represent signals.

1.3.4 Power of a Signal

Time average of the total energy is called the power of a signal.

Definition 1.4 **Power of a signal** for continuous time and discrete time signals are defined as follows:

$$P(x(t)) = \lim_{T\to\infty} \frac{1}{2T} \int_{-T}^{T} |x(t)|^2 dt = \lim_{T\to\infty} \frac{E(x(t))}{2T} \tag{1.14}$$

and

$$P[x[n]] = \lim_{N\to\infty} \frac{1}{2N+1} \sum_{-N}^{N} |x[n]|^2 = \lim_{N\to\infty} \frac{E[x[n]]}{2N+1}, \tag{1.15}$$

respectively.

Definition 1.5 A signal x is called a **power signal,** if its power is nonzero and finite, i.e.,

$$P(x) \neq 0 \quad \text{and} \quad P(x) < \infty. \tag{1.16}$$

Note that, when both the power and energy of a signal are infinite, the signal is neither a power nor an energy signal. In practice, a power signal cannot exist in the real world because it would require a power source that operates for an infinite amount of time.

Question: Show that the periodic signals are power signals and the aperiodic signals, which are nonzero in a finite interval are energy signals.

INTERACTIVE

Explore elementary operations on signals @ https://384book.net/i0102

1.4 Operations on the Time Variable of Signals

Recall that signals can be represented by functions. Hence, arithmetic operations, including addition, division, and multiplication, can be systematically employed to generate novel signals from the existing ones.

Signals can be manipulated in many ways. One major way is to apply operations on the time variable. In other words, we change the domain of the function that represents the signal and investigate the range of the function with respect to the newly defined domain. These operations can be categorized under four headings:

1) Time shift,
2) Time reverse,
3) Time scale,
4) Time shift and scale.

These operations enable us to define new signals, based on elementary functions, such as periodic functions, to represent some real-life signals.

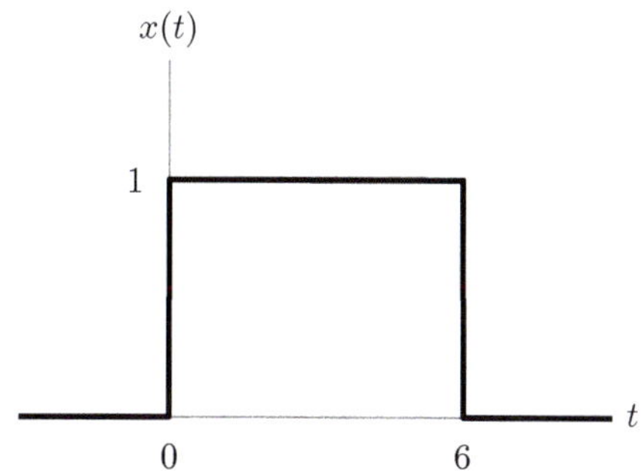

Figure 1.10 The plot of a continuous time pulse signal, which is nonzero in the interval [0, 6].

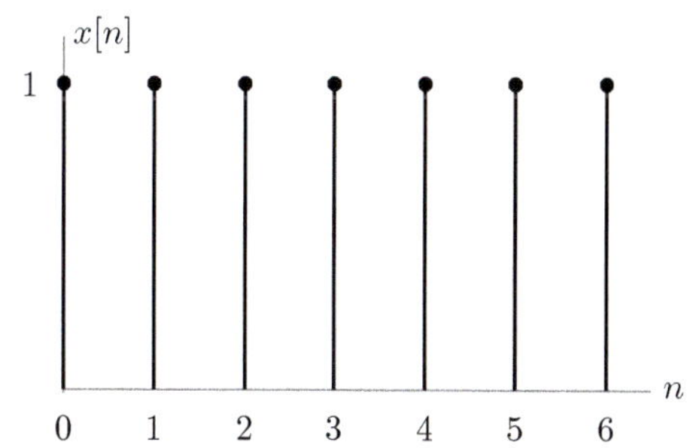

Figure 1.11 The plot of a discrete time pulse signal, which is nonzero in the interval [0, 6].

As an example, consider a simple signal, called a **pulse signal**, defined by the continuous time function in Figure 1.10. The analytical form of the pulse signal of Figure 1.10 is

$$x(t) = \begin{cases} 1, & \text{for } 0 \leq t \leq 6, \\ 0, & \text{otherwise.} \end{cases} \tag{1.17}$$

Discrete time version of this pulse signal is given in Figure 1.11.

The analytical form of the discrete time pulse signal of Figure 1.11 is given as follows:

$$x[n] = \begin{cases} 1, & \text{for } 0 \leq n \leq 6, \\ 0, & \text{otherwise.} \end{cases} \tag{1.18}$$

Let us define the operations on the time parameter of signals and apply these operations to the continuous and discrete time pulse functions, $x(t)$ and $x[n]$, defined in Equations (1.17) and (1.18), respectively.

1.4.1 Time Shift

Time shift operation for a continuous time signal replaces the time variable t of $x(t)$ by $t' = (t - T)$ to obtain $x(t') = x(t - T)$, where T is the amount of shift.

For the continuous time signal of Figure 1.10, time shift of $x(t)$ by the amount of T is given as:

$$x(t - T) = \begin{cases} 1, & \text{for } T \leq t \leq T + 6, \\ 0, & \text{otherwise.} \end{cases} \tag{1.19}$$

Similarly, time shift operation for a discrete time signal, replaces the time variable n of $x[n]$ by $n' = (n - N)$ to obtain $x[n'] = x[n - N]$, where N is the amount of shift.

For the discrete time signal of Figure 1.11, time shift of $x[n]$ by the amount of N is given as:

$$x[n - N] = \begin{cases} 1, & \text{for } N \leq n \leq N + 6, \\ 0, & \text{otherwise.} \end{cases} \tag{1.20}$$

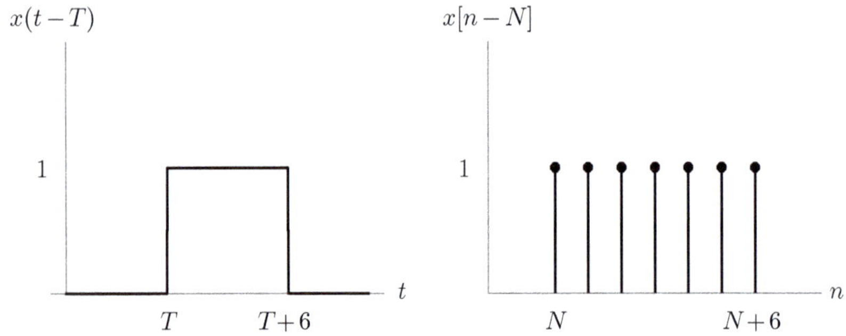

Figure 1.12 Shift of a continuous time pulse signal, $x(t)$, given in Figure 1.10 by the amount of $T > 0$ (a) and shift of a discrete time pulse signal, $x[n]$, given in Figure 1.11 by the amount of $N > 0$ (b).

Note that the amount of continuous time shift, T is a real number, whereas the amount of discrete time shift, N must be an integer number, for the time shift operation.

When $T > 0$, for continuous time signals, the time-shift operation **delays** the signal by T units in continuous time. The shift is toward the right on the time axis. Similarly, when $N > 0$, for discrete time signals, the signal is delayed N units in discrete time. Figure 1.12 shows the results of this time-shift operation.

For $T < 0$ and $N < 0$, the shift is toward the left on the time axis. This operation **advances** the signal by T units in continuous time and N units in discrete time. In practice, it is not possible to advance a signal in real time. We can only advance a signal if all of its previous values are recorded.

Question: Suppose that we are in a movie theater. What happens to the movie video, when we play it after a time shift operation for $T = 2$ hours?

If we assume that the start time of the movie is at $t = 0$, then there will be a delay of two hours for the movie to start.

1.4.2 Time Reverse

Time reverse operation of a continuous time signal changes the sign of the time variable t of the signal, $x(t)$, by $t' = -t$ to obtain $x(t') = x(-t)$.

For the signal of Figure 1.10, time reverse can be given as follows:

$$x(-t) = \begin{cases} 1, & \text{for } 0 \leq -t \leq 6, \\ 0, & \text{otherwise.} \end{cases} \tag{1.21}$$

Similarly, the time reverse operation of a discrete time signal changes the sign of the time variable n by $n' = -n$, as follows:

$$x[-n] = \begin{cases} 1, & \text{for } 0 \leq -n \leq 6, \\ 0, & \text{otherwise.} \end{cases} \tag{1.22}$$

Note that time reverse operation flips the signal with respect to the ordinate axis (Figure 1.13).

Question: Suppose that we are given a movie video. What happens to this video when we play it after the time reverse operation?

Time reverse operation makes the end of the signal, the start of the signal. In practice, there is no negative time. Thus, the reversed signal cannot start at $t = -6$. Hypothetically, if we assume that the movie starts at $t = -6$, we observe that the movie starts from the end and progresses toward the beginning, like Benjamin Button.

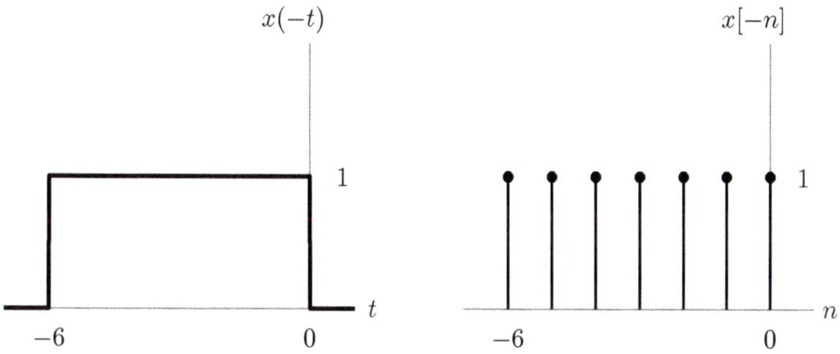

Figure 1.13 Time reversed versions of continuous time and discrete time pulse signals, given in Figures 1.10 and 1.11.

1.4.3 Time Scale

Time scale operation of a continuous time function replaces the time variable, t of $x(t)$ by $t' = at$ to obtain $x(t') = x(at)$, where a is a time scale parameter.

For the continuous function, $x(t)$ of Figure 1.10, time scale by the amount of a is given as follows:

$$x(at) = \begin{cases} 1, & \text{for } 0 \leq at \leq 6, \\ 0, & \text{otherwise.} \end{cases} \tag{1.23}$$

Time scale of a discrete time function by a **scalar,** a, replaces the time variable, n, by $n' = an$. For the discrete function, $x[n]$ of Figure 1.11, time scale by the amount of a can be given as follows:

$$x[an] = \begin{cases} 1, & \text{for } 0 \leq an \leq 6, \\ 0, & \text{otherwise.} \end{cases} \tag{1.24}$$

Time scale operation, either squishes or stretches a signal, depending on the value of the multiplicative factor, a. For $a > 1$, the signal becomes narrower, whereas for $0 < a < 1$ it gets wider.

Let us investigate the effect of the parameter a to the scaling process, in the following exercises.

Exercise 1.1 Given the continuous and discrete time signals of Figures 1.10 and 1.11, find and plot $x(2t)$ and $x[2n]$.

Solution

For the continuous time signal, $x(t)$, all we need to do is to replace t by $t' = 2t$, as follows:

$$x(2t) = \begin{cases} 1, & \text{for } 0 \leq t \leq 3, \\ 0, & \text{otherwise.} \end{cases} \tag{1.25}$$

For the discrete time signal, $x[n]$, we replace n by $n' = 2n$, as follows:

$$x[2n] = \begin{cases} 1, & \text{for } 0 \leq n \leq 3, \\ 0, & \text{otherwise.} \end{cases} \tag{1.26}$$

We can evaluate the discrete time function, $x[2n]$, in the nonzero interval, $0 \leq n \leq 3$, as follows:

For $n = 0 \rightarrow x[2n] = x[0] = 1$
For $n = 1 \rightarrow x[2n] = x[2] = 1$
For $n = 2 \rightarrow x[2n] = x[4] = 1$
For $n = 3 \rightarrow x[2n] = x[6] = 1$
For $n > 3 \rightarrow x[n] = 0.$

Figure 1.14 shows the plots of $x(2t)$ and $x[2n]$.

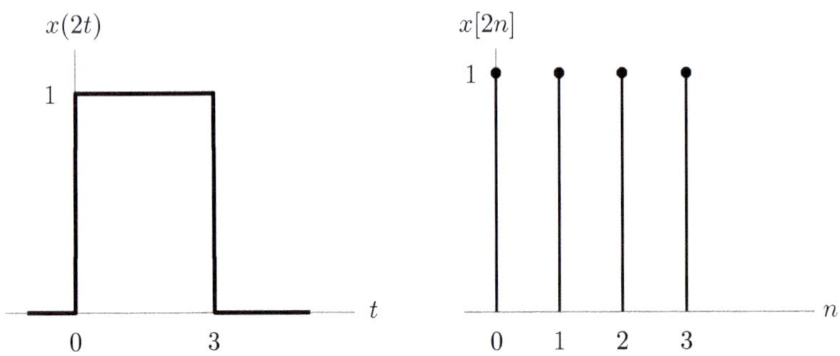

Figure 1.14 Scaled continuous time and discrete time signals for $a = 2$.

Note that, in the discrete time case, when $a > 1$, the scaled signal, $x[an]$ captures the original signal $x[n]$, at only each an time instance. Thus, we throw away the samples of the original signal between an and $a[n-1]$ time instances, for all n. This operation is called **decimation**.

Exercise 1.2 ***(Decimation)*** Given the discrete signal of Figure 1.11, find and plot $x[4n]$.

Solution

Let us evaluate the function $x[4n]$ for all possible values of n.

For $n < 0 \rightarrow x[4n] = x[0] = 0$
For $n = 0 \rightarrow x[4n] = x[0] = 1$
For $n = 1 \rightarrow x[4n] = x[4] = 1$
For $n \geq 2 \rightarrow x[n] = 0$.

Note that the decimation process by a factor of $a = 4$, keeps the values of the original function at $x[0]$ and the values at every $4n$, skipping the 3 values between $4n$, and $4[n-1]$, for all n. Therefore, it squishes the original signal by decimating with a factor of $a = 4$. The signal $x[4n]$ is plotted in Figure 1.15.

A good example of decimation with $a > 1$, corresponds to increasing the playing speed of a video recording.

Exercise 1.3 Given the continuous time signal of Figure 1.10, find and plot $x(t/2)$.

Solution

For the continuous time signal, $x(t)$, we replace t by $t' = t/2$, as follows:

$$x(t') = x(t/2) = \begin{cases} 1, & \text{for } 0 \leq t \leq 12, \\ 0, & \text{otherwise.} \end{cases} \tag{1.27}$$

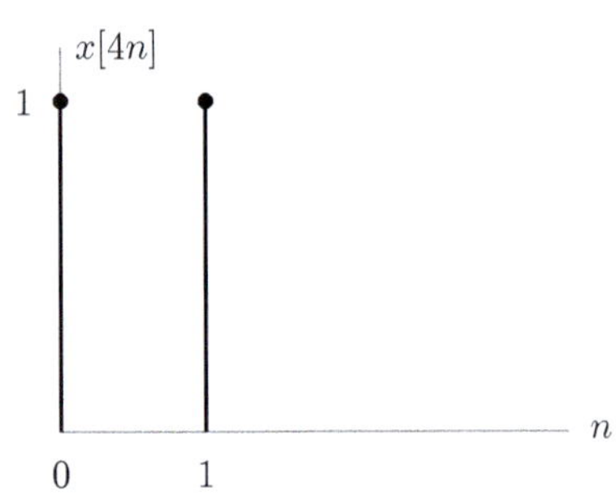

Figure 1.15 The original discrete time signal is decimated by a factor of $a = 4$.

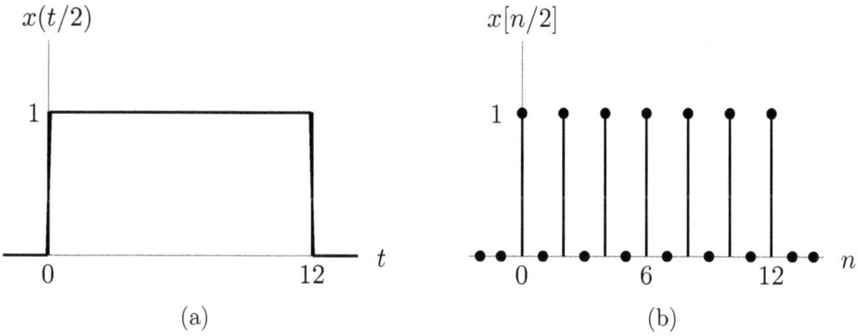

Figure 1.16 Time-scaled continuous and discrete time pulse signals, given in Figures 1.10 and 1.11, for $a = \frac{1}{2}$. (a) Notice that in the continuous time case, we simply stretch the function. (b) On the other hand, in discrete time case, we assign 0 values for each inserted time point. This operation is called expansion.

Note that for $a < 1$, the scaled continuous signal $x(at)$ is the stretched version of the original continuous signal, $x(t)$. The plot is given in Figure 1.16a.

Exercise 1.4 ***(Expansion)*** Given the discrete time signal of Figure 1.11, find and plot $x[n/2]$.

Solution

For the discrete time signal, $x[n]$, we replace n by $n' = n/2$, as follows:

$$x[n'] = x[n/2] = \begin{cases} 1, & \text{for } 0 \leq n \leq 12, \\ 0, & \text{otherwise.} \end{cases} \tag{1.28}$$

In this exercise, the values of the original discrete time signal, $x[n]$, are placed at every an time instance in the scaled signal, $x[an]$. However, when $a < 1$, we stretch the signal by inserting extra time instances between the time instances of the original signal. It is customary to put zero values to the inserted time instances. This operation is called **expansion**. The plot of $x[n/2]$ can be found in Figure 1.16b.

In some practical applications, it is possible to estimate nonzero values for the inserted time instances of the stretched function $x[an]$, using the past and future known values of the inserted points. For example, we can take the average of the closest known values of an inserted time sample and assign it the average. This process is called **interpolation**.

A good example of interpolation with $a < 1$, corresponds to slowing down the play speed of a video recording.

Question: Suppose that we are given a speech recording. How does it sound, when we scale the recording by $a < 1$ and $a > 1$?

For the values of $a > 1$, the decimation process skips some values of the signal resulting in a speed-up in the recording. Thus, the person speaks faster. For the values, $a < 1$, the interpolation process adds extra points in between the samples of the recording. Thus, the person speaks slower.

1.4.4 Time Scale and Shift

We can combine the time scale and time shift operations, to obtain the signal $x(at - b)$ from the continuous time signal, $x(t)$ and the signal $x[an - b]$ from the discrete time signal, $x[n]$.

For the continuous function, $x(t)$ of Figure 1.10, time scale and shift by the amount of a and b, can be given as follows:

$$x(at - b) = \begin{cases} 1, & \text{for } 0 \leq at - b \leq 6, \\ 0, & \text{otherwise.} \end{cases} \tag{1.29}$$

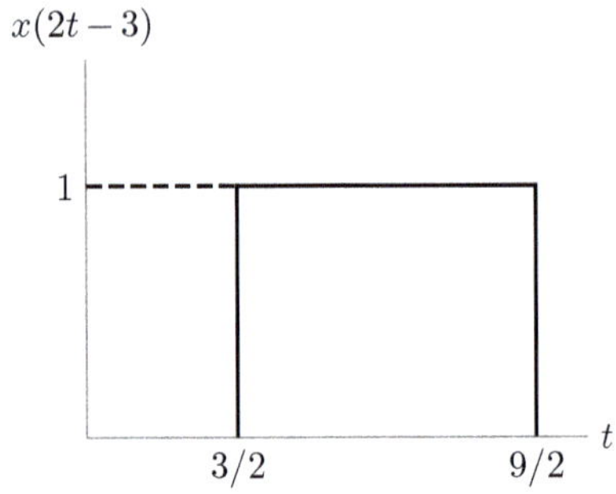

Figure 1.17 The plot of shifted and squished continuous time signal, $x(2t - 3)$.

For example, when we take $a = 2$ and $b = 3$, the continuous time signal is both squished and shifted (Figure 1.17) as follows:

$$x(2t-3) = \begin{cases} 1, & \text{for } \frac{3}{2} \leq t \leq \frac{9}{2}, \\ 0, & \text{otherwise.} \end{cases} \tag{1.30}$$

For the discrete function, $x[n]$ of Figure 1.11, time scale and shift of $x[2n - 3]$, is given as follows:

$$x[2n-3] = \begin{cases} 1, & \text{for } 2 \leq n \leq 4, \\ 0, & \text{otherwise.} \end{cases} \tag{1.31}$$

Note that the function, $x[2n - 3]$, is undefined for noninteger values of the domain, defined by $[2n - 3]$. In order to satisfy the inequality constraints, we have to round or truncate the upper and lower bounds of the domain of the function to integer values.

Exercise 1.5 Given the following continuous time signal:

$$x(t) = \begin{cases} t, & \text{for } 0 \leq t \leq 3, \\ 0, & \text{otherwise.} \end{cases} \tag{1.32}$$

Find and plot $x(2t - 3)$.

Solution

Let us define a new time variable $t' = 2t - 3$ and replace t by t' in Equation (1.32)

$$x(2t-3) = \begin{cases} 2t-3, & \text{for } 0 \leq 2t-3 \leq 3 \\ & \left(\text{for } \frac{3}{2} \leq t \leq 3\right), \\ 0, & \text{otherwise.} \end{cases} \tag{1.33}$$

A practical approach for the combined time shit and scale operation is to shift the function first, then apply the scaling operation to the shifted signal. Thus, we shift the function $x(t)$ by the amount of $b = 3$ and then scale the shifted function $x(t - 3)$ by the amount of $a = 2$. The shifted and scaled signal is depicted in Figure 1.18.

Exercise 1.6 Given the following discrete time signal:

$$x[n] = \begin{cases} n, & \text{for } 0 \leq n \leq 3, \\ 0, & \text{otherwise.} \end{cases} \tag{1.34}$$

Find and plot $x[2n - 3]$.

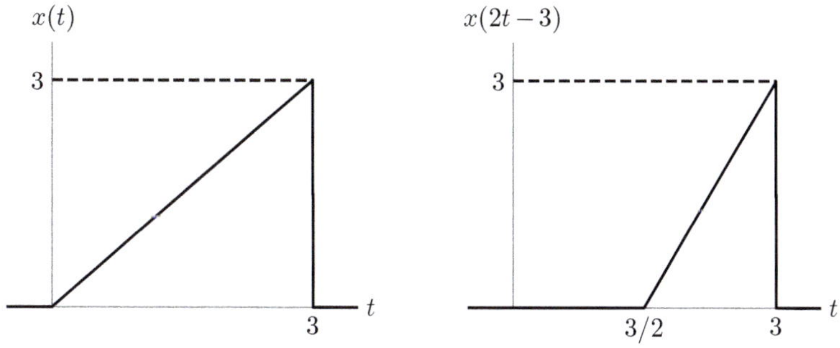

Figure 1.18 Plot of $x(t)$ and $x(2t-3)$.

Solution

As we did in the previous example, we define a new time variable $n' = 2n - 3$ and replace n by n' in Equation (1.34):

$$x[2n-3] = \begin{cases} 2n-3, & \text{for } 0 \le 2n-3 \le 3 \\ & \text{(for } 3/2 \le n \le 3\text{)}, \\ 0, & \text{otherwise.} \end{cases} \tag{1.35}$$

As in the continuous time case, first, we shift the signal to the right by the amount of $b = 3$, then apply the scaling operation to the shifted signal by the amount of $a = 2$. However, the lower bound of the domain is not integer-valued, and the signal is not defined for $x[3/2]$. Thus, we round the lower bound of the domain of the function to the nearest integer and define the nonzero interval in $2 \le n \le 3$ as follows:

$$x[2n-3] = \begin{cases} 2n-3, & \text{for } 2 \le n \le 3, \\ 0, & \text{otherwise.} \end{cases} \tag{1.36}$$

This fact can be observed from Figure 1.19.

Exercise 1.7 Given the following discrete time signal:

$$x[n] = \begin{cases} 2, & \text{for } n = -2, \\ -1, & \text{for } n = -1, \\ -3, & \text{for } n = 1, \\ 4, & \text{for } n = 2, \\ 0, & \text{otherwise.} \end{cases} \tag{1.37}$$

Find and plot the signal $x[3n-4]$.

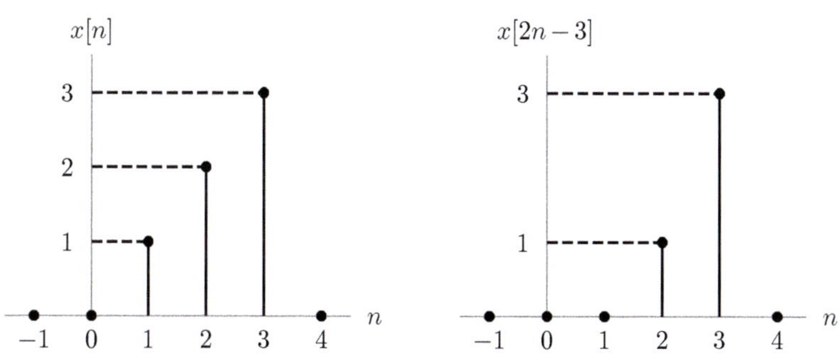

Figure 1.19 Plot of $x[n]$ and $x[2n-3]$.

Solution

First, we shift the signal to the left by $b = 4$:

$$x[n-4] = \begin{cases} 2, & \text{for } n = 2, \\ -1, & \text{for } n = 3, \\ -3, & \text{for } n = 5, \\ 4, & \text{for } n = 6, \\ 0, & \text{otherwise.} \end{cases} \tag{1.38}$$

Then, we scale the shifted signal $x[n - 4]$ by $a = 3$ to obtain the shifted and scaled signal, $x[3n - 4]$. For this purpose, we replace n by $n' = 3n$ in Equation (1.38). Since the noninteger values of n is not defined, for $3n = 2$ and $3n = 5$, the values of $x[3n - 4]$ disappear at $n = 2/3$ and $n = 5/3$. Finally, we get the following shifted and squished signal:

$$x[3n-4] = \begin{cases} -1, & \text{for } n = 1, \\ 4, & \text{for } n = 2, \\ 0, & \text{otherwise.} \end{cases} \tag{1.39}$$

The plots of the original function, $x[n]$, its shifted version, $x[n - 4]$, and then the scaled version, $x[3n - 4]$ are shown in Figure 1.20. Note that, scaling the function by a factor of $a = 3$ squishes the function, omitting the values of the original function at $n = -2$ and $n = 1$.

The aforementioned definitions and examples reveal that the operations on the time parameter of a signal require a little care for the discrete time functions, since we deal with integer arithmetic in the time variable, n.

Question: Suppose that we apply scaling and shifting processes at the same time to a movie video, represented by $x[n]$, where n is measured by seconds. What happens to the video for $x[2n - 3600]$ and for $x[0.5n + 3600]$?

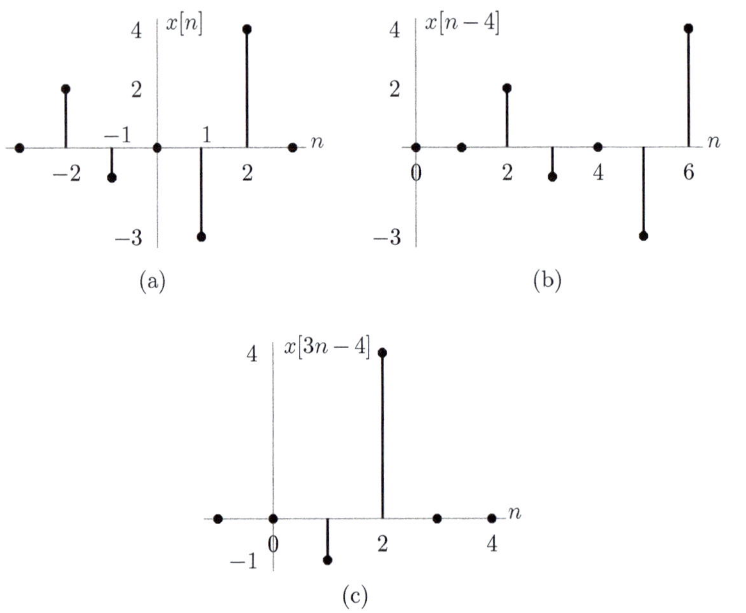

Figure 1.20 Plots of (a) $x[n]$, (b) $x[n - 4]$, and (c) $x[3n - 4]$ in Exercise 1.7.

For the signal, $x[2n - 3600]$, the movie starts with an hour delay and plays twice as fast as the original. For the signal, $x[0.5n + 3600]$, the movie starts an hour early and plays twice as slow as the original.

INTERACTIVE Explore operations on the time variable of signals @ https://384book.net/i0103

1.5 Signals with Symmetry Properties

Loosely speaking, symmetry of an object is a transformation that leaves certain properties of the object unchanged. For example, the number plate of a car does not change, when we change its location. If we drive a car, we may change its location, but not the plate of the car. Therefore, the number plate has a location translation symmetry.

Symmetry is a crucial property in many fields of mathematics, arts, and science. The study of symmetries in physics started with Noether's Theorem, which allows us to derive conserved quantities of physics from symmetries of the laws of nature. Loosely speaking, Noether's theorem states that every conservation law has a corresponding continuous symmetry transformation. For example, the conservation of energy arises from the time translation symmetry. In other words, time translation symmetry results in the law of conservation of energy.

There is a rigorous definition of symmetry in group theory, over the symmetry groups. A **symmetry group** is defined as the group of all transformations under which the objects of the group are invariant.

Although we do not know the exact reason(s), we find aesthetic values in symmetry. One reason may be the wide range of symmetry types found in natural objects. For example, the reflection symmetry of the human body, the rotational symmetry of flower petals, and the hexagonal symmetry of the honeycomb are aesthetically pleasing. In fact, asymmetry of an object may be an indication of a disease or a problem in the natural world.

Another reason for aesthetic values in symmetry may be attributed to the working principles of the human brain. It is well known that objects with some type of symmetry require relatively less storage space, compared to asymmetrical objects. For example, a square shape has 90° rotation symmetry and reflection symmetry. In order to store a square, all we need to know is the length of its edge. On the other hand, an amorphous shape with no symmetry requires to store all data points on the boundary. As an information processing and storing device, our brain creates a model of the physical world surrounding us from the sensory stimuli. The human brain compresses the world model by extracting a wide range of symmetry transformations and updates them continuously throughout our lives. Thus, symmetry brings efficiency to process information in our brain, which creates an aesthetically pleasing feeling.

The history of art is full of painters, architects, designers, and composers, who use the concept of symmetry in their artwork. A pioneering artist in this field, known for his symmetrical lithographs is M.C. Escher, who is inspired by the highly symmetrical geometric decorations of the Alhambra Palace. Escher displayed the beauty of mathematics in his lithographs without any formal mathematics education (Figure 1.21). His lithographs were sources of inspiration to many mathematicians, who work on tessellations. The famous mathematician R. Penrose was fascinated by the art of Escher, when he visited one of Escher's exhibitions and created highly symmetrical impossible tilings, one of which is shown in Figure 1.22.

Figure 1.21 *Snakes* by the Dutch artist M.C. Escher, which depicts rotational symmetry.[4] Source: The M.C. Escher Company B.V/https://mcescher.com/gallery/most-popular/#iLightbox[galleryimage1]/30/last accessed March 09, 2024.

Figure 1.22 A Penrose tiling with fivefold symmetrical two different rhombi.[5] Source: National Institute of Standards and Technology/Public domain.

Figure 1.23 shows the divine art of Alhambra Palace, which inspired many artists and mathematicians for over 800 years.

WATCH

Learn more about the symmetric decorations of the Alhambra Palace @ https://384book.net/v0105

4 https://mcescher.com/gallery/most-popular/#iLightbox[gallery_image_1]/30.
5 https://www.nist.gov/image/penrose-tiling.

Figure 1.23 Symmetric tiling Alhambra Palace.[6] Source: javiindy/Adobe Stock.

WATCH Learn more about the symmetric art of Escher @ https://384book.net/v0106

There are many forms of symmetry, specifically in group theory and representation theory. In this book, we deal with signals with specific forms of symmetry. First, we study two crucial classes of functions, namely

1) signals represented by periodic functions,
2) signals represented by even and odd functions.

Then, in the next chapter, we study the functions, which can be considered as the basic building blocks of a large class of functions. Interestingly, the symmetry functions enable us to represent a wide range of signals by rigorous mathematical models, in a compact way, as we shall see throughout this book.

WATCH Learn more about the mathematics of symmetry @ https://384book.net/v0107

Let us briefly study the basic properties of periodic signals, and even and odd signals in the following subsections.

1.5.1 Periodic Signals

Signals are called periodic if the representing function repeats itself at every finite interval, called the period. Periodic functions are symmetric with respect to periodic translation. In other words, periodic functions are translation invariant functions, where the translation of the signal by the amount of its period gives the same signal.

In this book, we use periodic functions intensively, to represent complicated signals, even the aperiodic and asymmetric ones, in continuous time and discrete time cases. Periodic functions open the door to the new vector spaces, where a large class of functions are represented in terms of periodic functions.

6 https://www.alhambra.info/img/interiores/4.jpg.

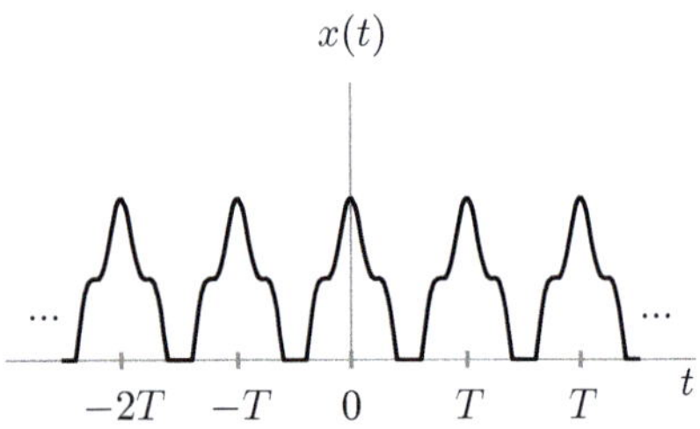

Figure 1.24 A continuous time periodic signal with the fundamental period, T. This function is symmetric with respect to reflection and translation by the amount of the period.

1.5.1.1 Continuous Time Periodic Signals

A continuous time signal, $x(t)$, is periodic if there exists a **finite and nonzero real value**, $T \in \mathbb{R}$, such that,

$$x(t) = x(t + T). \tag{1.40}$$

Definition 1.6 **Fundamental period**, $T_0 \in \mathbb{R}$, is defined as the smallest positive real number, in which the function repeats itself as follows:

$$x(t) = x(t + T_0), \tag{1.41}$$

and it is measured by a time unit, such as hours, minutes, or seconds. Figure 1.24 shows an example continuous time periodic signal.

Based on the fundamental period, T_0, there are two more measures for periodic signals:

- **Angular Frequency**: $\omega_0 = \frac{2\pi}{T_0}$, measured by rad/s.
- **Fundamental Frequency**: $f_0 = \frac{1}{T_0}$, measured by Hertz (cycle/s).

Exercise 1.8 Find the fundamental period of the following signals:

a) $x(t) = \cos t$,
b) $x(t) = \cos(\omega_0 t)$.

Solution
Recall from Calculus:

a) $x(t) = \cos t = \cos(t + 2\pi)$. Thus, the fundamental period is $T_0 = 2\pi$.
b) $x(t) = \cos(\omega_0 t) = \cos(\omega_0 (t + \frac{2\pi}{\omega_0}))$. Thus, the fundamental period is $T_0 = \frac{2\pi}{\omega_0}$.

Note that the parameter, ω_0, which multiplies the time variable, corresponds to angular frequency. We can, then, obtain the fundamental frequency, T_0 by using the relationship between the angular frequency and period. For $\omega_0 = 1$, the period of the cosine function is 2π.

Exercise 1.9 Plot the following signal and find its fundamental period:

$$x(t) = A\cos(\omega_0 t - K). \tag{1.42}$$

Solution
We need to find the smallest period T_0, such that

$$x(t) = A\cos(\omega_0 t - K) = x(t + T_0). \tag{1.43}$$

This equation is satisfied, when $T_0 = 2\pi/\omega_0$.

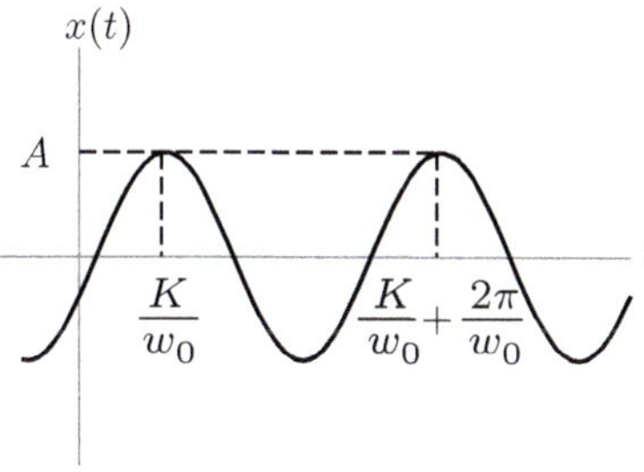

Figure 1.25 The continuous time cosine signal, $x(t) = A\cos(\omega_0 t - K)$, with amplitude A and period $T = 2\pi/\omega_0$. For $A = 1$ and $K = 0$ the plot is reduced to $x(t) = \cos\omega_0 t$ of Exercise 1.8.

For the signal $x(t) = A\cos(\omega_0 t - K)$, the parameters A, K and ω_0 have special names. The parameter A is called the **amplitude** of the signal. The parameter K is called the **phase** of the signal, $\omega_0 = 2\pi/T$, is called the **angular frequency.** It is plotted in Figure 1.25.

1.5.1.2 Discrete Time Periodic Signals

A discrete time signal, $x[n]$, is periodic if there exists a **finite and nonzero integer value**, $N \in \mathbb{I}$, such that

$$x[n] = x[n + N]. \tag{1.44}$$

An example is given in Figure 1.26.

Definition 1.7 The smallest integer, $N_0 \in \mathbf{I}$, which satisfies $x[n] = x[n + N_0]$ is called the **fundamental period**.

Based on the fundamental period, N_0, of a discrete time periodic function, we can define:

- **Angular Frequency**: $\Omega_0 = \frac{2\pi}{N_0}$, measured by rad/s.
- **Fundamental Frequency**: $f_0 = \frac{1}{N_0}$, measured by Hertz (cycle/s).

Exercise 1.10 Find the fundamental period of the following discrete time signal,

$$x[n] = A\cos(\Omega_0 n - K), \tag{1.45}$$

where K is an integer.

Solution

For periodicity, we need to find the fundamental period N_0, which satisfies,

$$x[n] = x[n + N_0]. \tag{1.46}$$

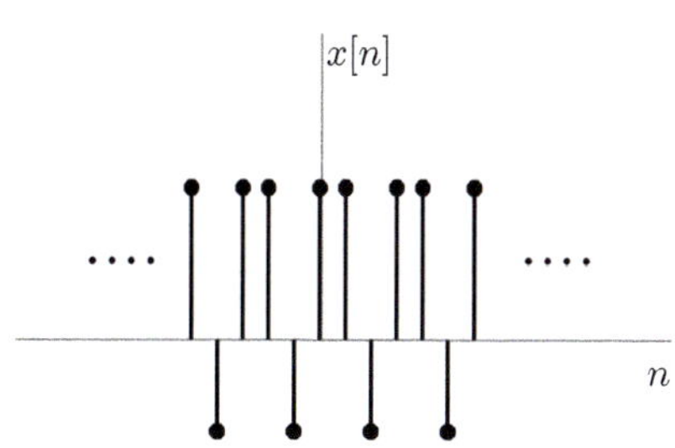

Figure 1.26 A discrete time periodic signal with fundamental period $N_0 = 3$.

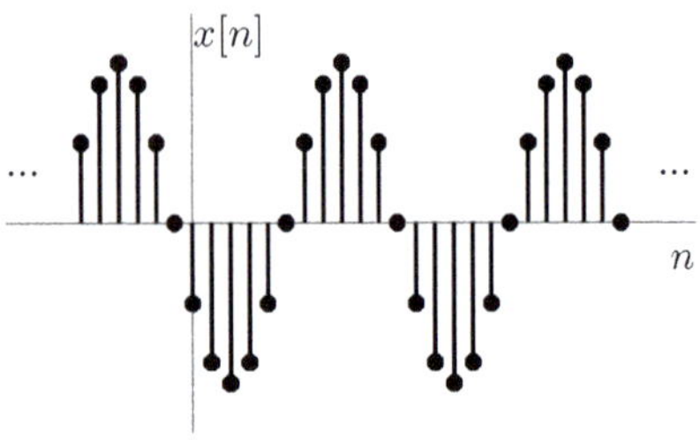

Figure 1.27 Plot of $x[n] = A\cos(\Omega_0 n - K)$, for $\Omega_0 = 2\pi/12$ and $K = 4$.

Since $x[n]$ is a discrete time signal, N_0 must be an integer. Thus, we need to find the fundamental period,

$$N_0 = \frac{2\pi}{\Omega_0}m, \tag{1.47}$$

where m is the smallest integer which makes N_0 an integer.

Note that, the angular frequency, Ω_0 is an irrational number, obtained by dividing 2π with an integer number. This integer divider corresponds to the fundamental period, N_0.

For example, if $\Omega_0 = \frac{\pi}{6} = \frac{2\pi}{12}$, then, $N_0 = 12$. This signal is plotted in Figure 1.27. On the other hand, if $\Omega_0 = \phi$, where ϕ is a real number, then $x[n]$ is not periodic at all.

Exercise 1.11 Is the following signal periodic? If yes, find the period.

$$x[n] = \sin\left(\frac{6\pi}{7}n + 1\right). \tag{1.48}$$

Solution
We need to find the smallest integer value, N_0, which satisfies the following equation:

$$x[n] = x[n + N_0] = \sin\left(\frac{6\pi}{7}(n + N_0) + 1\right) = \sin\left(\frac{6\pi}{7}n + \frac{6\pi}{7}N_0 + 1\right). \tag{1.49}$$

Here, $\frac{6\pi}{7}N_0$ must be equal to $2\pi m$, where m is the smallest integer, satisfying,

$$\frac{6\pi}{7}N_0 = 2\pi m, \; N_0 = \frac{7m}{3}. \tag{1.50}$$

The smallest integer, which satisfies Equation (1.50), is $m = 3$. Then, the fundamental period is $N_0 = 7$. Therefore, this signal is periodic.

1.5.2 Even and Odd Signals

Another set of signals with symmetry properties is the set represented by even and odd functions.

Even functions have reflection symmetry. In other words, they are invariant to flipping around the vertical axis (Figure 1.28). Mathematically, a signal is called even if

$x(t) = x(-t)$, for continuous time signals,
$x[n] = x[-n]$, for discrete time signals.

Odd functions have rotation symmetry. In other words, they are invariant to a rotation of 180° around the origin (Figure 1.29). Mathematically, a signal is called odd if

$x(t) = -x(-t)$, for continuous time signals,
$x[n] = -x[-n]$, for discrete time signals.

Exercise 1.12 An important family of functions, which is widely used in digital signal processing (DSP) technology is the parabola. Simply, a parabola is defined as the trajectory generated by

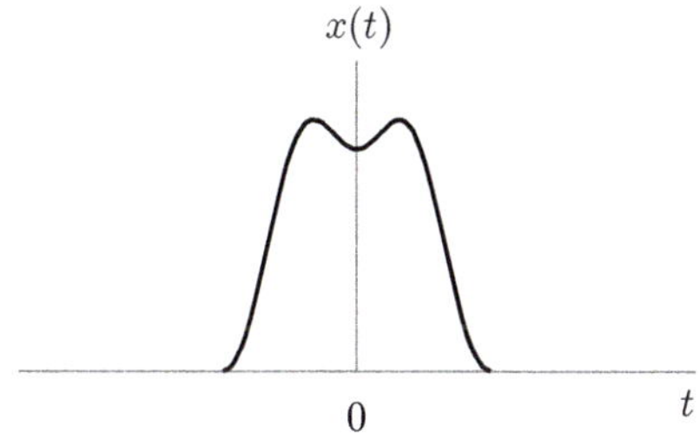

Figure 1.28 An even signal has reflection symmetry about the vertical axis. In other words, even functions are invariant to reflection, when they are flipped around the vertical axis.

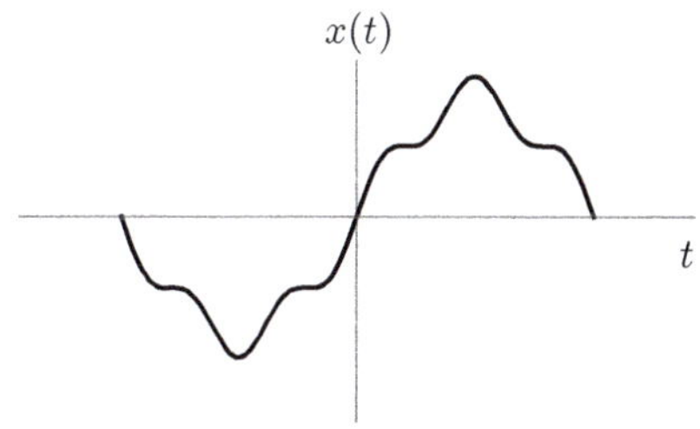

Figure 1.29 An odd signal has rotation symmetry about the origin. In other words, odd functions are invariant to rotation, when they are rotated 180° around the origin.

equidistant points from **a focal point and a fixed line**. It is represented by the following analytical form:

$$x(t) = at^2, \quad \forall a \neq 0. \tag{1.51}$$

Plot the parabolas for $a = 1$ and $a = 2$. Show that it is an even function.

Solution

Replacing t by $-t$, we get, $x(-t) = at^2 = x(t)$. Plots are given in Figure 1.30.

Exercise 1.13 Another important family of functions in DSP is called hyperbola. Simply, hyperbola is defined as the trajectory generated by a moving point, such that the difference of the distances from **two fixed focal points** is always constant. Hyperbola is represented by the following analytical form:

$$x(t) = a/t, \quad \forall a > 0. \tag{1.52}$$

Plot the hyperbolas for $a = 1$ and 2. Show that it is an odd function.

Solution

Replacing t by $-t$, we get $x(-t) = a/-t = -x(-t)$. Plots are given in Figure 1.31.

Exercise 1.14 Can a function be both even and odd? In other words, is there a function, which is symmetric with respect to both the vertical axis and the origin? If yes, give an example.

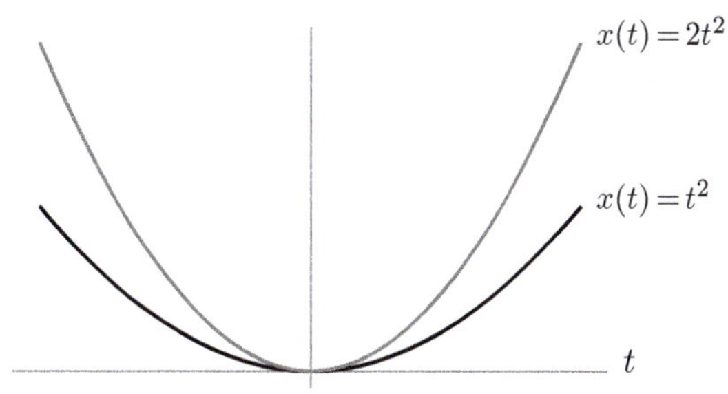

Figure 1.30 Plot of parabolas for $a = 1, 2$.

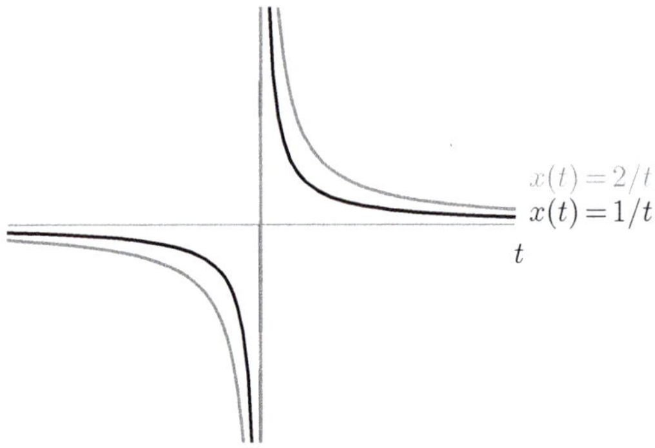

Figure 1.31 Plot of hyperbolas for $a = 1, 2$.

Solution

A function, $x(t)$ is even and odd if it satisfies both of the following equalities:

$$x(t) = x(-t) = -x(-t). \tag{1.53}$$

The only function, which satisfies Equation (1.53) is $x(t) = 0$. In other words, when there is no signal at all, the function is doubly symmetric.

Proposition 1.1 Any signal can be represented by its even and odd components, as follows:

$$x(t) = \text{Odd}\{x(t)\} + \text{Even}\{x(t)\}, \tag{1.54}$$

where

$$\text{Odd}\{x(t)\} = \frac{1}{2}(x(t) - x(-t)), \tag{1.55}$$

$$\text{Even}\{x(t)\} = \frac{1}{2}(x(t) + x(-t)). \tag{1.56}$$

The proof of this proposition follows from the addition of even and odd parts of the function.

The decomposition of a signal into its even and odd parts reveals an important property of the functions: Even if a function does not possess any type of symmetry property, it can be represented by the decomposition of functions with symmetry property.

Exercise 1.15 Given the plot of a continuous time signal, $x(t)$, in Figure 1.32, plot its even and odd parts.

Solution

Even and odd parts of $x(t)$ can be found using Equations (1.55) and (1.56). These are shown in Figure 1.33.

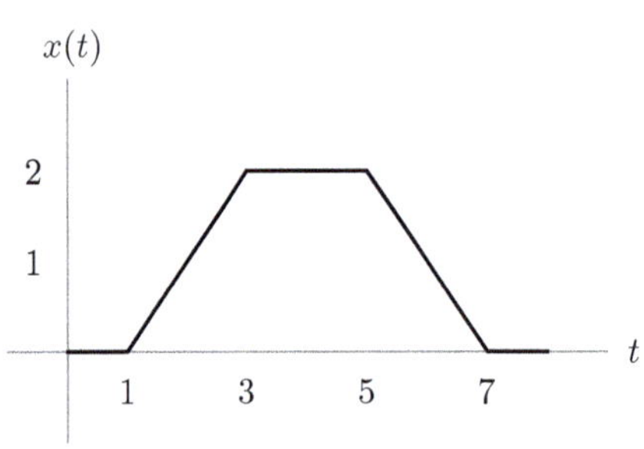

Figure 1.32 An arbitrary continuous time signal, $x(t)$.

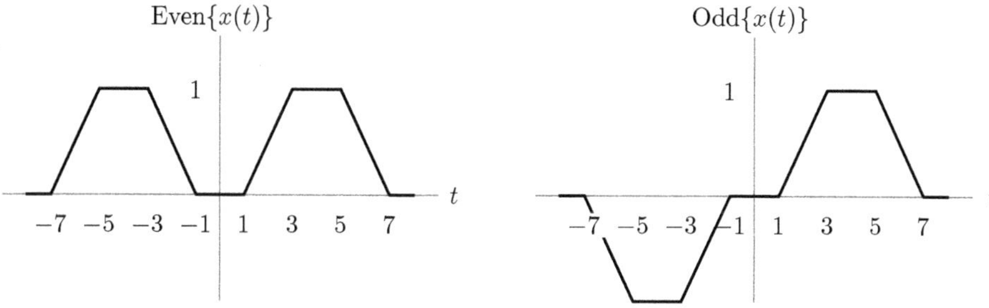

Figure 1.33 The odd and even parts of $x(t)$ of Figure 1.32.

Exercise 1.16 Find and plot the even and odd parts of the following function:

$$x(t) = t^3 - t + 1 \tag{1.57}$$

Solution

Using the definition of even and odd parts of functions, given earlier, we find $\text{Even}\{x(t)\} = 1$ and $\text{Odd}\{x(t)\} = t^3 - t$. Plots are given in Figure 1.34.

The definitions of even and odd functions can be trivially extended to the discrete time signals. Mathematically, a discrete time signal is even if

$$x[n] = x[-n] \tag{1.58}$$

and it is odd if

$$x[n] = -x[-n]. \tag{1.59}$$

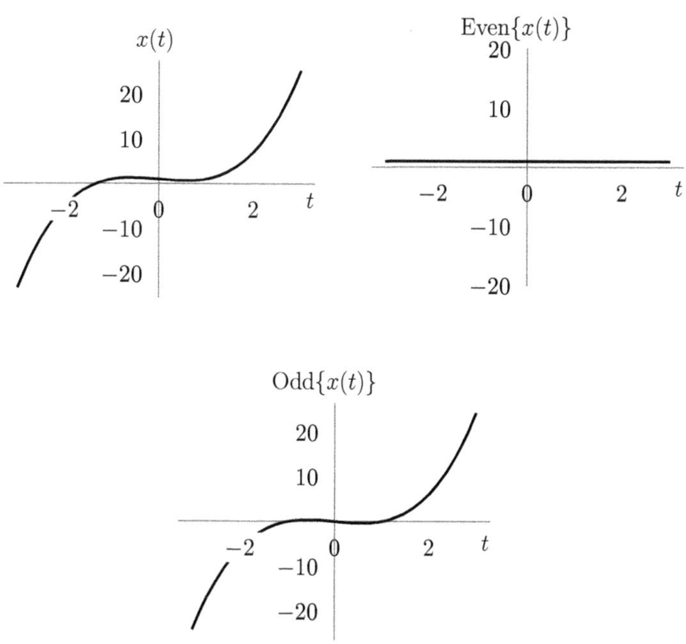

Figure 1.34 Plot of $x(t) = t^3 - t + 1$, and its even and odd parts.

Symmetry groups of functions is a very interesting field of mathematics, which is beyond the scope of this book.

INTERACTIVE

Decompose signals into their even and odd parts @ https://384book.net/i0104

1.6 Complex Signals Represented by Complex Functions

Most of the natural signals can be represented by functions with real- or integer-valued domains and ranges. For example, we perceive the objects around us by a world model, created in our brain, through the visible light reflected from the objects, in real-valued time domain. Although we define the domain and the range of the functions in the space of real numbers or integers, there are more compact and efficient ways of representing the functions in some other vector spaces spanned by a set of complex basis functions. Before diving into these elegant vector spaces, let us briefly overview the space of complex numbers.

Complex numbers are first introduced by Gerolamo Cardono, in his book, Ars Magna, in 15th century. It appears in a wide range of problems in engineering, science and mathematics. For example, when we need to find a solution to $x^2 + 1 = 0$, the space of real numbers, $\mathbb{R}$, does not enable us to provide the value for the variable, x. To find the solution to these types of equations, an **imaginary number**,

$$j = \sqrt{-1} \tag{1.60}$$

is introduced to define a new vector space of complex numbers, $\mathbb{C}$, where an extra imaginary dimension is added to the real number space.

Despite the historical nomenclature, imaginary numbers are not imaginary or nonsense at all. They are essential in many aspects of the scientific description of the natural and human-made world. Imaginary numbers can be considered as an extension of the one-dimensional vector space of real numbers, $\mathbb{R}$, to two-dimensional space of complex numbers, $\mathbb{C}$, called the **complex plane**. While the first dimension quantifies the amount of real value, the second dimension quantifies the imaginary value of a complex number.

Complex numbers can be equivalently represented in Cartesian and polar coordinate systems, as described in the following text.

1.6.1 Complex Numbers Represented in Cartesian Coordinate System

A **complex number** is represented in a Cartesian coordinate system $\mathbb{C}$ as follows:

$$z = x + jy, \tag{1.61}$$

where $x = \text{Re}\{z\}$, is called the **real part** and $y = \text{Im}\{z\}$ is called the **imaginary part** of the complex number z. Note that complex numbers form a two-dimensional vector space spanned by the standard basis vectors,

$$c_1 = [1 \quad 0] \quad \text{and} \quad c_2 = [0 \quad j]. \tag{1.62}$$

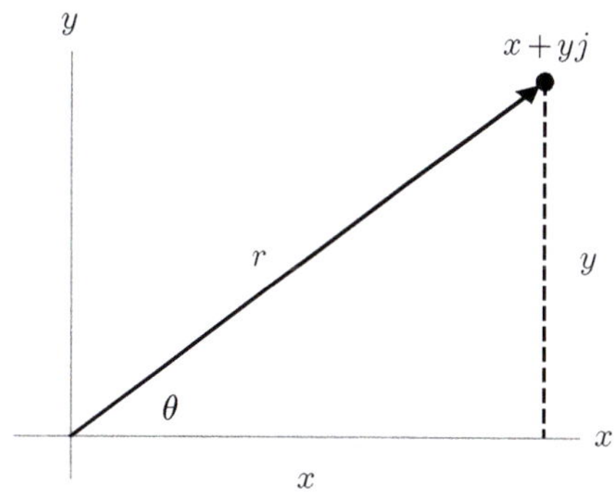

Figure 1.35 Complex plane, where the real numbers are augmented by imaginary numbers.

Thus, any complex number z can be represented in terms of the linear combination of c_1 and c_2, as follows:

$$z = xc_1 + yc_2. \tag{1.63}$$

Since one of the basis vectors has the imaginary number, basis vectors span the complex plane. The complex plane $\mathbb{C}$ reduces to the space of real numbers, $\mathbb{R}$, for $\forall y = \text{Im}\{z\} = 0$. It reduces to the space of purely imaginary numbers, $\mathbb{I}$, for $\forall x = \text{Re}\{z\} = 0$. The complex plane is illustrated in Figure 1.35.

Arithmetic operations, such as addition, subtraction, division and multiplication of complex numbers are defined by considering the fact that the imaginary number has a square,

$$j^2 = -1. \tag{1.64}$$

Reflection symmetry with respect to the real axis is called the **complex conjugate** of the complex number z,

$$z^* = x - jy. \tag{1.65}$$

Note that the multiplication of a complex number with its complex conjugate gives a real number,

$$z \cdot z^* = x^2 + y^2, \tag{1.66}$$

corresponding to the square of the Euclidean norm, which is measured by the length of the vector z:

Exercise 1.17 Given the following complex numbers,

$$z_1 = 1 + 2j \quad \text{and} \quad z_2 = 2 + 3j, \tag{1.67}$$

Find and plot the following.

a) $z_1 + z_2$,
b) $z_1 \cdot z_2$,
c) $z_1 \div z_2$.

Solution

a) We apply vector addition in two-dimensional space of complex numbers, as follows:

$$z_1 + z_2 = 3 + 5j. \tag{1.68}$$

Geometrically speaking, the addition operation translates one of the vectors by the amount of the other vector. Note that the addition operation is commutative.

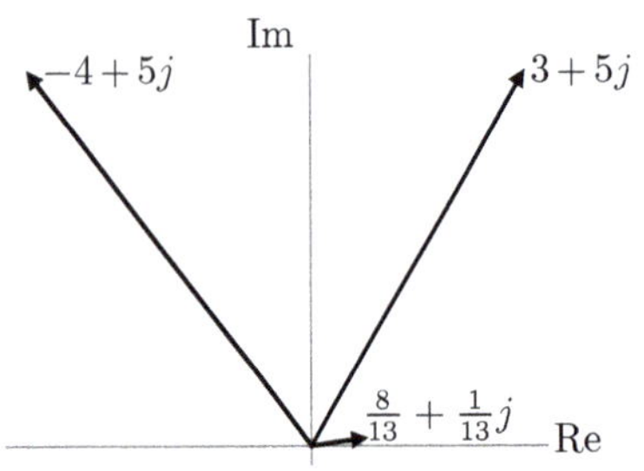

Figure 1.36 Plots of three complex numbers in Exercise 1.17.

b) Multiplication operation is associative and distributive, resulting in

$$z_1 \cdot z_2 = (1 + 2j)(2 + 3j) = 2 + 7j + 6j^2 = -4 + 7j. \tag{1.69}$$

c) For division operation, first, we multiply the dividend and the divisor by the complex conjugate of the divisor to ensure a real number at the denominator. Then, we apply multiplication operation to the numerator and divide both the real and imaginary part of the numerator by a real number obtained in the denominator, as follows:

$$z_1 \div z_2 = \frac{1 + 2j}{2 + 3j} = \frac{(1 + 2j)(2 - 3j)}{13} = \frac{8}{13} + \frac{1}{13}j. \tag{1.70}$$

The plots of these complex numbers can be found in Figure 1.36.

When we represent the complex numbers in the polar coordinate system, in the next subsection, we shall observe that the multiplication and division of two complex numbers correspond to rotation of the complex numbers.

1.6.2 Complex Numbers Represented in Polar Coordinate System and Euler's Number

A complex number, z, can be considered as a vector in the complex plane, represented by the length of the vector and the angle between the vector and x axis by defining the following relations:

$$x = \text{Re}\{z\} = r\cos\theta \tag{1.71}$$

and

$$y = \text{Im}\{z\} = r\sin\theta, \tag{1.72}$$

where $r = |z|$, is called the **magnitude** and $\theta = \angle z$ is called the **phase** of the complex number z. Then, the complex number, z, can be equivalently represented by

$$z = x + jy = r(\cos\theta + j\sin\theta). \tag{1.73}$$

There is an important number in mathematics called **Euler's number**, discovered by the great mathematician Jacob Bernoulli, while he was trying to compute the compound interest of a bank account from the following series:

$$e = \lim_{n\to\infty}\left(1 + \frac{1}{n}\right)^n = \sum_{n=0}^{\infty}\frac{1}{n!} \approx 2.71828182\ldots. \tag{1.74}$$

Like the irrational number π, Euler's number, e, has a great impact in mathematics. Euler's number has a wide range of interesting properties. One very important implication of Euler's number is to relate the trigonometric functions, such as cosine and sine of an angle, in the complex plane, through **Euler's formula**, as stated in the following proposition.

WATCH — Learn more about the Euler number @ https://384book.net/v0108

Proposition 1.2 ***Euler's Formula*** For any real number θ,

$$e^{j\theta} = \cos\theta + j\sin\theta. \tag{1.75}$$

Proof: We expand $e^{j\theta}$ to Maclaurin series (Taylor series around $\theta = 0$):

$$e^{j\theta} = 1 + j\theta + \frac{(j\theta)^2}{2!} + \frac{(j\theta)^3}{3!} + \frac{(j\theta^4)}{4!} + \cdots. \tag{1.76}$$

We expand $\cos\theta$ and $j\sin\theta$ to Maclaurin series, and add them up:

$$\cos\theta + j\sin\theta = \left(1 - \frac{\theta^2}{2!} + \frac{\theta^4}{4!} - \frac{\theta^6}{6!} + \cdots\right) + j\left(\theta - \frac{\theta^3}{3} + \frac{\theta^5}{5!} - \frac{\theta^7}{7!} + \cdots\right). \tag{1.77}$$

By a simple arrangement, we can show that the right-hand side of Equations (1.76) and (1.77) are the same. Therefore,

$$e^{j\theta} = \cos\theta + j\sin\theta. \tag{1.78}$$

Proof: (Alternative proof) Let us define a function $f(\theta)$ as the ratio of the right-hand side of the Euler's formula to the left-hand side:

$$f(\theta) = \frac{\cos\theta + j\sin\theta}{e^{j\theta}}. \tag{1.79}$$

Showing that $f(\theta) = 1$ would prove the Euler's formula. Let us take the derivative of $f(\theta)$:

$$\begin{aligned} f'(\theta) &= -je^{-j\theta}(\cos\theta + j\sin\theta) + e^{-j\theta}(-\sin\theta + j\cos\theta) \\ &= -je^{-j\theta}\cos\theta + je^{-j\theta}\cos\theta - j^2e^{-j\theta}\sin\theta - e^{-j\theta}\sin\theta = 0. \end{aligned} \tag{1.80}$$

Since $f'(\theta) = 0$, we conclude that $f(\theta)$ is a constant function. Further, when we evaluate $f(0)$, we get 1. So, we conclude that $f(\theta) = 1$.

WATCH — Euler's formula explained in simple group theory @ https://384book.net/v0109

Interestingly, for $\theta = \pi$, the Euler's formula becomes

$$e^{j\pi} = -1. \tag{1.81}$$

This beautiful equation contains two irrational numbers of infinite length, namely, e and π and the imaginary number j, on the left-hand side. And it is simply equal to -1.

Euler's formula enables us to represent any complex number, $z = x + jy$, in the polar coordinate system, equivalently;

$$z = re^{j\theta}, \tag{1.82}$$

where $r \in \mathbb{R}$ is the magnitude and $0 < \theta < 2\pi$ is the phase of the complex number. The relationship between the two coordinate systems is established by

$$x = r\cos\theta \quad \text{and} \quad y = r\sin\theta \tag{1.83}$$

and

$$r^2 = x^2 + y^2 \quad \text{and} \quad \theta = \arctan \frac{\text{Im}\{z\}}{\text{Re}\{z\}}. \tag{1.84}$$

Exercise 1.18 Consider the following complex numbers, given in polar coordinates:

$$z_1 = 2e^{j\pi/2} \quad \text{and} \quad z_2 = 2e^{j\pi}. \tag{1.85}$$

Compute the following:

a) $z_1 + z_2$
b) $z_1 \cdot z_2$
c) $z_1 \div z_2$

Solution

a) For addition operation in complex arithmetic, we need to use Cartesian coordinate representation. Using the Euler's formula, we get

$$\begin{aligned} z_1 + z_2 &= 2(\cos\pi/2 + j\sin\pi/2) + 2(\cos\pi + j\sin\pi) \\ &= (2\cos\pi/2 + 2\cos\pi) + j(2\sin\pi/2 + 2\sin\pi) = -2 + 2j. \end{aligned} \tag{1.86}$$

Next, we use Euler's formula to convert the Cartesian representation to the polar representation. Then, the magnitude is

$$r = \sqrt{\text{Re}\{z\}^2 + \text{Im}\,\{z\}^2} = \sqrt{x^2 + y^2} = \sqrt{8} = 2\sqrt{2}, \tag{1.87}$$

and the phase is

$$\theta = \arctan \frac{\text{Im}\{z\}}{\text{Re}\{z\}} = \arctan(-2/2) = -\pi/4. \tag{1.88}$$

Thus, the polar representation becomes

$$z_1 + z_2 = 2\sqrt{2}e^{-j\pi/4}. \tag{1.89}$$

Recall from the Cartesian representation of complex numbers, the addition operation translates the complex numbers with respect to the other one.

b) For the multiplication of complex numbers in the polar coordinate system, we multiply the magnitudes and add the phases in the exponent, as follows:

$$z_1 \cdot z_2 = 4e^{j(\pi+\pi/2)} = 4e^{j(3\pi/2)}. \tag{1.90}$$

Notice that the multiplication operation scales the amplitude and rotates one of the complex number by the amount of the phase of the other, in the clockwise direction.

c) For division of complex numbers, we divide the magnitudes and subtract the phases in the exponent,

$$z_1 \div z_2 = e^{j(\pi/2-\pi)} = e^{-j\pi/2}. \tag{1.91}$$

Notice that the division operation divides the amplitude of the dividend by that of the divider. It rotates the dividend in counterclockwise direction by the phase of the divider.

In summary, arithmetic operations on complex numbers have very nice geometric interpretations.

Exercise 1.19 Find the Cartesian representation of the following complex number:

$$z = e^{j\pi/2} + j. \tag{1.92}$$

Solution
Using the Euler's formula, we get

$$z = e^{j\pi/2} + j = \cos \pi/2 + j \sin \pi/2 + j = 2j. \tag{1.93}$$

Since the real part is 0, the number, $z = e^{j\pi/2} + j$, is purely imaginary.

1.6.3 Complex Functions

Based on the aforementioned introduction to complex numbers, we can define a complex function, as follows:

$$X : z \rightarrow X(z), \tag{1.94}$$

where the domain of the function is the complex variable $z \in \mathbb{C}$ and the range is $X(z) \in \mathbb{C}$.

Then, a complex function can be represented by its real and imaginary parts, in Cartesian coordinate system, as follows:

$$X(z) = \text{Re}\{X(z)\} + j\text{Im}\{X(z)\}. \tag{1.95}$$

Similarly, a complex function can be represented in polar coordinate system as follows:

$$X(z) = |X(z)|e^{j\Theta}, \tag{1.96}$$

where the **magnitude** and **phase** of the complex function, $X(z)$ is

$$|X(z)| = \sqrt{\text{Re}^2\{X(z)\} + \text{Im}^2\{X(z)\}} \tag{1.97}$$

and

$$\Theta = \angle X(z) = \tan^{-1} \frac{\text{Im}\{X(z)\}}{\text{Re}\{X(z)\}}, \tag{1.98}$$

respectively.

Arithmetic operations on complex functions are trivial extensions of the complex numbers, as depicted in the following exercises.

Exercise 1.20 Given the following complex functions, in Cartesian coordinate system:

$$X_1(z) = 1 + jz \quad \text{and} \quad X_2(z) = 2 + jz, \tag{1.99}$$

find the results of the following arithmetic operations:

a) $X_1(z) + X_2(z)$,
b) $X_1(z) \cdot X_2(z)$,
c) $X_1(z) \div X_2(z)$.

Solution
a) We apply vector addition in the Cartesian form of the functions in two-dimensional space of complex numbers, as follows:

$$X_1(z) + X_2(z) = 3 + j2z. \tag{1.100}$$

b) We apply multiplication operation in two-dimensional space of complex numbers as follows:

$$X_1(z) \cdot X_2(z) = 2 + 3jz + j^2z^2 = (2 - z^2) + 3jz, \tag{1.101}$$

where the real part

$$\mathrm{Re}\{X(z)\} = 2 - z^2 \tag{1.102}$$

and the imaginary part

$$\mathrm{Im}\{X(z)\} = 3z \tag{1.103}$$

of $X(z)$ are functions of the complex variable z.

c) For division, we simply multiply both divider and the dividend with the complex conjugate of the divider. This simple trick makes the divider a real number, reducing the vector-to-vector division to a scalar multiplication of a vector, as follows:

$$X_1(z) \div X_2(z) = \frac{1 + jz}{2 + jz} = \frac{(1 + jz)(2 - jz)}{4 - j^2z^2} = \frac{2 + z^2}{4 + z^2} + \frac{1}{4 + z^2}jz. \tag{1.104}$$

Exercise 1.21 Given the following complex functions, in polar coordinate system:

$$X_1(z) = ze^{j\pi} \quad \text{and} \quad X_2(z) = e^{j\pi z}, \tag{1.105}$$

find the results of the following arithmetic operations:

a) $X_1(z) + X_2(z)$,
b) $X_1(z) \cdot X_2(z)$,
c) $X_1(z) \div X_2(z)$.

Solution

When the complex functions are rather complicated, addition and subtraction operations are mostly done in Cartesian coordinate system, whereas multiplication and division are relatively easier in polar coordinate system, as we demonstrate in the following solutions.

a) Simplifying the addition of two functions in the polar form is not possible. Thus, we convert the polar form of the complex functions, $X_1(z)$ and $X_2(z)$ into Cartesian form representation by employing Euler's formula:

$$X_1(z) = ze^{j\pi} = z(\cos\pi + j\sin\pi) = -z \tag{1.106}$$

and

$$X_2(z) = e^{j\pi z} = \cos\pi z + j\sin\pi z. \tag{1.107}$$

Using vector addition, we get the addition of two functions in Cartesian coordinate system, as follows:

$$X_1(z) + X_2(z) = (\cos\pi z - z) + j\sin\pi z. \tag{1.108}$$

The magnitude of the addition is

$$|X_1(z) + X_2(z)| = \sqrt{(\cos\pi z - z)^2 + \sin^2\pi z}. \tag{1.109}$$

The phase of the addition is

$$\angle(X_1(z) + X_2(z)) = \tan^{-1}\frac{\sin\pi z}{\cos\pi z - z}. \tag{1.110}$$

Then, the polar representation of the addition becomes

$$X_1(z) + X_2(z) = \sqrt{(\cos \pi z - z)^2 + \sin^2 \pi z} e^{j\tan^{-1} \frac{\sin \pi z}{\cos \pi z - z}}. \quad (1.111)$$

b) We apply multiplication operation in polar coordinate system, as follows:

$$X_1(z) \cdot X_2(z) = ze^{j\pi} \cdot e^{j\pi z} = ze^{j\pi(1+z)}, \quad (1.112)$$

c) We apply division operation in polar coordinate system, as follows:

$$X_1(z) \div X_2(z) = ze^{j\pi} \div e^{j\pi z} = ze^{j\pi(1-z)}. \quad (1.113)$$

The simple aforementioned examples show that the arithmetic operations on complex functions result in real and imaginary parts in the Cartesian coordinate system and result in the magnitude and phase in polar coordinate system, as functions of a complex variable, $z = x + jy = |z|e^{j\theta}$, where the relationship between the polar and Cartesian representation is given by

$$x = |z| \cos \theta, \quad y = |z| \sin \theta,$$

$$|z| = \sqrt{x^2 + y^2}, \quad \theta = \tan^{-1} \frac{y}{x}.$$

As we shall see throughout this book, complex functions have very interesting symmetry properties, such that they simplify the signals and systems for a wide range of problems that deals with real functions.

1.7 Chapter Summary

What are systems? What are signals? How are they related to each other? What are the examples of systems in nature? What are the examples of human-made systems? How can we represent signals and systems mathematically, so that we can investigate the laws, which governs the natural systems based on the signals they generate? How can we design man-made systems? How can we operate on signals? What are the specific types of signals, which enable us to model a wide range of general signals? What is a complex signal?

In this chapter, we provide introductory answers to the aforementioned questions. We define a system as a unified collection of interrelated and interdependent parts, governed by a set of invariant laws. We define signals as anything that serves to indicate or communicate information. Signals can be considered as the measurements of our varying observations about a system and/or its parts. Depending of the nature of the underlying physical phenomena, signals can take different forms called, continuous time, discrete time or digital signals. The human body, a single cell, a molecule are some examples of natural systems. Computers, vehicles, telecommunication devices are some examples of human-made systems.

A wide range of systems can be represented by a set of equations and/or algorithms. On the other hand, a wide range of signals can be represented by functions. Thus, we can use mathematical tools available in calculus, linear algebra, differential equations and algorithms to model and analyze signals and systems.

Some specific types of functions, which represent signals, can be used as the building blocks of a general class of signals. These functions possess symmetry properties. A very popular set of functions, which have time translation symmetry are the periodic functions. Also, even signals have reflection symmetry and odd signals have rotation symmetry. Functions, which have some type of symmetry, can be used to represent asymmetric signals.

Some signals have domain and range of complex numbers. In order to represent this type of signals, we use complex functions, where the domain and range is represented by two-dimensional complex numbers. Complex functions are very powerful mathematical objects in solving many real-life problems, such as designing digital systems.

Problems

1.1 Consider the plot of the signal $x(t)$ given in Figure P1.1.

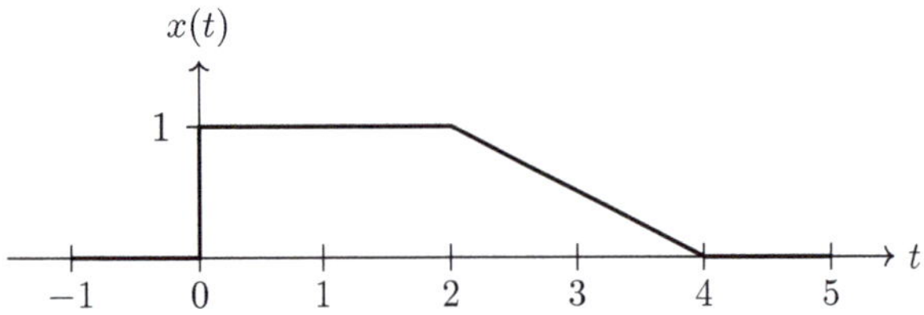

Figure P1.1

a) Find the analytical expression of this function.
b) Find and plot the following functions:
   i) $3x(4t + 8)$
   ii) $x(-4t + 8)$
   iii) $x(-4t - 8)$

1.2 Consider the plot of the signal $x(t)$ given in Figure P1.2.

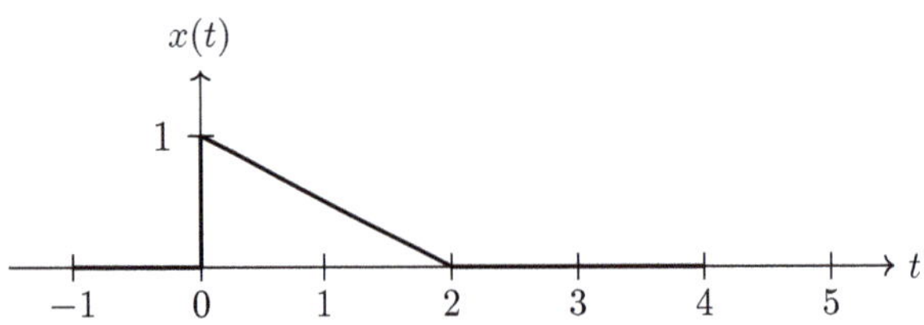

Figure P1.2

a) Find an analytical expression for this figure.
b) Find and plot $x\left(\frac{t}{2} + 1\right)$.
c) Find and plot $x(2t + 1)$.

1.3 Consider the plot of the signal $x[n]$, given in Figure P1.3.
a) Find an plot $x[1 - n]$.
b) Find and plot $x[2n + 2]$.
c) Find and plot $y[n] = x[2n + 2] + x[1 - n]$.

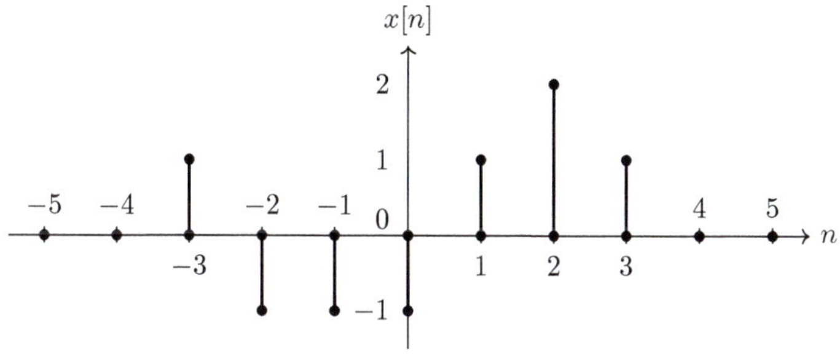

Figure P1.3

1.4 Consider the plot of the odd signal $x(t)$, given in Figure P1.4.

a) Find an analytical expression for this signal using the symmetry property.

b) Find and plot $y(t) = 2x(2t - 3)$. Is this an odd function?

c) Find and plot $y(t) = 2x(2t)$. Is this an odd function?

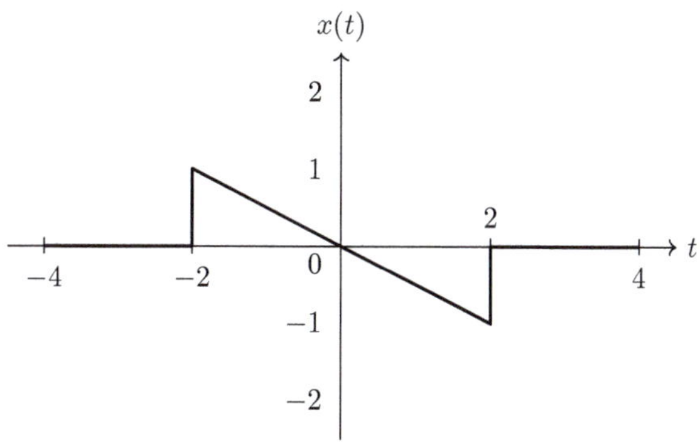

Figure P1.4

1.5 A continuous time signal $x(t)$ is given in Figure P1.5. Find the analytical expression of each of the following signals. Plot them all.

a) $y(t) = x\left(\frac{1}{2}t - 2\right)$

b) $y(t) = x(1 - 2t)$

c) $y(t) = x(2t)$

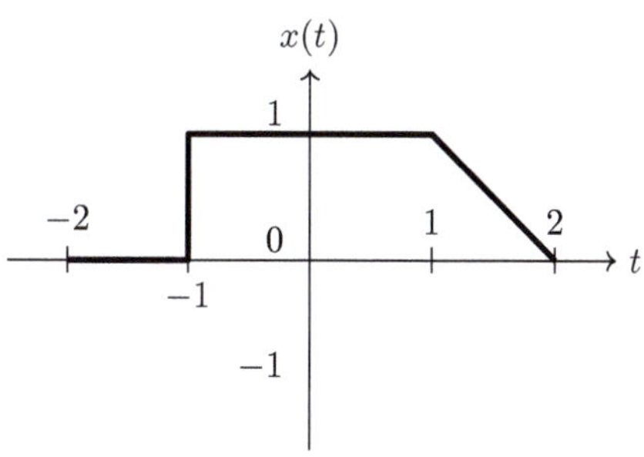

Figure P1.5

1.6 A discrete time even signal $x[n]$ is shown in Figure P1.6. Find the analytical expression for each of the following signals and plot them all.

a) Find an analytical expression of this signal using the symmetry property.

b) $y[n] = x[2 - n]$

c) $y[n] = x\left[\frac{1}{2}n + 1\right]$

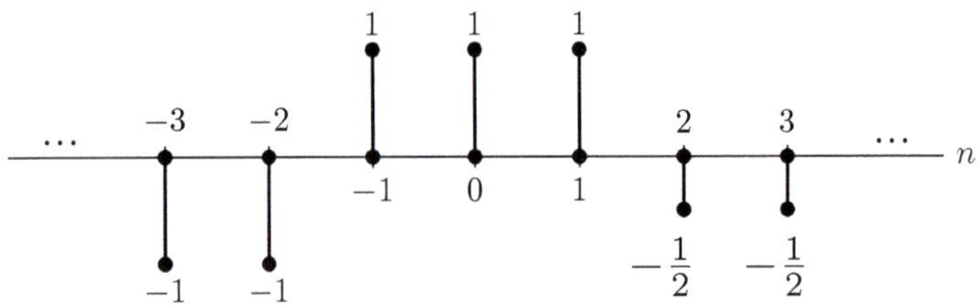

Figure P1.6

1.7 Determine whether or not the following signals are periodic. Determine the fundamental period of the periodic functions.

a) $x_1[n] = \cos\left(5\frac{\pi}{2}n\right)$

b) $x_2[n] = \sin(5n)$

c) $x_3(t) = 5\sin\left(4t + \frac{\pi}{3}\right)$.

1.8 Consider the following signal:

$$x(t) = \frac{\sin\left(\left(m + \frac{1}{2}\right)t\right)}{2\sin\left(\frac{t}{2}\right)},$$

where m is a natural number. Find the fundamental period of this signal.

1.9 Find and plot the even and odd parts of the continuous time signal given in Figure P1.9.

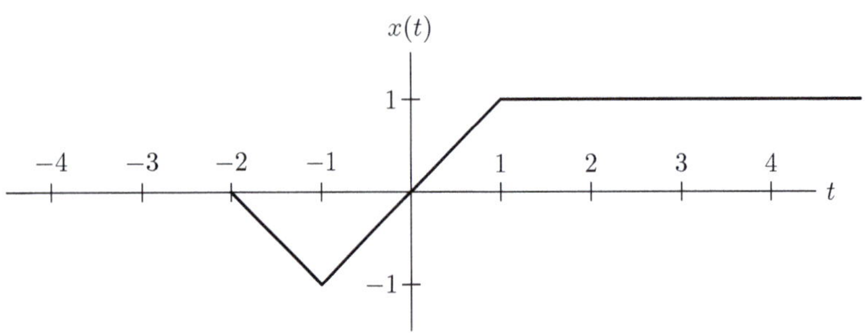

Figure P1.9

1.10 Determine whether or not each of the following continuous time signals is odd, even or neither.

a) $x(t) = \cos(3\pi t)$

b) $x(t) = 2e^{j(3t-2)}$
c) $x(t) = \sin(2\pi) + 3\cos(\pi)$
d) $x(t) = -2t \cdot \cos\left(\frac{1}{4}\pi\right)$.

1.11 Given the following odd signal:

$$x[n] = \sin\left(\frac{3}{4}\pi n\right),$$

show that

$$\sum_{n=-\infty}^{+\infty} x[n] = 0.$$

1.12 Given the following signals:

$$x_1[n] = \cos(n) \quad \text{and} \quad x_2(n) = \sin(n),$$

show that $y[n] = x_1[n]x_2[n]$ is an odd signal.

1.13 Determine whether the following functions are **periodic, even, odd, power**, and **energy** signals. If they are periodic, find the fundamental period.
a) $x(t) = \cos\left(4t + \frac{\pi}{9}\right)$.
b) $x[n] = 2\sin\left[\frac{\pi}{3}n\right]$.

1.14 Consider the following complex number:

$$z = \frac{\sqrt{2} + \sqrt{2}j}{2 + 2\sqrt{3}j}.$$

a) Find the real and imaginary part of this number.
b) Find the magnitude and phase of this number.

1.15 Consider the following complex number:

$$z = e^{5j} + e^{7j}.$$

a) Find the magnitude and phase of this number.
b) Find the real and imaginary part of this number.

1.16 Consider the following complex number:

$$z = (\sqrt{3} - \sqrt{3}j)^{40}.$$

a) Find the magnitude and phase of this number.
b) Find the real part and the imaginary part of this number

1.17 Solve the following and show all solution steps in detail. Simplify your results as much as possible to the format: $a + jb$, where a and b are real numbers.
a) $z = \frac{1}{4} - \frac{1}{3}j \quad \Rightarrow \quad \frac{1}{z} = ?$

b) $z_1 = 4 - 3j,\ z_2 = -j \quad \Rightarrow \quad |z_1 \cdot \bar{z}_2 + z_2 \cdot \bar{z}_1| = ?$

c) $z = \dfrac{(1+\sqrt{3}j)^2(2-j)}{(1+2j)^3} \quad \Rightarrow \quad |z| = ?$

1.18 Evaluate the following integrals and show all solution steps in detail.

a) $\displaystyle\int t^2 e^{-(\sqrt{2}+\sqrt{2}i)t}\,dt$

b) $\displaystyle\int e^{-t}\cos\left(\frac{\pi}{4}t + \frac{\pi}{6}\right)\,dt.$

1.19 Find the real and imaginary parts of the following complex functions, where

$$z = x + jy.$$

a) $X(z) = \cos z$
b) $X(z) = (z+1)\sin z$
c) $X(z) = z^3 + 5z - 1.$

1.20 Find the magnitude and phase of the following complex functions, where

$$z = |z|e^{j\theta}.$$

a) $X(z) = \sin z$
b) $X(z) = z^2 \cos z$
c) $X(z) = e^{jz} + e^{3jz}.$

1.21 Take the following integrals:

a) $\displaystyle\int_0^{1+j} e^z dz$

b) $\displaystyle\int_0^{1+j} z^3 dz.$

1.22 Write a computer program to plot the **even** and **odd parts** of a discrete time signal $x[n]$. Your program takes the signal and the starting index(s_i) of the signal as input. For example, let's say $x[n] = [1, 6, 8, 9]$ and $s_i = 3$, then $x[3] = 1, x[4] = 6, x[5] = 8, x[6] = 9$ and $x[n] = 0$ for other n values.
You should add your codes and the outputs for the given three input files (sine_part_a.csv,[7] shifted_sawtooth_part_a.csv,[8] and chirp_part_a.csv[9]) to your solution. The first element in the files is the starting index and remaining ones are the elements of the signal.

1.23 Write a computer program to plot the shifted and scaled version $x[an + b]$ of a discrete time signal $x[n]$. Your program takes the signal and the starting index(s_i) of the signal as input. Differently from part a, you should also take a and b values as input.

7 https://384book.net/resources/sine_part_a.csv.
8 https://384book.net/resources/shifted_sawtooth_part_a.csv.
9 https://384book.net/resources/chirp_part_a.csv.

You should add your codes and the outputs for the given three input files (sine_part_b.csv,[10] shifted_sawtooth_part_b.csv,[11] and chirp_part_b.csv[12]) to your solution. The first element in the files is the starting index, the second element is the value of a, the third element is the value of b and the remaining ones are the elements of the signal.
You should write your code in `Python` and no library is allowed other than `matplotlib.pyplot`.

10 https://384book.net/resources/sine_part_b.csv.
11 https://384book.net/resources/shifted_sawtooth_part_b.csv.
12 https://384book.net/resources/chirp_part_b.csv.

2

Basic Building Blocks of Signals

> With a bucket of LEGO toys, we can construct many objects, as long as we can imagine. With a set of simple functions, we can construct many complicated signals, as long as we speak the language of mathematics!

In Chapter 1, we introduced the general concepts about the signals and systems, together with some mathematical background needed for the rest of the book. As we mentioned before, we may attempt to model and analyze natural objects, such as the human brain or we may design and implement man-made objects, such as a computer vision system.

Unfortunately, the available mathematical tools are quite short to model all classes of systems and signals. However, it is possible to put a set of "valid" assumptions and rules to decompose a complicated system into interconnectable subsystems for modeling a large class of systems.

In this chapter, we explore some basic building blocks of signals, which enable us to design and implement mathematically tractable and realizable models with systems approach.

2.1 LEGO Functions of Signals

There are some basic functions, which can be used to represent a large variety of signals, such as speech and music. Just like playing with LEGO toys, we can play with the basic functions to construct relatively more complicated signals, using the available mathematical tools. Most of these functions have symmetry properties. In other words, they are invariant with respect to some type of transformations. In Chapter 1, we saw the class of periodic, even, and odd functions. These functions were symmetric with respect to translation, reflection, and rotation, respectively. In this section, we shall investigate additional basic functions, namely

1) exponential functions,
2) the unit impulse function,
3) the unit step function.

Later in this book, we shall see that linear combinations of the basic functions can be used to represent a large class of signals.

Signals and Systems: Theory and Practical Explorations with Python,
First Edition. Fatoş Tunay Yarman Vural and Emre Akbaş.

Companion website: www.wiley.com/go/signalsandsystems

2.2 King of the Functions: Exponential Function

One of the most important functions in mathematics is the exponential function. It is widely used to model natural phenomena, such as the growth of cancer cells or uncontrolled forest fires. It is, also, extensively used in man-made systems, for example in the "softmax layer" of artificial neural networks, or the nonlinear activation functions.

Loosely speaking, an exponential function is a function, in which its derivative (the amount of increase or decrease) with respect to its variable is constantly proportional to itself.

Formally speaking, the exponential function, in continuous time is defined as follows:

$$x(t) = Ce^{\alpha t}, \tag{2.1}$$

whereas the exponential function, in discrete time is defined as follows:

$$x[n] = Ce^{\alpha n}, \tag{2.2}$$

where

$$e = \sum_{n=0}^{\infty} \frac{1}{n!} \approx 2.71828182\ldots \tag{2.3}$$

is a transcendental number, called, the Euler number.

There are two parameters of an exponential function; the amplitude, C, and the parameter of exponent, α. Both parameters can be real or complex numbers. Depending on the parameters of the exponential function, we shall investigate two types of exponential functions, namely **real exponential function** and **complex exponential function**, for both **continuous time** and **discrete time signals**.

2.2.1 Real Exponential Function

A real exponential function has real parameters, $C \in \mathbb{R}$ and $\alpha \in \mathbb{R}$. Real exponential functions can be used to represent continuous time or discrete time signals.

2.2.1.1 Continuous Time Real Exponential Function

A **continuous time real exponential function** is a monotonically increasing function, when $\alpha > 0$ and it is a monotonically decreasing function, when $\alpha < 0$, as shown in Figure 2.1.

When both parameters, $C = \alpha = 1$, then we obtain **natural exponential function**, as follows:

$$x(t) = e^{t}. \tag{2.4}$$

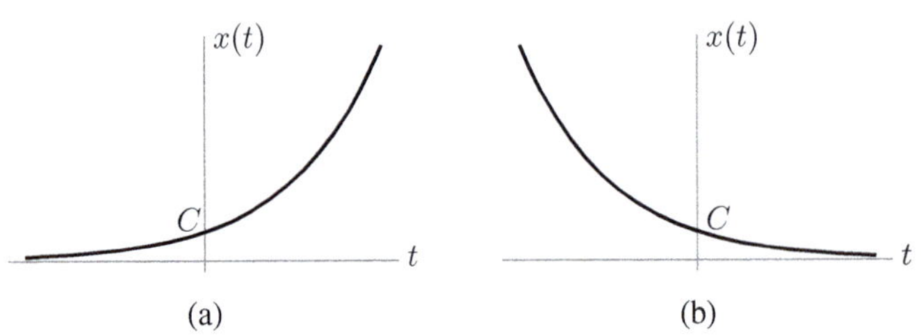

Figure 2.1 Continuous time exponential, $x(t) = Ce^{\alpha t}$, (a) for $\alpha > 0$ and (b) $\alpha < 0$.

The inverse of the natural exponential function is called the **natural logarithm**, which provides the variable t, as follows:

$$t = \ln x(t). \tag{2.5}$$

The natural exponential function has special importance in mathematics. It is the only function whose derivative and integral are the same as the function itself, as formally stated in the following Lemma.

Lemma 2.1 The first derivative of the natural exponential function is e^t itself:

$$\frac{de^t}{dt} = e^t, \tag{2.6}$$

and the integral of the real exponential in $(0, t)$ is itself:

$$\int_0^t e^t dt = e^t. \tag{2.7}$$

Proof: In order to find the derivative of e^t, we simply use the definition of the derivative,

$$\frac{de^t}{dt} = \lim_{h\to 0}\frac{e^{t+h} - e^t}{h} = \lim_{h\to 0}\frac{e^t e^h - e^t}{h} = e^t \lim_{h\to 0}\frac{e^h - 1}{h}. \tag{2.8}$$

The limit term in the right-hand side of Equation (2.8) is equal to 1, because, applying the L'Hopital rule

$$\lim_{h\to 0}\frac{e^h - 1}{h} = \lim_{h\to 0} e^h = 1. \tag{2.9}$$

Thus,

$$\frac{de^t}{dt} = e^t. \tag{2.10}$$

To show that the integral of the real exponential is itself, we simply take the derivative of both sides of Equation (2.7) with respect to t. The fundamental theorem of calculus gives us

$$\frac{d}{dt}\int_0^t e^t dt = e^t, \tag{2.11}$$

which is equivalent to the derivative of the right-hand side of Equation (2.7).

The aforementioned lemma shows that continuous time real exponentials, with base e are symmetric functions with respect to the derivative and integration operations.

Definition 2.1 **The general exponential function** for any parameter, β, is defined as follows:

$$x(t) = \beta^t. \tag{2.12}$$

The inverse of the general exponential function is defined by base-β logarithm, as follows:

$$t = \log_\beta x(t). \tag{2.13}$$

Motivating Question: Why do we call e^t as natural exponential?

The natural exponential function enables us to study all kinds of general exponential functions. Because the exponential function for any base parameter, $\beta \in \mathbb{R}$, can be written in terms of the natural exponential, with base-e, as shown by the following Lemma.

Lemma 2.2 Any general form of an exponential function can be written in terms of the natural exponential function, e^t, i.e.,

$$x(t) = A\beta^t = Ae^{\alpha t}. \tag{2.14}$$

Proof: Note that for any real number $a \in \mathbb{R}$,

$$a = e^{\ln(a)}. \tag{2.15}$$

Thus,

$$\beta^t = e^{\ln(\beta^t)} = e^{(t \ln \beta)}. \tag{2.16}$$

Then, for any A and β,

$$x(t) = A\beta^t = Ae^{\alpha t}, \tag{2.17}$$

where $\alpha = \ln \beta$.

Exercise 2.1 Show that the following algebraic properties hold for the real exponential function:

a) $e^{\alpha t+\beta} = e^{\alpha t} e^{\beta}$
b) $e^{-t} = 1/e^t$
c) $e^{\alpha t-\beta} = \frac{e^{\alpha t}}{e^{\beta}}$
d) $e^{\alpha t} = (e^t)^{\alpha}$, $\forall \alpha$ is rational.

Solution

a) Take the logarithm of the left-hand side of the equation,

$$\ln(e^{\alpha t+\beta}) = \alpha t + \beta = \ln e^{\alpha t} + \ln e^{\beta} = \ln(e^{\alpha t} e^{\beta}).$$

Recall that $\ln t$ is a monotonically increasing function. Thus, it is one-to-one and onto. Then,

$$e^{\alpha t+\beta} = e^{\alpha t} e^{\beta}.$$

b) Note that the 0th power of any real number is 1. Then,

$$e^0 = 1 = e^{(t-t)} = e^t e^{-t}.$$

Therefore, $e^{-t} = 1/e^t$.

c) Let's rephrase the left-hand side of the equation

$$e^{\alpha t-\beta} = e^{\alpha t+(-\beta)} = e^{\alpha t} e^{-\beta} = e^{\alpha t}/e^{\beta},$$

d) For $\alpha = n$ is integer,

$$e^{nt} = e^{(t+t+\ldots+t)} = e^t e^t e^t \ldots e^t = (e^t)^n.$$

For $\alpha = n/m$ is rational,

$$e^{\frac{n}{m}t} = (e^{\frac{1}{m}t})^n = ((e^t)^{\frac{1}{m}})^n = (e^t)^{\frac{n}{m}}.$$

2.2.1.2 Discrete Time Real Exponential Function

Discrete time real exponential function is an exponential function, which is only defined at every integer value of time instance.

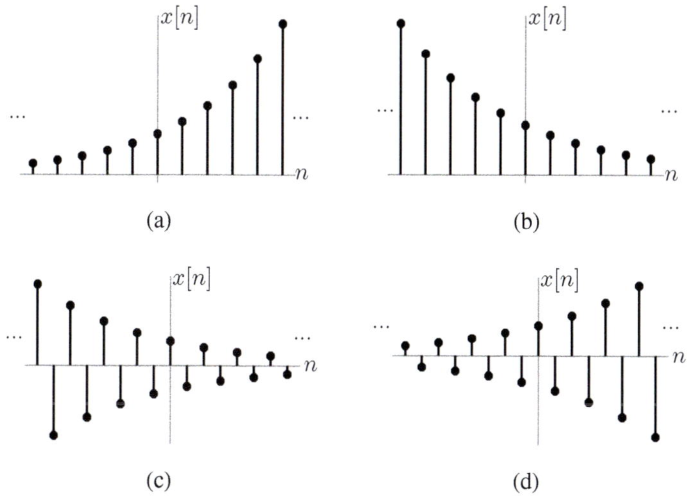

Figure 2.2 Discrete time real exponential function for (a) $\beta > 1$, (b) $0 < \beta < 1$, (c) $-1 < \beta < 0$, and (d) $\beta < -1$.

The analytic form of discrete time real exponential function is

$$x[n] = C\beta^n = Ce^{\alpha n}, \tag{2.18}$$

where $\beta = e^{\alpha}$, α, β are real-valued numbers and n is integer-valued variable. Discrete time real exponential function can be represented by a sequence of real numbers, where each value of the sequence is generated by Equation (2.18) for $n = 0, \pm1, \pm2, \pm3, \ldots$ The shape of the function depends on the value of the β or $\alpha = \ln \beta$ parameter. There are four different forms of a **discrete real exponential function** (see Figure 2.2):

- For $\beta > 1$, it increases monotonically,
- For $0 < \beta < 1$, it decreases monotonically,
- For $-1 < \beta < 0$, it alternates, while its absolute value decreases monotonically,
- For $\beta < -1$, it alternates, while its absolute value increases monotonically.

When we work with the discrete time exponential functions, we always keep in mind that the time variable n takes only integer values. For example, the logarithm of the discrete time exponential function, $x[n] = \beta^n$,

$$n = \ln_{\beta} x[n], \tag{2.19}$$

exits only at integer values of n.

2.2.2 Complex Exponential Function

When the parameter, α, of the exponential functions for a continuous time signal,

$$x(t) = Ce^{\alpha t}, \tag{2.20}$$

and a discrete time signal,

$$x[n] = Ce^{\alpha n}, \tag{2.21}$$

is a **complex number**, in other words,

$$\alpha = a + j\omega_0, \tag{2.22}$$

then we call them **complex exponential functions**. Adding an extra imaginary dimension to the parameter, α, at the exponent change the behaviour of the exponential function, as we shall see in the Chapter 8.

In the context of this book, we focus on a special form of complex functions, where α is purely imaginary. In other words, $a = 0$.

2.2.2.1 Continuous Time Complex Exponential Functions

We define the continuous time complex exponential function as follows:

$$x(t) = Ce^{j\omega_0 t}. \tag{2.23}$$

Note that the time variable is always a real number, in all of the functions, $t \in \mathbb{R}$. The amplitude, C, may or may not be a complex number in complex exponential. The Euler formula, introduced in Chapter 1, enables us to represent the continuous time complex exponential function in terms of trigonometric functions in the complex plane:

$$x(t) = Ce^{j\omega_0 t} = C(\cos \omega_0 t + j \sin \omega_0 t). \tag{2.24}$$

Continuous time complex exponential function has a very crucial symmetry property: It is periodic! This property has a special importance in representing signals in function spaces.

Let us prove the periodicity property of continuous time complex exponential in the following Lemma.

Lemma 2.3 Given a continuous time complex exponential function, $x(t) = e^{j\omega t}$, there exists a finite, nonzero value, $T \in \mathbb{R}$, such that,

$$e^{j\omega_0 t} = e^{j\omega_0 (t+T)}. \tag{2.25}$$

Proof: Let us split the complex exponential on the right-hand side of the equation into two expressions:

$$e^{j\omega_0 (t+T)} = e^{j\omega_0 t} e^{j\omega_0 T}. \tag{2.26}$$

In order to satisfy the periodicity property of complex exponential, we need,

$$e^{j\omega_0 T} = 1. \tag{2.27}$$

Using the Euler formula, we get,

$$e^{j\omega_0 T} = \cos \omega_0 T + j \sin \omega_0 T. \tag{2.28}$$

When the period of the complex exponential function satisfies,

$$T = k\frac{2\pi}{|\omega_0|}. \tag{2.29}$$

we can write,

$$e^{j\omega_0 T} = 1 = \cos \omega_0 \frac{2\pi k}{|\omega_0|} + j \sin \omega_0 \frac{2\pi k}{|\omega_0|}. \tag{2.30}$$

When k is an integer, the imaginary part of Equation (2.30) cancels and that of the real part becomes 1. Thus, a complex exponential function is periodic, with the fundamental period,

$$T = \frac{2\pi}{|\omega_0|}. \tag{2.31}$$

Learn more about the Euler formula, which explains complex exponential as rotation @ `https://384book.net/0201`

2.2.2.2 Harmonically Related Complex Exponential

Harmony is an important universal concept in arts and sciences. In music, harmony means simultaneously occurring harmonically related sounds of musical instruments and voices. In painting, color harmony refers to combining the colors of different frequencies in a way that is harmonious to the human eye (Figure 2.3).

Musical tunes, human voice, photographs, and paintings are all measurable signals, which can be represented by functions. Thus, the aesthetic values of harmony can be quantified to a certain extent by using the mathematics of harmony.

In the context of this book, we investigate the mathematical properties of harmonically related functions (signals), specifically, harmonically related complex exponential. Interestingly, harmonically related complex exponentials provide us with a set of orthogonal basis functions, which span a vector space, called function space. As we shall study in Chapter 6, in a function space, a large class of functions can be represented in terms of weighted summation of harmonically related complex exponentials.

Motivating Question: What do harmonically related frequencies or functions mean in mathematics?

A complex variable in polar coordinates,

$$z = e^{j\omega_0}, \tag{2.32}$$

Figure 2.3 A painting, generated by Adobe Firefly using color harmony with fundamental colors of red and orange. Source: AI-generated image created in Adobe Firefly.

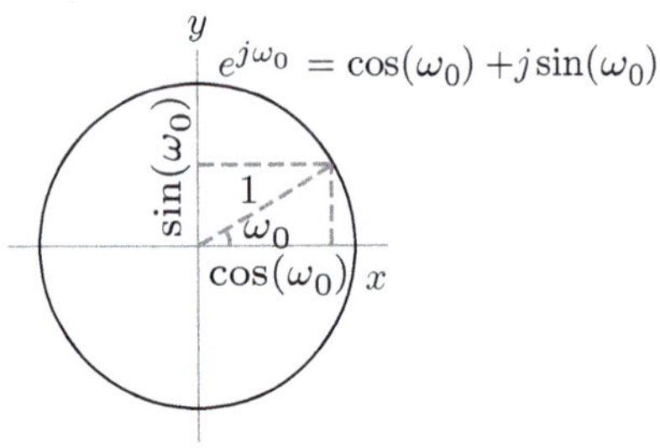

Figure 2.4 Periodic motion represented by a complex exponential function, in continuous time: Unit Circle on a complex plane, with $r = 1$ and $z = x + jy = e^{j\omega_0}$.

moves on a unit circle with radius $r = 1$, as we change the angle, ω_0, in the interval of $0 \leq \omega_0 \leq 2\pi$, in the complex plane (Figure 2.4).

The complex exponential function in continuous time,

$$x(t) = e^{j\omega_0 t} = \cos\omega_0 t + j\sin\omega_0 t, \tag{2.33}$$

includes a time dimension, t, which is perpendicular to the complex plane. Thus, like the complex variable z, complex exponential rotates on a unit circle as a function of time, drawing a spiral along the time axis (Figure 2.5).

WATCH

Visualization of continuous time complex exponential @ https://384book.net/0202

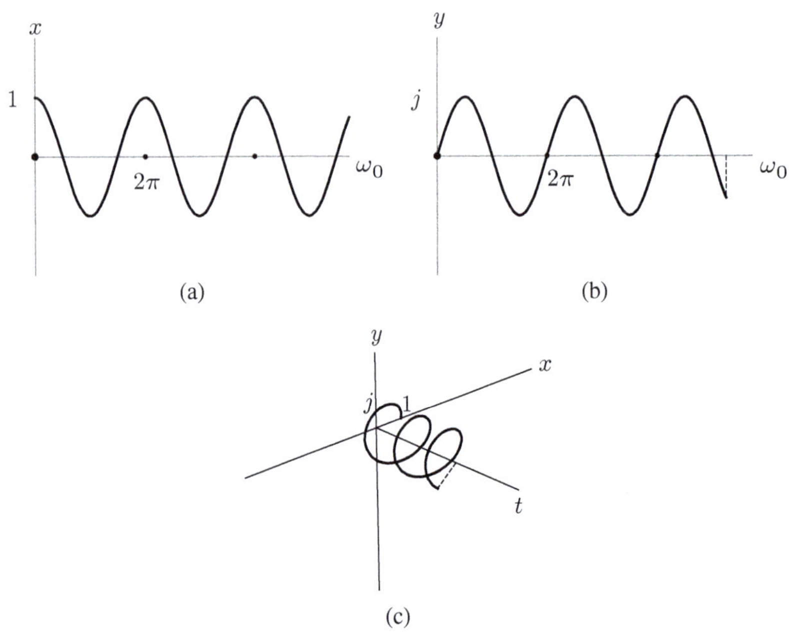

Figure 2.5 Periodic motion represented by a complex exponential function, in continuous time: (a) real part of complex exponential function, $\mathrm{Re}\{e^{j\omega_0 t}\} = \cos\omega_0 t$, (b) imaginary part of complex exponential function, $\mathrm{Im}\{e^{j\omega_0 t}\} = \sin\omega_0 t$, and (c) complex exponential function, $x(t) = e^{j\omega_0 t}$.

The real part, $\mathrm{Re}\{x(t)\} = \cos\omega_0 t$ and an imaginary part, $\mathrm{Im}\{x(t)\} = \sin\omega_0 t$ of the complex exponential, are, also periodic with the fundamental period,

$$T_0 = \frac{2\pi}{\mid \omega_0 \mid}, \tag{2.34}$$

as shown in Figure 2.4. The fundamental period, T_0 is, then, the amount of time seconds to complete one tour around the unit circle.

Motivating Question: What if we multiply the angular frequency ω_0 by an integer value k? The fundamental period becomes,

$$T_0 = \frac{2\pi}{k|\omega_0|}. \tag{2.35}$$

In Equation (2.35), the angular frequency, ω_0, is increased by a factor of k, while the fundamental period is decreased by a factor of k. Then, the complex exponential function completes a cycle in a shorter time on the unit circle, drawing spirals at a faster rate, as we increase the integer value k.

Definition 2.2 **Harmonically related sets of complex exponentials** are defined as:

$$\phi_k(t) = e^{jk\omega_0 t}, \quad \forall k = 0, \mp 1, \mp 2, \mp 3, \dots \tag{2.36}$$

Harmonically related complex exponential functions have angular frequencies, called **harmonics**, which are integer multiples of the angular frequency,

$$\omega_k = k\omega_0 \quad \forall k = 0, \mp 1, \mp 2, \mp 3, \dots \tag{2.37}$$

Definition 2.3 Given a periodic real-time signal, $x(t)$ with a fundamental frequency f_0, the **harmonically related periodic signals** are defined as the set of all signals,

$$x_k(t), \quad \forall k = 0, \mp 1, \mp 2, \mp 3, \dots, \tag{2.38}$$

where each $x_k(t)$, has a fundamental frequency of kf_0, called **harmonics**.

Harmonically related periodic functions have the fundamental frequencies, which are the integer multiple of the fundamental frequency, f_0 of a function, $x(t)$. Like the concept of symmetry, harmonically related signals have aesthetic values. For example, behind the art of composing music, there is a science of musical harmony for combining the harmonically related musical tunes. Similarly, great painters use color harmony theory, where they use harmonically related colors. Nature, as a whole, bears infinitely many harmonically related signals, which we can partially observe all over the universe. A popular example is the harmonically related electromagnetic waveforms, which make the observable universe.

WATCH

Learn more about the fundamental frequency and its harmonics in music @ `https://384book.net/0203`

Exercise 2.2 Consider the superposition of the following complex exponential signal

$$x(t) = \frac{1}{2}(e^{j\omega_0 t} + e^{-j\omega_0 t}) \tag{2.39}$$

a) Find its trigonometric form.
b) Find its angular frequency and the fundamental period.
c) Find and plot the superposition of the first and second harmonics of $x(t)$, given as follows:

$$x_1(t) + x_2(t), \tag{2.40}$$

for $\omega_0 = \pi/2$.

Solution

a) Use the Euler formula, which relates the complex exponential to the trigonometric functions,

$$e^{j\omega_0 t} = \cos\omega_0 t + j\sin\omega_0 t. \tag{2.41}$$

Then,

$$x(t) = \frac{1}{2}(\cos\omega_0 t + \cancel{j\sin\omega_0 t} + \cos\omega_0 t - \cancel{j\sin\omega_0 t}) = \cos\omega_0 t. \tag{2.42}$$

b) Angular frequency is ω_0 and the fundamental period is $T_0 = 2\pi/\omega_0$.
c) Recall that the kth harmonic of $x(t)$ is defined as:

$$x_k(t) = \frac{1}{2}(e^{jk\omega_0 t} + e^{-jk\omega_0 t}). \tag{2.43}$$

Superposition of the first two harmonics of $x(t)$ for $\omega_0 = 1$ is

$$x_1(t) + x_2(t) = cos\omega_0 t + \cos 2\omega_0 t = \cos t + \cos 2t. \tag{2.44}$$

The plot of $x_1(t) + x_2(t)$ is given in Figure 2.6.

Exercise 2.3 Show that trigonometric functions, $x(t) = \cos\omega_0 t$ and $x(t) = \sin\omega_0 t$, can be represented in terms of the complex exponentials, $e^{j\omega_0 t}$.

Solution

Recall the Euler formula

$$e^{j\omega_0 t} = \cos\omega_0 t + j\sin\omega_0 t, \tag{2.45}$$

Similarly,

$$e^{-j\omega_0 t} = \cos\omega_0 t - j\sin\omega_0 t, \tag{2.46}$$

Thus,

$$\cos\omega_0 t = \frac{e^{j\omega_0 t} + e^{-j\omega_0 t}}{2}, \tag{2.47}$$

$$\sin\omega_0 t = \frac{e^{j\omega_0 t} - e^{-j\omega_0 t}}{2j}. \tag{2.48}$$

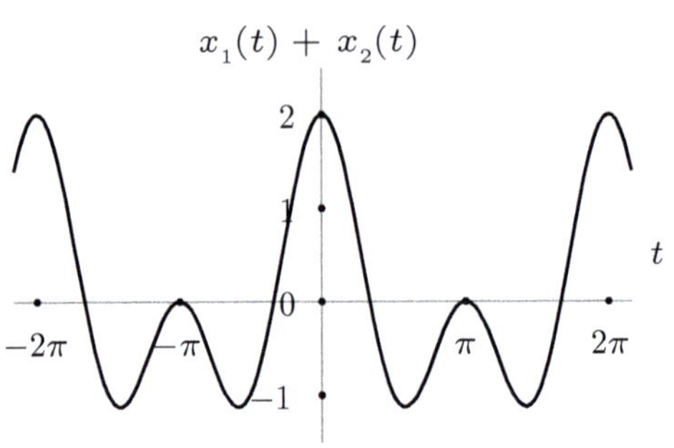

Figure 2.6 Plot of the superposition of the first two harmonics of $x(t)$.

Note that, although sine and cosine functions are real functions, they can be uniquely represented by the superposition of harmonically related complex exponentials, for $k = -1$ and 1.

Exercise 2.4 Consider the following function:

$$x(t) = e^{2j\omega_0 t} + e^{4j\omega_0 t}.$$

a) Find the magnitude and phase of this function.
b) Find the real part and the imaginary part of this function.

Solution

a) Recall that a complex function, represented in polar form,

$$x(t) = C(t)e^{j\theta(t)}.$$

has the magnitude $C(t)$ and the phase $\theta(t)$. Taking the average of the exponents and defining the complex exponential with the average exponent, we get,

$$x(t) = e^{3j\omega_0 t}(e^{j\omega_0 t} + e^{-j\omega_0 t}) = (2\cos\omega_0 t)e^{3j\omega_0 t}.$$

Thus, the magnitude is $C(t) = 2\cos\omega_0 t$ and the phase is $\theta(t) = 3\omega_0 t$.

b) Recall that a complex function in Cartesian form is represented by,

$$x(t) = \text{Re}\{x(t)\} + j\text{Im}\{x(t)\}.$$

Using the Euler formula, we obtain,

$$x(t) = \cos 2\omega_0 t + \cos 4\omega_0 t + j(\sin 2\omega_0 t + \sin 4\omega_0 t).$$

Thus the real part is

$$\text{Re}\{x(t)\} = \cos 2\omega_0 t + \cos 4\omega_0 t,$$

and the imaginary part is

$$\text{Im}\{x(t)\} = \sin 2\omega_0 t + \sin 4\omega_0 t.$$

2.2.2.3 Complex Exponential Function for Discrete Time Signals

Discrete time complex exponential function is an exponential function, which is only defined at every integer value of time instance.

The analytic form of discrete time complex exponential function is

$$x[n] = Ae^{j\omega_0 n}. \tag{2.49}$$

Most of the properties of the discrete time complex exponential functions are very similar to that of the continuous time exponential functions, except that n can only take integer values. This brings a serious constraint to the fundamental period, N_0, which has to be an integer.

Euler formula, introduced in Chapter 1, is also applicable to discrete time complex exponential functions to represent it in terms of trigonometric functions:

$$x(t) = Ae^{j\omega_0 n} = A(\cos\omega_0 n + j\sin\omega_0 n). \tag{2.50}$$

Motivating Question: Is discrete complex exponential periodic?
Not always!

As we mentioned earlier, the discrete time complex exponential is periodic if there exists an integer value N, which satisfies the following condition:

$$e^{j\omega_0(n+N)} = e^{j\omega_0 n} = e^{j\omega_0 N} e^{j\omega_0 n}. \tag{2.51}$$

Equation (2.51) is valid, if

$$e^{j\omega_0 N} = 1. \tag{2.52}$$

To satisfy the periodicity property of the discrete time complex exponentials, we need to find an integer period, N, satisfying $\omega_0 N = 2k\pi$.

For the fundamental period, we need to find the smallest integer value k, such that $N_0 = \frac{2\pi}{\omega_0}k$ is an integer.

Exercise 2.5 Is the following signal periodic?

$$e^{-jn} = \cos n - j \sin n$$

Solution

The angular frequency of the function is $\omega_0 = 1$. The period is

$$N = 2k\pi/\omega_0 = 2k\pi. \tag{2.53}$$

Due to the irrational number π, the period N cannot be an integer for any value of k. Therefore, the function e^{-jn} is not periodic!

Exercise 2.6 Is the following signal periodic?

$$e^{-j\pi n} = \cos \pi n - j \sin \pi n, \tag{2.54}$$

Solution

Angular frequency is π. The period is

$$N = 2k\pi/\pi. \tag{2.55}$$

For $k = 1$ the fundamental period is, $N_0 = 2$. Yes, it is periodic!

Exercise 2.7 Find the fundamental period of the following discrete time signals:

a) $x[n] = e^{(\pi/3)n} - e^{(\pi/4)n}$
b) $x[n] = e^{(2\pi/3)n} - e^{(\pi/4)n}$

Solution

a) The function $x[n]$ has two components: $x_1[n] = e^{(\pi/3)n}$ and $x_2[n] = e^{(\pi/4)n}$. The fundamental period of $x_1[n]$ is $N_1 = 6$ and that of $x_2[n]$ is $N_2 = 8$. The first component of x[n] repeats itself with period 6, and the second component with 8. Then, the combined signal will repeat itself with period $N = 24$, which is the least common multiple of 6 and 8.
b) As in part (a), $x[n]$ has two components: $x_1[n] = e^{(2\pi/3)n}$ and $x_2[n] = e^{(\pi/4)t}$. The fundamental period of $x_1[n]$ is $N_1 = 3$ and that of $x_2[n]$ is $N_2 = 8$. This time 8 is not integer multiple of 3. Thus, the fundamental period of $x[n]$ is the least common multiple of 3 and 8, which is $N = 24$.

INTERACTIVE

Explore the complex exponential signal @ https://384book.net/i0201

2.3 Unit Impulse Function

The simplest basic building block of the functions is the unit impulse function. If we play it as a sound, it would sound like an explosion for a very short time interval.

In the following, we shall investigate the discrete time and continuous time unit impulse functions.

2.3.1 Discrete Time Unit Impulse Function or Dirac-Delta Function

Discrete time unit impulse function is a real function, defined as:

$$\delta[n] = \begin{cases} 1 & \text{for } n = 0, \\ 0 & \text{otherwise.} \end{cases} \tag{2.56}$$

Discrete time unit impulse function has only one nonzero value at the origin, which is equal to 1 (Figure 2.7).

Exercise 2.8 Find the value of the following function:

$$x[n] = (n + 2)\delta[n].$$

Solution
The value of this function is nonzero at $n = 0$, only. Thus,

$$x[n] = 2\delta[n].$$

Exercise 2.9 Simplify the function:

$$x[n] = (n - 2)\delta[n - 2].$$

Solution
The value of this function is non-zero at $n = 2$, only. Thus,

$$x[n] = 0.$$

In general, we can write

$$x[n]\delta[n - n_0] = x[n_0]\delta[n - n_0]. \tag{2.57}$$

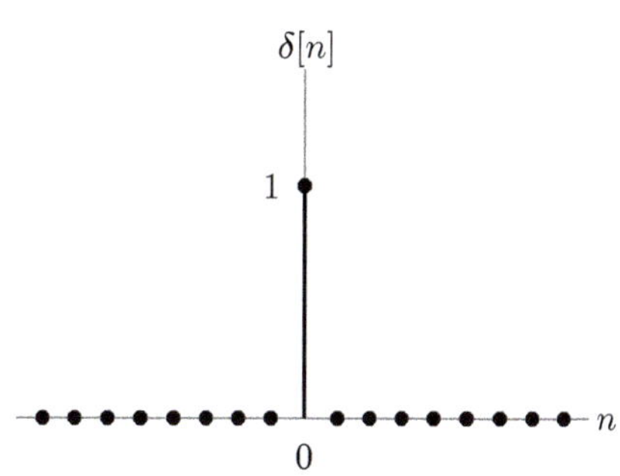

Figure 2.7 Discrete time unit impulse function. We put a small dot at every point to show that the height is either zero or finite value of 1, at $n = 0$.

2.3.2 Continuous Time Unit Impulse Function

Loosely speaking, a continuous time impulse function is a real-valued function with an infinite height and zero width and it integrates to one. Continuous time impulse function can be imagined as a really big instantaneous explosion.

The formal definition of the continuous time unit impulse function requires the concept of limit. First, we define a function, $\delta_\Delta(t)$ as:

$$\delta_\Delta(t) = \begin{cases} \frac{1}{\Delta} & \text{for } 0 < t < \Delta, \\ 0 & \text{otherwise.} \end{cases} \tag{2.58}$$

Note that the area under $\delta_\Delta(t)$ is equal to 1:

$$\int_{-\infty}^{\infty} \delta_\Delta(t)dt = 1. \tag{2.59}$$

The plot of $\delta_\Delta(t)$ is given in Figure 2.8.

Unit impulse function in continuous time is defined as the limit of the $\delta_\Delta(t)$ function, as follows:

$$\delta(t) = \lim_{\Delta \to 0} \delta_\Delta(t). \tag{2.60}$$

This function has a peculiar shape, with zero width and infinite height. However, the area under this function is finite

$$\int_{-\infty}^{\infty} \delta(\tau)d\tau = 1. \tag{2.61}$$

The unit impulse, $\delta(t)$, is plotted as an arrow of height 1 (Figure 2.9).

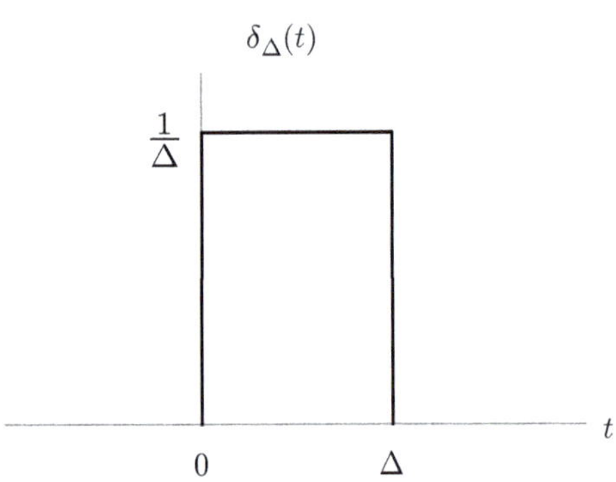

Figure 2.8 The plot of $\delta_\Delta(t)$ function. The width of this function is Δ and the height is $1/\Delta$.

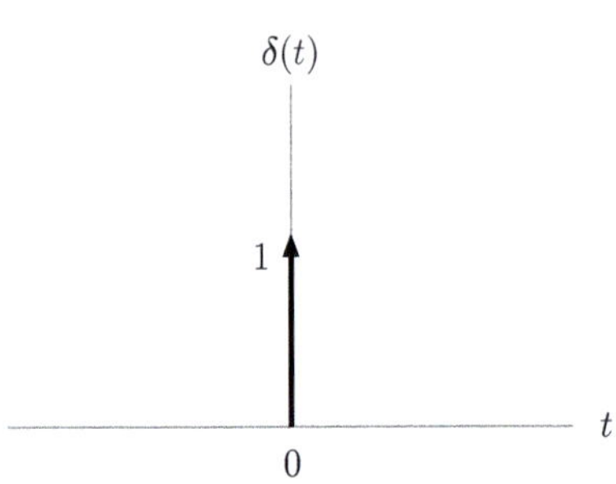

Figure 2.9 Schematic representation of the unit impulse function in continuous time. We put an arrow to the function to indicate that the height is infinity.

Exercise 2.10 The plot is given in Figure 2.10. Plot the superposition of two shifted continuous time unit impulse functions,

$$x(t) = \delta(t-1) + \delta(t-2). \tag{2.62}$$

Solution

Figure 2.10 Plot of $x(t) = \delta(t-1) + \delta(t-2)$.

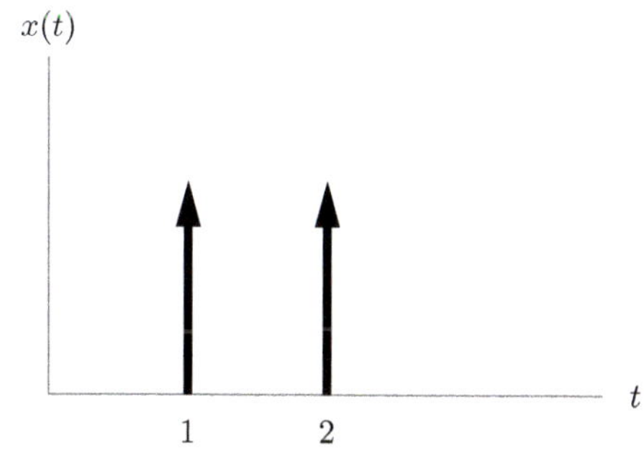

Multiplication of a function by the continuous time impulse function requires to take the limit. For sufficiently small Δ, we can write the following approximation:

$$x(t)\delta_\Delta(t) \approx x(0)\delta_\Delta(t). \tag{2.63}$$

Taking the limit of both sides, we obtain

$$\lim_{\Delta\to 0} x(t)\delta_\Delta(t) = \lim_{\Delta\to 0} x(0)\delta_\Delta(t). \tag{2.64}$$

Thus,

$$x(t)\delta(t) = x(0)\delta(t). \tag{2.65}$$

Similarly, we can show that

$$x(t)\delta(t-t_0) = x(t_0)\delta(t-t_0). \tag{2.66}$$

Exercise 2.11 Find the value of the following continuous time function:

$$x(t) = \int_{-\infty}^{\infty} (t+2)\delta(t)dt.$$

Solution
Let us first evaluate the integrand:

$$(t+2)\delta(t) = 2\delta(t).$$

Thus,

$$x(t) = 2\int_{-\infty}^{\infty} \delta(t)dt = 2. \tag{2.67}$$

2.3.3 Comparison of Discrete Time and Continuous Time Unit Impulse Functions

Discrete time impulse function is a bounded and realizable function, whereas the continuous time counterpart is unbounded and, thus unrealizable. Continuous time impulse function is a somewhat hypothetical function.

When we deal with the continuous time unit impulse function, we prefer to take its integral after the operations, such as multiplication or addition with other functions. Otherwise, the operations are not mathematically tractable.

Note that we could give the definition of discrete time unit impulse function by a simple analytical equation. However, we had to use limit or integral operation to define the continuous time unit impulse function. In mathematics, when we do not have a direct definition of a mathematical object, we provide some indirect definitions using mathematical tools. These second type of definitions is called **operational definitions**.

2.4 Unit Step Function

Unit step function is a real function, which is always 0 for negative values of its argument and always 1 for the positive values of its argument.

2.4.1 Discrete Time Unit Step Function

Discrete time unit step function is analytically defined as follows:

$$u[n] = \begin{cases} 1 & \text{for } n \geq 0 \\ 0, & \text{otherwise.} \end{cases} \tag{2.68}$$

It is customary to put little dots at the top of each bar of magnitude one and zero to show that the function exists at only integer values and it is bounded for all values of n, as shown in Figure 2.11.

2.4.2 Relationship Between the Discrete Time Unit Step and Unit Impulse Functions

There is an interesting relationship between the discrete time unit step and unit impulse functions: One of them can be used to represent the other. Mathematically speaking,

$$u[n] = \sum_{k=0}^{\infty} \delta[n-k], \tag{2.69}$$

and

$$\delta[n] = u[n] - u[n-1]. \tag{2.70}$$

The first equation reveals that the discrete time unit step function is nothing but the superposition of the discrete time shifted unit impulse functions with equal weights, for all positive values of k. In other words, when we add the shifted versions of unit impulse functions, $\delta[n], \delta[n-1], \delta[n-2], \ldots,$ we obtain the unit step function (Figure 2.12).

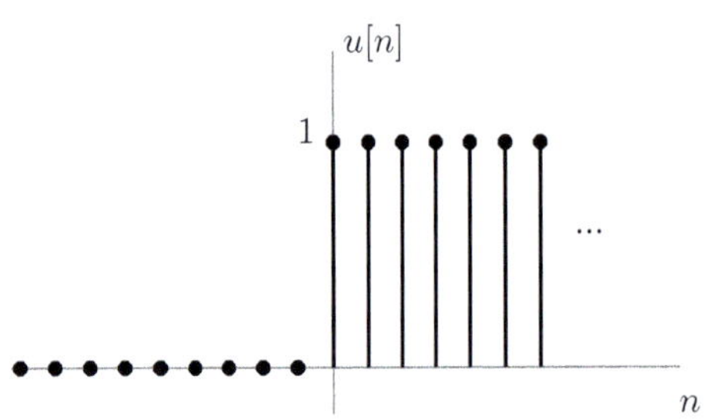

Figure 2.11 The plot of the discrete time unit step function, $u[n]$.

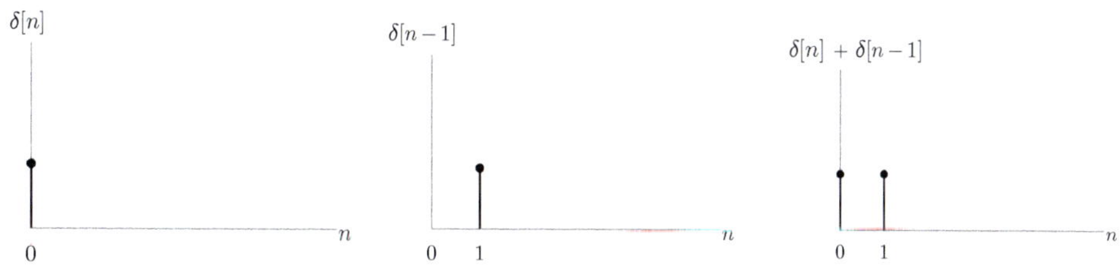

Figure 2.12 Addition of the unit impulse and its shifted version, $\delta[n-1]$. It is possible to generate the unit step function by adding the shifted impulse functions, $\delta[n-k]$, for all k.

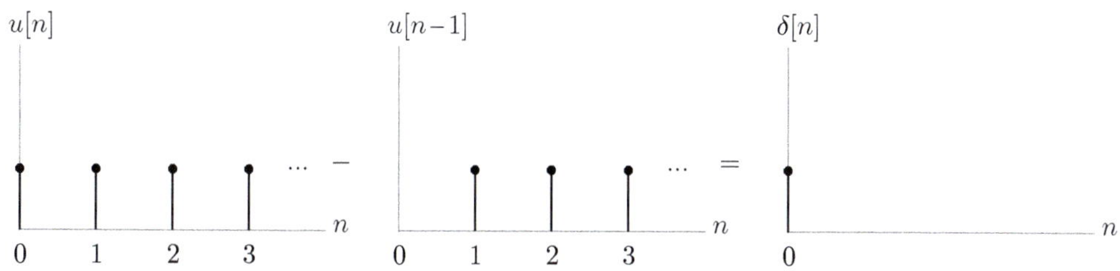

Figure 2.13 We can also generate the unit impulse function by subtracting the shifted unit step function from the unit step function, $\delta[n] = u[n] - u[n-1]$.

The second equation reveals that if we subtract the infinitely many impulses of unit step function from its shifted version, they all cancel out and we simply get the value of the unit step function at the origin, which is $\delta[n]$ (Figure 2.13).

Motivating Question: Consider any discrete time bounded function $x[n]$. Can we represent this signal by the superposition of the shifted discrete time impulse functions?

The answer is yes! Interestingly, the bounded functions, which do not have a closed analytical form, can be represented by the weighted summation of shifted impulses.

When we multiply the value of the discrete time function $x[n]$ by a shifted impulse $\delta[n-k]$, we get,

$$x[n]\delta[n-k] = \begin{cases} x[k] & \text{for } n = k \text{ (since } \delta[n-k] = 1 \text{ only when } n = k), \\ 0 & \text{otherwise.} \end{cases} \tag{2.71}$$

If we sum all the values of $x[k]$, we get $x[n]$,

$$x[n] = \sum_{k=-\infty}^{\infty} x[k]\delta[n-k]. \tag{2.72}$$

This equation reveals that we can represent any **bounded** discrete time function, $x[n]$, analytically by using the weighted summation of the shifted impulse functions, $\delta[n-k]$, where the weights correspond to the amplitude of the function at the point k.

Exercise 2.12 Find an analytical expression for the signal plotted in Figure 2.14.

Solution

This function does not have any closed-form representation. However, we can represent it by the superposition of the shifted impulse function,

$$x[n] = \sum_{k=-1}^{3} x[k]\delta[n-k], \tag{2.73}$$

where $x[k]$ is the value of $x[n]$ at point k.

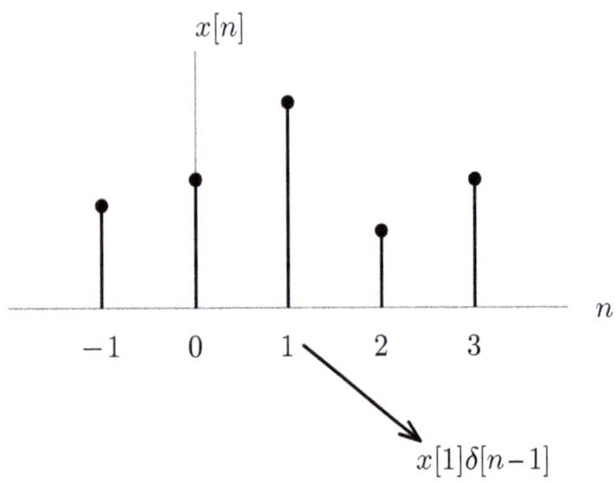

Figure 2.14 A discrete time signal, which has nonzero values in the interval $-1 \leq n \leq 3$.

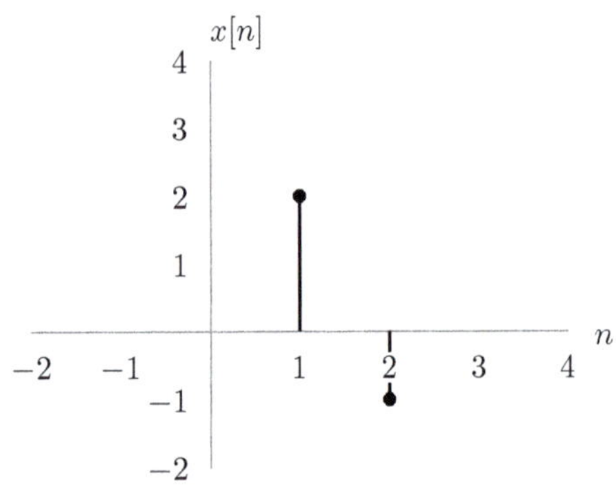

Figure 2.15 A function, which consists of two shifted impulse functions.

Exercise 2.13 Find an analytical expression for the signal plotted in Figure 2.15.

Solution

We can represent this signal as the superposition of two discrete time impulse function:

$$x[n] = 2\delta[n-1] - \delta[n-2]. \tag{2.74}$$

2.4.3 Continuous Time Unit Step Function

Continuous time unit step function (Figure 2.16) is defined as follows:

$$u(t) = \begin{cases} 1 & \text{for } t \geq 0, \\ 0 & \text{otherwise.} \end{cases} \tag{2.75}$$

The above definition of the continuous time unit step function reveals that there is a discontinuity at $t = 0$.

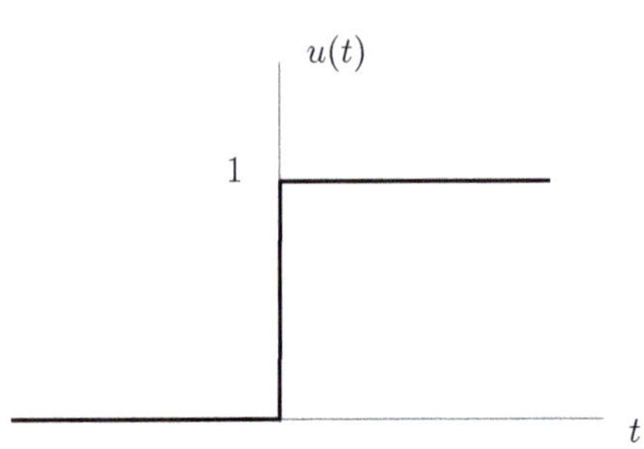

Figure 2.16 Plot of continuous time unit step function, $u(t)$. Note that this function has a discontinuity at the origin.

2.4.4 Comparison of Discrete Time and Continuous Time Unit Step functions

Recall the analytical form of the continuous time and discrete time unit step functions, as follows:

$$u(t) = \begin{cases} 1 & \text{for } t \geq 0, \\ 0 & \text{otherwise,} \end{cases} \tag{2.76}$$

and

$$u[n] = \begin{cases} 1 & \text{for } n \geq 0, \\ 0 & \text{otherwise.} \end{cases} \tag{2.77}$$

All we need to do is to replace the continuous variable t with the discrete variable n. Their analytical forms are the same. However, the discrete unit step function is **undefined** between the two integer values of n.

2.4.4.1 Relationship Between the Continuous Time Unit Step and Unit Impulse Functions

Continuous time unit step and unit impulse functions are related to each other with derivatives and integrals. Recall that we obtain the continuous time impulse function by taking the limit of the $\delta_\Delta(t)$ function, as follows:

$$\delta(t) = \lim_{\Delta \to 0} \delta_\Delta(t). \tag{2.78}$$

Recall, that the area under the impulse function was obtained by integrating it,

$$\int_{-\infty}^{\infty} \delta(\tau) d\tau = 1. \tag{2.79}$$

We use the relationship between the well-behaved function $\delta_\Delta(t)$ (Figure 2.8) and the unit step function,

$$\delta_\Delta(t) = \frac{u(t) - u(t-\Delta)}{\Delta} \qquad \text{(Figure 2.17 and 2.18).} \tag{2.80}$$

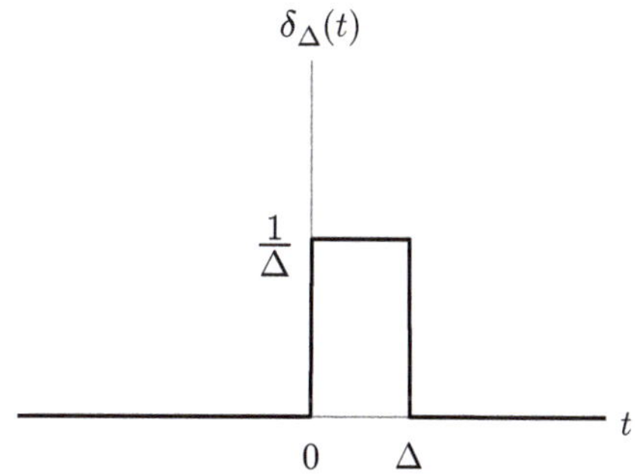

Figure 2.17 The well-behaved function $\delta_\Delta(t)$ approaches to the unit impulse function in the limit as $\Delta \to 0$.

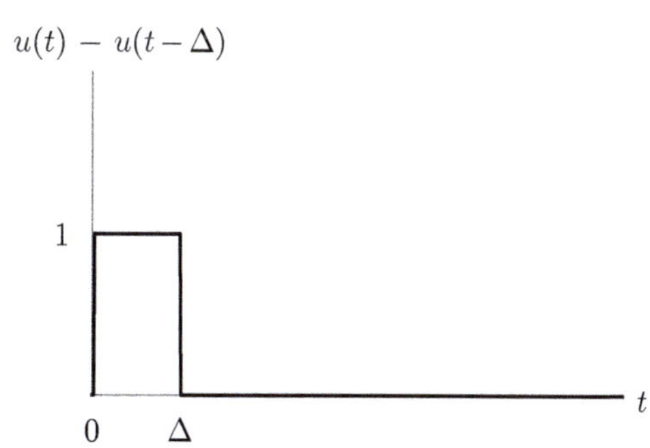

Figure 2.18 The well-behaved function $\delta_\Delta(t)$ can be written in terms of two unit step functions, as $\delta_\Delta(t) = \frac{u(t)-u(t-\Delta)}{\Delta}$.

Taking the limit as $\Delta \to 0$, we get,

$$\delta(t) = \lim_{\Delta \to 0} \delta_\Delta(t) = \lim_{\Delta \to 0} \frac{u(t) - u(t-\Delta)}{\Delta} = \frac{du(t)}{dt}. \tag{2.81}$$

Thus, we find the relationship between the unit impulse and unit step functions for the continuous time signals as follows:

$$\delta(t) = \frac{du(t)}{dt}. \tag{2.82}$$

If we integrate both sides of Equation (2.82), we obtain,

$$u(t) = \int_{-\infty}^{t} \delta(\tau) d\tau. \tag{2.83}$$

Comparing the relationship between the unit impulse function and unit step function in discrete time and continuous time signals, we observe that

- **Difference** operation in discrete time is replaced by the **differentiation** operation in continuous time,
- **Sum** operation in discrete time is replaced by the **integral** operation in continuous time.

These relationships are summarized in Table 2.1.

INTERACTIVE Explore the relation between the unit impulse and the unit step functions @ `https://384book.net/i0202`

Motivating Question: Consider any continuous time bounded function $x(t)$. Can we represent this signal by the weighted integral of continuous time impulse functions?

Suppose that we are given a continuous time bounded signal $x(t)$. Suppose, also, that we multiply $x(t)$ by $\delta_\Delta(t)$. What do we get? Assuming that Δ is very small, we get a rectangular function, where the width is Δ and height is,

$$x(0)\delta_\Delta(t). \tag{2.84}$$

When we shift the function $\delta_\Delta(t)$ by τ and multiply it by $x(\tau)$, we get another rectangular function, where the width is Δ and the height is,

$$x(\tau)\delta_\Delta(t-\tau). \tag{2.85}$$

Table 2.1 Relationship between the unit step and unit impulse functions.

	Continuous time	**Discrete time**
Obtain u from δ	Integration: $u(t) = \int_{-\infty}^{t} \delta(\tau) d\tau$	Summation: $u[n] = \sum_{k=0}^{\infty} \delta[n-k]$
Obtain δ from u	Differentiation: $\delta(t) = \frac{du(t)}{dt}$	Difference: $\delta[n] = u[n] - u[n-1]$

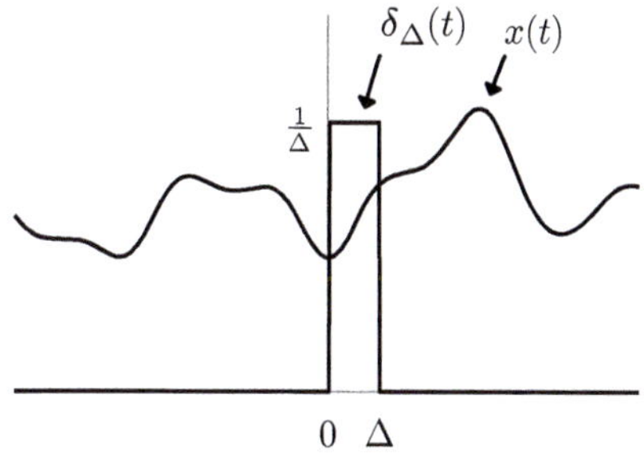

Figure 2.19 We can slide the function $\delta_\Delta(t)$, all over the function $x(t)$, as we multiply and integrate them to obtain a relatively coarse representation of $x(t)$.

This is illustrated in Figure 2.19. Now let us integrate $x(\tau)\delta_\Delta(t-\tau)$, in the interval of $-\infty < \tau < \infty$, to obtain an approximation of $x(t)$, as follows:

$$x_\Delta(t) = \int_{-\infty}^{\infty} x(\tau)\delta_\Delta(t-\tau)d\tau. \tag{2.86}$$

Now let's take the limit of Equation (2.86):

$$\lim_{\Delta\to 0} x_\Delta(t) = x(t) = \lim_{\Delta\to 0} \int_{-\infty}^{\infty} x(\tau)\delta_\Delta(t-\tau)d\tau = \int_{-\infty}^{\infty} x(\tau)\delta(t-\tau)d\tau. \tag{2.87}$$

$$x(t) = \int_{-\infty}^{\infty} x(\tau)\delta(t-\tau)d\tau. \tag{2.88}$$

Similar to the discrete case, we can represent any continuous time, bounded signal, $x(t)$, in terms of the weighted integral of shifted impulse functions, $\delta(t-\tau)$, where the weights correspond to the value of the function at the shift point, τ.

Exercise 2.14 Show that

$$\int_{-\infty}^{\infty} x(\tau)\delta(\tau)d\tau = x(0).$$

Solution

In the aforementioned integral, let us replace $\delta(\tau)$ by

$$\delta_\Delta(\tau) = \begin{cases} \frac{1}{\Delta} & \text{for } 0 < \tau < \Delta, \\ 0 & \text{otherwise.} \end{cases}$$

When Δ is very small, we approximate the integral as follows:

$$\int_{-\infty}^{\infty} x(\tau)\delta_\Delta(\tau)d\tau \approx x(0).$$

Take the limit,

$$\lim_{\Delta\to 0} \int_{-\infty}^{\infty} x(\tau)\delta_\Delta(\tau)d\tau = \int_{-\infty}^{\infty} x(\tau)\delta(\tau)d\tau = x(0)\int_{-\infty}^{\infty} \delta(\tau)d\tau = x(0).$$

Exercise 2.15 Show that

$$\int_{-\infty}^{\infty} e^{\tau}\delta(\tau)d\tau = 1.$$

Solution

Using the result of the previous example,

$$\int_{-\infty}^{\infty} x(\tau)\delta(\tau)d\tau = x(0)$$

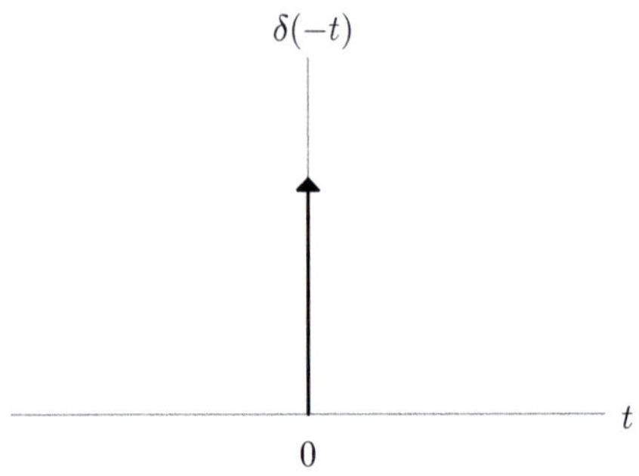

Figure 2.20 The continuous time unit impulse function is even.

and noting that $e^t = 1$, for $t = 0$, we obtain,

$$\int_{-\infty}^{\infty} e^{\tau}\delta(\tau)d\tau = e^0 = 1.$$

Exercise 2.16 Show that the impulse function is even, in other words, $\delta(t) = \delta(-t)$.

Solution
Note that $\delta(t)$ is symmetric with respect to the y-axis. Thus, it is even (Figure 2.20). This directly implies that

$$\delta(t) = \delta(-t).$$

Exercise 2.17 Find an analytical expression for signal in Figure 2.21.

Solution
We can use the superposition of shifted superposition of unit step functions, as follows:

$$x(t) = 2u(t-1) - u(t-2) - u(t-5). \tag{2.89}$$

Exercise 2.18 Find an analytical expression for the signal in Figure 2.22.

Solution
We can use the shifted superposition of unit step functions, as follows:

$$x(t) = u(t+2) + u(t-1) - u(t-5) - u(t-6). \tag{2.90}$$

This exercise reveals that a bounded piece-wise constant function can be represented by the superposition of continuous time unit step functions.

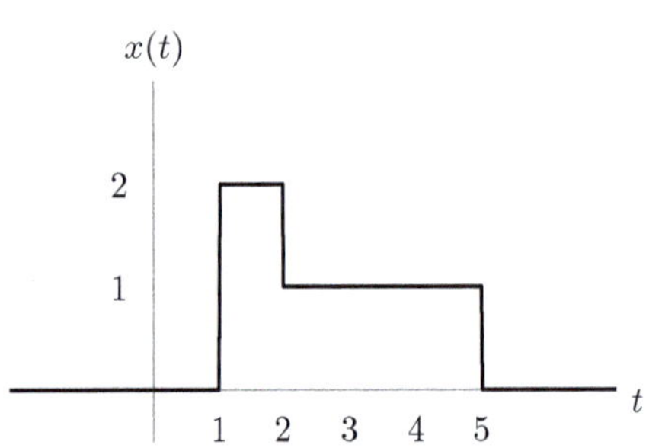

Figure 2.21 A piecewise constant function, which does not have a compact analytical form.

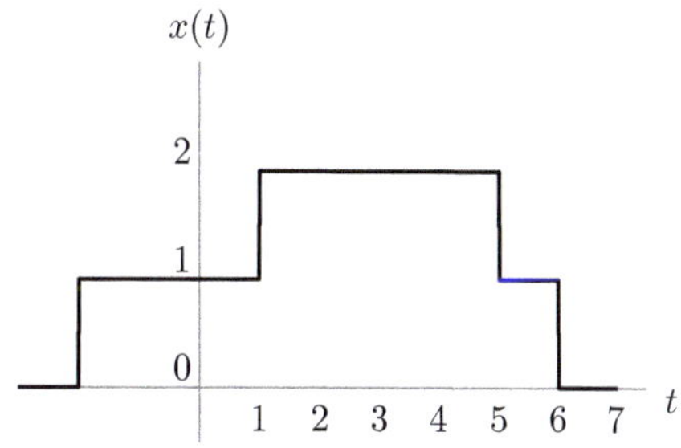

Figure 2.22 A bounded piece-wise constant function.

2.5 Chapter Summary

Can we represent simple signals by some basic functions? Is it possible to use these basic signals to represent more complicated ones? How can we manipulate a signal to generate a signal of the desired form?

In this chapter, we introduced basic building blocks of signals, which have well-defined analytical forms, namely exponential functions, unit step functions, and unit impulse functions. We introduced both continuous time and discrete time counterparts of these functions. We also investigate the methods for manipulating them by changing the time variable.

Throughout this book, we shall use and manipulate these basic functions to model and analyze the signals, even the complicated ones. Among these basic functions, we pay special attention to harmonically related complex exponential functions, which will be used as basis of a vector space to represent natural and man-made signals in a very efficient and compact way.

Problems

2.1 Consider the following real exponential function:

$$x(t) = e^{-0.5t}u(t).$$

a) Find and plot $x(2t-4)$.

b) Find and plot the derivative

$$\frac{dx(t)}{dt}.$$

c) Find and plot the integral

$$y(t) = \int_0^t x(\tau)d\tau.$$

d) Find and plot the integral

$$y(t) = \int_0^t x(2\tau-4)d\tau.$$

2.2 Find and plot the real and imaginary parts of the following continuous time signals:

a) $x_1(t) = e^{j\pi/2}\cos(2t+2\pi)$

b) $x_2(t) = 4e^{-2t}\sin(3t+2\pi)$

c) $x_3(t) = 2je^{(-20+60j)t}$

2.3 Find and plot the magnitude and phases of the following discrete time signals:
a) $x_1[n] = e^{j\pi/2}\cos(2n)$
b) $x_2[n] = 4e^{-2n}\sin(3n + 2\pi)$
c) $x_3[n] = 2je^{(-20+j60)n}$

2.4 Determine if the following signals are periodic or not, for each periodic signal determine its fundamental period:
a) $x_1(t) = 4e^{j20t}$
b) $x_2(t) = e^{(-3+j2)t}$
c) $x_3[n] = e^{j9\pi n}$

2.5 Determine the fundamental period of the following signal:

$$x(t) = 4\cos(5t + 2) - 2\sin(10t - 1).$$

2.6 Determine the fundamental period of the following signal:

$$x[n] = 13 + e^{j8\pi n/3} - e^{j4\pi n/5}.$$

2.7 Consider the superposition of the following periodic signal:

$$x(t) = \sin\frac{2\pi}{3}t + 2\cos\frac{\pi}{2}t. \tag{2.91}$$

a) Find its exponential form.
b) Find its angular frequency and the fundamental period.
c) Find and plot the first and second harmonics of $x(t)$.

2.8 Given a continuous time signal $x(t) = e^{-t}$, sketch the following functions:
a) $x(t - 2)$
b) $x(2t - 2)$
c) $x(t)[\delta(2t + 4)]$
d) $[x(2t) + x(t)]u(t)$

2.9 Given a discrete time signal $x[n] = u[n]$, sketch the following functions:
a) $x[n - 5]$
b) $x[2n + 2]$
c) $x[n]u[2n]$
d) $x[n] + (-1)^n x[n]$
e) $x[n + 2]\delta[n + 2]$

2.10 State what kind of symmetry or symmetries the following function has:

$$x[n] = u[n + 1] - u[n - 1].$$

2.11 Express and plot the following discrete time function in terms of shifted impulse functions:

$$x[n] = \begin{cases} n & 0 \le n \le 2 \\ -n & -2 \le n < 0 \\ 0 & \text{otherwise.} \end{cases}$$

2.12 Let $x(t) = x(t + 6)$ be a continuous time periodic function, represented by the following analytical expression in one full period:

$$x(t) = \begin{cases} t & |t| < 2 \\ 0 & 2 \leq |t| \leq 4. \end{cases}$$

a) Plot this function.
b) Find and plot $x(2t)$.
c) What is the fundamental period of $x(2t)$.

2.13 Consider the following discrete time periodic function, where $x[n] = x[n + 4]$. The function is defined in one enumeratefull period, as follows:

$$x(t) = \begin{cases} 2 & 0 \leq t \leq 2 \\ -2 & 2 < t < 4. \end{cases}$$

a) Find the derivative $\frac{dx(t)}{dt}$ and represent it by the sum of shifted impulse functions.
b) Find the integral $\int_0^t x(\tau)d\tau$.

2.14 Show that

$$\delta(at) = \frac{1}{|a|}\delta(t).$$

2.15 Plot each of the functions given below:
a) $x(t) = 2\delta(2t)$
b) $x(t) = \delta(-2t) + \delta(2t)$
c) $x(t) = \delta(2t + 1)$

2.16 Suppose that $x(t)$ is a continuous time function, show that $x(t)\delta(t) = x(0)\delta(t)$. Find and plot $x(t)\delta(t - 2)$, for $x(t) = 2t^2$.

2.17 Given the following shifted discrete time signal:

$$x[n + 3] = 4 - \sum_{k=5}^{\infty} \delta[n - 2k],$$

find a closed form for $x[n]$ in terms of a shifted and scaled unit step function.

2.18 Solve the following integrals:
a) $\int_{-\pi/4}^{\pi/4} \cos(2t)u(t)dt$
b) $\int_{-10}^{10} \cos(4\pi t)[\delta(t + 5) + \delta(t - 3)]\, dt$
c) $\int_{-6}^{6} \sin(6\pi t)u(t - 1)dt$

2.19 Consider the plot of the signal $x[n]$, given in Figure P2.1.
a) Find an analytical expression of this figure using shifted impulse functions.
b) Find and plot $y[n] = x[2n + 2] + x[1 - n]$.
c) Find $y[n] = x[2n + 2] + x[1 - n]$ signal in terms of the shifted impulse functions.

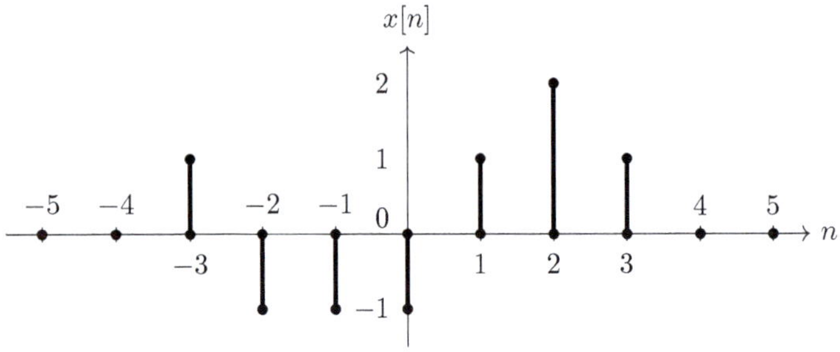

Figure P2.1

3

Basic Building Blocks and Properties of Systems

We are the basic building blocks of the cosmos to know itself.
Adopted from Carl Sagan "We Are A Way For The Cosmos To Know Itself."

In Chapter 2, we studied the basic building blocks of signals and their mathematical representations as functions. In this chapter, we shall study the basic building blocks of systems and their representations by equations.

We shall combine the basic blocks of a system, called subsystems, in various forms to represent relatively more complicated systems. We also study the properties of systems, which provide us with a framework to model, design, and implement a wide range of systems, using the available mathematical tools.

3.1 Representation of Systems by Equations

A system can be considered as a mapping, which transforms the input signal(s) to the output signal(s). Thus, a system is a functional operator in which signals are transformed into other signals.

Suppose that a system, represented by a model, h, receives a set of input signals to generate a set of output signals, as shown in the block diagram representation of Figure 3.1.

In order to represent a system by a mathematical model, we need to establish a relationship between the input signal, $x(\cdot)$ and output signal, $y(\cdot)$ in the following general form,

$$y(\cdot) = h(x(\cdot)). \tag{3.1}$$

In Equation (3.1), h stands for the model of the system. It can be an **algebraic equation, a differential equation, an integral equation, a nonlinear equation**, etc. We use a loose notation, $(\cdot)$ to cover both continuous and discrete time systems.

In most practical problems, the model of the system, h, is not available, thus, we represent it with a black box. However, we can observe a set of input–output pairs, $\{x_i(\cdot), y_i(\cdot)\}_{i=1}^{n}$, which enables us to find a model h, satisfying,

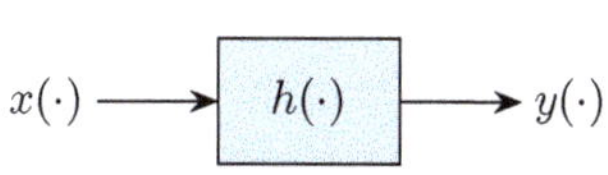

Figure 3.1 Black box representation of a system, where h represents a functional operator, which maps an input signal to an output signal. Dot notation is used to indicate both continuous time and discrete time input signal $x(\cdot)$, output signal $y(\cdot)$, and the model, $h(\cdot)$.

Signals and Systems: Theory and Practical Explorations with Python,
First Edition. Fatoş Tunay Yarman Vural and Emre Akbaş.

Companion website: www.wiley.com/go/signalsandsystems

$$
\begin{aligned}
h(x_1(\cdot)) &= y_1(\cdot) \\
h(x_2(\cdot)) &= y_2(\cdot) \\
\dots\dots\dots & \\
h(x_n(\cdot)) &= y_n(\cdot).
\end{aligned}
\tag{3.2}
$$

As we mentioned before, the model of the system, h, should be invariant under all of our varying input–output observations, $\{x_i(\cdot), y_i(\cdot)\}_{i=1}^{n}$.

Exercise 3.1 Consider a system represented by the following equation:

$$y(t) = Ax(t). \tag{3.3}$$

Find the outputs of this system for the following inputs:

a) $x_1(t) = \cos(\omega_0 t)$,
b) $x_2(t) = e^{j\omega_0 t}$.

Solution
The model, h, for this system is simply a multiplier, A, of an algebraic equation. The input signal is multiplied by the constant parameter, A, to generate the corresponding output signal. Thus, the output function is the constant multiple of the input function.

a) $y_1(t) = A\cos(\omega_0 t)$.
b) $y_2(t) = Ae^{j\omega_0 t}$.

Note that we may feed different inputs to the same system. The corresponding output obeys the rules governed by the system equation. No matter what the input is, the system equation, which relates the input and output signals remains invariant.

There are also some systems, where the input and output pairs change the system model h. This type of systems, called model-aware systems, is beyond the scope of this book.

3.2 Interconnection of Basic Systems: Series, Parallel, Hybrid, and Feedback Control Systems

When a system, such as a cellular phone or a computer, is very complicated, it is not possible to find a single equation, which we can put into the black box of Figure 3.1. Instead, we may consider its subsystems and the relationships among the subsystems. In such cases, we may represent a system by collection of subsystems, which are interrelated by signals. The connections among the subsystems can be in various forms, such as series connection, parallel connection, hybrid connection, or loopy connection. Let us exemplify the type of connections among the subsystems of a large system.

3.2.1 Series Systems

As an example, let us consider modeling the human visual system. It has many components, which is highly difficult to model by a single equation. Fortunately, we can split it into two subsystems, namely the eye and the brain. We represent each subsystem with a black box. The input to the first box (the eye) is the light signal, which generates a set of neural signals at the output. The output of the eye component is continuously fed to the brain component (second box) as the input. The output of the brain component can be one of the wide ranges of cognitive processes,

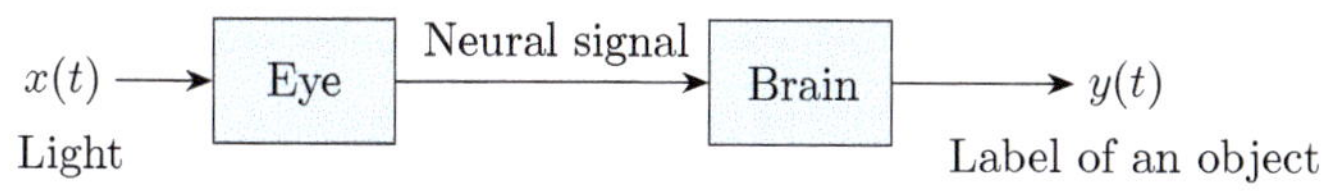

Figure 3.2 Human visual system, represented by the **series connection** of two black box subsystems: eye and brain. In this representation, the light, reflected from the objects, is the input signal to the eye. After processing this input signal, the eye outputs a set of neural signals. The input to the brain is this set of signals. They are processed in various regions of the brain to generate the final output signal, which consists of a high-level decision, such as the label or location of an object.

such as perceiving colors and shapes, recognizing objects, and interpreting scenes (Figure 3.2). These types of connections, where the outputs of the subsystems are fed to the input of another subsystem sequentially, are called **series representation**. Instead of finding a single equation to represent the human visual system, we can find two relatively more tractable equations, one of which represents the eye and the other represents the brain.

Note that the eye as a subsystem is still too complicated to be represented by a single equation. Thus, it needs to be partitioned into further subsystems, such as the eye lens, retina, blind spot, eye muscles, etc. Similarly, we can partition the brain components into anatomical regions, which are responsible in vision.

In summary, we can partition a complicated system into as many subsystems as we need, until we get a mathematically tractable representation for each subsystem. However, we need to also define the signals, which establish the relationships among the subsystems, considering the input and output signal of each subsystem.

3.2.2 Parallel Systems

As a second example, let us consider the human audio-visual system. The auditory system receives the sound waves as input, whereas the visual system receives the light. These two different types of signals are separately processed in auditory and visual systems. Then, the outputs of both systems are combined in anatomic regions of our brain, which are responsible for the audio-visual process. Finally, a set of cognitive processes can be generated, such as creating and storing a world model in our brain. As we did in the previous example, we can construct subsystems for the auditory system as the ear and the auditory regions of the brain. We can further represent the ear and the auditory regions by subsystems until we obtain a mathematically tractable model for the audio-visual system. Note that we should also establish the relationships by defining the input and output signals among the subsystems

In this case, the auditory and visual systems receive and generate different input–output pairs. Then, the outputs of the subsystems are merged by a system component, such as an adder or multiplier (Figure 3.3). These types of models are called **parallel systems**.

3.2.3 Hybrid Systems

It is more realistic to represent the audio-visual system by further decomposing it into smaller subsystems. In this case, we need to develop parallel and series connections to form a **hybrid representation**, as shown in Figure 3.4. This block diagram approximates the human audio-visual system better than the coarse parallel representation of Figure 3.3. It does not only split the auditory and visual systems into two blocks, but it also adds an additional brain block, which processes the added signals together. It is possible to add more blocks to obtain finer representations of the human

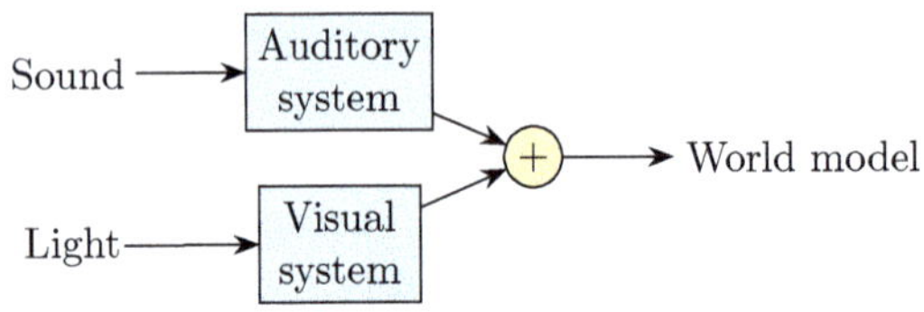

Figure 3.3 Block diagram representation of the audio-visual system, as two parallel black box subsystems. Each subsystem receives different inputs. While the eye subsystem receives light and outputs a set of neural signals, the ear (auditory system) receives sound and outputs a different set of neural signals. These two sets of signals are added and processed in some anatomical regions of the brain to generate an output signal, which is the perceived world.

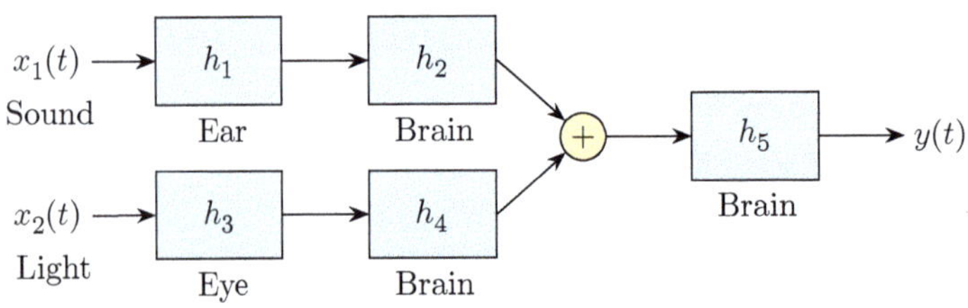

Figure 3.4 Block diagram representation of the audio-visual system, as a combination of parallel and series subsystems. This representation is called **hybrid representation**.

audio-visual system. This example shows that there is more than one block diagram representation for the same physical phenomenon.

In real-life applications for designing or modeling a system, we may need hundreds or thousands of subsystems, hybridly interconnected with each other and working in harmony to achieve a certain goal or to serve to some other systems.

3.2.3.1 Feedback Control Systems

Some systems feed their generated output back to the system as an input to evaluate or adjust their functionalities. For example, the thermostat of a heating system measures the temperature of a house. Then, this output temperature is fed back to the system, as an input. If the temperature exceeds a certain threshold value, the system halts for a while. These types of interconnected systems are called feedback control systems. A simple feedback control system is depicted in Figure 3.5.

In feedback control systems, each subsystem receives a set of signals, called input and emits a set of signals, called output. An output of a subsystem is fed as the input to another subsystem. These signals glue all the subsystems to generate the overall system.

Systems approach enables us to model and implement complex human-made systems as a collection of subsystems interconnected by signals. Good examples include cars, airplanes, robots,

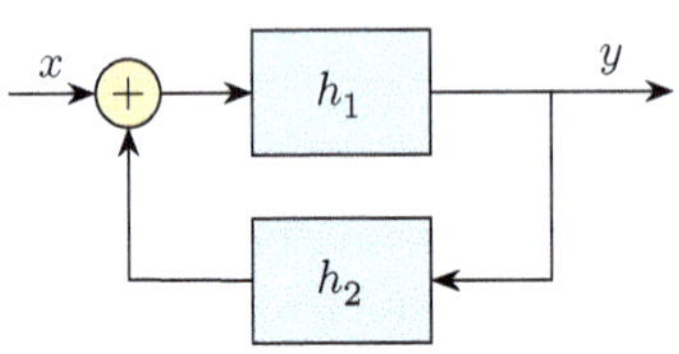

Figure 3.5 Block diagram representation of a feedback control system.

telecommunication networks, etc. However, modeling the natural systems may get complicated in many cases. For example, the basic building blocks of natural systems are the simple cells. There are several representations of the simple cells, each of which requires several dozens of subsystems to approximately model and implement by circuit elements. Considering the fact that there are approximately 2 billion cells in a spider, modeling it with a systems approach is quite difficult, because of the scalability problem.

3.2.3.2 An Example of System Modeling: Neurons as a Subsystem of Human Brain

Despite the great effort in many fields of science and engineering, the available methods for representing the cognitive activities of the human brain are too short to decipher the underlying complex structure. Nevertheless, let us give a try to model the human brain.

Neurons can be considered as the basic building blocks of the human brain. Thus, they can be considered as billions of subsystems of the brain, massively interconnected by neural signals. A single neuron receives multiple inputs, through the dendrites and outputs a single signal through an axon.

When a neuron receives a set of inputs from the dendrites, these neural signals are processed in the neuron by a bunch of highly complicated electrochemical activities. Finally, the neuron

- either stays silent, when the electrochemical processes generate a weak signal or
- fires a single output, when the electrochemical processes generate a signal which is above a certain threshold.

The output signal, generated by a neuron is conveyed to the dendrites of the other neurons by synaptic connections, as inputs.

WATCH

Learn more about how signal travels through a neuron @ https://384book.net/v0301

A simple mathematical model for a neuron was proposed by Frank Rosenblatt in 1957. In this representation, neuron is defined as a system, which receives multiple inputs, x_i, $\forall i = 1, \ldots, n$. Then, each input is multiplied by a weight w_i, and the linear combination of the inputs,

$$\mathbf{x} = \sum_{i=1}^{n} w_i x_i, \tag{3.4}$$

are fed to a function $f(\mathbf{x})$ to generate an output,

$$y = f(\mathbf{x}). \tag{3.5}$$

$f(\cdot)$ is a nonlinear function, typically a unit step function defined as:

$$f(\mathbf{x}) = \begin{cases} 1, & x \geq 0, \\ 0, & x < 0. \end{cases} \tag{3.6}$$

The block diagram representation of this model is given in Figure 3.6.

If we organize many of these simple artificial neurons layer by layer, we can construct an artificial neural network. The weights w_{ij} are adjusted by an optimization algorithm in such a way that the label of the input sample is predicted at the output neuron. Figure 3.7 depicts an example artificial

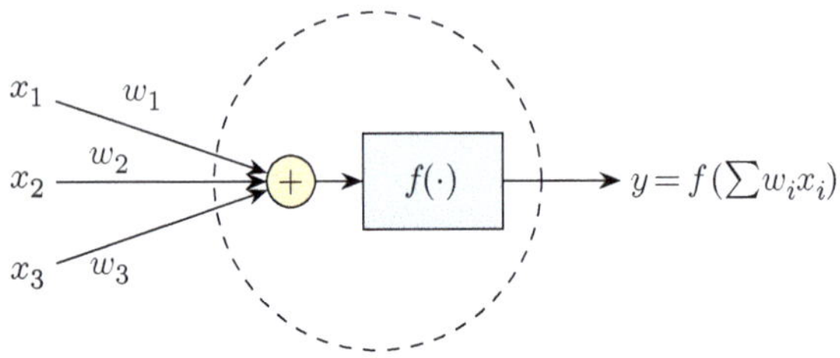

Figure 3.6 An artificial neuron. It first takes the linear combination of the input signals to obtain $\mathbf{x} = \sum w_i x_i$. This signal is fed to a function $f(\mathbf{x})$ to generate the output of the neuron, y. The circle with a dashed line shows the artificial neuron.

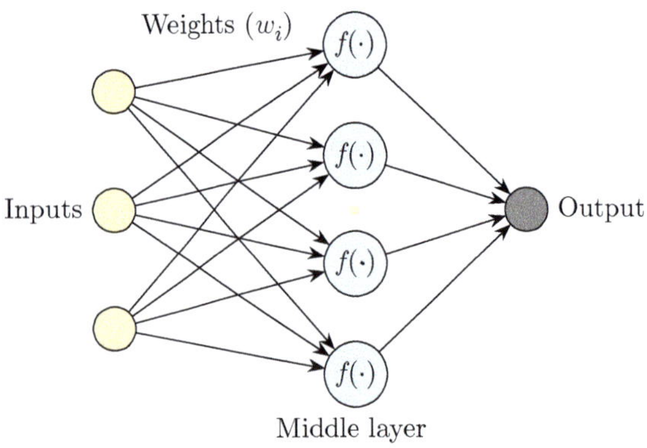

Figure 3.7 An artificial neural network, with one hidden layer, obtained by the interconnection of eight artificial neurons. The three neurons at the first layer receive the input values of a sample. The neurons in the middle layer compute the weighted linear combination of the inputs as $\sum w_i x_i$, and then compute the output as $f(\sum w_i x_i)$. The output of each neuron is fed to the next layer as the input. In this example, there is a single output neuron, which receives the weighted combination of the outputs of the hidden layer and passes it through a loss function to predict the label of the input sample.

neural network with three layers: an input layer (leftmost three neurons), a hidden layer (four neurons in the middle), and an output layer (a single rightmost neuron). Depending on the application domain, we can design an artificial neural network with as many neurons and layers as we need.

The artificial neural networks, briefly mentioned earlier, do not get even close to represent the human brain. However, they are widely used in artificial intelligence and machine learning systems. Design of these networks for a specific goal, such as designing large language models, detecting an object in a video or diagnosing a disease from an X-ray, is beyond the scope of this book.

3.3 Properties of Systems

Studying the systems with respect to some quantifiable properties, simplifies the modeling, design, and implementation problems. Let us investigate the basic properties of systems under six headings:

1) Memory
2) Causality
3) Invertibility
4) Stability
5) Time invariance
6) Linearity

These properties establish a mathematically tractable framework to model and design systems. Once we decide on what type of properties are required for a system, we restrict the system equation to satisfy the desired properties.

In the following subsections, we provide the definitions of the system properties enriched with simple examples.

3.3.1 Memory

Memory is a cognitive process of humans and some animals to store, retain and retrieve the information perceived by sensory stimuli.

The definition of memory is slightly different in system theory. A **system is memoryless** if the present value of the output depends only on the present value of the input. A memoryless system operates on the current value of the input to generate a current value of the output. Otherwise, the system has memory. **A system with memory** can store and retrieve past or future values of the input values. Therefore, the present value of the output $y(t)$ can be expressed in terms of the **past or future** values of an input for the systems with memory.

Let us give the formal definitions of systems with and without memory. In the following definitions, we suppose that a system is represented by a model h.

Definition 3.1 A continuous time system, represented by the model h is **memoryless**, if the model h relates the present value of the input $x(t)$ to the present value of the output $y(t)$, as follows:

$$y(t) = h(x(t)). \tag{3.7}$$

Similarly, a discrete time system, represented by the model h is memoryless, if the model h relates the present value of the input $x[n]$ to the present value of the output $y[n]$, as follows:

$$y[n] = h[x[n]]. \tag{3.8}$$

Definition 3.2 A continuous time system, represented by the model h **has memory**, if the model h relates the past and/or future values of the input $x(t)$ to the present value of the output $y(t)$, as follows:

$$y(t) = h(x(t - t_0)), \quad t_0 \neq 0. \tag{3.9}$$

Similarly, a discrete time system, represented by the model h has memory, if the model h relates the past and/or future values of the input $x[n]$ to the present value of the output $y[n]$, as follows:

$$y[n] = h[x[n - n_0]], \quad n_0 \neq 0. \tag{3.10}$$

For $t_0 > 0$ or $n_0 > 0$, the present value of the output of the system depends on the past value of the input. For $t_0 < 0$ or $n_0 < 0$, the present value of the output of the system depends on the future value of the input. Thus, the formal definition of memory extends beyond its everyday meaning. In the cognitive process, memory is confined to recalling the past, not predicting the future.

Exercise 3.2 Is this system memoryless?

$$y(t/3) = x(t). \tag{3.11}$$

Solution

No! This system has memory. For example, for $t = 3$, the output value depends on the future value of the input, $y(1) = x(3)$.

Exercise 3.3 ***(Accumulator)*** Consider a discrete time system represented by the following equation:

$$y[n] = \sum_{k=-\infty}^{n} x[n]. \tag{3.12}$$

Does this system have memory?

Solution

This system has memory. It accumulates all the past values of the input to generate an output. Thus, it remembers all the past values of the input.

Exercise 3.4 Consider a continuous time system, represented by the following equation:

$$y(t) = x^2(t) + 1. \tag{3.13}$$

Does this system have memory?

Solution

This is a memoryless system. For all values of t, the present values of the output depend on just the present values of the input. Specifically, the system receives the current value of an input at time t. It adds 1 to the square of the input to generate the output at the same time.

3.3.2 Causality

The memory property of systems can be counterintuitive. In systems with memory, the output may depend not only on past values but also on future input values. In other words, the system remembers both the past and the future values. This is contradictory in most physical systems, which are causal.

Causality is a fundamental concept in many fields of science and philosophy. The major assumption of causality is that the response of a system is caused by stimuli. In systems, output signals are caused by input signals. In other words, responses cannot come before the signals they are responding to.

The concept of memory can be further restricted to define causal systems, where the output $y(t)$ depends on the past and present values of the input. If the present value of the input depends on the future value of the input, then this system is called **noncausal**. In real-life problems, we have many examples of noncausal systems. For example, the prediction systems are nonausal. An aircraft pilot defines the route at a present time based on the future weather forecast.

Definition 3.3 A continuous time system is **causal** if

$$y(t) = h(x(t - t_0)), t_0 \geq 0, \tag{3.14}$$

and a discrete time system is **causal** if

$$y[n] = h[x[n - n_0]], n_0 \geq 0,$$

Note that a causal system may or may not be memoryless. A memoryless system is always causal.

Definition 3.4 **A system is non causal**, if $\exists\, t_0 < 0$ such that

$$y(t) = h(x(t - t_0)). \tag{3.15}$$

A noncausal system has always memory.

Exercise 3.5 Is the following system causal and/or memoryless?

$$y(t) = \frac{dx(t)}{dt}. \tag{3.16}$$

Solution

The answer follows from the definition of derivatives:

$$y(t) = \frac{dx(t)}{dt} = \lim_{\Delta t \to 0} \frac{x(t) - x(t - \Delta t)}{\Delta t}. \tag{3.17}$$

As $\Delta t \to 0$, the system becomes memoryless. Because the current value of the output depends only on the current value of the input. It is also causal.

Exercise 3.6 ***(Averaging)*** Is the following system causal?

$$y[n] = \frac{1}{2N+1} \sum_{k=-N}^{N} x[n-k]. \tag{3.18}$$

Solution

This system takes the average of past and future values of the input signal, defined in a time interval $(-N, N)$. Thus, it is noncausal.

3.3.3 Invertibility

In a general form, a system receives an input signal and emits an output signal, according to the model, h, satisfying the following equation:

$$y(\cdot) = h(x(\cdot)). \tag{3.19}$$

In some practical applications, we can find a model h and measure the output, $y(\cdot)$, without observing the input which causes $y(\cdot)$. As an example, consider the speech signal, generated by the vocal track. There are several reliable methods, which model the subsystems of the vocal track, such as the throat, teethes, tongue, mouth, nasal cavities, etc. With a simple recorder, we can easily record a speech signal. However, we do not have access to the input signal of the vocal track, which is the air in the lung pushed by the diaphragm. It may be important to study the properties of the input air, which generates speech signal at the output of the vocal track. Can we compute the input function, $x(\cdot)$, from the output function, $y(\cdot)$, using the model h?

Motivating Question: Suppose that we are given the model, h, of a system. Suppose also that we can observe the output, but not the input of the system. Is it possible to obtain the input signal for any observed output signal?

This is only possible if the system is invertible. When we concatenate the system, represented by a model h, to its inverse system, h^{-1}, in a series connection, we obtain the input to the original system, h, at the output of the inverse system, h^{-1} (Figure 3.8).

Definition 3.5 Given a continuous time system, $y(t) = h(x(t))$, if $\exists$ a **unique** h^{-1} so that

$$x(t) = h^{-1}(y(t)), \tag{3.20}$$

then, the system is **invertible**.

Definition 3.6 Given a discrete time system, $y[n] = h(x[n])$, if $\exists$ a **unique** h^{-1} for all possible inputs, so that

$$x[n] = h^{-1}[y[n]], \tag{3.21}$$

then, the system is **invertible**.

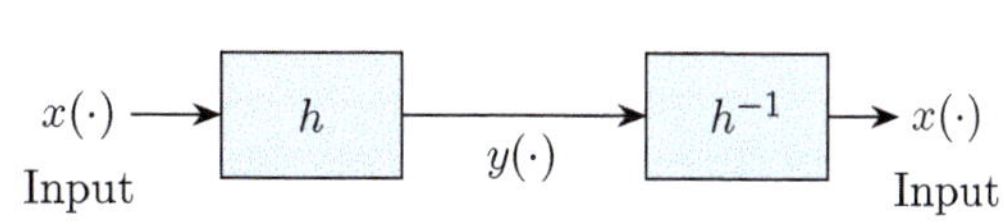

Figure 3.8 When we cascade the system, h with its inverse h^{-1}, the input of the system is obtained at the output of the inverse system.

Exercise 3.7 Is the following continuous time system invertible?

$$y(t) = \cos x(t). \tag{3.22}$$

Solution

In order to show that the system is **not invertible**, it is sufficient to find two different inputs, which generate the same output. This violates the uniqueness assumption of the inverse model, making the system noninvertible.

Take, for example, two different constant signals, $x_1(t) = \pi/2$ and $x_2(t) = -\pi/2$, as input. Both of them generate the same output:

$$y_1(t) = \cos x_1(t) = 0 = \cos x_2(t) = y_2(t).$$

The inverse is not unique. Thus, the system is not invertible.

Exercise 3.8 Is the following continuous time system invertible?

$$y(t) = 0.5x(t) + 3. \tag{3.23}$$

Solution

Yes, this system is invertible. The inverse of this system is uniquely obtained by solving Equation (3.23) for $x(t)$ as follows:

$$x(t) = h^{-1}(y(t)) = 2y(t) - 6. \tag{3.24}$$

The input signal $x(t)$ can be uniquely obtained from the output signal $y(t)$. Thus, the system is invertible.

Exercise 3.9 Is the following discrete time system invertible?

$$y[n] = 0.5x^2[n] + 3. \tag{3.25}$$

Solution

The inverse of this system can be obtained by solving the aforementioned equation for $x[n]$ as follows:

$$x[n] = \pm\sqrt{2y[n] - 6}. \tag{3.26}$$

Since there are two distinct inputs to generate the same output, there is not a unique inverse for the model h. Thus, the system is not invertible.

Exercise 3.10 Is the following discrete time system invertible?

$$y[n] = x[n-2]. \tag{3.27}$$

Solution

Yes, this system is invertible. The inverse of this system is uniquely obtained by,

$$h^{-1}[y[n]] = y[n+2] = x[(n+2) - 2] = x[n]. \tag{3.28}$$

Note that finding the inverse of a system may not be possible for many systems, even if there exists a unique inverse.

3.3.4 Stability

Stability is an important performance characteristic of a system. It assures the controllability of the output signal generated by the system for **all** the input signals. A stable system is robust to imperfections and unexpected changes at the input. It bears self-correcting mechanisms to bring the output of the system into equilibrium (Figure 3.9).

Most of the ecological systems are stable in their environment. They keep the prey and predator populations under an equilibrium. However, when we manipulate an internal parameter of an ecological system by an external source, we may create an unstable system. As an example, a small industrial waste slightly changes the internal characteristics of an ecological water system. This external perturbation, even if it is small, may spoil the equilibrium between the preys and predators, which may result in destroying most of the population, including the water-cleaning agents. In this case, we do not only lose important species, but we may lose control over water pollution, creating an extremely toxic environment.

An unstable system lacks the ability to control inputs, leading to explosively large outputs. This results in an uncontrollable system incapable of self-restoration, potentially causing catastrophic outcomes for certain inputs.

Loosely speaking, a stable system generates a bounded output for all possible bounded inputs. This type of stability is called BIBO (bounded input, bounded output) stability.

Definition 3.7 A continuous time signal is **bounded** if there exists a finite number $b \in \mathbb{R}$ such that for all $t \in \mathbb{R}$,

$$|x(t)| \leq b. \tag{3.29}$$

Definition 3.8 A discrete time signal is **bounded** if there exists a finite number, $b \in \mathbb{R}$ such that for all $n \in \mathbb{I}$,

$$|x[n]| \leq b. \tag{3.30}$$

Definition 3.9 A continuous time system, represented by a model h, is **BIBO stable** if for any bounded input function $x(t)$, the corresponding output function,

$$y(t) = h(x(t)), \tag{3.31}$$

is also bounded. In other words, for all inputs $x(t) < b$ there exists a finite number $b' \in \mathbb{R}$ such that the output $y(t)$ remains below b', i.e., $y(t) < b'$.

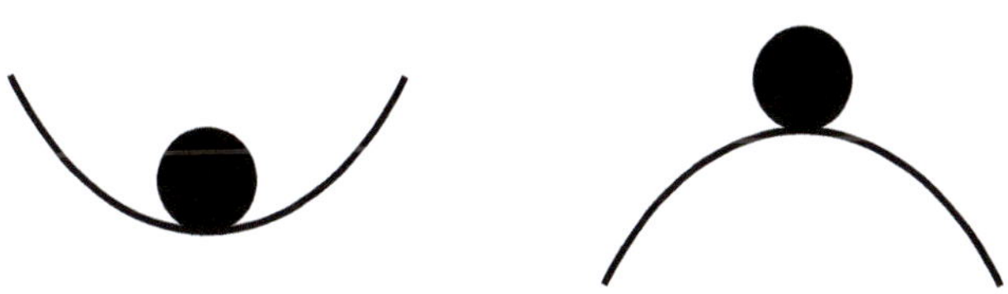

(a) (b)

Figure 3.9 **Stable system** (a): When we put a glass billiard ball to the side of a bowl, it will swing for a while, then after a certain period of time it will reach an equilibrium at the bottom of a bowl. **Unstable system** (b): If we put the billiard ball on top of an upside-down bowl, its falling speed will increase, uncontrollably; apparently it will fall down the bowl and crash.

Definition 3.10 A discrete time system represented by a model h, is **BIBO stable** if for any bounded input function $x[n]$, the corresponding output function,

$$y[n] = h[x[n]], \tag{3.32}$$

is also bounded. In other words, for all n there exists a finite number $b' \in \mathbb{R}$ such that $y[n] \leq b'$.

Definition 3.11 If a system does not satisfy the BIBO stability condition, then, it is called **unstable**.

Exercise 3.11 Is the following system stable?

$$y[n] = \sum_{k=-\infty}^{n} x[k]. \tag{3.33}$$

Solution

In order to show the stability property of a system, we select a bounded input, such that $|x[n]| < b$ for all values of n and show that the corresponding output is also bounded. A popular bounded signal is the unit step function, where $u[n] = 1$ for all values of $n \geq 0$. Suppose that we feed the unit step function, a bounded signal, as input: $x[n] = u[n]$. The output is

$$y[n] = \sum_{k=-\infty}^{n} u[k] = (n+1)u[n]. \tag{3.34}$$

The output signal of Equation (3.34) is unbounded as $n \to \infty$. Thus, this system is unstable!

Exercise 3.12 Is the following system stable?

$$y(t) = Ae^{x(t)}, \tag{3.35}$$

where A is a bounded parameter.

Solution

Suppose that we feed a bounded input signal to this system. In other words, there exists a finite number b, such that, $|x(t)| < b$. Then, we have $y(t) < Ae^{b}$. Since the right-hand side of this inequality is bounded, the output $y(t)$ is also bounded. Thus, this system is BIBO stable.

3.3.5 Time Invariance

Time invariance is a crucial system property, which makes the behavior of the system independent of time. A system is time invariant when a time shift at the input generates the same time shift at the output. In other words, no matter when you feed an input signal to the system, you receive the same corresponding output.

As an example, consider a piano as a system. While we play it, we press a set of keys at the input and we hear a set of tunes at the output. No matter when we play it, as long as we press the same set of keys, we hear the same tunes. Thus, a piano is a time-invariant system. On the other hand, if a system is time-varying, the dynamics of this system change with time. A good example of time-varying system is human behavior. We may have a diverse set of moods to communicate with our friends at different times.

Below is the formal definitions of time invariance for continuous and discrete time systems.

Definition 3.12 A continuous time system, $y(t) = h(x(t))$, is **time invariant** if for all $t_0 \in \mathbb{R}$ and for all input signals, $x(t)$, the corresponding output satisfies,

$$y(t - t_0) = h(x(t - t_0)). \tag{3.36}$$

Definition 3.13 A discrete time system, $y[n] = h[x[n]]$, is **time invariant** if for all $n_0 \in \mathbb{Z}$ and for all input signals, $x[n]$, the corresponding output satisfies,

$$y[n - n_0] = h[x[n - n_0]]. \tag{3.37}$$

A system is **time varying** if it is not time invariant.

Exercise 3.13 Is the following continuous time system time invariant?

$$y(t) = \cos x(t) + \sin x(t). \tag{3.38}$$

Solution

Suppose that we feed an input signal, shifted by t_0: $x_1(t) = x(t - t_0)$. The corresponding output becomes

$$y_1(t) = \cos x_1(t) + \sin x_1(t) = \cos x(t - t_0) + \sin x(t - t_0) = y(t - t_0). \tag{3.39}$$

We obtain the same amount of shift at the output. Thus, the system is time invariant.

Exercise 3.14 Is the following system time invariant?

$$y[n] = x^2[n] + nx[n]. \tag{3.40}$$

Solution

No! Since for input $x_1[n] = x[n - n_0]$, we have

$$y_1[n] = x_1^2[n] + nx_1[n] = x^2[n - n_0] + nx[n - n_0], \tag{3.41}$$

which is not equal to

$$y[n - n_0] = x^2[n - n_0] + [n - n_0]x[n - n_0]. \tag{3.42}$$

Thus, this system is time varying.

3.3.6 Linearity and Superposition Property

In general, linearity is a relationship between two mathematical objects, which are proportional to each other. In our context, the mathematical objects are functions, which represent input and output signals.

For example, a continuous time system h is linear when the input signal is proportional to the output signal, as follows:

$$y(t) = ax(t), \tag{3.43}$$

where $a \neq 0$ is a constant parameter.

Definition 3.14 ***(Superposition Property)*** Suppose that a continuous time system, represented by a model, $y(t) = h(x(t))$, is fed by two different inputs, $x_1(t)$ and $x_2(t)$. The corresponding outputs are $y_1(t) = h[x_1[n]]$ and $y_2(t) = h[x_2[n]]$.

Superposition property is satisfied, when the following equality hold:

$$y(t) = a_1 y_1(t) + a_2 y_2(t) = h(a_1 x_1(t) + a_2 x_2(t)), \tag{3.44}$$

where a_1 and a_2 are arbitrary numbers.

Similarly, suppose that a discrete time system, represented by a model, $y[n] = h[x[n]]$, is fed by two different inputs, $x_1[n]$ and $x_2[n]$. The corresponding outputs are $y_1[n] = h[x_1[n]]$ and $y_2[n] = h[x_2[n]]$.

Superposition property holds iff,

$$y[n] = a_1 y_1[n] + a_2 y_2[n] = h[a_1 x_1[n] + a_2 x_2[n]]. \tag{3.45}$$

Definition 3.15 A system is called **linear** if and only if it satisfies the superposition property. In other words, the superposition of two different inputs yields the same superposition at the output:

$$y_1(\cdot) = h(x_1(\cdot)) \text{ and } y_2(\cdot) = h(x_2(\cdot)) \iff a_1 y_1(\cdot) + a_2 y_2(\cdot) = h(a_1 x_1(\cdot) + a_2 x_2(\cdot)), \tag{3.46}$$

where $(\cdot)$ shows a generic notation for both continuous time variable t and discrete time variable n (Figure 3.10).

Exercise 3.15 Are the following continuous and discrete time systems linear?

$$y(t) = ax(t) + b, \tag{3.47}$$

and

$$y[n] = ax[n] + b. \tag{3.48}$$

Solution

Let us omit the time variable $(\cdot)$, for simplicity. Suppose that, the input x_1 generates the output $y_1 = ax_1 + b$ and the input x_2 generates the output $y_2 = ax_2 + b$. Superposition of two inputs,

$$x = a_1 x_1 + a_2 x_2, \tag{3.49}$$

does not generate the same superposition of the output,

$$y = a_1 y_1 + a_2 y_2 = a_1(ax_1 + b) + a_2(ax_2 + b) \neq a(a_1 x_1 + a_2 x_2) + b. \tag{3.50}$$

Thus, superposition property **does not** hold. This system does not satisfy the linearity property of Definition 3.15.

However, if we take the difference of the system equations with different input–output pairs, the additive term b vanishes and the difference becomes linear. Formally, for the continuous time case, the system equation for two different pairs of input–output are

$$y_1(t) = ax_1(t) + b, \tag{3.51}$$

$$y_2(t) = ax_2(t) + b. \tag{3.52}$$

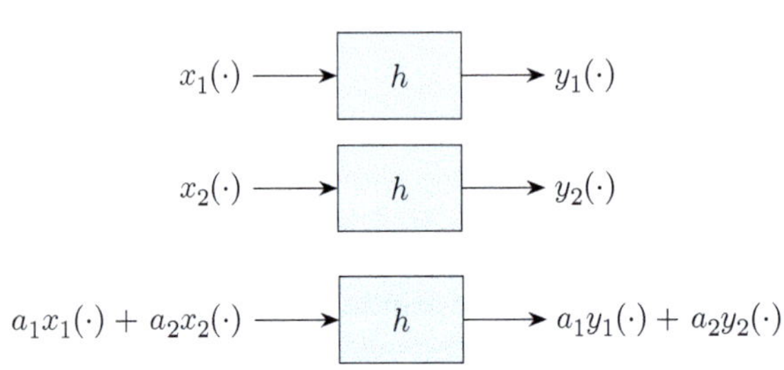

Figure 3.10 Linearity property. Given that $y_1(\cdot) = h(x_1(\cdot))$ and $y_2(\cdot) = h(x_2(\cdot))$, a linear system represented by $h(\cdot)$ satisfies $a_1y_1(\cdot) + a_2y_2(\cdot) = h(a_1x_1(\cdot) + a_2x_2(\cdot))$.

We define a new input–output pair, $x_3(t)$ and $y_3(t)$, which are the differences of the input–output pair of Equations (3.51) and (3.52), as follows:

$$y_3(t) = y_1(t) - y_2(t) = a(x_1(t) - x_2(t)), \tag{3.53}$$

$$y_3(t) = ax_3(t), \tag{3.54}$$

which is linear.

Similar derivations are applied for the discrete time case, where we feed two different inputs to receive the following outputs:

$$y_1[n] = ax_1[n] + b, \tag{3.55}$$

$$y_2[n] = ax_2[n] + b. \tag{3.56}$$

We define a new input–output pair, $x_3[n]$ and $y_3[n]$, which are the differences of the input–output pairs of Equations (3.55) and (3.56), as follows:

$$y_3[n] = y_1[n] - y_2[n] = a(x_1[n] - x_2[n]). \tag{3.57}$$

Hence,

$$y_3[n] = ax_3[n], \tag{3.58}$$

which is linear.

Definition 3.16 A system is called **incrementally linear** if the difference of the system equations for different input–output pairs are linear.

The difference between linear and incrementally linear systems is similar to the difference between linear and affine transformation in linear algebra.

Exercise 3.16 Is the following continuous time system linear, time invariant, causal, invertible, stable, and memoryless?

$$y(t) = tx(t). \tag{3.59}$$

Solution

Linearity: Suppose we feed two inputs, $x_1(t)$ and $x_2(t)$, to the system. The corresponding outputs will be

$$x_1(t) \rightarrow y_1(t) = tx_1(t), \tag{3.60}$$

$$x_2(t) \rightarrow y_2(t) = tx_2(t). \tag{3.61}$$

Output for the superposition of the two inputs, $x_3(t) = a_1x_1(t) + a_2x_2(t)$ is

$$y_3(t) = t(a_1x_1(t) + a_2x_2(t)) = a_1y_1(t) + a_2y_2(t). \tag{3.62}$$

Superposition property holds. Thus, the system is linear.

Time Invariance: Let us shift the input of the system by t_0 to define a new input, $x_1(t) = x(t - t_0)$. Then, the corresponding output becomes

$$y_1(t) = tx_1(t) = tx(t - t_0) \neq y(t - t_0) = (t - t_0)x(t - t_0). \tag{3.63}$$

The shift for the input $x(t - t_0)$ does not give the same amount of shift at the output. Thus, the system is not time invariant.

Memory and Causality: Present values of the output depend on the present values of the input. For all t, the output is $y(t) = h(x(t)) = tx(t)$. The system is memoryless. Therefore, it is causal.

Invertibility: For $t = 0$, it is not invertible. However, for $t \neq 0$, it is invertible, where the inverse can be obtained from the following equation:

$$x(t) = \frac{1}{t}y(t) \quad \text{for} \quad t \neq 0. \tag{3.64}$$

Stability: The system is not stable because a bounded input does not generate bounded output. For any bounded input $x(t) = B$,

$$\lim_{t \to \infty} y(t) = tx(t) = tB = \infty. \tag{3.65}$$

Exercise 3.17 Is the following discrete time system linear, time invariant, memoryless, causal, invertible, and stable?

$$y[n] = \left(\cos \frac{\pi}{2} n\right) x[n], \tag{3.66}$$

Solution

In the right-hand side of the system equation, there is a multiplicative factor, which depends on the time variable n. Therefore, the system equation can be written as $y[n] = A(n)x[n]$ where $A(n) = \cos \frac{\pi}{2} n$.

Linearity: Suppose we feed two inputs, $x_1[n]$ and $x_2[n]$, to the system. The corresponding outputs will be

$$x_1[n] \to y_1[n] = A(n)x_1[n], \tag{3.67}$$

$$x_2[n] \to y_2[n] = A(n)x_2[n]. \tag{3.68}$$

Output for the superposition of the two inputs, $x_3[n] = a_1x_1[n] + a_2x_2[n]$ is

$$y_3[n] = A(n)(a_1x_1[n] + a_2x_2[n]) = a_1y_1[n] + a_2y_2[n]. \tag{3.69}$$

Superposition property holds. Thus, the system is linear.

Time Invariance: Let us shift the input of the system by the time amount of n_0 to define a new input, $x_1[n] = x[n - n_0]$. Then, the corresponding output becomes,

$$y_1[n] = A(n)x_1[n] = A(n)x[n - n_0] \neq y[n - n_0] = A(n - n_0)x[n - n_0]. \tag{3.70}$$

The shift for the input $x[n - n_0]$ does not give the same amount of shift at the output. Thus, the system is not time invariant.

Memory and Causality: Present values of the output depend on the present values of the input. For all n, the output is

$$y[n] = h[x[n]] = \left(\cos \frac{\pi}{2} n\right) x[n]. \tag{3.71}$$

The system is memoryless. In addition, it is causal.

Invertibility: Let us solve this equation for $x[n]$:

$$x[n] = \frac{y[n]}{\cos \frac{\pi}{2} n}. \tag{3.72}$$

Note that, $x[n] \to \infty$ for all odd values of n. Thus, the system is not invertible.

Stability: The system is stable because a bounded input generates bounded output. For any bounded input $x[n] = b$,

$$y[n] = \left(\cos \frac{\pi}{2} n\right) b < \infty, \tag{3.73}$$

is bounded.

3.4 Basic Building Blocks of Systems and Their Properties

Recall that we introduced the basic building blocks of signals, such as trigonometric functions, exponential functions, unit impulse, and unit step functions. We could use these functions as the basic building blocks of more complicated functions. Similarly, we can define basic building blocks or system components for systems. These simple components can be used as subsystems to build a more complicated system.

There are dozens of basic building blocks in system theory, for both continuous time and discrete time systems. The type of the basic components varies depending on the application domain. For example, if we design an electric circuit, the basic building blocks of the system are the circuit elements, such as resistors, capacitors, inductors, etc.

In this book, we shall focus on some generic basic building blocks, for modeling a wide range of discrete time and continuous time systems, as described in the following text.

3.4.1 Scalar Multiplier

Scalar multiplier simply multiplies an input signal by a scalar number. This component is used for both discrete time and continuous time systems, as follows:

$$y(\cdot) = Ax(\cdot). \tag{3.74}$$

It is easy to show that scalar multiplier is a **memoryless, linear, time invariant, stable**, and **invertible** system component, where the present value of the output is the scaled version of the present value of the input (Figure 3.11).

3.4.2 Adder

An adder simply adds multiple inputs to generate a single output. This component, too, is used for both discrete time and continuous time systems, as follows:

$$y(\cdot) = x_1(\cdot) + x_2(\cdot) + \cdots + x_n(\cdot). \tag{3.75}$$

This is a **memoryless, linear, time invariant, stable, and noninvertible** system component (Figure 3.12).

3.4.3 Multiplier

A multiplier multiplies all the inputs to generate the output. It is used for both discrete time and continuous time systems, as follows:

$$y(\cdot) = x_1(\cdot) \times x_2(\cdot) \times \cdots \times x_n(\cdot). \tag{3.76}$$

A multiplier is a **memoryless, nonlinear, time invariant, stable, and noninvertible** system component, which multiplies multiple inputs (Figure 3.13).

Figure 3.11 Schematic representations of a scalar multiplier. Both representations are used for scalar multipliers.

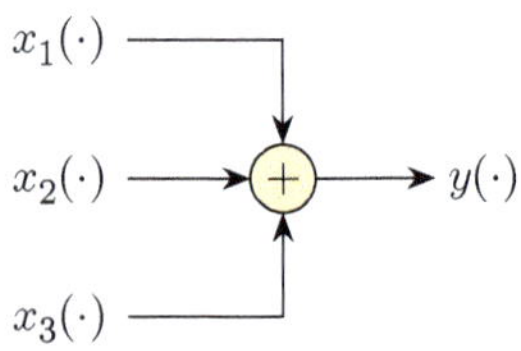

Figure 3.12 An adder, which adds three inputs, to generate output $y(\cdot) = x_1(\cdot) + x_2(\cdot) + x_3(\cdot)$.

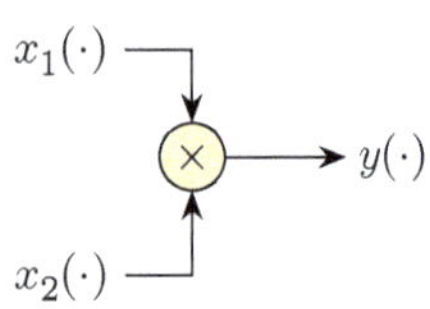

Figure 3.13 Schematic representation of a multiplier, which multiplies two inputs to generate output $y(\cdot) = x_1(\cdot) \times x_2(\cdot)$.

3.4.4 Integrator

Integrator is used in the **continuous time** systems, which takes the integral of the input for all of the past values, until the present time t, as follows:

$$y(t) = \int_{-\infty}^{t} x(\tau)d\tau. \tag{3.77}$$

An integrator is **linear, time invariant, invertible**, and **causal** system with **memory** (Figure 3.14).

3.4.5 Differentiator

Differentiator is used in the **continuous time** systems, which takes the derivative of the input to generate the output, as follows:

$$y(t) = \frac{dx(t)}{dt}. \tag{3.78}$$

A differentiator is a **memoryless, linear, time invariant, stable**, and **noninvertible** system component (Figure 3.15).

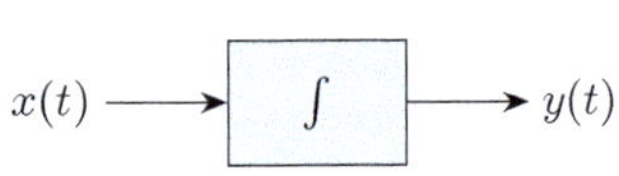

Figure 3.14 Schematic representation of an integrator.

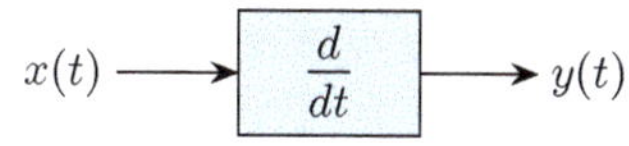

Figure 3.15 Schematic representation of a differentiator.

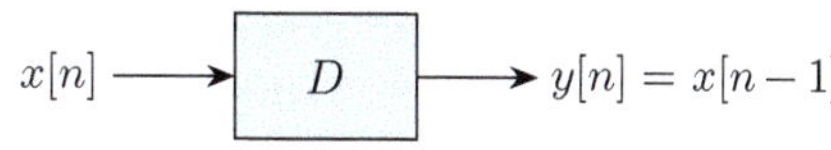

Figure 3.16 Schematic representation of the unit delay operator.

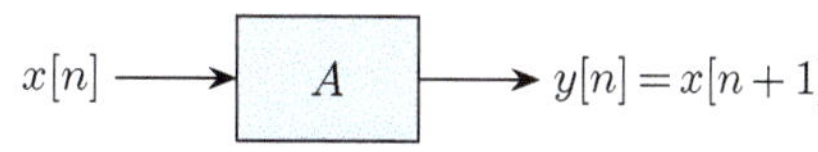

Figure 3.17 Block diagram representation of the system, $y[n] = x[n + 1]$, represented by the unit advance operator A.

3.4.6 Unit Delay Operator

Unit delay operator is used in **discrete time** systems. It is related to the discrete version of the differentiator (i.e., $y[n] = x[n] - x[n - 1]$). Unit delay operator is defined as follows:

$$y[n] = x[n - 1]. \tag{3.79}$$

It is easy to show that the unit delay operator is a **linear, time invariant, causal, invertible**, and **stable** system with **memory** (Figure 3.16).

3.4.7 Unit Advance Operator

Unit advance operator is used in **discrete time** systems. It is defined as:

$$y[n] = x[n + 1]. \tag{3.80}$$

The unit advance operator is a **linear, time invariant, invertible**, and **stable** system with **memory** (Figure 3.17). However, it is **noncausal**.

Exercise 3.18 Find a block diagram representation for the following system:

$$y[n] = x[n] - x[n - 1]. \tag{3.81}$$

Is this system memoryless? Is this system causal?

Solution

This system requires an adder with a minus sign and a unit delay operator, as shown in Figure 3.18.

The system has memory. Since, e.g., the output $y(1)$ depends on $x(0)$.

The system is causal, because the present value of the output does not depend on the future values of the input.

Exercise 3.19 Find the system equation of the block diagram representation of Figure 3.19. Is this system linear? Is this system invertible? If yes, find its inverse.

Solution

From the given block diagram, we can write the system equation as

$$y(t) = ax(t) + b.$$

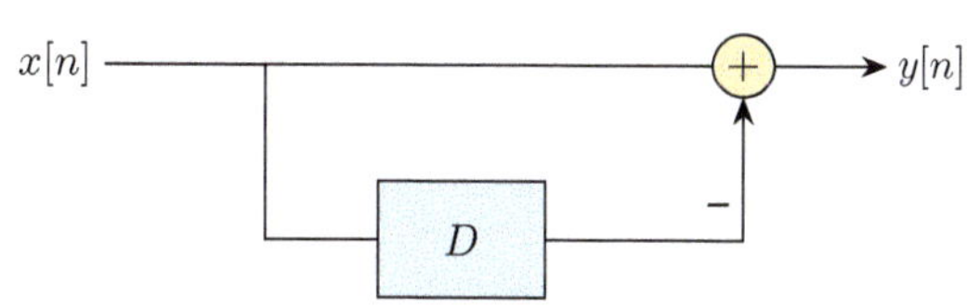

Figure 3.18 Block diagram representation of the system, $y[n] = x[n] - x[n - 1]$.

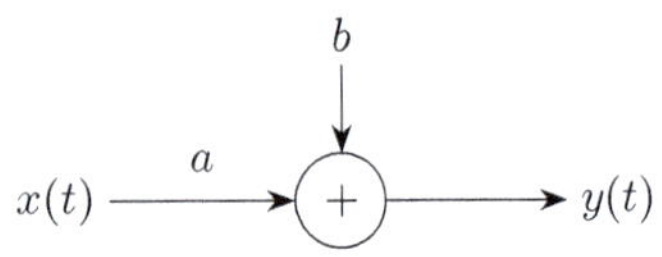

Figure 3.19 Block diagram of the system in Exercise 3.19.

This system does not satisfy the superposition property because when the input is $x(t) = a_1x_1(t) + a_2x_2(t)$, the corresponding output is not $y(t) = a_1y_1(t) + a_2y_2(t)$:

$$y(t) = a[a_1x_1(t) + a_2x_2(t)] + b \neq a_1y_1(t) + a_2y_2(t). \tag{3.82}$$

Although this system is not strictly linear, it bears many properties of a linear system, such as invertibility. Recall that we call these systems as **incrementally linear**.

The inverse of the incrementally linear system can be directly obtained from the system equation as follows:

$$x(t) = \frac{y(t) - b}{a} \quad \text{for} \quad a \neq 0. \tag{3.83}$$

There exists a unique input, $x(t)$ for every output $y(t)$. Thus it is invertible (Figure 3.20).

Exercise 3.20 Find the equation for the system represented by the block diagram of Figure 3.21.

Solution

This is a feedback control system, where the adder receives two inputs: $x(t)$ and $y(t)$. The output of the adder must be $\frac{dy(t)}{dt}$ so that when it is integrated, it outputs $y(t)$. Thus, this system can be represented by the following differential equation:

$$\frac{dy(t)}{dt} - y(t) = x(t). \tag{3.84}$$

This system is represented by a first-order differential equation. It is possible to represent a system with cascaded integrators and adders, by higher order differential equations. We shall explore the properties of the systems, represented by differential equations, in Chapter 5.

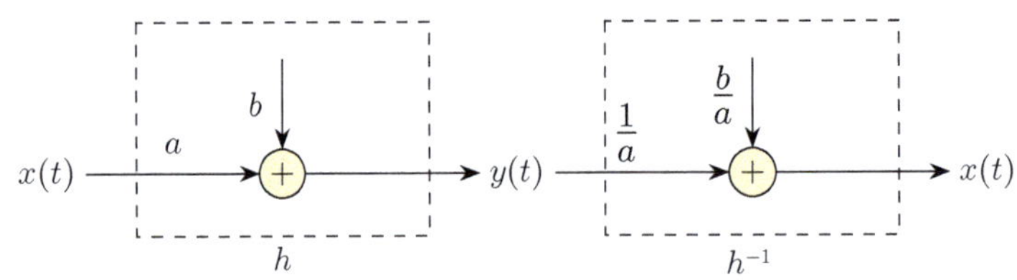

Figure 3.20 Cascaded representation of an incrementally linear system (given in Exercise 3.19), represented by the model, h and its inverse, h^{-1}.

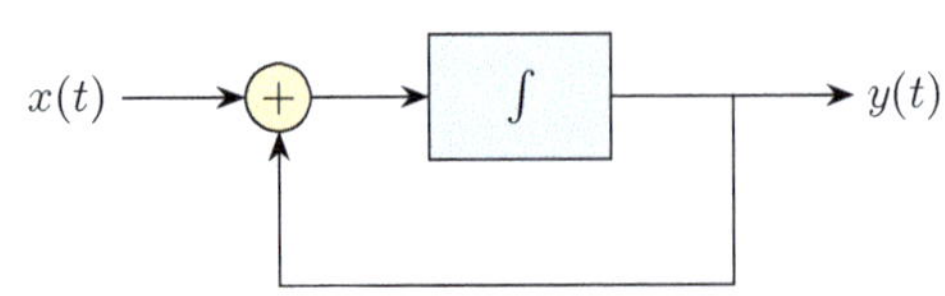

Figure 3.21 Block diagram of the system in Exercise 3.20.

3.5 Chapter Summary

How can we represent a discrete time or a continuous time system? Is it possible to decompose a complicated system into a set of interrelated subsystems, so that we can model and design a large class of natural and human-made systems? How do we categorize systems with respect to predefined properties?

In this chapter, we define a system as a mapping between the input signal(s) and output signal(s). Depending on the behavior of the system, this mapping can be represented by an algebraic equation, differential equation, or integral equation.

In most applications, finding a single equation to represent a system is not possible. In order to represent complicated systems, we decompose them into a set of subsystems, each of which is represented by an equation. The interrelations among the subsystems are established by the input and output signals of each subsystem.

We can combine subsystems, in various forms, called, **series, parallel, hybrid and feedback control systems**. We also study the properties of systems, which provide us with a framework to model, design and implement a wide range of systems, using the available mathematical tools. In case we can identify some of these properties, namely, **memory, causality, invertibility, stability, time invariance**, and **linearity,** it is possible to represent systems in more compact and precise models.

Finally, it is possible to define a set of basic building blocks for both discrete time and continuous time systems. Combining the simple building blocks, such as **adders, multiplies, integrators**, and **differentiators** for continuous time systems, and **unit advance and unit delay operators** for the discrete time systems enable us to model and design a large class of systems.

Problems

3.1 Consider two discrete time subsystems $S1$ and $S2$, defined by the following difference equations:

$$S1:\ y_1[n] = 4x_1[n] + 2x_1[n-1],$$
$$S2:\ y_2[n] = y_1[n-2].$$

Suppose that $S1$ and $S2$ are connected in series to form an overall system, S, as shown in the following:

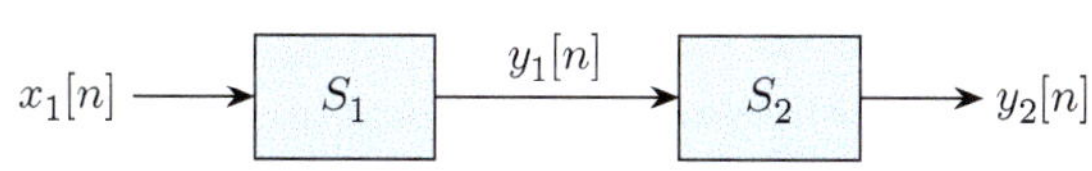

a) Find the difference equation for the overall system, which relates the input $x[n] = x_1[n]$ and output $y[n] = y_2[n]$.
b) Would the system equation you obtain in part a be different if the order of the series connection of $S1$ and $S2$ is reversed? In other words, is the series connection of sub systems commutative? Verify your answer.
c) Is the overall system S linear? Show if the superposition property holds or not.
d) Is the overall system S time invariant? Verify your answer.

3.2 Consider two continuous time subsystems $S1$ and $S2$, defined by the following differential equation:

$$S1:\ y_1(t) = 4x(t) + 2\frac{dx(t)}{dt},$$
$$S2:\ y_2(t) = \frac{dx(t)}{dt}.$$

Suppose that these $S1$ and $S2$ are connected in parallel to form an overall system S with the same input $x(t)$ and with an output $y(t) = y_1(t) + y_2(t)$, as shown in the following:

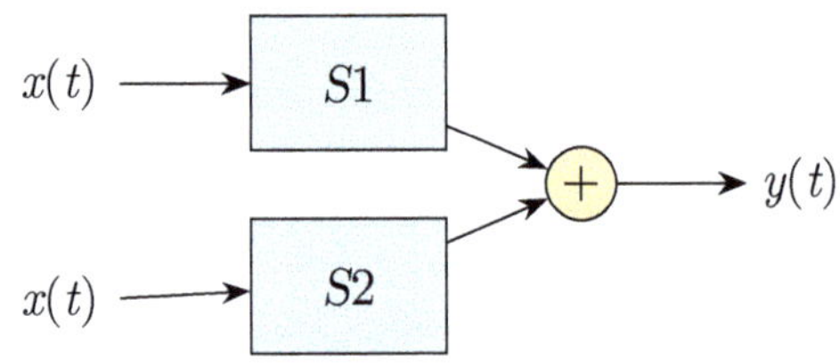

a) Find the differential equation, which relates the input $x(t)$ and the output $y(t)$ for the overall system S.
b) Is the overall system S linear?
c) Is the overall system S time invariant?
d) If $S1$ and $S2$ were interchanged in the parallel configuration, would it make any difference to the overall system S?

3.3 Consider a discrete time system, represented by the following difference equations:

$$y[n] = x[n]\sin\left[\frac{\pi}{2}n\right].$$

a) Is this system stable?
b) Is this system invertible?
c) Is this system causal?
d) Is this system linear?

3.4 Consider the continuous time systems, represented by the following equations. Are these systems linear and time invariant? Verify your answers.
a) $y(t) = 2tx(t+1)$
b) $y(t) = x(t)\sin(t-1)$
c) $y(t) = 2\delta(t)$
d) $y(t) = x(2t^2)$

3.5 Consider the discrete time systems, represented by the following equations. Are these systems linear and time invariant? Verify your answers.
a) $y[n] = x[2n]\cos[\pi n]$
b) $y[n] = x[n^2 - 1]$
c) $y[n] = 2x[n-1] + x[2n-2]$
d) $y[n] = x^2[2n+2]$

3.6 Given the following equations for discrete time systems, check if the properties of **memory, stability, causality, linearity, invertibility,** and **time invariance** hold. Verify your answers for each system and for each property.

a) $y[n] = 2x[n^2]$
b) $y[n] = (x[n-100] + x[100-n])\sin 5n$
c) $y[n] = \delta[n]x[2n]$
d) $y[n] = x\left[\frac{n}{3}\right]$
e) $y[n] = \begin{cases} 0, & n < 0 \\ x[n+4] & n \geq 0 \end{cases}$

3.7 Given the following equations for continuous time systems, check if the properties of **memory, stability, causality, linearity, invertibility,** and **time invariance** hold. Verify your answers for each system and for each property.
a) $y(t) = \int_{-\infty}^{5t} x(2\tau)\,d\tau$
b) $y(t) = \frac{d(x(t)\sin(3t))}{dt}$
c) $y(t) = x\,(2t+3)$
d) $y(t) = \int_{-5t}^{\infty} 2x(\tau)d\tau$
e) $y(t) = \frac{dx(2t)}{dt}$

3.8 Given the following system equations, check if the properties of **memory, stability, causality, linearity, invertibility,** and **time invariance** hold. Verify your answers for each system and each property.
a) $y(t) = x(t-5)$
b) $y(t) = \left(\frac{\sin(2t)}{t}\right)^2 x(t)$
c) $y(t) = (3t + \cos t)x(t)$
d) $y[n] = \sum_{k=-\infty}^{n} x[k]$
e) $y(t) = \frac{d(x(t)) + \sin(\cos(t))}{dt}$

3.9 Consider a discrete time system S, represented by the following difference equation:

$$y[n] = x[n]h[n+1] + 2h[n].$$

a) If $h[n] = C$ for all n and C is a constant, show that S is time invariant.
b) If $h[n] = n$, show that S is not time invariant.
c) Is this system linear for $h[n] = n$?

3.10 Consider the following statements and determine if they are always true, always false or neither. Justify your answers.
a) The parallel connection of two linear time-invariant systems is itself a time-invariant system.
b) The series connection of two causal and linear systems is itself causal and linear.

c) A system consisting of a causal and linear system and a nonlinear and time varying connected serially is not causal or linear.

3.11 Consider the subsystems $S1$, $S2$, and $S3$, where the inputs and outputs are represented by the following equations:

$$S1: \; y_1[n] = \begin{cases} 0, & n \text{ even}, \\ x[n]^2, & n \text{ odd} \end{cases}$$

$$S2: \; y_2[n] = 2y_1[n/2],$$

$$S3: \; y[n] = y_2[n+2] + \frac{1}{4}y_2[n].$$

Suppose that these systems are connected in series, as shown in the following block diagram:

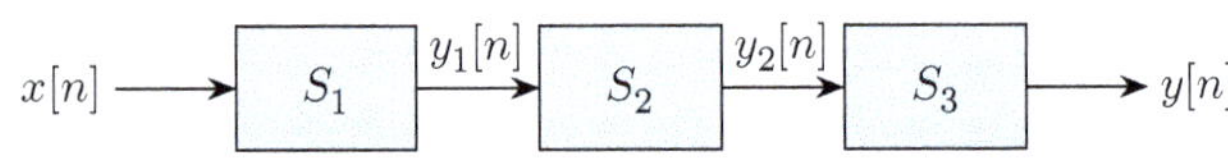

a) Find the equation, which represent the overall system S, which relates the overall input $x[n]$ to the overall output $y[n]$.
b) Are the subsystems, $S1, S2$, and $S3$ linear, time invariant, causal, invertible, and stable? Verify your answer for each subsystem.
c) Is the overall system S linear, time invariant, causal, invertible, and stable? Verify your answer for S.

3.12 Show that if the input of a time-invariant system is periodic, then the corresponding output is also periodic for both continuous and discrete time systems.

3.13 Consider a time-invariant system with an aperiodic input. Is the output signal always aperiodic? Verify your answer by giving examples.

3.14 Consider the block diagram representation of a causal feedback control system

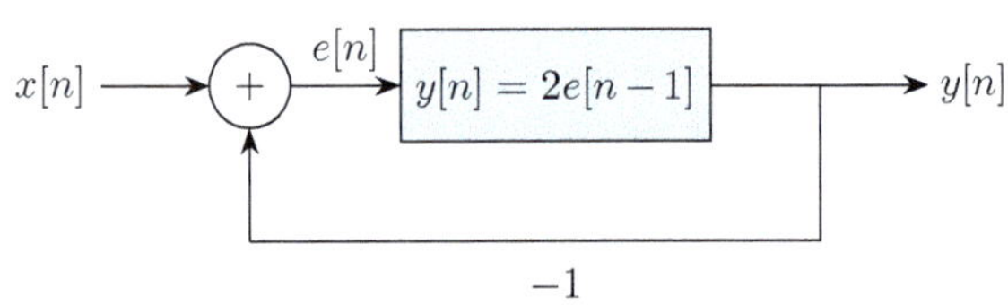

a) Find and plot the output for $x[n] = \delta[n] + \delta[n-2]$.
b) Find and plot the output given $x[n] = u[n] - u[n-4]$.

3.15 Are the following systems invertible? Find the inverse systems if they are invertible.
a) $y[n] = 2x[n^2]$
b) $y(t) = \sin x(t)$
c) $y[n] = x[n+1]x[n-1]$
d) $y(t) = \dfrac{dx(t)}{dt}$

e) $y[n] = \begin{cases} x[2n], & n \geq 0 \\ x[n-2], & n < 0 \end{cases}$

f) $y(t) = \cos(3t)x(t)$

3.16 Find a block diagram representation for the continuous time system, represented by the following equation:

$$\frac{d^2y(t)}{dt} + ay(t) = \frac{dx(t)}{dt}.$$

3.17 Find a block diagram representation for the discrete time system, represented by the following equation:

$$y[n-2] + 0.5y[n-1] + 0.6y[n] = x[n].$$

3.18 Consider the following continuous time system equation, where the input is represented by the superposition of derivatives of the output:

$$\sum_{k=0}^{N} a_k \frac{d_k y(t)}{dt^k} = x(t).$$

a) Find a block diagram representation of this system.

b) For $x(t) = 0$, show that if s_0 satisfies the following algebraic equation:

$$p(s) = \sum_{k=0}^{N} a_k s^k = 0,$$

then, $Ae^{s_0 t}$ satisfies the system equation given above, where A is an arbitrary complex constant number.

3.19 Consider the system equation given as follows:

$$\frac{d^3y(t)}{dt^3} + 4\frac{d^2y(t)}{dt^2} + 5\frac{dy(t)}{dt} + 2y(t) = x(t).$$

a) Find a block diagram representation of this system.

b) For $x(t) = 0$, find a solution to this equation in the form $Ae^{s_0 t}$.

3.20 Consider the following discrete time system equation, where the input is equal to the superposition of differences of the output:

$$\sum_{k=0}^{N} a_k y[n-k] = x[n].$$

a) Find a block diagram representation of this system.

b) For $x[n] = 0$, show that, if z_0 is a solution of the equation

$$p(s) = \sum_{k=0}^{N} a_k s^{-k} = 0,$$

then, Az_0^n is a solution of the aforementioned difference equation, where A is an arbitrary constant.

3.21 Given a discrete time system, represented by the following equation:

$$y[n] - 3y[n-1] + 2y[n-2] = x[n].$$

a) Find a block diagram representation of this system.
b) For $x[n] = 0$, find a solution to this equation in terms of Az_0^n.

4

Representation of Linear Time-Invariant Systems by Impulse Response and Convolution Operation

"The universe is nonlinear. But straight lines are really nice."

D. C. Vural

In Chapter 1, we mentioned about the systems approach, where we noted that the whole is more than the sum of its parts, in a wide range of systems. However, this inequality becomes an equality, if a system is linear, holding the superposition property. In this case, the whole system can be represented by a set of subsystems interconnected with each other by signals. The behavior of the linear systems can be studied by modeling their subsystems and the signals relating to the subsystems. On the other hand, if a system is not linear, the interactions among the subsystems may generate unpredictable outputs, which cannot be studied by the analysis of the subsystems. This behavior prevents us from modeling the nonlinear systems as a collection of interrelated subsystems. Thus, linear models are very important, in the sense that they enable us to break down the systems into a bunch of mathematically tractable components.

Time invariance is a type of symmetry, where the representation of a system does not change in time. In continuous time systems, a time-invariant system performs the same operation at any time t and at time $(t - t_0)$, for any $t_0 \neq 0$. Similarly, in discrete time systems, the system behaves the same at time n and at time $[n - n_0]$, for any $n_0 \neq 0$. This property simplifies the modeling and analysis of systems.

A large class of systems can be represented by linear, time-invariant (LTI) systems, which enable us to find a unique model for a system by establishing the relationship between the input and the output signals. In this chapter, we shall study the discrete time and continuous time LTI systems.

Our goal is to find an equation, which relates the input $x(\cdot)$ to the corresponding output $y(\cdot)$ to represent an LTI system.

Suppose that we observe an input–output pair, $x(\cdot)$ and $y(\cdot)$, of an LTI system. In Chapter 2, we showed that it is possible to represent any function in terms of the weighted integral of shifted impulse functions $\delta(t - \tau)$ for continuous time systems. Similarly, we can represent any function in terms of weighted summation of shifted impulses, $\delta[n - k]$ for discrete time signals, as shown in Figure 4.1.

For example, suppose that a continuous time, LTI system receives an input signal, $x(t) = e^{-j\omega t}$, and outputs a unit step signal, $y(t) = u(t)$. These signals can be represented by the weighted integral

Signals and Systems: Theory and Practical Explorations with Python,
First Edition. Fatoş Tunay Yarman Vural and Emre Akbaş.

Companion website: www.wiley.com/go/signalsandsystems

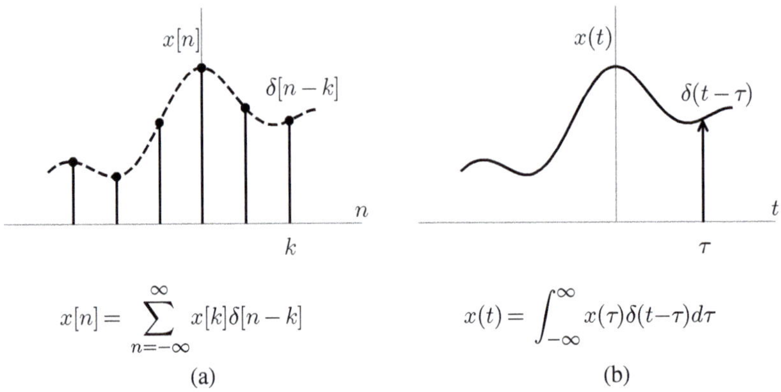

Figure 4.1 Representation of discrete time signals in terms of weighted summation of shifted impulse functions (a) and representation of continuous time signals in terms of the weighted integral of shifted impulse functions (b).

of shifted impulse functions. Mathematically,

$$x(t) = \int_{-\infty}^{\infty} e^{-j\omega\tau}\delta(t-\tau)d\tau = e^{-j\omega t}. \tag{4.1}$$

$$y(t) = \int_{-\infty}^{t} \delta(t-\tau)d\tau = x(t) = u(t). \tag{4.2}$$

Motivating Question: Can we find a relationship between $x(t)$ and $y(t)$, using Equations (4.1) and (4.2) to represent an LTI system by a single equation?

In this chapter, we answer this question by defining a new operation, called **convolution** for both continuous time and discrete time systems. This operation relates the input–output pair of an LTI system through a specific response, called the **impulse response**.

4.1 Representation of LTI Systems by Impulse Response

Suppose that we feed an input signal $x(\cdot)$ and observe a signal $y(\cdot)$ at the output of a black box of Figure 4.2. What is it in the box? The answer lies in the very important concept, called the **impulse response.**

Definition 4.1 **Impulse response** is the signal $y(\cdot) = h(\cdot)$ of an LTI system, observed at the output when the input signal is a unit impulse function, $x(\cdot) = \delta(\cdot)$ (Figure 4.3).

Motivating Question: Why is impulse response very important? Because it uniquely represents an LTI system! In other words, impulse response uniquely relates any input to the corresponding output of an LTI system.

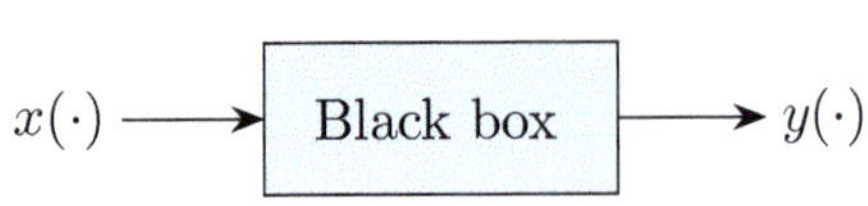

Figure 4.2 If we observe the input $x(\cdot)$ and the corresponding output $y(\cdot)$, can we find a unique model $h(\cdot)$ for the "black box"?

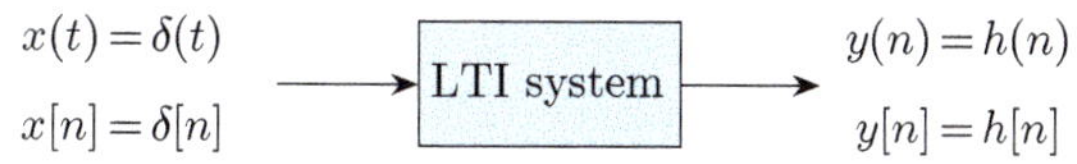

Figure 4.3 Impulse response is the response of an LTI system to a unit impulse function. A discrete time LTI system generates the impulse response $h[n]$ for the unit impulse input $\delta[n]$, whereas a continuous time LTI system generates the impulse response $h(t)$ for the unit impulse input $\delta(t)$.

4.1.1 Representation of Discrete Time Linear Time-Invariant Systems by Impulse Response

Suppose that we feed a discrete time unit impulse, $\delta[n]$ to an LTI system and at the output, we measure the impulse response, $h[n]$, as shown in Figure 4.4. Next, we shift the impulse response at the input and feed the shifted impulse response, $\delta[n-1]$. Since the system is time invariant, we obtain the same shift at the output and obtain the shifted impulse response, $h[n-1]$. If we keep sifting the impulse functions, $\delta[n-k]$ for all k and feeding it as an input to an LTI, we obtain the shifted versions of the impulse responses, $h[n-k]$. Now, let us multiply the shifted impulse function by the kth component of a general signal $x[n]$, to feed $x[k]\delta[n-k]$ at the input. Since the system is linear, at the output, we get $x[k]h[n-k]$.

Finally, let us superpose the input signals, $x(k)h[n-k]$ for all k, as follows:

$$x[n] = \sum_{k=-\infty}^{\infty} x[k]\delta[n-k], \tag{4.3}$$

and feed it as the input signal. Since the system is linear and time invariant, the superposition of the shifted impulses outputs the same superposition of the shifted impulse responses, $x[k]h[n-k]$, as follows:

$$\boxed{y[n] = \sum_{k=-\infty}^{\infty} x[k]h[n-k]}, \tag{4.4}$$

as shown in Figure 4.4. This equation is called the **convolution summation**.

Basically, this equation states that the output of a discrete time LTI system is equal to the weighted superposition of its shifted impulse responses. The weights are the value of the input signal at the

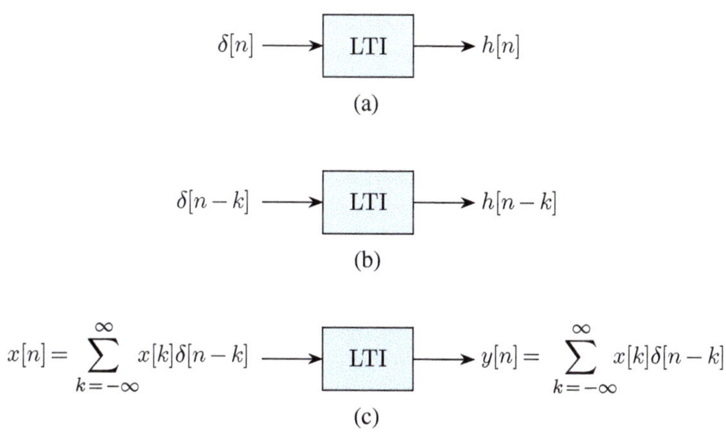

Figure 4.4 Relationship between the input and output signal of a discrete time LTI system: (a) response of a system, $y[n] = h[n]$ to a unit impulse function, $\delta[n]$. (b) Since the system is time invariant, shifted unit impulse, $\delta[n-k]$, generates a shifted impulse response, $h[n-k]$. (c) Since the system is linear, we can superpose all the shifted impulses at the input and obtain the same superposition at the output.

point of the shift. Thus, convolution summation relates a general input $x[n]$ to a general output $y[n]$ through the impulse response, providing us with an equation for representing an LTI system.

4.1.2 Representation of Continuous Time Linear Time-Invariant System

We can use the aforementioned methodology to derive a relationship between the input and output signals of the continuous time LTI systems. This time, we feed a continuous time unit impulse input, $\delta(t)$ to an LTI system and at the output, we measure the impulse response, $h(t)$, as shown in Figure 4.5.

Next, we shift the impulse response at the input, and feed the shifted impulse response, $\delta(t - \Delta t)$, as in Figure 4.5. Since the system is time invariant, we obtain the same shift at the output to obtain the shifted impulse response, $h(t - \Delta t)$. If we keep sifting the impulse responses, $\delta(t - k\Delta t)$ for all k and keep feeding it as an input to an LTI, we obtain the shifted versions of the impulse responses, $h(t - k\Delta t)$.

Now, let us multiply the shifted impulse function by $x(k\Delta t)$, to feed $x(k\Delta t)\delta(t - k\Delta t)$ at the input. Since the system is linear, at the output of the system, we get $x(k\Delta t)h(t - k\Delta t)$.

Next, let us superpose all the input signals to obtain a new signal,

$$x_{\Delta t}(t) = \sum_{k=-\infty}^{\infty} x(k\Delta t)\delta(t - k\Delta t), \tag{4.5}$$

and feed it as the input signal. Since the system is linear, the superposition of the shifted impulse functions at the input results in the superposition of the shifted impulse responses at the output, as follows:

$$y_{\Delta t}(t) = \sum_{k=-\infty}^{\infty} x(k\Delta t)h(t - k\Delta t). \tag{4.6}$$

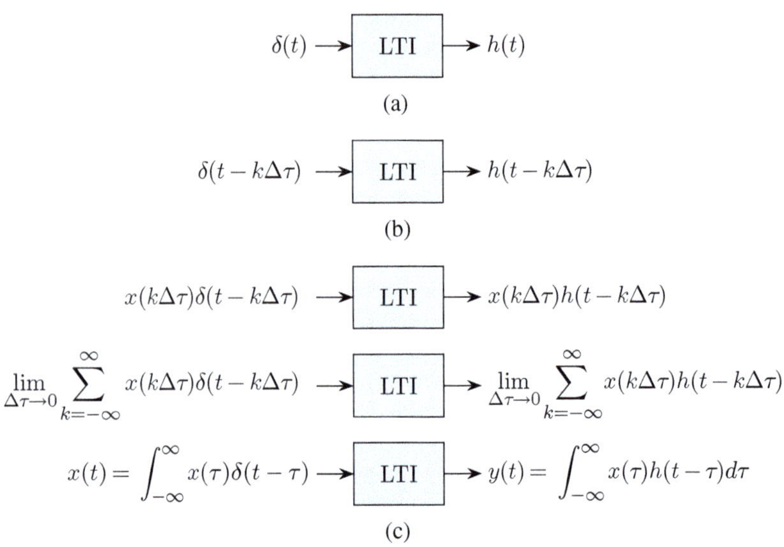

Figure 4.5 Relationship between the input and output signal of a continuous time LTI system: (a) Response of a system, $y(t) = h(t)$ to a unit impulse function, $\delta(t)$. (b) since the system is time invariant, shifted unit impulse, $\delta(t - \Delta t)$, generates a shifted impulse response, $h(t - \Delta t)$, (c) Since the system is linear, we can superpose all of the shifted impulses at the input and obtain the same superposition at the output.

If we take the limits with respect to time, the summation operation of Equations (4.6) converges to the integral operations:

$$x(t) = \lim_{\Delta t \to 0} x_{\Delta t}(t) = \int_{-\infty}^{\infty} x(\tau)\delta(t-\tau)d\tau, \tag{4.7}$$

$$y(t) = \lim_{\Delta t \to 0} y_{\Delta t}(t) = \int_{-\infty}^{\infty} x(\tau)h(t-\tau)d\tau, \tag{4.8}$$

where $\lim_{\Delta t \to 0} x_{\Delta t}(t) \to x(t)$ and $\lim_{\Delta t \to 0} y_{\Delta t}(t) \to y(t)$. Thus, we have

$$\boxed{y(t) = \int_{-\infty}^{\infty} x(\tau)h(t-\tau)d\tau}, \tag{4.9}$$

as shown, in Figure 4.5. This equation is called **convolution integral**.

Basically, the convolution integral states that the output of a continuous time LTI system can be uniquely obtained from the weighted integral of its shifted impulse responses. The weights are the values of the input signal $x(t)$ at the point of shift, $t = \tau$, of the impulse response. Thus, the convolution integral relates a general input $x(t)$ to a general output $y(t)$ through the impulse response, providing us an equation for representing a continuous time LTI system.

Later in this book, we shall see that given an input–output pair of a continuous time or discrete time LTI system, we can uniquely find its impulse response. Since impulse response uniquely defines an LTI system, it is possible to model the black box of an LTI system from an observed input–output signal pair.

Convolution integral and convolution sum uniquely identify continuous time and discrete time LTI systems, respectively (Figure 4.6). Thus, the convolution operation, for continuous time LTI systems,

$$y(t) = \int_{-\infty}^{\infty} x(\tau)h(t-\tau)d\tau = x(t) * h(t), \tag{4.10}$$

and for discrete time systems,

$$y[n] = \sum_{-\infty}^{\infty} x[k]h[n-k] = x[n] * h[n], \tag{4.11}$$

are considered as system equations. We use "$*$" as the shorthand notation for the convolution operation.

Remark 4.1 In continuous time LTI systems, the operand functions and the output of the convolution integral are all **continuous time functions**. Similarly, in discrete time LTI systems, the operand functions and the output of the convolution summation are all **discrete time functions**.

Question: Show that convolution operation satisfies the following properties:

1) **Commutativity**: Convolution operation is commutative:

$$x(\cdot) * h(\cdot) = h(\cdot) * x(\cdot). \tag{4.12}$$

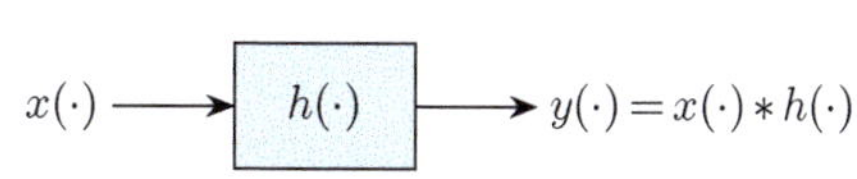

Figure 4.6 Representation of LTI system by its impulse response, $h(\cdot)$.

Formally speaking,

$$y(t) = \int_{-\infty}^{\infty} x(\tau)h(t-\tau)d\tau = \int_{-\infty}^{\infty} x(t-\tau)h(\tau)d\tau, \tag{4.13}$$

$$\left\{x(\cdot) \to \boxed{h(\cdot)} \to y(\cdot)\right\} = \left\{h(\cdot) \to \boxed{x(\cdot)} \to y(\cdot)\right\}. \tag{4.14}$$

Commutativity property of convolution operation reveals that we can replace the input signal with the impulse response of an LTI system without changing the output.

2) **Associativity:** Convolution operation is associative:

$$x(\cdot) * [h_1(\cdot) * h_2(\cdot)] = [x(\cdot) * h_1(\cdot)] * h_2(\cdot). \tag{4.15}$$

This property reveals that for an input signal $x(\cdot)$, passing it through $h_1(\cdot)$ and $h_2(\cdot)$ in a series connection, and passing it through a single system represented by $h_1(\cdot) * h_2(\cdot)$ would yield the same output signal $y(\cdot)$ (Figure 4.7).

Together with commutativity, associativity property reveals that in the series connection, the order of the subsystems does not matter (Figure 4.7).

3) **Distributivity:** Convolution operation is distributive:

$$x(\cdot) * [h_1(\cdot) + h_2(\cdot)] = x(\cdot) * h_1(\cdot) + x(\cdot) * h_2(\cdot). \tag{4.16}$$

The distributivity property of convolution operation reveals that the addition of convolutions of an input signal with two different impulse responses is equivalent to the convolution of added impulse responses with the same input. This property reduces a parallel system with two subsystems into a system with one component, which has two added impulse responses of the parallel system (Figure 4.8).

In summary, the properties of convolution operation allow us to interchange the input signals and impulse responses of subsystems without changing the output to a particular input of an overall system.

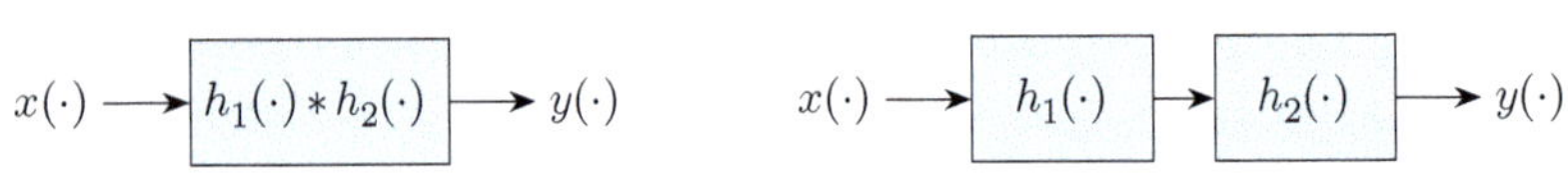

(a) (b)

Figure 4.7 Associativity of convolution. Provided that $h_1(\cdot)$ and $h_2(\cdot)$ are impulse responses of two LTI systems, the system shown in (a) is equivalent to the system shown in (b). Additionally, the commutativity property enables us to switch the order of $h_1(.)$ and $h_2(.)$.

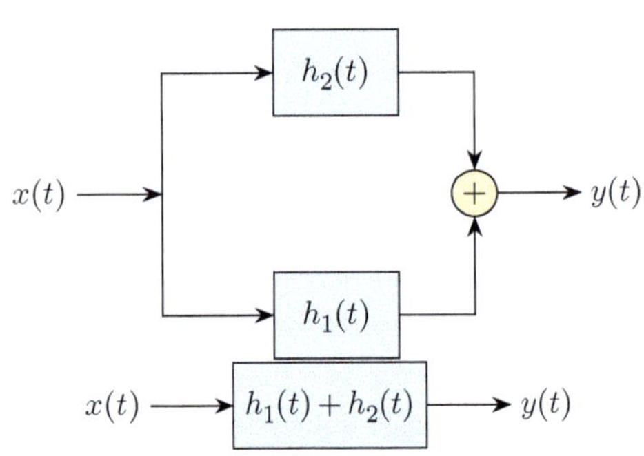

Figure 4.8 Distributivity of convolution. Two block diagrams are equivalent.

Motivating Question: What does convolution operation do in a real-life system, which is fed by a particular input? When we convolute an input signal $x(\cdot)$ with an impulse response, $h(\cdot)$, what do we measure at the output signal?

Let us further study the convolution operation to understand the meaning and implications on systems.

4.1.3 Convolution Operation in Continuous Time

When we convolute two functions, $x(t)$ and $h(t)$, we apply the following steps:

Step 1: Change the time variable to the dummy variable of integral to get $h(\tau)$.
Step 2: Time reverse the dummy variable of integral, τ, to obtain the reversed impulse response, $h(-\tau)$.
Step 3: Shift the impulse response, by the time variable t, to obtain reversed and shifted impulse response, $h(t-\tau)$.
Step 4: Multiply the shifted impulse response with the input signal, $x(\tau)$ to get $h(t-\tau)x(\tau)$.
Step 5: Find the area under the multiplication, using the integration operation for each translation of the time variable t,

$$y(t) = \int_{-\infty}^{\infty} x(\tau)h(t-\tau)d\tau = x(t) * h(t). \tag{4.17}$$

Since the convolution operation is commutative, in the aforementioned steps we can replace the signal $x(t)$, by the impulse response, $h(t)$. Formally,

$$y(t) = \int_{-\infty}^{\infty} x(\tau)h(t-\tau)d\tau = \int_{-\infty}^{\infty} x(t-\tau)h(\tau)d\tau. \tag{4.18}$$

Notice that in the convolution operation, there are two variables. The variable τ is the dummy variable of integration. After we take the integral of $x(\tau)h(t-\tau)$, it disappears. The time variable t translates the reversed impulse response $h(-\tau)$ all over the input function $x(\tau)$. Due to the commutativity property of the convolution operation, we can replace the impulse response with the input.

INTERACTIVE

Explore convolution @ https://384book.net/i0401

Loosely speaking, convolution operation measures the degree of similarity of two functions, $x(t)$ and $h(-t)$ as a function of time. The convolution operation is intensively used in designing a wide range of systems, including signal, video and image processing, and machine learning.

Impulse response, which represents an LTI system, takes different names, depending on the application domains.

- In signal processing, the impulse response is called **filter** because it passes only the parts of the input signal, which resemble the impulse response and suppresses the rest. Like a water filter, which keeps the chemicals and only passes the clean part of polluted water, a filter purifies the signal. For example, one can design a filter for a microphone to suppress the noise of the environment and just pass the clean speech or music. In other words, the convolution operation modifies the input signal depending on the shape of the filter function.

- In image and video processing, impulse response is called **mask** because it detects an object in an image. When we design a mask with the characteristics of an object, the convolution of an image signal with a mask, outputs high values in the regions that resemble the object. Thus, it is possible to identify the location or class of an object at the output of the convolution operation.
- In machine learning, it is called **model** because it is very handy in designing a learning algorithm for object detection and classification. In this case, a network architecture, called convolutional neural network, learns the impulse responses to detect and/or classify an object.

Designing a problem-specific LTI system is beyond the scope of this book.

Exercise 4.1 Suppose that we are given the following impulse response and the input signal,

$$h(t) = e^{-2t}u(t), \quad x(t) = e^{-t}u(t). \tag{4.19}$$

Find and plot the output of this LTI system.

Solution
We follow the steps of convolution operation, given earlier and take the convolution integral, as follows:

$$\begin{aligned} y(t) &= \int_{-\infty}^{\infty} x(t-\tau)h(\tau)d\tau \\ &= \int_{0}^{t} e^{-(t-\tau)}e^{-2\tau}d\tau \\ &= e^{-t}\int_{0}^{t} e^{-\tau}d\tau \\ &= e^{-t}(1-e^{-t})u(t). \end{aligned} \tag{4.20}$$

INTERACTIVE

Explore convolution of two exponential functions @ https://384book.net/i0402

Exercise 4.2 Find the output of an LTI system, for any input, $x(t)$, when the impulse response is an impulse input, $h(t) = \delta(t)$.

Solution
Setting $h(t) = \delta(t)$ in the convolution integral, we get

$$y(t) = x(t) * h(t) = \int_{-\infty}^{\infty} x(\tau)\delta(t-\tau)d\tau = x(t). \tag{4.21}$$

Thus, $\delta(t)$ acts as an identity function in convolution operation (Figure 4.9).

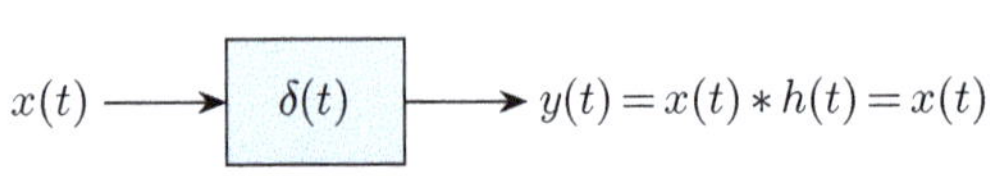

Figure 4.9 When the impulse response of an LTI system is $\delta(t)$, the system acts as an identity system. The output is equal to the input.

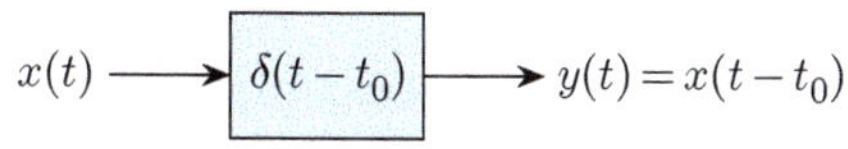

Figure 4.10 When the impulse response of an LTI system is $\delta(t-t_0)$, the system delays its input signal by t_0.

Exercise 4.3 Find the output of an LTI system for any input, $x(t)$, when the impulse response is the shifted impulse function, $h(t) = \delta(t - t_0)$.

Setting $h(t) = \delta(t - t_0)$ in the convolution operation, we get

$$y(t) = x(t) * \delta(t - t_0) = \int_{-\infty}^{\infty} x(\tau)\delta(t - t_0 - \tau)d\tau = x(t - t_0). \tag{4.22}$$

The output is just the time-shift version of the input. Thus, $\delta(t - t_0)$ acts as a delay operator (Figure 4.10).

Exercise 4.4 Find the response of an LTI system represented by the impulse response, $h(t) = u(t)$, when the input is $x(t) = u(t)$.

Solution

We need to evaluate

$$y(t) = \int_{-\infty}^{\infty} u(\tau)u(t - \tau)d\tau. \tag{4.23}$$

Since the unit step function $u(t)$ equals to 1 for $t > 0$, the integral operand is non-zero for $\tau > 0$ and $t - \tau > 0$, hence, for $0 < \tau < t$. Then, the output is

$$y(t) = \int_{-\infty}^{\infty} u(\tau)u(t - \tau)d\tau = \int_{0}^{t} d\tau = tu(t). \tag{4.24}$$

We multiplied t with $u(t)$ in Equation (4.24) because $y(t) = 0$ for $t < 0$. This convolution is illustrated in Figure 4.11.

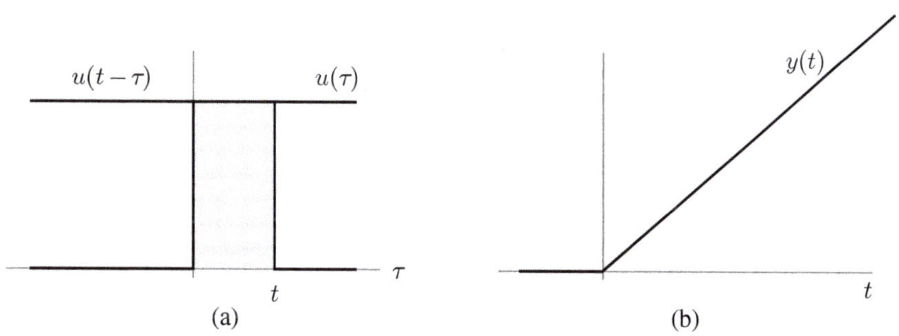

Figure 4.11 (a) Convolution of two unit step functions. Prior to integration, we multiply $u(\tau)$ by $u(t - \tau)$. The shaded area shows the overlap. Note that as we change the time variable t between $(-\infty, \infty)$, we translate one of the unit step functions over the other one. For $t < 0$, there is no overlap. Thus the output signal $y(t) = 0$ for $t < 0$. As we increase the time variable for $t > 0$, we get more and more overlap between the two unit step functions. Thus, the convolution operation outputs a monotonically increasing function $y(t)$, which is a ramp. (b) The result of convolution, $y(t) = tu(t)$.

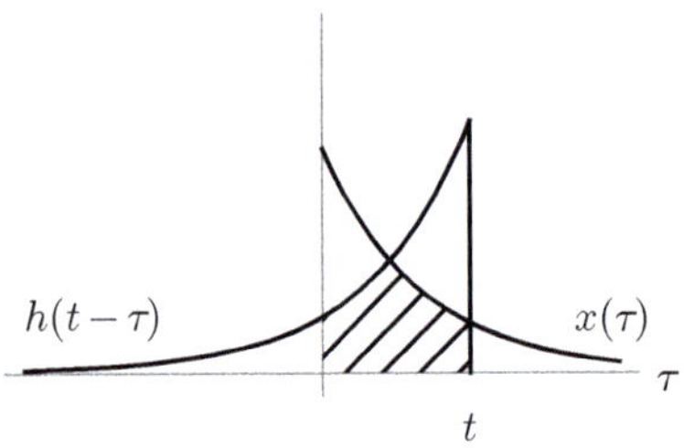

Figure 4.12 While we convolute two exponential functions with negative decays, we reverse one of the exponentials and translate it over the other one as we change the time variable $-\infty < t < \infty$.

Exercise 4.5 Find the response of an LTI system, represented by the impulse response, $h(t) = e^{\lambda_2 t}u(t)$ for the input, $x(t) = e^{\lambda_1 t}u(t)$. An example is illustrated in Figure 4.12.

Solution

This is the general form of convolution of two exponential functions:

$$\begin{aligned} y(t) &= \int_0^t e^{\lambda_2(t-\tau)}e^{\lambda_1\tau}d\tau \\ &= e^{\lambda_2 t}\int_{-0}^t e^{\tau(\lambda_1-\lambda_2)}d\tau \\ &= \frac{e^{\lambda_2 t}}{\lambda_1-\lambda_2}\, e^{\tau(\lambda_1-\lambda_2)}\Big|_0^t \\ &= \frac{e^{\lambda_1 t}-e^{\lambda_2 t}}{\lambda_1-\lambda_2}u(t). \end{aligned}$$

Note that there is no overlap between $x(\tau) = e^{\lambda_1\tau}u(\tau)$ and $h(t-\tau) = e^{\lambda_2(t-\tau)}u(t-\tau)$ for $t < 0$. Thus, $y(t) = 0$ for $t < 0$; hence, we have $u(t)$ as the multiplier. It increases as we increase $t > 0$, until the tails get sufficiently small. Then it starts to decrease.

Exercise 4.6 Find the convolution of the following input signal $x(t)$ and impulse response $h(t)$ given as follows to obtain the output signal, $y(t) = x(t) * h(t)$:

$$x(t) = \begin{cases} 1 & \text{for } -1 \le t \le 2 \\ 0 & \text{otherwise.} \end{cases} \tag{4.25}$$

$$h(t) = \begin{cases} 1 & \text{for } -2 \le t \le 2 \\ 0 & \text{otherwise.} \end{cases} \tag{4.26}$$

Solution

The input signal and impulse response are both nonzero in a finite interval. Thus, the output of the convolution integral will be also nonzero in a finite interval. In order to find the nonzero interval of the output, we add the lower limits and upper limits of the two functions. In this particular example $y(t) \neq 0$ for $-3 \le t \le 4$.

When we evaluate the convolution integral, as the time variable changes in $(-\infty, \infty)$ the interval of the nonzero overlap, thus, the limits of the integral change. This requires to evaluate the integral for all different limits of the integral as t varies. Since there are no overlaps, for $t < -3$ and $t > 4$, $y(t) = 0$.

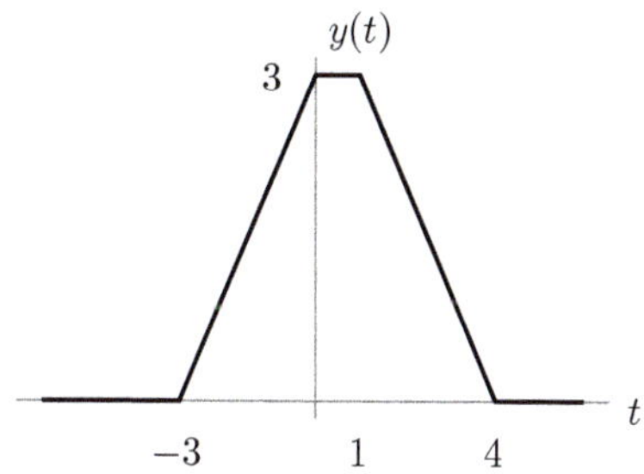

Figure 4.13 Result of the convolution operation $y(t) = x(t) * h(t)$, in Exercise 4.6.

Considering the convolution formula,

$$y(t) = \int_{-\infty}^{\infty} x(\tau)h(t-\tau)d\tau, \tag{4.27}$$

$x(\tau)$ and $h(t-\tau)$ will

- partially overlap for $-3 \leq t < 0$,
- fully overlap for $0 \leq t < 1$,
- partially overlap for $1 \leq t < 4$.

We need to evaluate the convolution integral for these three regions separately.

For $-3 \leq t < 0$, we need to find the boundaries of the integral, i.e., the bounds on the integration variable τ. From the definitions of x and h, we know that $-1 \leq \tau \leq 2$ and $t-2 \leq \tau \leq t+2$. Imposing the constraint $-3 \leq t < 0$ on these two inequalities restricts τ to $-1 \leq \tau \leq t+2$. We have

$$y(t) = \int_{-1}^{t+2} d\tau = t+3 \quad \text{for} \quad -3 \leq t < 0. \tag{4.28}$$

For $0 \leq t < 1$; $-1 \leq \tau \leq 2$ and $t-2 \leq \tau \leq t+2$ restricts τ to $-1 \leq \tau \leq 2$:

$$y(t) = \int_{-1}^{2} d\tau = 3 \quad \text{for} \quad 0 \leq t < 1. \tag{4.29}$$

For $1 \leq t < 4$, $-1 \leq \tau \leq 2$ and $t-2 \leq \tau \leq t+2$ restricts τ to $t-2 \leq \tau \leq 2$:

$$y(t) = \int_{t-2}^{2} d\tau = -t+4 \quad \text{for} \quad 1 \leq t < 4. \tag{4.30}$$

Overall, the output $y(t)$ is a trapezium, which shows that the maximum resemblance between the input signal and the impulse response is for $0 \leq t < 1$ (Figure 4.13).

4.1.4 Convolution Operation in Discrete Time Systems

For discrete time signals and systems, instead of convolution integral, we apply the convolution summation. The properties and meaning of the convolution sum are very similar to that of the convolution integral. Thus, we do not repeat them in this section. All we need to do is to replace the signal and the impulse response with their discrete time counterparts, $x[n]$ and $h[n]$, and use the convolution sum,

$$y[n] = \sum_{k=-\infty}^{\infty} x[k]h[n-k]. \tag{4.31}$$

Let us go over some exercises to see the similarities and distinctions between the continuous time and discrete time convolution.

Exercise 4.7 Find the output of a discrete time LTI system, represented by the impulse response, $h[n] = \alpha^n u[n]$, $0 < \alpha < 1$, when the input is $x[n] = u[n]$.

Solution

When we deal with infinite limits of the convolution summation, we need to take into account of the nonzero intervals of the input function and the impulse response. In this example, the convolution sum is as follows:

$$y[n] = \sum_{k=-\infty}^{\infty} u[k]\alpha^{n-k}u[n-k], \tag{4.32}$$

where $u[k]$ is nonzero for $k \geq 0$, and $u[n-k]$ is nonzero for $n-k \geq 0$. These two constraints yield $0 \leq k \leq n$. The convolution sum becomes

$$y[n] = \sum_{k=0}^{n} \alpha^{n-k} = \sum_{k'=0}^{n} \alpha^{k'}. \tag{4.33}$$

The closed form of the summation in the right-hand side of Equation (4.33) can be written as follows:

$$y[n] = \sum_{k'=0}^{n} \alpha^{k'} = \frac{1-\alpha^{n+1}}{1-\alpha}. \tag{4.34}$$

The discrete time functions, $x[n]$, $h[n]$, and $y[n]$ are illustrated in Figure 4.14.

Exercise 4.8 Find the output of an LTI system, represented by the impulse response, $h[n] = \delta[n - n_0]$, for any input $x[n]$.

Solution

This LTI system simply time-shifts the input by an integer amount of n_0 (Figure 4.15), as follows:

$$y[n] = x[n] * h[n] = x[n] * \delta[n-n_0] = x[n-n_0]. \tag{4.35}$$

Exercise 4.9 Find the impulse response of the discrete time system, represented by the following difference equation:

$$y[n] = x[n] - x[n-1]. \tag{4.36}$$

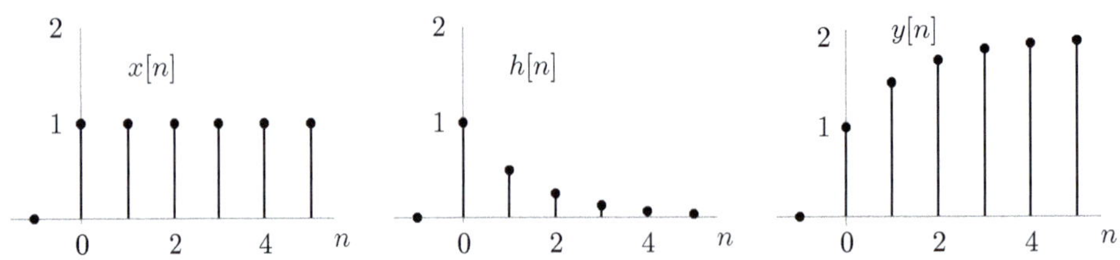

Figure 4.14 Output $y[n]$ of a discrete time LTI system, represented by the impulse response, $h[n] = 0.5^n u[n]$, when the input is $x[n] = u[n]$.

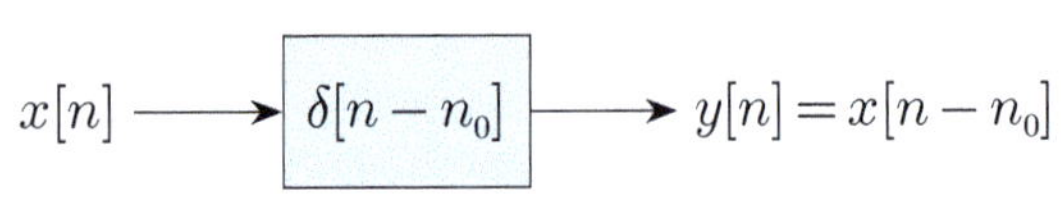

Figure 4.15 When the impulse response of a discrete time LTI system is $\delta[n - n_0]$, the system delays its input signal by n_0.

Solution

We replace the input by the unit impulse function. Then, the corresponding output is the impulse response:

$$h[n] = \delta[n] - \delta[n-1]. \tag{4.37}$$

Exercise 4.10 Find the output $y[n]$ of a discrete time system represented by the impulse response,

$$h[n] = u[n+1] - u[n-2], \tag{4.38}$$

when the input is

$$x[n] = u[n] - u[n-2]. \tag{4.39}$$

Solution

Both the input and impulse response can be represented by shifted impulse functions, as follows:

$$x[n] = \delta[n] + \delta[n-1], \tag{4.40}$$

$$h[n] = \delta[n-1] + \delta[n] + \delta[n+1]. \tag{4.41}$$

Convolution of $x[n]$ and $h[n]$,

$$y[n] = (\delta[n] + \delta[n-1]) * (\delta[n-1] + \delta[n] + \delta[n+1]), \tag{4.42}$$

yields

$$y[n] = \delta[n+1] + 2\delta[n] + 2\delta[n-1] + \delta[n-2]. \tag{4.43}$$

Here we used the distributivity property of convolution and the facts that (i) convolution with $\delta[n]$ does not change the input, (ii) convolution with $\delta[n-n_0]$ shifts the input by n_0.

4.1.5 Cross-correlation and Autocorrelation Operations

Convolution of two functions involves a time reverse operation together with the translation over time in one of the functions, for both continuous time and discrete time LTI systems, given as follows:

$$y(t) = \int_{-\infty}^{\infty} x(\tau)h(t-\tau)d\tau = x(t) * h(t), \tag{4.44}$$

$$y[n] = \sum_{-\infty}^{\infty} x[k]h[n-k] = x[n] * h[n]. \tag{4.45}$$

However, in many practical applications, the filter, representing the system can be designed in a reversed form or it can be an even function. In this case, time reverse step of the convolution operation becomes redundant. Furthermore, the signals can be complex functions, which makes the time reverse operation more difficult in real and imaginary axes. Thus, we can define a slightly simple version of convolution operation called cross-correlation.

Definition 4.2 **Cross-correlation or correlation operation** between two continuous time signals $x(t)$ and $h(t)$ is defined as:

$$y(t) = \int_{-\infty}^{\infty} x^*(\tau)h(t+\tau)d\tau = x(t) \star h(t), \tag{4.46}$$

where $x^*(t)$ indicates the complex conjugate of $x(t)$ and $\bigstar$ indicates the correlation operation. When the signals are represented by real functions, their complex conjugates become the same and complex conjugate operations disappear.

Note that the integral of Equation (4.46) approaches to ∞ for power signals. It is customary to normalize the cross-correlation functions for the power signals as

$$y(t) = \lim_{T\to\infty} \frac{1}{T} \int_{-T/2}^{T/2} x^*(\tau)h(t+\tau)d\tau = x(t) \bigstar h(t). \tag{4.47}$$

Similarly, cross-correlation or correlation operation between two discrete time signals $x[n]$ and $h[n]$ is defined as:

$$y[n] = \sum_{k=-\infty}^{\infty} x[k]^* h[n+k] = x[n] \bigstar h[n]. \tag{4.48}$$

For the discrete time power signals, the cross-correlation function is normalized as

$$y[n] = \lim_{N\to\infty} \frac{1}{2N+1} \sum_{k=-N}^{N} x[k]^* h[n+k] = x[n] \bigstar h[n]. \tag{4.49}$$

Cross-correlation operation measures the degree of containment of one signal in the other signal. Since it avoids the time reverse operation, it is preferred in many application domains. If one of the signals is even, convolution and correlation yields the same result.

Exercise 4.11 Find the cross-correlation between the following continuous time complex exponential functions with different harmonics:

$$x(t) = e^{jk\omega_0 t} \quad \text{and} \quad h(t) = e^{jl\omega_0 t} \quad \text{for } k \neq l. \tag{4.50}$$

Solution

Since this is a periodic signal, it is a power signal. Thus, cross-correlation function is defined as:

$$\begin{aligned} y(t) &= \lim_{T\to\infty} \frac{1}{T} \int_{-T/2}^{T/2} x^*(\tau)h(t+\tau)d\tau \\ &= \lim_{T\to\infty} \frac{1}{T} \int_{-T/2}^{T/2} e^{-jk\omega_0\tau} e^{jl\omega_0(t+\tau)} d\tau \\ &= \lim_{T\to\infty} \frac{1}{T} e^{jl\omega_0 t} \int_{-T/2}^{T/2} e^{-j\omega_0\tau(k-l)} d\tau. \end{aligned} \tag{4.51}$$

Using the Euler formula, we obtain

$$y(t) = \lim_{T\to\infty} \frac{1}{T} e^{jl\omega_0 t} \int_{-T/2}^{T/2} (\cos(\omega_0\tau(k-l)) - j\sin(\omega_0\tau(k-l)))d\tau = 0. \tag{4.52}$$

As long as $k \neq l$, sine and cosine functions remain periodic, hence, the integral of Equation (4.52) becomes 0. Therefore, for $k \neq l$, the complex exponentials with different harmonics are not contained in each other. These types of signals are called **orthogonal**.

Exercise 4.12 Find the cross-correlations between the following two toy digital signals.

$$x[n] = -3\delta[n] + 2\delta[n-1] - 1\delta[n-2] + \delta[n-3], \tag{4.53}$$

$$h[n] = -\delta[n] - 3\delta[n-2] + 2\delta[n-3]. \tag{4.54}$$

Solution

The cross-correlation of these to signals is obtained from the definition:

$$y[n] = \sum_{k=-\infty}^{\infty} x[k]h[n+k]. \tag{4.55}$$

For $n \leq -4$, $y[n]$ evaluates to 0, since the two signals, $x[k]$ and $h[k+n]$, do not overlap.

For $n = -3$, the last nonzero element of $x[k]$ overlaps with the first nonzero element of $h[k-3]$, which yields $y[-3] = -1$.

As we increase n, we obtain $y[-2] = 1$, $y[-1] = -5$, $y[0] = 8$, $y[1] = -8$, $y[2] = 13$ and $y[3] = -6$. For $n \geq 4$, the two signals do not overlap; therefore, $y[n] = 0$ for $n \geq 4$. Overall, $y[n]$ can be expressed as:

$$y[n] = -\delta[n+3] + \delta[n+2] - 5\delta[n+1] + 8\delta[n] - 8\delta[n-1] + 13\delta[n-2] - 6\delta[n-3].$$

Definition 4.3 Auto-correlation operation of a continuous time signal $x(t)$ is defined as:

$$y(t) = x(t) \star x(t) = \int_{-\infty}^{\infty} x^*(\tau)x(t+\tau)d\tau. \tag{4.56}$$

Similarly auto-correlation operation of a discrete time signal $x[n]$ is defined as:

$$y[n] = x[n] \star x[n] = \sum_{k=-\infty}^{\infty} x^*[k]x[n+k]. \tag{4.57}$$

Auto-correlation operation measures the correlation of a signal with a lagged copy of itself as a function of the lag. It measures the similarity between different time instances of a function.

Time series analysis of signals, such as speech, medical signals, seismographic signals, etc. intensively uses the auto-correlation functions. Following is a very simple example of auto-correlation of a finite-duration signal.

INTERACTIVE

Explore cross-correlation and auto-correlation @ `https://384book.net/i0403`

Exercise 4.13 Find the auto-correlation function of the signal given as follows:

$$x[n] = \begin{cases} -1 & \text{for} \quad t = 0, \\ 2 & \text{for} \quad t = 1, \\ 1 & \text{for} \quad t = 2. \end{cases} \tag{4.58}$$

Solution

We need to evaluate the discrete time auto-correlation given in Equation (4.57).

For $n < -2$ and $n > 2$, $x[k]$ and $x[k+n]$ do not overlap; hence, $y[n] = 0$.

For other values of n, we have $y[-2] = -1$, $y[-1] = 0$, $y[0] = 6$, $y[1] = 0$ and $y[2] = -1$. Overall, we can express the result as:

$$y[n] = -\delta[n+2] + 6\delta[n] - \delta[n-2].$$

4.2 Properties of Impulse Response for LTI Systems

LTI systems are of special importance in system theory due to many reasons. First of all, they can be represented by a unique function, called impulse response. Secondly, a relatively complicated system can be represented by parallel, series, hybrid, or loopy connections of many simple LTI subsystems. Thus, it is easy to develop mathematically tractable methods to design, analyze, and model natural and human-made objects using LTI systems. Even if the system is not inherently linear and/or time invariant, it is possible to define manifolds, where the system is piecewise linear and locally time invariant.

In Section 4.1, we showed that an LTI system can be uniquely represented by its impulse response. Let us investigate the properties and behavior of the impulse response for memory, causality, invertibility, and stability.

4.2.1 Impulse Response of Memoryless LTI Systems

Recall that a system is memoryless if the present value of the output depends **only** on the present value of the input. Recall also that an LTI system has a system equation, where the input is proportional to the output, as follows:

$$y(\cdot) = Kx(\cdot). \tag{4.59}$$

If we replace the input signal with the unit impulse function, we obtain the impulse response of a **memoryless LTI system** at the output:

$$\begin{aligned} h(t) &= K\delta(t), \forall K, \quad \text{for continuous time systems,} \\ h[n] &= K\delta[n], \forall K, \quad \text{for discrete time systems.} \end{aligned} \tag{4.60}$$

For all other cases, where $h(t) \neq K\delta(t)$, the **LTI system has memory**.

When an LTI system is memoryless, it simply multiplies an input by a constant factor K. Thus, a memoryless LTI system is a scalar multiplier.

4.2.2 Impulse Response of Causal LTI Systems

Recall that if a system has memory, it may be causal or noncausal. A system with memory is causal if the present value of the output depends on the **past and/or present** value of the input. Recall also that all the memoryless systems are causal.

When a system is LTI, the convolution sum and integral ranges between $(-\infty, \infty)$. The definition of causality is violated for negative values of the time variable. Thus, an LTI system is causal if the limits of the convolution summation and convolution integral ranges between $(0, \infty)$, as follows:

$$y[n] = \sum_{k=0}^{\infty} x[n-k]h[k], \qquad \text{for discrete time systems,} \tag{4.61}$$

$$y(t) = \int_{0}^{\infty} x(t-\tau)h(\tau)d\tau, \qquad \text{for continuous time systems.} \tag{4.62}$$

Furthermore, the impulse response of a causal LTI system satisfies the following condition:

$$\begin{aligned} h[n] &= 0 \quad \text{for} \quad n < 0, \\ h(t) &= 0 \quad \text{for} \quad t < 0. \end{aligned} \tag{4.63}$$

Thus, the impulse response of a causal system is all zero values for negative values of time instances.

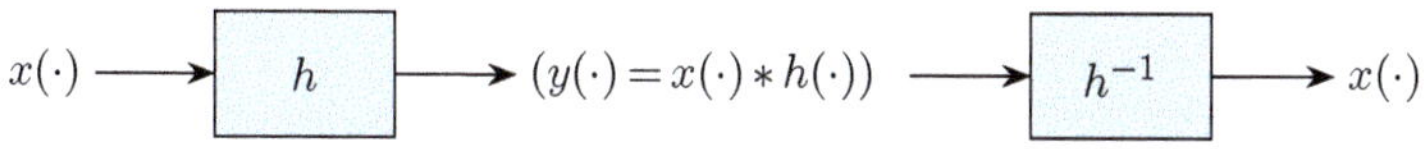

Figure 4.16 If the output of an LTI system represented by an impulse response $h(\cdot)$, for input signal $x(\cdot)$, is fed to the inverse of the system, represented by $h^{-1}(\cdot)$, we obtain the original signal $x(\cdot)$ back.

4.2.3 Inverse of Impulse Response for LTI Systems

In general, a system represented by a model $y(\cdot) = h(x(\cdot))$, is invertible if there exists a unique inverse model h^{-1}, such that

$$x(\cdot) = h^{-1}(y(\cdot)). \tag{4.64}$$

In particular, an LTI system, represented by $y(\cdot) = x(\cdot) * h(\cdot)$, is invertible if there exists a unique inverse impulse response, $h^{-1}(\cdot)$, such that

$$x(\cdot) = h^{-1}(\cdot) * y(\cdot), \tag{4.65}$$

which is illustrated in Figure 4.16.

Motivating Question: How can we find $h^{-1}(t)$ and $h^{-1}[n]$, for continuous time and discrete time LTI systems, respectively?

This is not an easy task for a general impulse response function. It requires an operation called deconvolution, which is beyond the scope of this book.

However, we can find a relationship between an impulse response $h(\cdot)$ and its inverse, $h^{-1}(\cdot)$ by convoluting both sides of system equation with $h^{-1}(\cdot)$, as follows:

$$y(\cdot) * h^{-1}(\cdot) = x(\cdot) * h(\cdot) * h^{-1}(\cdot). \tag{4.66}$$

Recall that the convolution of a function by a unit impulse function is the function itself, i.e.,

$$x(t) = x(t) * \delta(t). \tag{4.67}$$

Thus, for

$$x(\cdot) = y(\cdot) * h^{-1}(\cdot), \tag{4.68}$$

we need

$$h(\cdot) * h^{-1}(\cdot) = \delta(\cdot). \tag{4.69}$$

Thus, we seek an inverse impulse response such that the convolution of the impulse response and its inverse results in a unit impulse function.

In summary, the relationship between an impulse response and its inverse is

$$h(t) * h^{-1}(t) = \delta(t), \quad \text{for continuous time systems,} \tag{4.70}$$

$$h[n] * h^{-1}[n] = \delta[n], \quad \text{for discrete time systems,} \tag{4.71}$$

which satisfies

$$x(t) = h^{-1}(t) * y(t), \quad \text{for continuous time systems,} \tag{4.72}$$

$$x[n] = h^{-1}[n] * y[n], \quad \text{for discrete time systems.} \tag{4.73}$$

Exercise 4.14 Consider an LTI system represented by the following equation:

$$y(t) = x(t - t_0). \tag{4.74}$$

a) Find the impulse response, $h(t)$.
b) Find the inverse of impulse response, $h^{-1}(t)$.
c) Does this system have memory?
d) Is this system causal?

Solution

a) In order to find the impulse response, we replace the input with the unit impulse function. The corresponding output is the impulse response. Thus, the impulse response is

$$h(t) = \delta(t - t_0), \tag{4.75}$$

b) In order to find the inverse of the impulse response, we need to satisfy

$$h(t) * h^{-1}(t) = \delta(t). \tag{4.76}$$

Equation (4.74) is satisfied, when

$$\delta(t - t_0) * \delta(t + t_0) = \delta(t). \tag{4.77}$$

Thus,

$$h^{-1}(t) = \delta(t + t_0). \tag{4.78}$$

c) This system has memory for $t_0 \neq 0$.
d) This system is causal, for $t_0 \geq 0$ because the present value of the output depends only on the past values. Otherwise, it is noncausal.

Exercise 4.15 Consider an LTI system, called accumulator, represented by the following equation:

$$y[n] = \sum_{k=-\infty}^{n} x[k]. \tag{4.79}$$

a) Find the impulse response, $h(t)$.
b) Find a block diagram representation of this system, using an adder and a unit delay operator.
c) Find the inverse of the impulse response, $h^{-1}(t)$.
d) Does this system have memory?
e) Is this system causal?

Solution

a) We replace the input with the unit impulse function. The corresponding output is the impulse response. Thus the impulse response is

$$h[n] = \sum_{k=-\infty}^{n} \delta[k] = u[n], \tag{4.80}$$

where $u[n]$ is the unit step function.
b) We can use a simple mathematical trick to find a closed-form equation to represent this system by taking the difference between $y[n]$ and $y[n-1]$, to obtain a more compact form of the system

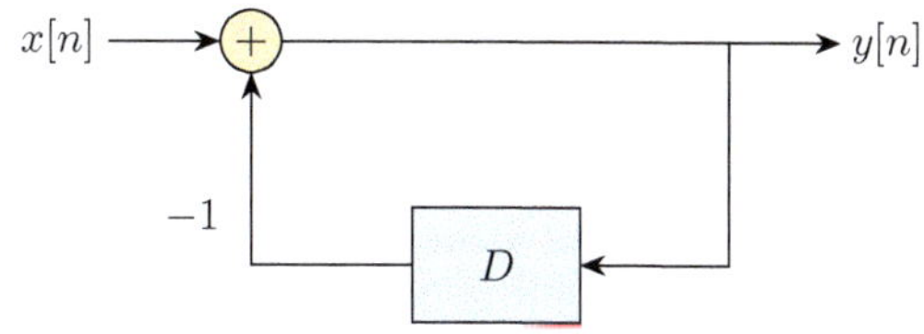

Figure 4.17 Block diagram for the LTI system represented by $y[n] - y[n-1] = x[n]$ in Exercise 4.15.

equation as follows:

$$y[n] - y[n-1] = x[n]. \tag{4.81}$$

This LTI system can be represented by the feedback control system illustrated in Figure 4.17.

c) The inverse of the impulse response of $h[n] = u[n]$ should satisfy

$$h[n] * h^{-1}[n] = \delta[n]. \tag{4.82}$$

We need to find a function $h^{-1}[n]$, such that when it is convoluted by $u[n]$, the output is the impulse function. We know that

$$u[n] - u[n-1] = \delta[n] \tag{4.83}$$

and we know that $\delta[n-1] * u[n] = u[n-1]$. Thus,

$$h^{-1}[n] = \delta[n] - \delta[n-1]. \tag{4.84}$$

Note that finding the inverse of an impulse response requires some heuristics. In this particular example, we found an inverse function $h^{-1}[n]$ to satisfy

$$\underbrace{u[n]}_{h[n]} * \underbrace{(\delta[n] - \delta[n-1])}_{h^{-1}[n]} = \delta[n], \tag{4.85}$$

and

$$y[n] * \underbrace{(\delta[n] - \delta[n-1])}_{h^{-1}[n]} = x[n]. \tag{4.86}$$

d) This is a causal system because $h[n] = 0$ for $n < 0$.

e) This system has memory because $h[n] \neq K\delta[n]$.

Exercise 4.16 Consider the LTI system represented by the following equation,

$$y(t) = x(t-2) + x(t+2). \tag{4.87}$$

a) Find the impulse response of this system.

b) Is this system memoryless?

c) Is this system causal?

Solution

a) Impulse response of this LTI system is

$$h(t) = \delta(t-2) + \delta(t+2). \tag{4.88}$$

b) This system has memory because $h(t) \neq K\delta(t)$.

c) This system is noncausal because $h(t) = \delta(t-2) + \delta(t+2) \neq 0$, for $t < 0$.

4.2.4 Impulse Response of Stable LTI Systems

Recall that a system is called stable, if a bounded input signal generates a bounded output signal. Let us apply this definition to the convolution sum and convolution integral.

For a discrete time system, when the input is bounded, in other words,

$$|x[n]| < B, \tag{4.89}$$

we need that the convolution summation results in a bounded output. Mathematically speaking,

$$|x[n]| < B \Rightarrow |y[n]| < B \sum_{k=-\infty}^{\infty} |h(k)|, \tag{4.90}$$

$$\Rightarrow \sum_{k=-\infty}^{\infty} |h(k)| < \infty. \tag{4.91}$$

When a discrete time function, $h[n]$, satisfies the inequality of Equation (4.91), it is called **absolutely summable function.**

For a continuous time system, when the input is bounded, $|x(t)| < B$, we need that the convolution integral results in a bounded output. Mathematically speaking,

$$|x(t)| < B \Rightarrow |y(t)| < B \int_{-\infty}^{\infty} |h(\tau)| d\tau, \tag{4.92}$$

$$\Rightarrow \int_{-\infty}^{\infty} |h(\tau)| d\tau < \infty. \tag{4.93}$$

When a continuous time function, $h(t)$, satisfies the inequality of Equation (4.93), it is called **absolutely integrable function.**

Exercise 4.17 Consider a discrete time LTI system represented by the impulse response $h[n] = u[n]$.

a) Find the system equation, which gives the relationship between the input and the output.
b) Is this system memoryless?
c) Is this system causal?
d) Is this system stable?

Solution

a) The system equation of this LTI system can be obtained from the convolution summation:

$$y[n] = x[n] * u[n] = \sum_{k=-\infty}^{n} x[k]u[n-k] = \sum_{k=-\infty}^{n} x[k]. \tag{4.94}$$

For $y[n-1]$, the aforementioned equation can be written as:

$$y[n-1] = \sum_{k=-\infty}^{n-1} x[k]u[n-k] = \sum_{k=-\infty}^{n-1} x[k]. \tag{4.95}$$

Subtracting Equation (4.95) from Equation (4.94) side by side yields

$$y[n] - y[n-1] = x[n], \tag{4.96}$$

b) This system has memory because $h[n] \neq K\delta[n]$.
c) This system is causal because $h[n] = 0$ for $n < 0$.

d) This system is unstable because for a bounded input $x[n] < B$, the output becomes unbounded due to the infinite summation.

Exercise 4.18 Consider a continuous time LTI system represented by the following equation:

$$y(t) = \int_0^3 x(t-\tau)d\tau, \tag{4.97}$$

a) Find the impulse response of this system.
b) Is this system memoryless?
c) Is this system causal?
d) Is this system stable?

Solution

a) We replace the input by the impulse function to obtain the impulse response as follows:

$$h(t) = \int_0^3 \delta(t-\tau)d\tau = u(t) - u(t-3). \tag{4.98}$$

b) This system has memory because $h(t) \neq K\delta(t)$.
c) This system is causal because $h(t) = 0$ for $t < 0$.
d) This system is stable because $h(t)$ is absolutely summable:

$$\int_{-\infty}^{\infty} |h(\tau)|d\tau = \int_0^3 u(\tau)d\tau < \infty. \tag{4.99}$$

WATCH

Learn more about the impulse response and the acoustics of instruments @ https://384book.net/v0401

4.2.5 Unit Step Response

The aforementioned analysis shows that at the heart of an LTI system lies the concept of impulse response, which is the response of an LTI system to a unit impulse function. In other words, given an LTI system, if we can measure its response to a unit impulse input, we can find its response to any input through the convolution operation. Similar to the impulse response, we can define a unit step response as defined in the following text.

Definition 4.4 **Unit Step Response** is the response of an LTI system to the unit step function.

For a discrete time system, replacing the input by the unit step function in convolution summation, we can obtain the discrete time **unit step response**, as follows:

$$s[n] = u[n] * h[n] = \sum_{k=-\infty}^{\infty} u[n-k]h(k) = \sum_{k=-\infty}^{n} h(k). \tag{4.100}$$

There is a one-to-one correspondence between the unit step response and impulse response. If we take the difference between the unit step responses of $s[n]$ and $s[n-1]$, we get

$$\boxed{h[n] = s[n] - s[n-1]}. \tag{4.101}$$

For a continuous time system, replacing the input by the unit step function in convolution integral, we can obtain the **unit step response** as:

$$s(t) = u(t) * h(t) = \int_{-\infty}^{\infty} u(t-\tau)h(\tau)d\tau = \int_{-\infty}^{t} h(\tau)d\tau. \tag{4.102}$$

There is a one-to-one correspondence between the continuous time unit step response and impulse response. If we take the derivative of the convolution integral of Equation (4.102), we get

$$\boxed{h(t) = \frac{ds(t)}{dt}}. \tag{4.103}$$

Exercise 4.19 Find the impulse response of a continuous LTI system if its unit step response is

$$s(t) = e^{-\alpha t}u(t). \tag{4.104}$$

Solution
Take the derivative of the unit step response with respect to t,

$$h(t) = \frac{ds(t)}{dt} = -\alpha e^{-\alpha t}u(t). \tag{4.105}$$

Look at the beautiful symmetry of the continuous time exponential function! When the unit step response is an exponential function, its impulse response is also an exponential function, scaled by the exponent.

Exercise 4.20 Find the impulse response of a discrete time LTI system if the unit step response is

$$s[n] = e^{-\alpha n}u[n]. \tag{4.106}$$

Solution
Take the first difference of the unit step response,

$$h[n] = s[n] - s[n-1] = e^{-\alpha n}u[n] - e^{-\alpha(n-1)}u[n-1] = \delta[n]. \tag{4.107}$$

Since, $u[n] - u[n-1] = \delta[n]$ and $e^0 = 1$, the result is simply $\delta[n]$.

4.3 An Application of Convolution in Machine Learning

Convolution and cross-correlation operations play a crucial role in computer vision and machine learning, particularly in tasks like visual recognition. Let us delve into a practical application of convolution in hand-written digit recognition.

Our objective is to develop a system utilizing a cross-correlation filter, with the filter weights learned through minimizing an error measure on the MNIST dataset.[1] This dataset comprises 60,000 28×28 grayscale images of hand-written digits from 0 to 9. Some examples from the dataset are illustrated in Figure 4.18.

Instead of tackling the multiclass problem, for simplicity, we will focus on a binary classification task, specifically recognizing the digit "3." Thus, we will treat the class "3" as the positive class and all other digits as the negative class. The goal of our system is to predict the class of its input image.

1 https://yann.lecun.com/exdb/mnist.

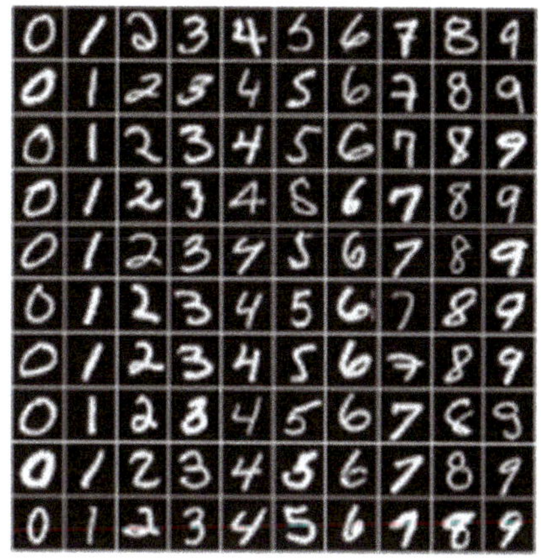

Figure 4.18 Sample images from the MNIST (Modified National Institute of Standards and Technology) dataset.

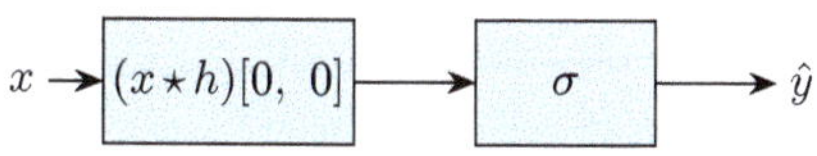

Figure 4.19 The system that takes an input image x (as a 28×28 matrix), processes it through a correlation filter h and applies the sigmoid function (σ) on the correlation output. Finally, the system produces $\hat{y}$, which should be close to 1 if the image belongs to the positive class, and to 0, otherwise.

Specifically, our system takes an image x, which is a 28×28 matrix, as input and processes it through a cross-correlation operation with a 28-by-28 filter denoted with h. Then, the system outputs a scalar number $\hat{y}_i$ which should be 1 if the input image belongs to the positive class, 0 otherwise. The system is illustrated in Figure 4.19.

In this chapter, we have defined the cross-correlation operations for one-dimensional signals. In two dimensions, it is defined as:

$$(x \star h)[i,j] = \sum_m \sum_n x[i+m, j+n]h[m,n]. \tag{4.108}$$

This operation slides the 28×28 filter h on the input signal x, and computes results for all possible locations, even if x and h overlap partially. However, in our system, we are only interested in $(x \star h)[0,0]$, which is the result of correlation when the filter and the input fully overlap. Subsequently, $(x \star h)[0,0]$ is passed through the sigmoid function defined as:

$$\sigma(x) = \frac{1}{1+e^{-x}}, \tag{4.109}$$

which squashes its input into the interval $[0,1]$.

"Training" this system means finding the filter h that solves this task in such a way that it makes a minimal amount of prediction errors on the examples of the training set. This error is measured by a "loss function." There are many different loss functions in machine learning, which are beyond the scope of this book. Here, we will use the "mean squared error (MSE)" loss function, which is easy to understand but not necessarily the best for the task. The MSE is defined as:

$$\text{MSE} = \frac{1}{N}\sum_{i=1}^{N}(\hat{y}_i - y_i)^2, \tag{4.110}$$

where $\hat{y}_i$ is the output of our system for input image x_i. This loss function achieves its minimal value, which is zero, when all the predictions ($\hat{y}_i$) are equal to their corresponding labels (y_i). To minimize MSE on the training set, we apply a method called the "gradient descent." In essence,

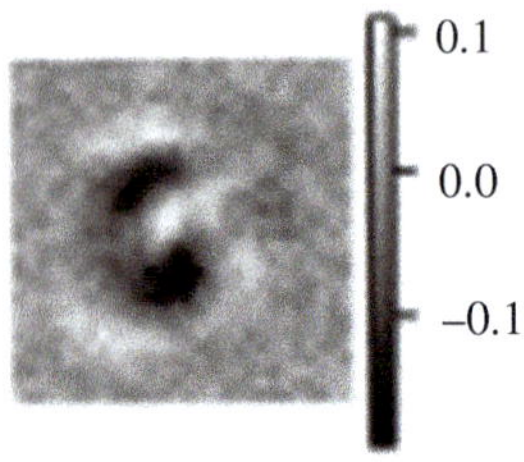

Figure 4.20 The optimal filter h learned on the MNIST training set by minimizing the MSE. The filter looks like the digit 3.

it involves iteratively adjusting the filter weights in the negative direction of the gradient of MSE with respect to the filter. Note that the gradient is a vector that points in the direction of the greatest rate of increase. In each iteration of training, the system learns from the discrepancies between its predictions ($\hat{y}_i$) and the actual labels (y_i). The cross-correlation filter gradually adapts to recognize distinctive features of the digit "3." After training, the optimal h looks like the digit "3" as shown in Figure 4.20.

It is important to note that this simplified explanation focuses on a binary classification task. Extending this approach to multiclass classification involves modifying the output layer, utilizing techniques like softmax activation and using different loss functions like cross-entropy. We provide the code for this example and its multiclass extension in the companion website of the book, accessible through the following link.

INTERACTIVE

A convolution (cross-correlation) example from machine learning @ `https://384book.net/i0404`

4.4 Chapter Summary

Can we find a unique equation, which relates the input–output pair of a continuous time and discrete time system? Can we define a function, which uniquely represents a continuous time and discrete time system? If so, can we study the basic properties of the systems using these representations?

The answers to the aforementioned questions are all "yes," provided that the system is LTI. Thus, linearity and time invariance are crucial properties to model and analyze systems. In this chapter, we introduced an interesting function, called impulse response, which uniquely represents an LTI system. Impulse response is defined as the output of an LTI system when the input is a unit impulse function. We define a new mathematical operation, called convolution, which relates any input–output pair of an LTI system through the impulse response. As an alternative to convolution, we also define correlation and auto-correlation operations, which are widely used in machine learning applications.

Finally, we studied the stability, memory, and invertibility properties of LTI systems using their mathematical representation with the impulse response.

Problems

4.1 Consider a discrete time LTI system, represented by the following impulse response:

$$h[n] = u[n + 2].$$

a) Find the output of the system for the following input:

$$x[n] = (0.5)^{(n-0.5)}u[n-2].$$

b) Is this system invertible? If yes, find its inverse.

c) Is this system Bounded Input, Bounded Output (BIBO) stable? Verify your answer.

4.2 Find the output, $y[n]$, of the system for the following input and impulse response plots:

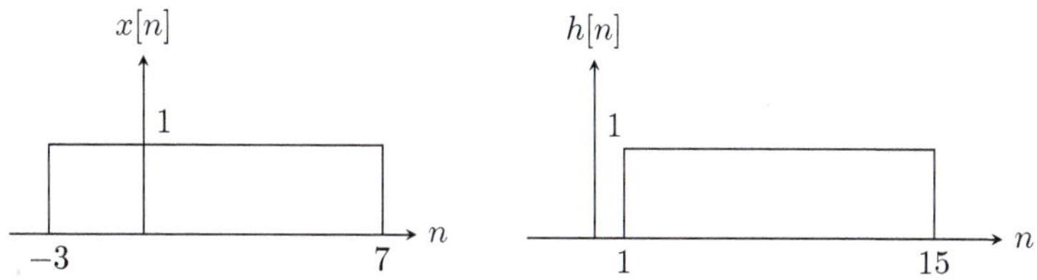

4.3 Consider the discrete time system, represented by the following system equation:

$$y[n] = \sum_{k=-\infty}^{\infty} x[k]g[n-2k],$$

where $g[n] = u[n+2] - u[n-2]$

a) Find and plot $y[n]$ for $x[n] = \delta[n]$.

b) Find and plot $y[n]$ for $x[n] = \delta[n-2]$.

c) Find and plot $y[n]$ for $x[n] = u[n]$.

4.4 Find and plot the convolution of the following continuous time signals:

$$x(t) = \delta(t+2) + \delta(t+1),$$

$$h(x) = \begin{cases} t+1 & \text{for } -1 \le t \le 1 \\ 2-t & \text{for } \quad 1 < t \le 3, \\ 0 & \text{otherwise.} \end{cases}$$

4.5 Given the input $x(t) = 2e^{-\alpha t}(u(t) - u(t-1))$ and the impulse response $h(t) = u(t/\alpha)$, where $0 < \alpha \le 1$

a) Find and plot $y(t) = x(t) * h(t)$.

b) Find $\dfrac{dx(t)}{dt} * h(t)$.

4.6 Given the input and impulse response of an LTI system,

$$x(t) = u(4-t) - u(8-t),$$

$$h(t) = e^{-4t}u(-t).$$

a) Find the output $y(t) = x(t) * h(t)$.

b) Find the function $g(t) = \frac{dx(t)}{dt} * h(t)$.

c) Find $g(t)$ in terms of $y(t)$.

4.7 Consider the output of a continuous time LTI system represented by the following convolution equation:

$$y(t) = e^{-t}u(t) * \sum_{k=-\infty}^{\infty} \delta(t-3k).$$

a) Find the output function $y(t)$ for $0 \le t < 3$, in terms of exponential function by taking the convolution operation.
b) What is a possible input and impulse response of this system?
c) When you switch the input with the impulse response, does the output function of this system change? Explain why?

4.8 Find and plot the following outputs of the discrete time LTI systems for the input $x[n] = \delta[n] + 2\delta[n-2] - 3\delta[n-4]$ and the impulse response $h[n] = 2\delta[n+2] + \delta[n-2]$:
a) $y_1[n] = x[n] * h[n]$
b) $y_2[n] = x[n+2] * h[n]$
c) $y_3[n] = x[n+2] * h[n-2]$.

4.9 Consider the two signals given as follows:

$$x[n] = 3^n u[-2n-1],$$
$$h[n] = u[1-n].$$

a) Find and plot the output of the cross-correlation operation between these functions.
b) Find and plot the output of the convolution operation between these functions.
c) Compare the results of parts (a) and (b).

4.10 Consider a discrete time causal LTI system, represented by the following difference equation:

$$y[n] = \frac{1}{5}x[n-1] + x[n].$$

a) Find the impulse response, $h[n]$, of the system.
b) Find the output, $y[n]$, for the input $x[n] = \delta[n-2]$.
c) Is this system BIBO stable?
d) Does this system have memory?
e) Is this system invertible? If yes, find its inverse.

4.11 Consider the following impulse responses, each of which represents a continuous time LTI system. Determine if these systems are BIBO stable. Verify your answers.
a) $h_1(t) = e^{-(1-2j)t}u(t-1)$
b) $h_2(t) = e^{-t}\cos(t)u(-t)$
c) $h_3(t) = e^{-t}\sin(t)u(-t)$

4.12 Consider the following impulse responses, each of which represents a discrete time LTI system. Determine if these systems are BIBO stable. Verify your answers.
a) $h_1[n] = \frac{4n}{\pi}\cos\left(\frac{\pi}{4}n\right)u[n]$,
b) $h_2[n] = 7^n u[7-n]$.

4.13 Consider a discrete time LTI system, represented by the following impulse response:

$$h[n] = (2n)\alpha^n u[n],$$

where $|\alpha| < 1$. Show that the step response of this system is

$$s[n] = 2\left(\frac{1-\alpha^n}{(1-\alpha)^2} - \frac{n\alpha^n}{1-\alpha}\right)u[n].$$

(*Hint*: Note that

$$\sum_{k=0}^{N}(k)\alpha^k = \frac{d}{d\alpha}\sum_{k=0}^{N}\alpha^k,$$

and

$$\sum_{k=0}^{N}\alpha^k = \frac{1-\alpha^{n+1}}{1-\alpha}.\Bigg)$$

4.14 Consider the discrete time signal

$$x[n] = \left(\frac{1}{3}\right)^{n-1}\{u[n+3] - u[n-3]\}.$$

Find an analytical expression for $x[n-k]$ as a piece-wise linear function.

4.15 Prove that convolution operation is commutative, associative and distributive:

a) $x(t) * h(t) = h(t) * x(t)$,
b) $x(t) * (h_1(t) * h_2(t)) = x(t) * (h_1(t)) * (h_2(t))$,
c) $x(t) * (h_1(t) + h_2(t)) = x(t) * h_1(t) + x(t) * h_2(t)$.

4.16 Does cross-correlation have the commutativity, associativity, and distributivity properties? Prove your answer.

4.17 Write a computer program to take discrete convolution of two signals. (You are not allowed to use any xx.convolve() function from any library.) Your function takes four inputs: the first signal $x[n]$, the starting index of the first signal s_i^x, the second signal $h[n]$ and the starting index of the second signal s_i^h (starting indexes and signals are in the same format as the ones in HW1) and returns the output signal $y[n]$ and the starting index of the output signal s_i^y.

a) (5 pts) Generate a shifted discrete impulse function $\delta[n-5]$ in the given signal form and plot the output function that is the result of your discrete convolution function when $x[n]$ = "the signal in signal.csv" and $h[n] = \delta[n-5]$. What is the effect of convolution with $\delta[n-5]$? Comment on that.
b) (10 pts) The N-point moving average filter is defined as follows:

$$h[n] = \begin{cases} \frac{1}{N} & \text{if } 0 \le n \le N-1 \\ 0 & \text{otherwise.} \end{cases}$$

Generate an N-point moving average filter $m[n]$ in the given signal form and plot four output functions that are the result of your discrete convolution function when $x[n]$ = "the signal in signal.csv" and $h[n] = m[n]$ by setting $N = 3, 5, 10, 20$. What is the effect of convolution with $m[n]$? What are the differences between different N values?

You should write your code in `Python 3`. You are not allowed to use any library other than **matplotlib.pyplot** and **numpy**.

5

Representation of LTI Systems by Differential and Difference Equations

"...Since Newton, mankind has come to realize that the laws of physics are always expressed in the language of differential equations."

Steven H. Strogatz

In Chapter 4, we studied a representation of linear time-invariant (LTI) systems by **impulse response**, which relates the input and output signals by convolution summation and convolution integral. We showed that the response of an LTI system to any arbitrary input, $x(\cdot)$, is completely characterized by the impulse response, $h(\cdot)$.

There are other mathematical tools to represent an LTI system. An important group of models involves **differential equations** for the continuous time systems and **difference equations** for discrete time systems. Differential and difference equations are one of the most fundamental tools for modeling systems. A differential equation links the rate of change of an input to that of the output in continuous time systems. Similarly, the difference equation links the present, future, and past values of input signal to that of the output signals, in discrete time systems.

WATCH Learn more about differential equations @ https://384book.net/v0501

There is a wide range of types of differential and difference equations. In this book, we shall only focus on *linear, constant-coefficient differential*, and *difference equations*. We shall study the differential and difference equation to relate an input signal $x(\cdot)$ to an output signal $y(\cdot)$ to represent an LTI system.

5.1 Linear Constant-Coefficient Differential Equations

A linear, constant-coefficient differential equation is given in the following general form:

$$\sum_{k=0}^{N} a_k \frac{d^k y(t)}{dt^k} = \sum_{k=0}^{M} b_k \frac{d^k x(t)}{dt^k}, \tag{5.1}$$

Signals and Systems: Theory and Practical Explorations with Python,
First Edition. Fatoş Tunay Yarman Vural and Emre Akbaş.

Companion website: www.wiley.com/go/signalsandsystems

where $\frac{d^k}{dt^k}$ shows the kth derivative of the input $x(t)$ and the output $y(t)$. The constant parameters $\{a_k, b_k\}$ show the degree of the contribution of each derivative to the behavior of the system. In most of the practical systems, $N \geq M$, where N is the order of the differential equation and M is the order of the derivative of $x(t)$.

In order to find an explicit relationship between the input and output pairs, we need to solve the differential equation. This is only possible if we can obtain N auxiliary (initial) conditions about the system,

$$y(t_0),\, \dot{y}(t_0) = \frac{dy(t_0)}{dt},\, \dots,\, y^{(N-1)}(t_0) = \frac{d^{(N-1)}y(t_0)}{dt}, \tag{5.2}$$

at a specific time instance t_0.

An important set of initial conditions can be obtained, when an LTI system is **initially at rest.**

Definition 5.1 A continuous time LTI system is **initially at rest** at time t_0, if the input and output pairs, $x(t), y(t)$, are zero for all $t < t_0$.

Formally speaking,

$$x(t) = 0 \ \text{ for } t < t_0 \Rightarrow y(t_0) = 0 \text{ for } t < t_0. \tag{5.3}$$

The condition of Equation (5.3) implies that a continuous time LTI system is initially at rest if,

$$x(t) = 0 \text{ for } t < t_0 \Rightarrow y(t_0) = \dot{y}(t_0) = \cdots = y^{N-1}(t_0) = 0 \text{ for } t < t_0. \tag{5.4}$$

In most practical applications, the initial time starts from $t_0 = 0$. For this reason, we assume that the system is initially at rest, when there is no input and output for $t < 0$. Thus, initial rest condition gives us the following initial conditions,

$$y(0) = \dot{y}(0) = \cdots = y^{N-1}(0) = 0. \tag{5.5}$$

Motivating Question: Why do we need initial conditions?

A differential equation does not give an explicit equation, which relates the input to the output, but, instead, it gives a relationship among the rates of changes of the input and output. There are infinitely many solutions to a differential equation, depending on the starting points of the system. These initial values shape up the explicit equation between the input and output, which is obtained by solving the differential equation. In order to obtain a unique solution to a differential equation, we need a set of initial conditions.

5.2 Representation of a Continuous Time LTI System by Differential Equations

Recall that the derivative of a function measures the rate of change of that function with respect to the variable of differentiation, which is typically the time, t in the context of this book. Similarly, the second derivative measures the rate of the change of change. As we increase the degree of the derivatives, we measure the change of the change of the change, etc. When it is possible to relate the rate of the changes of the input and output signals, we can represent this system by a differential equation.

Proposition 5.1 A linear constant coefficient differential equation, which is **initially at rest**

$$\sum_{k=0}^{N} a_k \frac{d^k y(t)}{dt^k} = \sum_{k=0}^{M} b_k \frac{d^k x(t)}{dt^k} \tag{5.6}$$

represents a continuous time, **causal LTI system**, when the input–output pair of this system is $x(t)$ and $y(t)$.

Verification of the proposition: In order to show that the system represented by a differential equation is linear, we need to check the superposition property. Mathematically speaking, given two input–output pairs, $x_1(t) \to y_1(t)$ and $x_2(t) \to y_2(t)$, superposition of the inputs must yield the same superposition at the output, as follows:

$$x_S(t) = A_1 x_1(t) + A_2 x_2(t) \to y_S(t) = A_1 y_1(t) + A_2 y_2(t), \quad \text{for} \quad t \geq 0, \tag{5.7}$$

where A_1 and A_2 are arbitrary scalar numbers.

When the system is initially at rest, the output and all of its derivatives are zero, until we feed a nonzero input to the system. Thus, the superposition property can be shown by simply replacing the input and output pairs, $x_1(t) \to y_1(t)$ and $x_2(t) \to y_2(t)$, in the equation, for $t \geq 0$, as follows:

$$\sum_{k=0}^{N} a_k \frac{d^k y_1(t)}{dt^k} = \sum_{k=0}^{M} b_k \frac{d^k x_1(t)}{dt^k}, \tag{5.8}$$

$$\sum_{k=0}^{N} a_k \frac{d^k y_2(t)}{dt^k} = \sum_{k=0}^{M} b_k \frac{d^k x_2(t)}{dt^k}. \tag{5.9}$$

When we multiply the first equation by A_1 and the second equation by A_2, we obtain

$$A_1 \sum_{k=0}^{N} a_k \frac{d^k y_1(t)}{dt^k} = A_1 \sum_{k=0}^{M} b_k \frac{d^k x_1(t)}{dt^k}, \tag{5.10}$$

$$A_2 \sum_{k=0}^{N} a_k \frac{d^k y_2(t)}{dt^k} = A_2 \sum_{k=0}^{M} b_k \frac{d^k x_2(t)}{dt^k}, \tag{5.11}$$

and we add them side by side to obtain,

$$\sum_{k=0}^{N} a_k \frac{d^k [A_1 y_1(t) + A_2 y_2(t)]}{dt^k} = \sum_{k=0}^{M} b_k \frac{d^k [A_1 x_1(t) + A_2 x_2(t)]}{dt^k}. \tag{5.12}$$

Thus, for any superposed input output pairs, $x_S(t) \to y_S(t)$, the equation,

$$\sum_{k=0}^{N} a_k \frac{d^k y_S(t)}{dt^k} = \sum_{k=0}^{M} b_k \frac{d^k x_S(t)}{dt^k} \tag{5.13}$$

is satisfied.

In the aforementioned derivations, we could freely move the constant parameters, A_1 and A_2, inside of the sum and derivation operators because both **summation and derivation are linear operators.** However, if the system is **not** initially at rest, there are some nonzero outputs even if the input values are zero for some T, $t < T$. In this case, the superposition property of the differential equation is not satisfied for the initial conditions.

Since the superposition property holds, an "initial-at-rest" system represented by a constant-coefficient linear differential equation is linear with respect to the input–output pairs, $x(t)$ and $y(t)$.

Linear constant coefficient differential equations are, also, time invariant, when the system is initially at rest, because, for an arbitrary input output pair, $x(t) \rightarrow y(t)$, a time shift by t_0 of the input generates the same shift at the output, for $t \geq 0$, i.e.,

$$\sum_{k=0}^{N} a_k \frac{d^k y(t - t_0)}{dt^k} = \sum_{k=0}^{M} b_k \frac{d^k x(t - t_0)}{dt^k}. \tag{5.14}$$

Defining a shifted time variable, $\tau = t - t_0$ and noting that $d\tau = dt$, we can show time invariance. Thus, the system represented by a linear constant coefficient differential equation, which is initially at rest, is time invariant, $x(t - t_0) \rightarrow y(t - t_0)$.

Recall that causality simply means that the response of a system at a given time, t, does not depend on the future of the input signal. Since the system is initially at rest, the output takes nonzero values starting from $t > t_0$. Considering the fact that the derivation operator is causal, the initial rest condition assures the causality condition of a linear constant coefficient differential equation.

Motivating Question: How do we estimate the parameters a_k and b_k, and the orders N and M to model a continuous time system by a differential equation or to model a discrete time system by a difference equation?

This requires a great amount of domain knowledge and experimentation. Sometimes it takes the effort of hundreds of scientists over the centuries, like quantum mechanical systems, represented by Schrodinger's equation. Sometimes it is not possible at all. Recently, there is a trend to estimate the parameters of the differential equations by using artificial neural networks. These machine learning tools are trained by the experimental data of the underlying physical phenomenon, obtained in the laboratory environments. It is possible to model and analyze biological, chemical, or physical systems by finding the representative differential or difference equation based on LTI differential or difference equations. However, these techniques are beyond the scope of this book.

5.3 Solving the Linear Constant Coefficient Differential Equations That Represent LTI Systems

When we model a continuous time LTI system by a differential equation, instead of the impulse response, we put the differential equation in the black box (Figure 5.1). This representation formalizes the dynamic interactions between the input and output pairs, with respect to the time variable.

In order to study the behavior of the system we need to find an explicit relationship between the input and output pairs, so that we can compute the output for a given input signal. When the system is nonlinear, finding the solution of the differential equation may not always be easy, or may not be possible at all. However, when the system is linear, time invariant, and causal, there are systematic

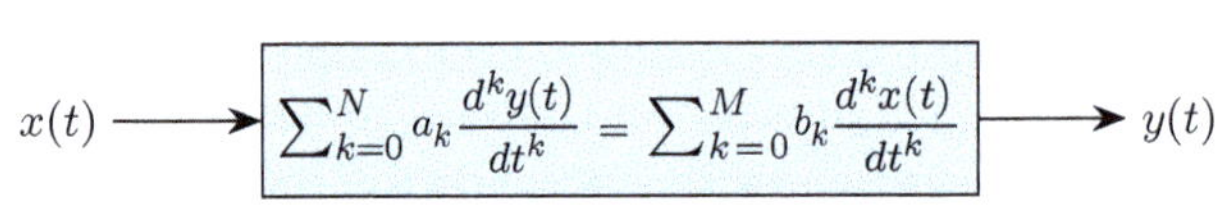

Figure 5.1 Representation of a continuous time LTI systems by differential equation.

methods for solving the corresponding differential equation. A detailed study of solutions, which covers all forms of input–output pairs is beyond the scope of this book. However, we shall focus on an important class of differential equations, which represent the LTI systems.

There are two solutions, which satisfy a linear constant coefficient differential equation:

1) **Particular solution**, which is the output, $y_p(t)$ of the system for a given input, $x(t)$.
2) **Homogeneous solution**, which is the solution of the differential equation of a system, when the input is $x(t) = 0$.

Recall that for an LTI system, if a set of input–output pairs satisfy the representative differential equation, then the superposition of all input–output pairs also satisfies this equation. Therefore, the general solution of a differential equation, which represents an LTI system, is obtained by superposing all the solutions. Mathematically,

$$y(t) = y_h(t) + y_p(t). \tag{5.15}$$

Let us study the methods for finding the solutions mentioned earlier.

5.3.1 Finding the Particular Solution

In System Theory, a particular solution of a linear constant coefficient differential equation is a unique response of the underlying system to a particular input, which satisfies the differential equation. Since the system is LTI, the analytical form of the particular solution, $y_p(t)$ must be similar to that of the input signal $x(t)$. Thus, a practical method for finding the particular solution is to assume that the particular solution has the general analytical form of the input. Then, we find the parameters of the particular solution, which satisfies the differential equation.

Motivating Question: How do we find the unique parameters of the particular solution, given its analytical form?

All we have to do is to take the derivatives of the assumed particular solution $y_p(t)$ for a given $x(t)$ and insert it into the differential equation. Then, solve this equation for the parameters of the particular solution to obtain a unique set of parameters.

Exercise 5.1 Find the particular solution of the following differential equation, when the input is $x(t) = t + 1$.

$$\frac{dy(t)}{dt} + 3y(t) = x(t). \tag{5.16}$$

Solution

The input is a line equation with slope and intercept equal to 1. Since the differential equation is linear with a constant coefficient, the particular solution should be another line equation with a different slope and intercept. Thus, it should be in the following analytical form:

$$y_p(t) = At + B.$$

The derivative of the particular solution is $\dot{y}_p(t) = A$. Let us insert the right-hand side of the particular solution and its derivative into the differential equation:

$$3At + (A + 3B) = t + 1.$$

Equating the coefficient of t and the constant term on both sides of the equation, we obtain

$$3A = 1 \quad \text{and} \quad A + 3B = 1.$$

Thus, $A = 1/3$ and $B = 2/9$, yielding

$$y_p(t) = \frac{1}{3}t + \frac{2}{9}.$$

In general, if the input is an nth-order polynomial, the particular solution is another nth-order polynomial with $n + 1$ arbitrary parameters,

$$y_p(t) = \sum_{k=0}^{n} A_k t^k. \tag{5.17}$$

The parameters are computed by inserting all the derivatives of $y_p(t)$ into the differential equation and finding the unique set of $\{A_k\}_{k=0}^{n}$.

Similarly, if the input is an exponential function, the particular solution is another exponential function with a different magnitude and parameter of the exponent. If the input consists of trigonometric functions, the particular solution consists of similar trigonometric functions with different amplitudes and frequencies, etc.

Note that the parameters of the particular solution depend on the parameters of the differential equation, a_k and b_k and the analytical form of the input signal, $x(t)$.

5.3.2 Finding the Homogeneous Solution

When the right-hand side of the differential equation is zero, the corresponding output $y_h(t)$ is represented by the following differential equation:

$$\sum_{k=0}^{N} a_k \frac{d^k y_h(t)}{dt^k} = 0. \tag{5.18}$$

This equation is called the **homogeneous equation.**

The solution to the homogeneous equation is called **homogeneous solution**. The homogeneous solution satisfies the differential equation for $x(t) = 0$. Due to the linearity property of the systems described by linear differential equations, it is trivial to show that

$$y(t) = y_h(t) + y_p(t) \tag{5.19}$$

also satisfies the differential equation.

Motivating Question: How do we find the solution of the homogeneous differential equation?

At this point, we shall use a crucial property of the exponential function. Recall that the kth derivative of an exponential function is the function itself, scaled by the kth power of the exponent. Formally,

$$\frac{d^k(e^{\beta t})}{dt^k} = \beta^k e^{\beta t}. \tag{5.20}$$

Suppose that a solution to the homogeneous differential equation is in the following form:

$$y_h(t) = Ce^{\beta t}, \tag{5.21}$$

where C is a nonzero constant. Then, using this form in the homogeneous Equation (5.18), we obtain

$$\sum_{k=0}^{N} a_k \beta^k C e^{\beta t} = C e^{\beta t} \sum_{k=0}^{N} a_k \beta^k = 0. \tag{5.22}$$

Since $Ce^{\beta t}$ cannot be zero, we get

$$\sum_{k=0}^{N} a_k \beta^k = 0, \tag{5.23}$$

which is an algebraic equation with N roots, i.e., $\beta_1, \beta_2, \ldots, \beta_N$. This fact reveals that the homogeneous solution is actually in the following form:

$$y_h(t) = \sum_{k=1}^{N} C_k e^{\beta_k t}, \tag{5.24}$$

with separate constants C_k for each root β_k. Since the system is linear, the superposition of all of the valid solutions for Equation (5.24) satisfies the homogeneous differential equation. Note that the exponential form of the homogeneous solution to Equation (5.21) converts the Nth-order homogeneous differential given by Equation (5.18) into an Nth-order algebraic equation given by Equation (5.23).

Finally, the linearity property enables us to add the particular and homogeneous solutions to obtain a **general solution** to the system:

$$y(t) = y_h(t) + y_p(t). \tag{5.25}$$

The method for solving a linear constant coefficient differential equation shows the beauty of the linearity property of the differential equation and the derivative property of the exponential function. Combining these two properties generates a very efficient and powerful method to solve linear constant coefficient differential equations:

i) Derivative property of the exponential function converts the problem of solving a homogeneous differential equation (when there is no input) into an algebraic equation.
ii) Linearity property enables us to superpose all of the valid solutions. In order to find the overall homogeneous solution, we superpose all of the valid homogeneous solutions. Then, we superpose the homogeneous and particular solutions to obtain the general solution.

5.3.3 Finding the General Solution

Equation (5.24) reveals that we can obtain infinitely many general solutions to a differential equation by changing the constant superposition weights, C_k. However, a model for an LTI system requires a unique explicit relationship between the output and input. Thus, we need additional information about the system. This additional information comes from some auxiliary conditions about the system, called initial conditions, at a specific time instance t_0, as follows:

$$y(t_0), \dot{y}(t_0) = \frac{dy(t_0)}{dt}, \ldots, y^{(N-1)}(t_0) = \frac{d^{(N-1)}y(t_0)}{dt}. \tag{5.26}$$

Recall that for this system to be a causal LTI system, it is to be initially at rest, i.e., for $t = 0$ the initial conditions are to be zero:

$$y(0) = \dot{y}(0) =, \ldots, y^{(N-1)}(0) = 0. \tag{5.27}$$

These N initial conditions are used to find a unique set of constant coefficients of C_k for $k = 1, \ldots, N$ in the general solution,

$$y(t) = \sum_{k=1}^{N} C_k e^{\beta^k t} + y_p(t). \tag{5.28}$$

Exercise 5.2 Given the following first-order differential equation,

$$\frac{dy(t)}{dt} + 2y(t) = x(t), \tag{5.29}$$

a) Find the particular solution, when the input is $x(t) = e^{-t}u(t)$.
b) Find the homogeneous solution.
c) Find the general solution for $y(t) = 0$, for $t \leq 0$.
d) Is this system initially at rest?
e) Is this system causal?

Solution

a) Since this equation is linear, the particular solution would be another exponential function in the following form:

$$y_p(t) = Ke^{-t}u(t). \tag{5.30}$$

Its derivative is $\dot{y}_p(t) = -Ke^{-t}u(t) + Ke^{-t}\delta(t)$. However, for practical purposes, we will use $\dot{y}_p(t) = -Ke^{-t}u(t)$ as explained in the following remark.

Remark 5.1 As we discussed in Chapter 2, the unit impulse $\delta(t)$ is operationally defined as the derivative of unit step function $u(t)$. The derivative of the discontinuity of the unit step function at $t = 0$ is not realizable as $\delta(t)$, with infinite magnitude and infinitesimally narrow interval. As such, impulse function is not actually a function, it is only tractable inside of an integral.

For practical purposes, we ignore the $Ke^{-t}\delta(t)$ term in $\dot{y}_p(t)$. This corresponds to considering the behavior of $\dot{y}_p(t)$ starting from $t = 0^+$ and ignoring its behavior right at $t = 0$.

Replacing $y_p(t)$ and $\dot{y}_p(t)$ in the differential equation and solving for K, we obtain $-K + 2K = 1$. Thus, $K = 1$ and $y_p(t) = x(t) = e^{-t}u(t)$.

b) A homogeneous solution for $x(t) = 0$ has the form: $y_h(t) = Ce^{\beta t}$. Its derivative is $\dot{y}_h(t) = C\beta e^{\beta t}$. Inserting $y_h(t)$ and $\dot{y}_h(t)$ into the homogeneous equation, we find the characteristic equation: $\beta + 2 = 0$. Thus, $\beta = -2$, which gives us the overall homogeneous equation as:

$$y_h(t) = Ce^{-2t}. \tag{5.31}$$

c) General solution is the superposition of the particular and homogeneous solutions:

$$y(t) = y_h(t) + y_p(t) = Ce^{-2t} + e^{-t}u(t). \tag{5.32}$$

The constant coefficient, C can be obtained from the initial condition, $y(0) = 0$, as $y(0) = C + 1 = 0$. Thus, $C = -1$. Since $y(t) = 0$ for $t \leq 0$, the general solution is

$$y(t) = \left(-e^{-2t} + e^{-t}\right) u(t). \tag{5.33}$$

d) This system is initially at rest because

$$x(t) = 0 \text{ for } t < 0 \Rightarrow y(t) = 0 \text{ for } t < 0. \tag{5.34}$$

e) Recall that for an LTI system to be causal, we need $h(t) = 0$ for $t < 0$. This condition is implied by the initial rest conditions given in (c) earlier. Thus, the system is causal.
The system is also memoryless because the present value of the output depends only on the present value of the input for all t.

Exercise 5.3 Consider the following first-order differential equation:

$$\frac{dy(t)}{dt} + 2y(t) = x(t), \tag{5.35}$$

with the initial conditions $y(-1) = 1$ for a particular input, $x(t) = e^{-t}u(t)$.

a) Find the general solution for this differential equation.
b) Is this system initially at rest?
c) Is this system causal?

Solution

a) The general solution is found from the previous example, as:

$$y(t) = y_h(t) + y_p(t) = Ce^{-2t} + e^{-t}u(t), \tag{5.36}$$

where the constant term C is to be obtained from a different initial condition, which is $y(-1) = 1$. Replacing this initial condition in Equation (5.36), we get,

$$y(-1) = Ce^{2} + eu(-1) = 1. \tag{5.37}$$

The second term on the right-hand side of this equation is zero. Then, the constant term becomes

$$C = e^{-2}. \tag{5.38}$$

Replacing the value of C in the general solution of Equation (5.36), we obtain

$$y(t) = e^{-2(t+1)} + e^{-t}u(t), \tag{5.39}$$

Comparison of Equations (5.33) and (5.39) shows that the initial conditions change the analytic form of the solution of the differential equations.

b) The input to the system, $x(t) = e^{-t}u(t)$, is zero for $t \leq 0$. The initial rest condition implies $y(t) = 0$ for $t \leq 0$. However, this is not true since $y(-1) = 1$. Therefore, the system is not initially at rest.

c) Since this is an LTI system, for $y(-1) = 1$, we need the convolution integral to evaluate to 1:

$$y(-1) = \int_{-\infty}^{\infty} x(\tau)h(-1-\tau) = 1. \tag{5.40}$$

We know that $x(\tau) = 0$ for $\tau \leq 0$. So, we must have $h(-1-\tau) \neq 0$ for $\tau > 0$, which violates the causality property for LTI systems. Recall that an LTI system is causal if its impulse response is zero for negative time values, that is, $h(t) = 0$ for $t < 0$.

Exercise 5.4 Given the following second-order differential equation:

$$\ddot{y}(t) + 3\dot{y}(t) + 2y(t) = \dot{x}(t). \tag{5.41}$$

a) Find the homogeneous solution for zero input, $x(t) = 0$.
b) Find the particular solution for the input,

$$x(t) = e^{\lambda t}u(t), \tag{5.42}$$

where λ is a parameter to determine the decay rate or the growth rate of the exponential.

c) Find the general solution for $\lambda = 1$ and assuming that the system is initially at rest with the initial conditions given here:

$$y(0) = \dot{y}(0) = 0. \tag{5.43}$$

Solution

a) The homogeneous solution can be obtained by assuming the following analytical form:

$$y_h(t) = e^{\beta t}, \tag{5.44}$$

where β is the parameter to be computed by taking the derivatives of $y_h(t)$,

$$\begin{aligned} y_h(t) &= e^{\beta t}, \\ \dot{y}_h(t) &= \beta e^{\beta t}, \\ \ddot{y}_h(t) &= \beta^2 e^{\beta t}. \end{aligned} \tag{5.45}$$

and inserting into the differential equations as follows:

$$(\beta^2 + 3\beta + 2)e^{\beta t} = 0. \tag{5.46}$$

Equation (5.46) gives a second-order characteristic equation in terms of β, with two roots:

$$\beta^2 + 3\beta + 2 = 0 \Rightarrow \beta_1 = -1, \beta_2 = -2. \tag{5.47}$$

Notice that for a second-order differential equation, we obtain two values of β. Thus, the homogeneous solution, for zero input response is the superposition of two exponential functions, e^{-t} and e^{-2t}, in the following form:

$$y_h(t) = C_1 e^{-t} + C_2 e^{-2t}. \tag{5.48}$$

b) Since the system is linear, we can assume that the particular solution has the following form:

$$y_p(t) = Kx(t) = Ke^{\lambda t}u(t). \tag{5.49}$$

Let us take the derivative of the particular solution and insert it into the differential equation to find the constant, K. Then, the differential equation becomes

$$(K\lambda^2 + 3K\lambda + 2K)e^{\lambda t}u(t) = \lambda e^{\lambda t}u(t). \tag{5.50}$$

Exponential function, $e^{\lambda t}u(t)$, cancels in both sides of the equation and we obtain an algebraic equation in terms of K, as follows:

$$K\lambda^2 + 3K\lambda + 2K = \lambda. \tag{5.51}$$

The constant value K can be obtained as:

$$K = \frac{\lambda}{\lambda^2 + 3\lambda + 2}. \tag{5.52}$$

Interestingly, the constant parameter K depends on the parameter λ, which is the decay or growth rate of the exponential function. Note that $\lambda = -1$ or -2 are also the roots of the characteristic equation and for these values of λ, $K \to \infty$. This is a degenerate solution. When the roots of the characteristic equation are the same as the decay values λ, the analytical form of the particular solution must take a different form to avoid degeneracy. The following exercise solves this problem.

c) In order to find the constant coefficients C_1 and C_2, we form the general solution, as follows:

$$y(t) = [C_1 e^{-t} + C_2 e^{-2t} + Ke^{\lambda t}]u(t). \tag{5.53}$$

Then, we apply the initial conditions to find the values of the constant parameters, C_1 and C_2. Assuming that the system is initially at rest, we obtain the following initial conditions:

$$y(0) = \dot{y}(0) = 0. \tag{5.54}$$

For $\lambda = 1$,

$$K = \frac{1}{6} \tag{5.55}$$

and the general solution becomes

$$y(t) = \left[C_1 e^{-t} + C_2 e^{-2t} + \frac{1}{6} e^{t}\right] u(t). \tag{5.56}$$

Finally, using the initial conditions, we obtain the values for C_1 and C_2, and the general solution, as follows:

$$\begin{aligned} C_1 + C_2 + \frac{1}{6} = 0,\ -C_1 - 2C_2 + \frac{1}{6} = 0 \\ C_1 = -\frac{1}{2},\ C_2 = \frac{1}{3} \\ y(t) = \left[-\frac{1}{2} e^{-t} + \frac{1}{3} e^{-2t} + \frac{1}{6} e^{t}\right] u(t). \end{aligned} \tag{5.57}$$

Exercise 5.5 Given the following second-order differential equation:

$$\ddot{y}(t) + 3\dot{y}(t) + 2y(t) = \dot{x}(t). \tag{5.58}$$

a) Find the homogeneous solution for zero input, $x(t) = 0$.
b) Find the particular solution for the input,

$$x(t) = e^{-\lambda t} u(t), \tag{5.59}$$

where $\lambda = -2$.
c) Find the general solution for $\lambda = -2$ and assuming that the system is initially at rest with the initial conditions given here:

$$y(0) = \dot{y}(0) = 0. \tag{5.60}$$

Solution

a) The homogeneous solution is the same as the previous example,

$$y_h(t) = C_1 e^{-t} + C_2 e^{-2t}. \tag{5.61}$$

b) One of the roots of the characteristic equation is the same as the decay rate $\lambda = -2$. In order to avoid the degenerate solution, we assume that the particular solution has the following analytical form:

$$y_p(t) = Kte^{-2t} u(t). \tag{5.62}$$

Let us take the derivatives of the particular solution with respect to t:

$$\dot{y}_p(t) = K(1 - 2t)e^{-2t} u(t), \tag{5.63}$$

$$\ddot{y}_p(t) = 4K(t - 1)e^{-2t} u(t), \tag{5.64}$$

and insert them into the differential equation to find the constant parameter, $K = 2$. Then, the particular solution is

$$y_p(t) = 2te^{-2t} u(t). \tag{5.65}$$

c) The general solution is the sum of homogeneous and particular solutions,

$$y(t) = [C_1 e^{-t} + C_2 e^{-2t} + 2te^{-2t}] u(t). \tag{5.66}$$

The constant parameters, C_1 and C_2 are obtained by substituting initial conditions,

$$y(0) = \dot{y}(0) = 0. \tag{5.67}$$

into the general solution;

$$y(0) = C_1 + C_2 = 0, \tag{5.68}$$

$$\dot{y}(0) = -C_1 - 2C_2 + 2 = 0. \tag{5.69}$$

Solving Equations (5.68) and (5.69), we find that $C_2 = -C_1 = 2$.

$$y(t) = 2[te^{-2t} - e^{-t} + e^{-2t} +]u(t). \tag{5.70}$$

Remark 5.2 The aforementioned simple examples show that finding the particular solutions to the linear constant coefficient differential equations requires heuristics to make an initial guess about the analytical form.

5.3.4 Transfer Function of a Continuous Time LTI System

Exponential inputs are of special importance for the LTI systems represented by

$$\sum_{k=0}^{N} a_k \frac{d^k y(t)}{dt^k} = \sum_{k=0}^{M} b_k \frac{d^k x(t)}{dt^k}. \tag{5.71}$$

Linearity property enables us to model the particular solution to an exponential input, as follows:

$$y_p(t) = Ke^{\lambda t}. \tag{5.72}$$

Differentiation property of the exponential function,

$$\frac{d^k y_p(t)}{dt^k} = K\frac{d^k e^{\lambda t}}{dt^k} = K\lambda^k e^{\lambda t} \tag{5.73}$$

enables us to obtain the value of K in terms of the parameters of the equation a_k, b_k, and λ. Inserting the derivatives of the particular solution into the differential equation, we obtain

$$K = H(\lambda) = \frac{\sum_{k=0}^{M} b_k \lambda^k}{\sum_{k=0}^{N} a_k \lambda^k}. \tag{5.74}$$

The coefficient of the particular solution, $K = H(\lambda)$, is called the **transfer function.**

Motivating Question: What is the meaning of the transfer function?

When the input is an exponential function, $x(t) = e^{\lambda t}$, the corresponding output,

$$y_p(t) = H(\lambda)e^{\lambda t}, \tag{5.75}$$

is just the scaled version of the exponential input, $x(t)$. The scaling factor is the transfer function, which is parameterized by the coefficient of the exponent, λ. Thus, the transfer function directly determines how much of the exponential input is transferred to the output (Figure 5.2).

When the input is an exponential function, $x(t) = e^{\lambda t}$, the general solution has the following form:

$$y(t) = \underbrace{\sum C_k e^{\beta_k t}}_{y_h} + \underbrace{H(\lambda)e^{\lambda t}}_{y_p} = \sum_{k=1}^{N} C_k e^{\beta_k t} + H(\lambda)x(t). \tag{5.76}$$

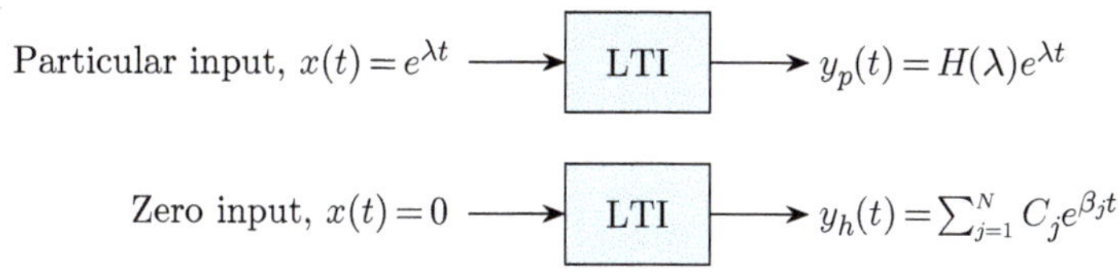

Figure 5.2 For an LTI system represented by a linear constant-coefficient differential equation, when the input is an exponential signal, $e^{\lambda t}$, the output is also an exponential signal scaled by the transfer function, $H(\lambda)$. When the input is the zero signal, the output is of the form $\sum C_j e^{\beta_j t}$. Since both the particular input and zero input satisfy the differential equation, the general solution, $y(t) = y_p(t) + y_h(t)$, is the superposition of the homogeneous and particular solutions.

Remark 5.3 The constant coefficients, C_k, of the homogeneous solution not only depend on the initial condition of the differential equation, but also depend on the particular solution of an input.

Exercise 5.6 Consider the following differential equation given:

$$\ddot{y}(t) + 3\dot{y}(t) + 2y(t) = \dot{x}(t). \tag{5.77}$$

a) Find the particular solution for $x(t) = \cos \omega_0 t$.
b) Find the homogeneous solution.
c) Find the general solution in terms of the constant coefficients of the homogeneous solution and the angular frequency ω_0.

Solution

a) Recall the Euler formula to represent cosine function in terms of complex exponential,

$$\cos \omega_0 t = \frac{e^{j\omega_0 t} + e^{-j\omega_0 t}}{2}. \tag{5.78}$$

We can directly use the result of the previous exercise by setting $\lambda = j\omega_0$ to obtain the transfer function as $H(j\omega_0)$.

When the input is $x(t) = e^{j\omega_0 t}$, the corresponding output is $y_p(t) = H(j\omega_0)e^{j\omega_0 t}$. When the input is $x(t) = e^{-j\omega_0 t}$, the corresponding output is $y_p(t) = H(-j\omega_0)e^{-j\omega_0 t}$.

If we superpose the two inputs as:

$$\cos \omega_0 t = \frac{e^{j\omega_0 t} + e^{-j\omega_0 t}}{2}, \tag{5.79}$$

we can obtain the output as the superposition of the two particular solutions as follows:

$$y_P(t) = \frac{1}{2}\left[H(j\omega_0)e^{j\omega_0 t} + H(-j\omega_0)e^{-j\omega_0 t}\right], \tag{5.80}$$

where

$$H(j\omega_0) = \frac{j\omega_0}{(j\omega_0)^2 + 3j\omega_0 + 2}. \tag{5.81}$$

This approach shows the power of the linearity property. Each term in the right-hand side of Equation (5.80) shows a subsystem of the overall system. The first subsystem receives,

$$x_1(t) = \frac{e^{j\omega_0 t}}{2} \tag{5.82}$$

and outputs $y_{p1}(t)$, whereas the second one receives

$$x_2(t) = \frac{e^{-j\omega_0 t}}{2} \tag{5.83}$$

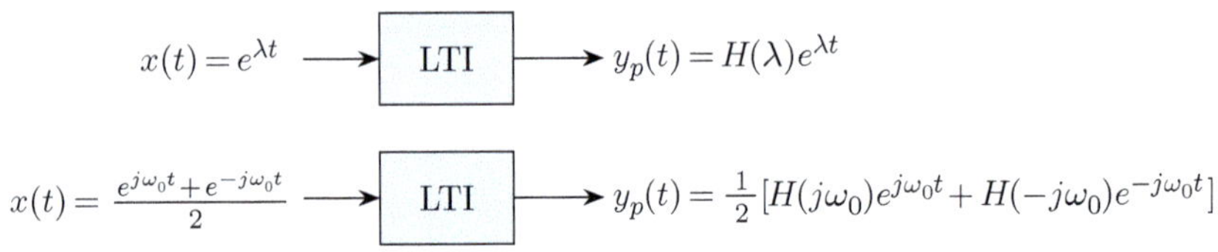

Figure 5.3 Using the superposition property to find the particular solution in Exercise 5.6

and outputs $y_{p2}(t)$. The overall particular solution is the superposition of the two responses,

$$y_p(t) = y_{p1}(t) + y_{p2}(t). \tag{5.84}$$

Note: As in the previous example, we use the superposition property to find the particular solution (Figure 5.3).

b) The homogeneous solution is the same as the previous example.

$$y_h(t) = C_1 e^{-t} + C_2 e^{-2t}. \tag{5.85}$$

c) The general solution is

$$y(t) = C_1 e^{-t} + C_2 e^{-2t} + y_p(t) = C_1 e^{-t} + C_2 e^{-2t} + \frac{1}{2}\left[H(j\omega_0)e^{j\omega_0 t} + H(-j\omega_0)e^{-j\omega_0 t}\right]. \tag{5.86}$$

Since the initial conditions are not given, there are infinitely many solutions, each of which depends on the constant coefficients, C_1 and C_2. We leave them as unknown parameters.

5.4 Linear Constant Coefficient Difference Equations

A linear constant coefficient difference equation is given in the following general form:

$$\sum_{k=0}^{N} a_k y[n-k] = \sum_{k=0}^{M} b_k x[n-k], \tag{5.87}$$

where $y[n-k]$ and $x[n-k]$ shows the kth difference, N is the order of the difference equation, M is order of the difference of $x[n]$, and $\{a_k, b_k\}$ are the constant parameters of the difference equation.

As in the continuous time systems, we can define the initial condition to satisfy the initial rest property of the system, as defined in the following text.

Definition 5.2 Discrete time LTI system is **initially at rest** at time n_0, if the input and output pairs, $x[n]$ and $y[n]$, are zero for all $n < n_0$. Formally speaking, a system is initially at rest, if

$$x[n] = 0 \text{ for } n < n_0 \Rightarrow y[n] = 0 \text{ for } n < n_0. \tag{5.88}$$

The initial rest condition is crucial for discrete time LTI systems represented by a difference equation. It simplifies finding the solution, which provides an explicit analytical expression for the output of the system for a given input.

5.4.1 Representation of a Discrete Time LTI Systems by Difference Equations

If we can relate the linear combinations of the past, present, and/or the future values of the input and output signals, then, we can represent a discrete time system by a difference equation. There is

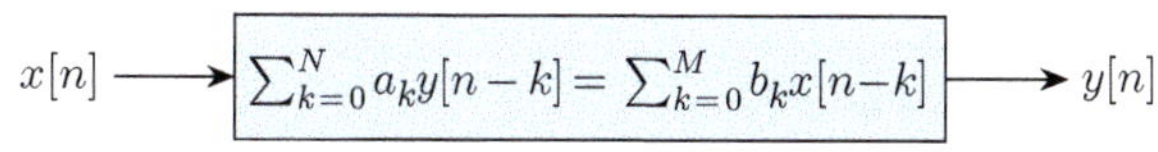

Figure 5.4 Representation of a discrete time LTI system with difference equation.

a wide range of applications of difference equations to model discrete time systems. For example, they are used in biomedical signal processing, for modeling the heart or brain signals. They are also used to model time series, where the present value of a discrete time function is related to its past values. Examples of time series data include speech signals, financial, economic, or demographic data, collected on a yearly, monthly or daily basis.

Proposition 5.2 A difference equation, with **initial rest conditions,**

$$\sum_{k=0}^{N} a_k y[n-k] = \sum_{k=0}^{M} b_k x[n-k], \tag{5.89}$$

represents a discrete time, **causal LTI system**, which relates the superposition of the input $x[n-k]$ to that of the output $y[n-k]$.

Since the verification of this proposition is a trivial extension of the continuous time case, it is omitted here. When we represent a discrete time causal LTI system with a difference equation, we can replace the impulse response with the difference equation, in the black box, as shown in Figure 5.4.

5.4.2 Solution to Linear Constant Coefficient Difference Equations

The solution to a difference equation can be obtained by using a recursive method. Suppose we normalize Equation (5.89) by dividing all the constant parameters to a_0 so that the coefficient of $y[n]$ is 1. Leaving the $y[n]$ alone on the left-hand side of Equation (5.89), we get,

$$y[n] = -\sum_{k=1}^{N} a_k y[n-k] + \sum_{k=0}^{M} b_k x[n-k]. \tag{5.90}$$

This is a recursive equation. Given the input, $x[n]$ and N initial conditions,

$$y[n_0], y[n_0-1], \ldots, y[n_0-N], \tag{5.91}$$

we can start from the initial conditions and iterate it for all possible values of $y[n]$. Let us solve an example.

Exercise 5.7 Consider the following difference equation, which represents a discrete time LTI system:

$$y[n] - \frac{1}{2} y[n-1] = x[n], \tag{5.92}$$

with the input, $x[n] = n^2 u[n]$ and initial condition, $y[-1] = 16$.

a) Find and plot the output, $y[n]$, using the recursive method.
b) Is this system initially at rest?

Solution

a) Let us leave $y[n]$ alone in the left-hand side of the equation,

$$y[n] = \frac{1}{2}y[n-1] + x[n]. \tag{5.93}$$

Then, using the initial condition and input, let us evaluate the values of $y[n]$ for all n, as follows:

$$\begin{aligned} y[0] &= 8 \\ y[1] &= \frac{1}{2}y[0] + 1 = 5 \\ y[2] &= \frac{1}{2}y[1] + 4 = 6.5 \\ y[3] &= \frac{1}{2}y[2] + 9 = 12.25 \\ &\vdots \end{aligned} \tag{5.94}$$

$y[n]$ is plotted in Figure 5.5.

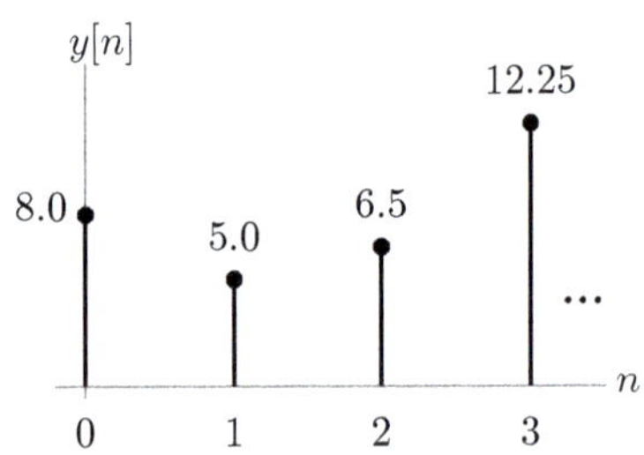

Figure 5.5 Solution of the difference equation of $y[n] = \frac{1}{2}y[n-1] + x[n]$ with initial condition, $y[-1] = 16$. Notice that the plot continues as we increase n.

b) This system is **not** initially at rest, since for $n = 0$, the input is $x[n] = 0$, however, the output is $y[0] = 8$.

Exercise 5.8 Consider the following equation, which represents a discrete time LTI system:

$$y[n] - \frac{1}{2}y[n-1] = x[n]. \tag{5.95}$$

Assuming that the system is initially at rest, with $y[-1] = 0$, find the output, when the input is $x[n] = u[n]$.

Solution

Leave $y[n]$ in the left-hand side of the equation,

$$y[n] = x[n] + 0.5y[n-1], \tag{5.96}$$

and solve it recursively:

$$\begin{aligned} y[0] &= x[0] + 0.5y[-1] = 1, \\ y[1] &= x[1] + 0.5y[0] = 1 + 0.5 = 1.5, \\ y[2] &= x[2] + 0.5y[1] = 1 + 0.5(1 + 0.5), \\ &\vdots \end{aligned} \tag{5.97}$$

In a more compact form, the output is

$$y[n] = \sum_{k=0}^{n} (0.5)^k u[n]. \tag{5.98}$$

Recall that

$$\sum_{k=0}^{n} \alpha^k = \frac{1 - \alpha^{n+1}}{1 - \alpha}. \tag{5.99}$$

Set $\alpha = 0.5$ to find a closed-form solution for the output

$$y[n] = [2 - 0.5^n]u[n]. \tag{5.100}$$

5.4.3 Transfer Function of a Discrete Time LTI System

As in the continuous time case, exponential inputs are also very important for discrete time LTI systems. An LTI system can be represented by the following difference equation:

$$\sum_{k=0}^{N} a_k y[n-k] = \sum_{k=0}^{M} b_k x[n-k]. \tag{5.101}$$

It can also be represented by its impulse response $h[n]$. Suppose that we feed the following exponential input to the system:

$$x[n] = e^{\lambda n}.$$

The corresponding output can be obtained by the convolution sum:

$$y[n] = x[n] * h[n] = \sum_{k=-\infty}^{\infty} e^{\lambda(n-k)} h[k] = e^{\lambda n} \left(\sum_{k=-\infty}^{\infty} h[k] e^{-\lambda k} \right). \tag{5.102}$$

Equation (5.102) reveals that the exponential input is directly passed to the output with a scaling factor,

$$H(e^{\lambda}) = \sum_{k=-\infty}^{\infty} h[k] e^{-\lambda k}, \tag{5.103}$$

which is called the **transfer function**.

The output to the exponential input $x[n] = e^{\lambda n}$ is then written as:

$$y[n] = e^{\lambda n} H(e^{\lambda}).$$

Inserting the value of the output $y[n]$ and the exponential input $x[n] = e^{\lambda n}$, in the difference equation, we obtain

$$\sum_{k=0}^{N} a_k H(e^{\lambda}) e^{\lambda(n-k)} = \sum_{k=0}^{M} b_k e^{\lambda(n-k)}. \tag{5.104}$$

Arranging this, we obtain the transfer function for discrete time LTI systems in terms of the parameters of the difference equation, as follows:

$$H(e^{\lambda}) = \frac{\sum_{k=0}^{M} b_k e^{-\lambda k}}{\sum_{k=0}^{N} a_k e^{-\lambda k}}. \tag{5.105}$$

Hence, when the input of a discrete time LTI system is an exponential function, the corresponding output is just the scaled version of the input, where the scaling factor $H(e^{\lambda})$ is called the **transfer function**.

5.5 Relationship Between the Impulse Response and Difference or Differential Equations

In Chapter 4, we mentioned that impulse response uniquely represents an LTI system through

- convolution integral for continuous time systems and
- convolution summation for the discrete time systems.

In this chapter, we have seen that it is possible to uniquely represent a LTI system by

- differential equation for continuous time systems and
- difference equation for discrete time systems.

Therefore, we should be able to obtain impulse response from a differential or difference equation. All we have to do is to replace the input by the impulse function and to replace the corresponding output by the impulse response. Then, we solve the differential equation, which represents the LTI system, to get an explicit analytical form for the impulse response.

Let us try to find the impulse response from a differential equation in the following example.

Exercise 5.9 Find the impulse response, $h(t)$ of the following first-order differential equation, which is initially at rest:

$$\dot{y}(t) + 3y(t) = x(t). \tag{5.106}$$

Solution

Recall that impulse response is the output of an LTI system, when the input is an impulse function, i.e., $x(t) = \delta(t)$. Thus, replacing the input by the unit impulse function in the differential equation, we obtain the following differential equation:

$$\dot{h}(t) + 3h(t) = \delta(t). \tag{5.107}$$

Let us solve this differential equation by finding the homogeneous solution and particular solution, then adding them to obtain the overall solution.

The homogeneous solution, $h_H(t)$, is obtained from the homogeneous equation,

$$\dot{h}_H(t) + 3h_H(t) = 0. \tag{5.108}$$

Solving it, for $h_H(t) = Ce^{\alpha t}$, we find

$$h_H(t) = Ce^{-3t}u(t). \tag{5.109}$$

Particular solution is obtained by replacing $x(t) = \delta(t)$ and integrating both sides of the differential equation between a little bit larger than 0, (+0), and a little bit smaller than 0, (−0), as follows:

$$\int_{-0}^{+0} dh(t) + 3\int_{-0}^{+0} h(\tau)d\tau = \int_{-0}^{+0} \delta(\tau)d\tau = 1. \tag{5.110}$$

The right-hand side of Equation (5.110) is equivalent to the operational definition of the impulse function,

$$\int_{-0}^{+0} \delta(\tau)d\tau = \int_{-\infty}^{\infty} \delta(\tau)d\tau = 1. \tag{5.111}$$

The first term on the left-hand side of Equation (5.110) is

$$\int_{-0}^{+0} dh(\tau) = h(+0) - h(-0). \tag{5.112}$$

Since the system is initially at rest, $h(-0) = 0$. Thus,

$$\int_{-0}^{+0} dh(\tau) = h(+0). \tag{5.113}$$

The second term in the left-hand side of Equation (5.110) integral approaches to 0,

$$\lim_{\substack{0^+ \to 0 \\ 0^- \to 0}} \int_{-0}^{+0} h(\tau)d\tau = 0. \tag{5.114}$$

Thus, the particular solution provides an auxiliary condition for the impulse response, as $h(+0) = 1$. This condition is used to find the constant parameter C of the homogeneous solution $h_H(t)$,

$$h_H(+0) = Ce^{-3\times 0} = C = 1. \tag{5.115}$$

Thus, the general solution of the differential equation is $h(t) = e^{-3t}u(t)$.

Remark 5.4 The particular solution just provides us with the initial condition as $h(+0) = 1$.

The aforementioned example to find the impulse response from the differential equation, can be expanded to a general differential equation as follows.

Given an Nth-order ordinary differential equation, which represents a system which is initially at rest,

$$\sum_{k=0}^{N} a_k y^{(k)}(t) = x(t), \quad \sum_{k=0}^{N} a_k h^{(k)}(t) = \delta(t). \tag{5.116}$$

The impulse response $h(t)$ of a continuous time system, which is initially at rest, is obtained by the following two steps:

Step 1: Find the homogeneous solution, $h_H(t) = \sum_{k=0}^{N} C_k e^{\alpha_k t}$.
Step 2: Obtain the constant coefficients, C_k, from the N initial conditions.

$$\begin{aligned} h(0^+) = \dot{h}(0^+) = \cdots = h^{(N-2)}(0^+) = 0 \quad \text{and} \\ h^{(N-1)}(0^+) = \frac{1}{a_N}. \end{aligned} \tag{5.117}$$

Recall that the derivative of the unit step response provides the impulse response of an LTI system. Therefore, finding the solution of a differential equation to a unit step response and taking the derivative of this solution also provides the impulse response, as exemplified in the following exercise.

Exercise 5.10 Given the following differential equation of an LTI system, which is initially at rest for $\dot{y}(0) = y(0) = 0$:

$$\ddot{y}(t) + 3\dot{y}(t) + 2y(t) = x(t), \tag{5.118}$$

a) Find the unit step response.
b) Find the impulse response.

Solution

a) The unit step response, $s(t)$ is obtained when the particular input to this equation is $x(t) = u(t)$. The equation becomes

$$\ddot{s}(t) + 3\dot{s}(t) + 2s(t) = u(t). \tag{5.119}$$

The homogeneous solution of Equation (5.118) has the following analytical form:

$$s_h(t) = e^{\alpha t}, \tag{5.120}$$

where the parameter α is determined by finding the roots of the characteristic equation. Inserting the right-hand side of $s_h(t)$ and its derivatives into Equation (5.119) gives the following characteristic equation:

$$\alpha^2 + 3\alpha + 2 = 0. \tag{5.121}$$

with two roots $\alpha_1 = -1$ and $\alpha_2 = -2$. Thus, the homogeneous solution, for zero input response has the following form:

$$s_h(t) = C_1 e^{-t} + C_2 e^{-2t}. \tag{5.122}$$

The particular solution has the following form:

$$s_p(t) = Ku(t) = \begin{cases} K, & \text{for } t \geq 0, \\ 0, & \text{otherwise.} \end{cases} \tag{5.123}$$

The derivatives of the particular solution are all zero, yielding

$$2K = 1 \quad \text{for} \quad t \geq 0. \tag{5.124}$$

The general solution is then,

$$s(t) = \left[C_1 e^{-t} + C_2 e^{-2t} + \frac{1}{2}\right] u(t). \tag{5.125}$$

The parameters C_1 and C_2 are obtained by using the initial conditions $\dot{s}(0) = s(0) = 0$, in Equation (5.125) as follows:

$$C_1 + C_2 + \frac{1}{2} = 0 \tag{5.126}$$

$$-C_1 - 2C_2 + \frac{1}{2} = 0 \tag{5.127}$$

Then, $C_1 = -1$ and $C_2 = \frac{1}{2}$. The unit step response is then,

$$s(t) = [-e^{-t} + \frac{1}{2}e^{-2t} + \frac{1}{2}]u(t). \tag{5.128}$$

b) The impulse response of the LTI system is obtained by taking the derivative of the unit step response:

$$h(t) = [e^{-t} - e^{-2t}]u(t) \tag{5.129}$$

Obtaining the impulse response for a discrete time LTI system from a difference equation is relatively easy compared to the continuous case, as shown in the following example.

Exercise 5.11 Find the impulse response for the following discrete time LTI system, which is initially at rest:

$$y[n] - 0.5y[n-1] = x[n], \tag{5.130}$$

where the initial rest condition is $h[-1] = 0$.

Solution

Set $y[n] = h[n]$ for $x[n] = \delta[n]$ to obtain

$$h[n] - 0.5h[n-1] = \delta[n]. \tag{5.131}$$

Using the recursion equation,

$$h[n] = 0.5h[n-1] + \delta[n], \tag{5.132}$$

with $h[-1] = 0$, we can obtain the values of $h[n]$ for all $n \geq 0$, as follows:

$$\begin{aligned} h[0] &= 0.5h[-1] + 1 = 1, \\ h[1] &= 0.5h[0] + 0 = \frac{1}{2}, \\ h[2] &= \frac{1}{2} \cdot \frac{1}{2} = \left(\frac{1}{2}\right)^2, \\ h[3] &= \frac{1}{2} \cdot \frac{1}{2} . \frac{1}{2} = \left(\frac{1}{2}\right)^3, \\ &\cdots \end{aligned} \tag{5.133}$$

Thus, the impulse response can be obtained in the following closed form:

$$h[n] = \left(\frac{1}{2}\right)^n u[n]. \tag{5.134}$$

Remark 5.5 The filter, $h[n] = \left(\frac{1}{2}\right)^2 u[n]$, has an infinite length of $0 \leq n \leq \infty$. For this reason, this is an **infinite impulse response (IIR) filter** (Figure 5.6).

Exercise 5.12 Find the discrete time impulse response corresponding to the following difference equation, when the system is initially at rest:

$$y[n] = \sum_{k=0}^{M} b_k x[n-k]. \tag{5.135}$$

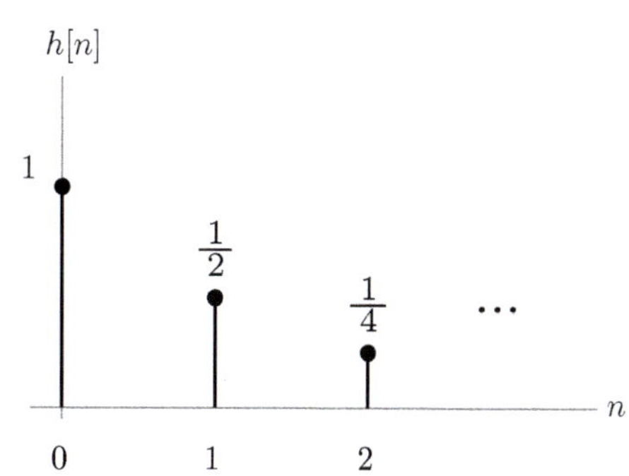

Figure 5.6 Impulse response of an infinite impulse response (IIR) filter.

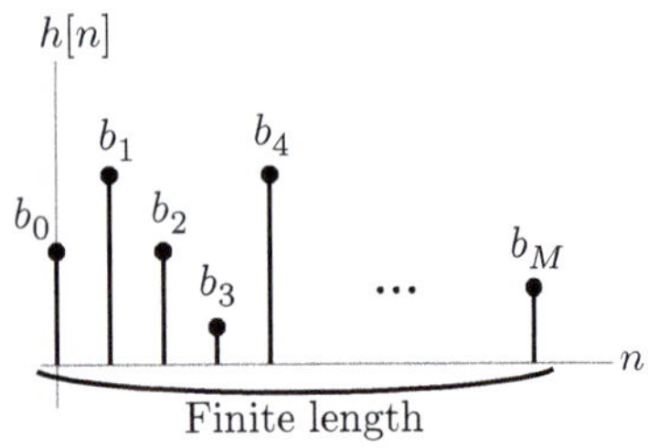

Figure 5.7 Impulse response of a finite impulse response (FIR) filter is the shifted superposition of the impulse functions, $\delta[n-k]$.

Solution

Let us set $y[n] = h[n]$ for $x[n] = \delta[n]$. Then, the impulse response is obtained as follows:

$$h[n] = \sum_{k=0}^{M} b_k \delta[n-k]. \tag{5.136}$$

Remark 5.6 The filter, represented by the impulse response of Equation (5.136) has only nonzero values in a finite interval, i.e.,

$$h[n] \neq 0 \quad \text{for} \quad 0 \leq n \leq M.$$

For this reason, it is a **finite impulse response (FIR) filter** (Figure 5.7).

Note that FIR filters are realizable in the physical environment by hardware. They are widely used in many application areas of signal processing to shape up a finite-length input signal.

5.6 Block Diagram Representation of Differential Equations for LTI Systems

In Chapter 3, we have seen that a system can be represented by a set of subsystems connected with each other by input and output signals. A subset of these components can be used to represent LTI systems, as summarized below.

An **adder** is used to add or subtract signals for both continuous time and discrete time systems and it is symbolized as shown in Figure 5.8.

A **scalar multiplier** multiplies its input signal by a scalar parameter for both continuous and discrete time systems, and it is symbolized as shown in Figure 5.9.

A **unit delay operator** is used to translate an input signal, $x[n]$, to obtain $y[n] = x[n-1]$ for **discrete time systems**, and it is symbolized as shown in Figure 5.10.

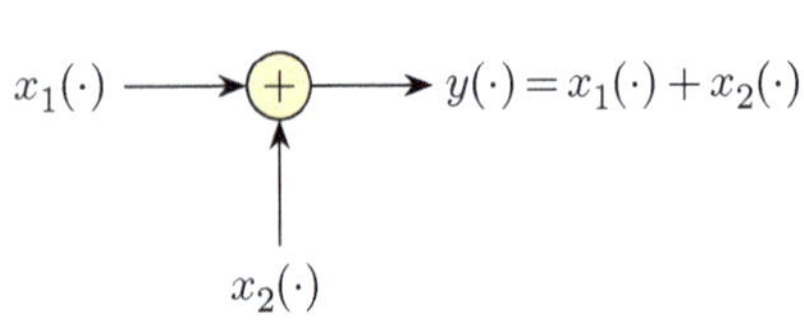

Figure 5.8 Schematic representation of an adder for two input signals.

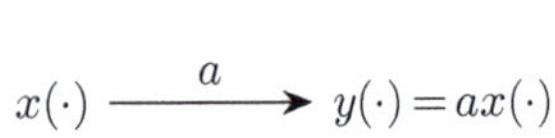

Figure 5.9 Schematic representation of a scalar multiplier.

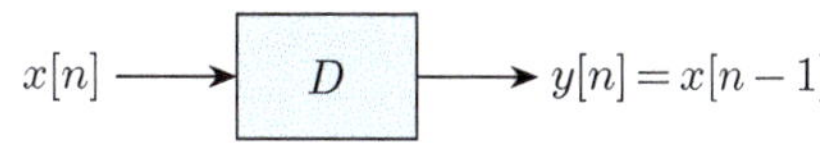

Figure 5.10 Schematic representation of unit delay operator.

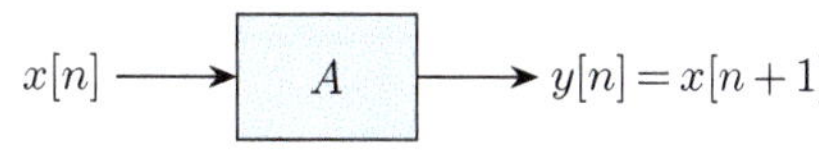

Figure 5.11 Schematic representation of the unit advance operator.

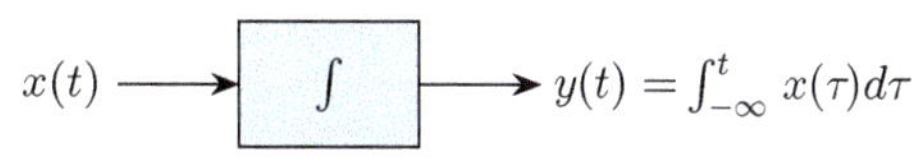

Figure 5.12 Schematic representation of an integrator.

A **unit advance operator** is used to translate an input signal, $x[n]$, to obtain $y[n] = x[n+1]$ for **discrete time systems**, and it is symbolized as shown in Figure 5.11.

An **integrator** integrates an input signal as:

$$y(t) = \int_{-\infty}^{t} x(\tau)d\tau \tag{5.137}$$

for **continuous time systems**, and it is symbolized shown in Figure 5.12.

A **differentiator** takes the derivative of an input signal as:

$$y(t) = \frac{dx(t)}{dt}. \tag{5.138}$$

for **continuous time systems**, and it is symbolized as shown in Figure 5.13.

In the following exercises, we find the block diagram representation of differential and difference equations to realize LTI systems. We also find the differential and difference equations, given block diagrams.

Exercise 5.13 Find a block diagram representation of the following differential equation:

$$\dot{y}(t) + ay(t) = bx(t). \tag{5.139}$$

Solution

Leave the highest order of the derivative on the left-hand side of the equation,

$$\dot{y}(t) = bx(t) - ay(t). \tag{5.140}$$

The block diagram representation requires an integrator and an adder to realize the first-order differential equation, as depicted in Figure 5.14.

Remark 5.7 When we draw a block diagram, traditionally, we always put the input to the left-hand side, and the output to the right-hand side.

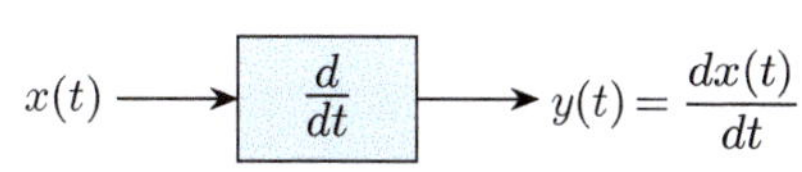

Figure 5.13 Schematic representation of a differentiator.

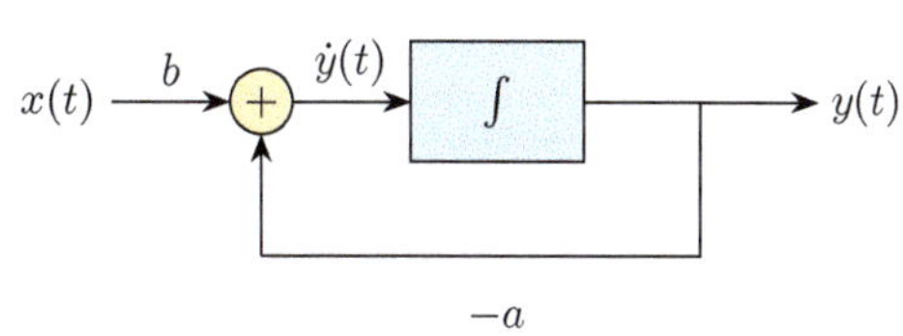

Figure 5.14 The output of the adder is $\dot{y}(t)$, which is equal to $bx(t) - ay(t)$. If we integrate $\dot{y}(t)$, we get the output signal $y(t)$.

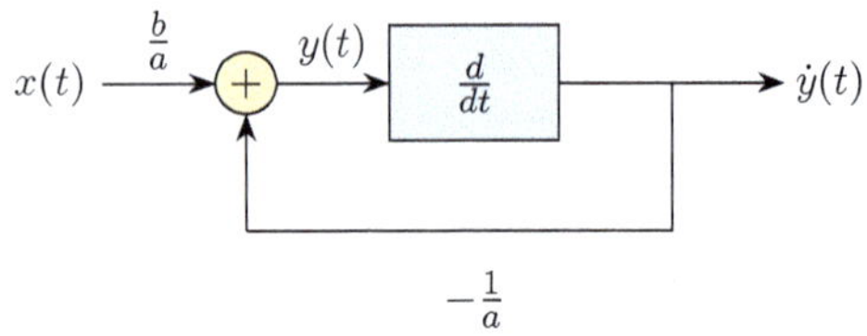

Figure 5.15 Realization of the differential equation, $\dot{y}(t) + ay(t) = bx(t)$ by a differentiator, an adder, and two scalar multipliers.

Remark 5.8 The block diagram representation is **not unique**. We could use differentiators instead of the integrator to represent the same system. Another block diagram representation can be obtained by leaving $y(t)$ on the left-hand side:

$$y(t) = \frac{b}{a}x(t) - \frac{1}{a}\dot{y}(t). \tag{5.141}$$

The corresponding block diagram is given in Figure 5.15.

Motivating Question: Which block diagram is better? Figure 5.14 or 5.15?

Integrators reduce the power consumption compared to the differentiators. Thus, the block diagram representation of Figure 5.14 is more efficient and cost effective compared to that of Figure 5.15.

The subsystems to be used in block diagram representations are a design issue, which is beyond the scope of this book.

Exercise 5.14 Find the block diagram representation of the following discrete time LTI system:

$$y[n] + ay[n-1] - by[n-2] = x[n-1]. \tag{5.142}$$

Solution

Leave $y[n]$ in the left-hand side of the equation, as follows:

$$y[n] = x[n-1] - ay[n-1] - by[n-2]. \tag{5.143}$$

Then, form the right-hand side of the equation using the adder and unit delay operators, as shown in Figure 5.16.

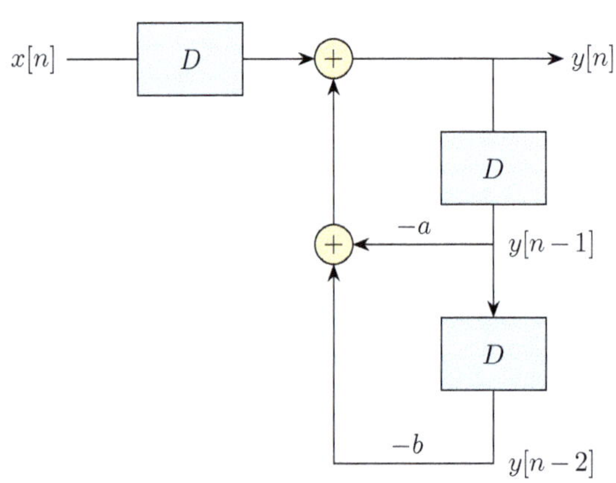

Figure 5.16 Block diagram representation of the difference equation, $y[n] + ay[n-1] - by[n-2] = x[n-1]$.

Exercise 5.15 Find the differential equation which is represented by the block diagram of Figure 5.17.

Figure 5.17 Block diagram representation of a differential equation in Exercise 5.15.

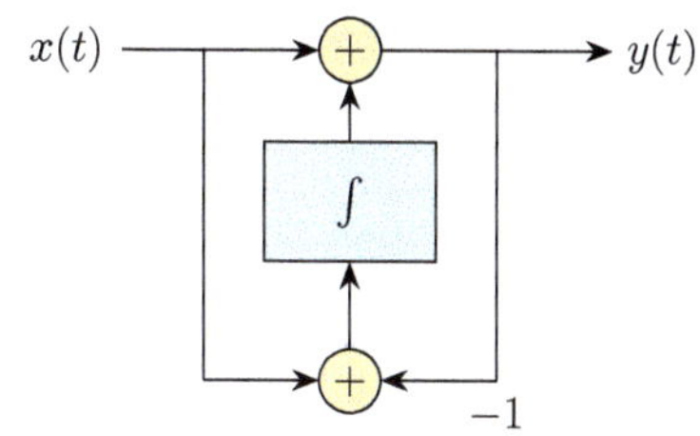

Solution

The output $y(t)$ is expressed as the integral of difference of $(x(t) - y(t))$ and added by $x(t)$:

$$y(t) = \int [x(t) - y(t)]dt + x(t). \tag{5.144}$$

Taking the derivative of both sides of Equation (5.144), we obtain,

$$\dot{y}(t) = x(t) - y(t) + \dot{x}(t) \tag{5.145}$$

or

$$\dot{y}(t) + y(t) = \dot{x}(t) + x(t). \tag{5.146}$$

Exercise 5.16 Find the difference equation that is represented by the block diagram of Figure 5.18.

Figure 5.18 Block diagram representation of a differential equation in Exercise 5.16.

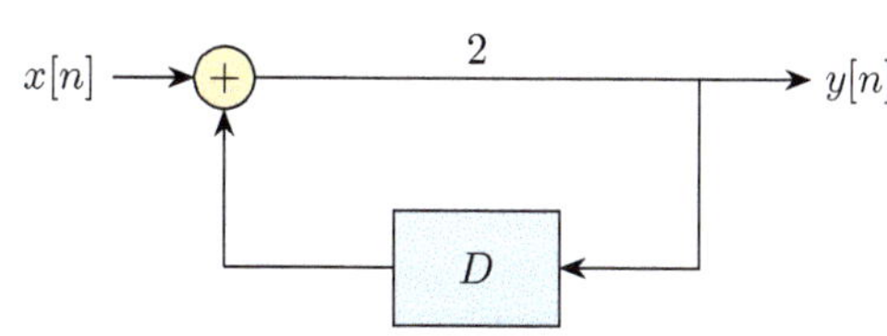

Solution

The output of the adder gets multiplied by 2, and becomes $y[n]$. Therefore, the output of the adder should be $\frac{1}{2}y[n]$. If we equate the inputs and the output of the adder, we obtain the difference equation:

$$\frac{1}{2}y[n] = x[n] + y[n-1]. \tag{5.147}$$

5.7 Chapter Summary

What is a differential equation? What is a difference equation? Can we use differential equations to represent continuous time systems? Can we use difference equations to represent discrete time systems?

A differential equation basically links the rate of change of an input to that of the output, in continuous time systems. Similarly, a difference equation relates the rate of change on the input–output pairs, when the time variable is integer-valued.

Both the difference and differential equations can be used to represent the dynamic changes of the systems. In particular, a linear and constant-coefficient differential equation, which satisfies the initial rest condition, uniquely describes a continuous time LTI system. Similarly, when the time variable is integer-valued, a linear and constant coefficient difference equation, which satisfies the initial rest condition, uniquely describes a discrete time LTI system. Thus, differential and difference equations are the mathematical objects to represent LTI systems.

There is a one-to-one correspondence between the impulse response and the differential equation, which represents a continuous time system. Similarly, there is a one-to-one correspondence between the impulse response and the difference equation, which represents a discrete time system.

In order to realize the LTI systems in practice, we can use simple components, such as adders, differentiators, integrators, unit delay, and unit advance operators. These components enable us to represent LTI systems with block diagrams and realize them in real-life applications.

Problems

5.1 The general form of an Nth-order homogeneous differential equation is given here:

$$\sum_{k=0}^{N} a_k \frac{d^k y(t)}{dt^k} = 0.$$

a) Find a solution to this differential equation in terms of the roots of the following algebraic equation:

$$\sum_{k=0}^{N} a_k s^k = 0.$$

b) How many different solutions can you obtain for this differential equation if the initial conditions are not specified?

5.2 A second-order homogeneous differential equation is given here:

$$\ddot{y}(t) + 6\dot{y}(t) + 9y(t) = 0.$$

a) Find the solution of this equation for the initial conditions, $y(0) = 0$ and $\dot{y}(0) = 18$.
b) Find the solution of this equation for the initial conditions, $y(0) = 0$ and $\dot{y}(0) = 0$.
c) Compare the solutions you obtained in parts a and b. Explain the effect of the initial conditions on the solutions.

5.3 Let us make a slight modification in Problem 5.2 as follows:

$$\ddot{y}(t) - 6\dot{y}(t) + 9y(t) = 0.$$

a) Find the solution of this equation for the initial conditions $y(0) = \dot{y}(0) = 0$.
b) Compare the solution you obtained in part a to that of Problem 5.2b. Explain the changes and no changes.

5.4 An initially at rest continuous time system is represented by the following first-order differential equation:

$$\dot{y}(t) + 2y(t) = x(t).$$

a) Find the output of the system, $y_1(t)$, for the input $x_1(t) = 3e^{3t}$.
b) Find the output of the system, $y_2(t)$, for the input $x_2(t) = 2e^{2t}$.
c) Find the output of the system, $y(t)$, for the input $x(t) = 6e^{3t} + 6e^{2t}$.
d) Find the output of the system, $y_3(t)$, for the input $x_3(t) = Ae^{2t}u(t)$.
e) Find the output of the system in terms of $y_3(t)$, which you calculated in part **d**, for the input $Ae^{2(t-T)}u(t-T)$.

5.5 A continuous time LTI system is represented by the following first-order differential equation:

$$\dot{y}(t) + 6y(t) = x(t),$$

where $x(t)$ is the input and $y(t)$ is the output of the system.
a) Find the output $y(t)$ of this system for the input $x(t) = e^{(3j-1)t}u(t)$.
b) What is the output of the system, $y(t)$, when the input is $\text{Re}\{x\}(t)$?
c) Find a transfer function of this system.

5.6 The transfer function of a continuous time LTI system is given as follows:

$$H(\lambda) = \frac{2\lambda}{\lambda^2 - 2\lambda + 1},$$

where the system is initially at rest.
a) Find the differential equation that represents this system.
b) Find the output of this system for $x(t) = 0$.
c) Find the output of this system for $x(t) = (2t+1)u(t)$.

5.7 The general form of an Nth-order homogeneous difference equation is given here:

$$\sum_{k=0}^{N} a_k y[n-k] = 0.$$

a) Find a solution to this difference equation in terms of the roots of the following algebraic equation:

$$\sum_{k=0}^{N} a_k z^k = 0.$$

b) How many different solutions can you obtain for this differential equation if the initial conditions are not specified?

5.8 A second-order homogeneous difference equation is given here:

$$y[n] - 2y[n-1] + y[n-2] = 0.$$

a) Find the solution of this equation for the initial conditions, $y[0] = 0, y[1] = 3$.
b) Find the solution of this equation for the initial conditions, $y[0] = 0, y[1] = 0$.
c) Compare the solutions you obtained in parts a and b. Explain the effect of the initial conditions on the solutions.

5.9 Let us make a slight modification in Problem 5.8, as follows:

$$y[n] - 2y[n-1] + y[n-2] = 0.$$

a) Find the solution of this equation for the initial conditions $y[0] = y[1] = 0$.
b) Compare the solution you obtained in part a to that of Problem 5.8b. Explain the changes and no changes.

5.10 A discrete time system is represented by the following difference equation:

$$y[n] = \frac{1}{3}y[n-1] + \frac{1}{9}x[n].$$

a) Does this system satisfy the conditions of initially rest for the initial condition $y[0] = 1$?
b) Is this system LTI? Verify your answer.
c) Find the transfer function of this system.

5.11 A discrete time LTI system, which is initially at rest, is represented by the following difference equation:

$$y[n] = \frac{1}{5}y[n-1] + 2x[n-2].$$

a) Find the impulse response of this system.
b) Find the transfer function of this system.
c) Find a block diagram representation of this system using the adders and unit delay operators.

5.12 A discrete time LTI system is represented by the following difference equation. Assume that the system is initially at rest.

$$y[n] + \frac{1}{2}y[n-1] + \frac{3}{20}y[n-2] = x[n]$$

Find the output, $y[n]$, of the system for the following input.

$$x[n] = \delta[n+2] + 2\delta[n+1] + 3\delta[n] + 2\delta[n-1] + 2\delta[n-2] + \delta[n-3].$$

5.13 A discrete time system is represented by the following difference equation;

$$y[n] = \frac{1}{2}y[n-1] + x[n],$$

where the system is initially at rest.
a) Find the homogeneous solution of the system.
b) Find the general solution of the system for the input $x[n] = (\frac{1}{4})^n u[n]$.
c) Find the impulse response of this system.

5.14 A discrete time causal LTI system consists of two subsystems, S_1 and S_2, given here:

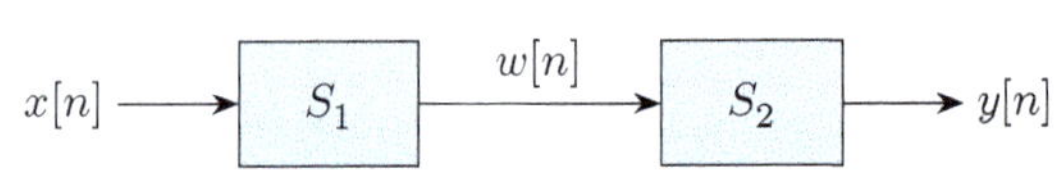

The subsystem S_1 is represented by the following difference equation:

$$w[n] = \frac{1}{3}w[n-1] + x[n].$$

The subsystem S_2 is represented by the following difference equation:

$$y[n] = \frac{2}{3}y[n-1] + \frac{1}{2}w[n].$$

a) Find the difference equation for the overall system, which relates the input $x[n]$ and output $y[n]$.
b) Draw a block diagram of the overall system, which receives the input $x[n]$ and outputs $y[n]$, using adders and unit delay operators.
c) Find and plot the impulse response of the overall system.

5.15 The following discrete system consists of three subsystems, with impulse responses, $h_1[n]$, $h_2[n]$, and $h_3[n] = h_2[n]$, respectively.

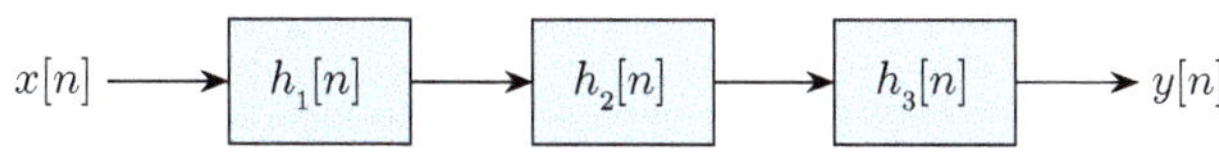

The impulse response $h_2[n]$ is given as follows:

$$h_2[n] = u[n] - u[n-2]$$

and the overall impulse response, $h[n]$, of the system is:

$$h[n] = \delta[n] + 5\delta[n-1] + 10\delta[n-2] + 11\delta[n-3] + 8\delta[n-4] + \delta[n-5].$$

a) Find and plot the impulse response $h_1[n]$.
b) Find and plot the output $y[n]$ for the input $x[n] = \delta[n] - \delta[n-1]$.

5.16 A discrete time LTI system is represented by the following impulse response:

$$h[n] = \left(\frac{1}{4}\right)^{n+1} u[n+3].$$

a) Find the output, $y[n]$, of the system for the following input:

$$x[n] = 4^n u[-n] + \left(\frac{1}{4}\right)^n u[n].$$

b) Find and plot the output for the input $x[n] = e^{\lambda n}$.
c) Find a transfer function of this system.
d) Is this system causal?

5.17 A discrete time LTI system is represented by the following difference equation:

$$y[n] = x[n] - 5y[n-1].$$

Assume that the system is initially at rest.
a) Find and plot the impulse response of this system.
b) Find and plot the output, $y[n]$ of the system for the input $x[n] = u[n]$.
c) Find and plot the output for the input $x[n] = e^{\lambda n}$.
d) Find a transfer function of this system.
e) Is this system causal?

5.18 A continuous time system S is represented by the impulse response $h(t)$.

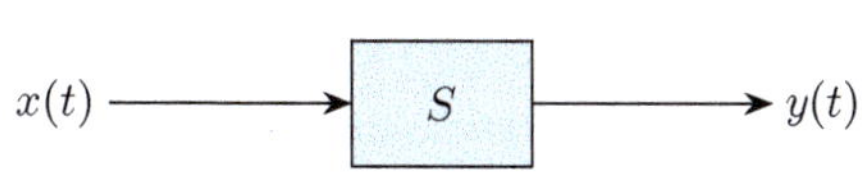

When the derivative $x_1(t) = \frac{dx(t)}{dt}$ of the input signal $x(t) = 2^{-3t}u(t-1)$ is fed to this system, the corresponding output becomes

$$y_1(t) = -3y(t) + e^{-2t}u(t).$$

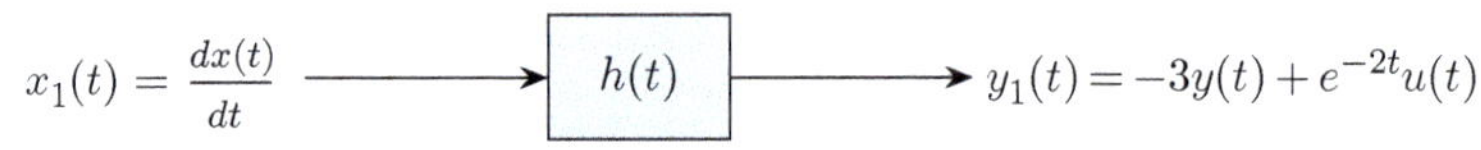

Find the impulse response of the system S.

5.19 A discrete time system, which is initially at rest, is represented by the following difference equation:

$$y[n] - \frac{1}{4}y[n-1] = x[n].$$

a) Find and plot the output $y[n]$ for $n = 0$ and for the input $x[n] = \delta[n]$.
b) Find and plot the impulse response, $h[n]$, for $n \geq 1$.

5.20 Find the impulse response, $h[n]$ of an initially at rest discrete time system, represented by the following difference equation:

$$y[n] - \frac{1}{4}y[n-1] = x[n] + 2x[n-1].$$

5.21 For the discrete time LTI system given here:

$$\sum_{k=0}^{N} y[n-k] = x[n],$$

find $y[0]$ for $x[n] = \delta[n]$.

5.22 Find the impulse responses of the causal LTI systems represented by the following difference equations:
a) $y[n] - \frac{1}{2}y[n-2] = x[n]$
b) $y[n] - \frac{1}{2}y[n-2] = x[n] + x[n-1]$
c) $y[n] - y[n-2] = x[n] - 2x[n-4]$
d) $y[n] - \frac{\sqrt{3}}{4}y[n-1] + \frac{1}{4}y[n-2] = x[n]$

5.23 A discrete time system is represented by the following difference equation:

$$y[n] = \frac{1}{3}y[n-1] + \frac{1}{2}x[n].$$

a) Find a block diagram representation of this system using adders and unit delay operators.
b) Find a block diagram representation of this system using adders and unit advance operators.

5.24 A discrete time LTI system is represented by the following block diagram:
a) Find the difference equation corresponding to the following block diagram.
b) Find a block diagram representation using unit advance operators and adders.

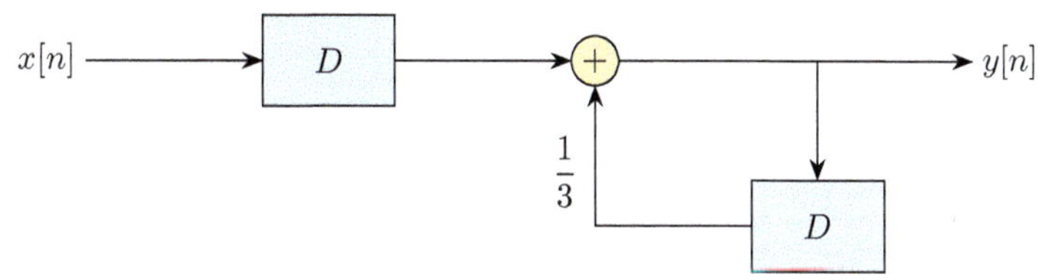

5.25 A continuous time causal LTI system is represented by the following differential equation:

$$y(t) = -\frac{1}{2}\dot{y}(t) + 4x(t).$$

a) Find a block diagram representation of this system using integrators and adders.
b) Find a block diagram representation of this system using differentiators and adders.

5.26 A continuous time LTI system is represented by the following block diagram:

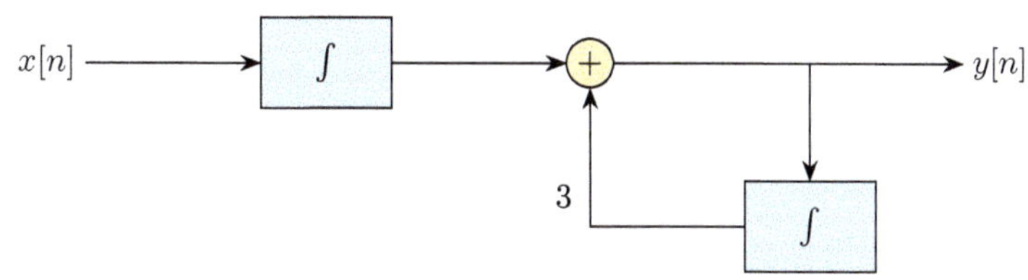

a) Find the differential equation corresponding to this system.
b) Find a block diagram representation of this system using adders and differentiators.

6

Fourier Series Representation of Continuous Time Periodic Signals

"The deep study of nature is the most fruitful source of knowledge!"

Jean Baptiste Joseph Fourier

We, humans, perceive the physical world mostly by dynamic changes of signals, such as light, sound, speech, and heat, perceived by sensory stimuli, in the time domain. Thus, it is natural to model signals and systems in the time domain.

Until now, we represented the **signals** as a function of time. In other words, the domain of the functions was time. In the time domain, we represent signals in terms of weighted integral and sum of basic functions, such, as **unit impulses**:

$$x(t) = \int_{-\infty}^{\infty} x(\tau)\delta(t-\tau)d\tau, \tag{6.1}$$

for continuous time signals, and,

$$x[n] = \sum_{k=-\infty}^{\infty} x[k]\delta[n-k], \tag{6.2}$$

for discrete time signals.

We represented linear time-invariant (**LTI**) systems with equations, which establish a relationship between the input and output signals of LTI systems using the **impulse response,** $h(\cdot)$ by convolution integral and sum:

$$y(t) = \int_{-\infty}^{\infty} x(\tau)h(t-\tau)d\tau, \tag{6.3}$$

for continuous time systems,

$$y[n] = \sum_{k=-\infty}^{\infty} x[k]h[n-k], \tag{6.4}$$

for discrete time systems.

Equivalently, we also represented a continuous time LTI system with a **linear constant coefficient differential equation,**

$$\sum_{k=0}^{N} a_k \frac{d^k y(t)}{dt^k} = \sum_{k=0}^{M} b_k \frac{d^k x(t)}{dt^k}, \tag{6.5}$$

Signals and Systems: Theory and Practical Explorations with Python,
First Edition. Fatoş Tunay Yarman Vural and Emre Akbaş.

Companion website: www.wiley.com/go/signalsandsystems

and a discrete time LTI system with a **linear constant coefficient difference equation,**

$$\sum_{k=0}^{N} a_k y[n-k] = \sum_{k=0}^{M} b_k x[n-k]. \tag{6.6}$$

All of the aforementioned functions and equations represent the signals and systems in terms of **time**.

Motivating Questions: Can we represent signals and systems in domains other than time, such that they provide different types of useful information about the underlying phenomena? How can we model our observations, such as the trajectory of celestial objects, to study the universe surrounding us? How do we represent a periodic motion? Is it possible to represent complicated periodic motions in terms of simple periodic functions? Is it possible to represent any function in terms of simple periodic motions at all? If this is ever possible, what is the relationship between this new representation and the time domain representation of the function?

Human curiosity has searched for answers to questions of this kind over the centuries of the history of science, as summarized in the next section.

6.1 History

At the heart of the Fourier analysis lies the **periodic motions and harmony**.

The very first question was asked by the Babylonians, around 1500 BCE, when they attempted to understand what was happening in the sky: Is it possible to model the motion of terrestrial objects, which repeats with some regularity?

In order to answer this question, they recorded the location of the sun and moon relative to time to predict their trajectories. Following the Babylonians, Indian mathematicians developed an early version of the periodic functions, called **Jiva,** in Sanskrit. In the 9th century, Muhammad ibn Musa Al-Khwarizmi, produced the first accurate tables of periodic motions, by improving Jiva, and named it as **Jaib**, which means bosom in Arabic. The term Jaib is translated in Latin as **Sinus**, meaning bosom or bay, to represent periodic motion. **Al-Khwarizmi** was, also, a pioneer in circular trigonometry. The scientists in medieval Islam, accomplished a series of studies for calculating the trajectories of celestial objects using the simple periodic functions. Among them, **Al-Biruni, Al-Farghani,** and **Al-Haytham,** used circular motion and trigonometry to formalize the periodic motion.

The next important question was asked by the European mathematicians in enlightenment: Is it possible to represent complicated periodic motions in terms of simple periodic functions?

Answering this question was possible by using the concept of **harmony**. In 18th century, **L. Euler** expressed a periodic motion of a string in terms of the linear combination of the harmonically related sinusoidal functions. However, **J. Bernoulli** and **J. L. Lagrange** argued that the representation of functions as a superposition of periodic waveforms was not possible, especially, when the function has sharp corners.

Finally, **Jean-Baptiste Joseph Fourier** claimed that any periodic function can be represented in terms of the superposition of harmonically related sine and cosine functions, today, known as **Fourier series** for continuous time functions. Then, he extended this idea to **any** continuous time aperiodic function and he wrote his pioneering paper, in 1807. Four referees examined this mind blowing work: **S.F. Lacroix, G. Mogne, P.S. Laplace**, and his advisor, **J. L. Lagrange**. Three of the committee members accepted the paper. However, his advisor, Lagrange rejected the paper and

Figure 6.1 Digital artwork showing colorful waves. Source: Eigens/Adobe Stock Photos.

it was not published. In 1822, Fourier published his theorems in a book called **Theory Analytique de La Chaleur (Heat Diffusion)**. Later, in 1829, **P. G. L. Dirichlet,** a student of Fourier, showed that under some conditions, Fourier's theorems are correct. These conditions are called Dirichlet conditions.

WATCH

Learn more about the life of J. B. Fourier @ https://384book.net/v0601

Loosely speaking Fourier's theorems together with Dirichlet condition states that

- Any periodic function, satisfying a set of conditions, called Dirichlet conditions, can be represented in terms of the superposition of **harmonically related waves**.
- Furthermore, any function, satisfying a set of conditions, called Dirichlet conditions, can be represented in terms of the superposition of **harmonically related waves**.

Motivating Question: But what is a wave?

Loosely speaking, waves are defined as propagating dynamic changes from an equilibrium of one or more quantities (Figure 6.1). Waves can be periodic, in which case those quantities oscillate repeatedly about an equilibrium value at some frequency.

Examples of waves include sound waves, light waves, radio waves, microwaves, water waves, stadium waves, earthquake waves, and waves on a string.

6.2 Mathematical Representation of Waves and Harmony

In mathematics, waves or waveforms are frequently represented by periodic functions, such as **sines, cosines,** and **complex exponentials**.

Recall that a complex exponential, represented by the Euler Formula,

$$\Phi_k(t) = e^{jk\omega_0 t} = \cos(k\omega_0 t) + j\sin(k\omega_0 t), \tag{6.7}$$

is a periodic function, where the angular frequency is $\omega_0 = \frac{2\pi}{T_0}$ and T_0 is the fundamental period. The set of all complex exponentials, $\Phi_k(t)$, for the integer values, $-\infty < k < \infty$, are called **harmonically related exponentials**.

In Equation (6.7), for each integer value of k, we define a complex exponential, $\Phi_k(t)$, with frequency, $k\omega_0$, called the kth harmonics. Complex exponentials are very interesting functions, which rotate on a unit circle in a complex plane, as a function of time. If we take the time as the third dimension, which is orthogonal to the complex plane, we can generate a sinusoidal signal in the time domain, as the complex exponential rotates in the complex plane. The speed of the rotation depends on the angular frequency, ω_0. We generate the kth harmonics of the complex exponential, $e^{j\omega_0 t}$, as follows:

$$\Phi_k(t) = e^{jk\omega_0 t}, \tag{6.8}$$

where the rotation speed of the exponential function on the unit circle becomes k times faster. Recall that the exponential functions with the integer multiples of angular frequencies, $k\omega_0$, are called the harmonically related complex exponentials. The motion of the complex exponential on the unit circle and the corresponding sinusoidal function is depicted in Figure 6.2.

Motivating Question: Can we represent a periodic function in terms of weighted summation of waves, such as sinusoidals or complex exponentials?

Mathematically speaking, given a periodic function, $x(t)$, our goal is to find a set of coefficients $\{a_k\}$ such that we can represent $x(t)$ by the superposition of harmonically related complex exponentials, as follows:

$$x(t) = \sum_{k=-\infty}^{\infty} a_k e^{jk\omega_0 t}. \tag{6.9}$$

Let us study the following simple example to see if this is ever possible.

Exercise 6.1 Can we represent the following signal in terms of the superposition of complex exponentials?

$$x(t) = \beta_1 \sin(2\pi t) + \beta_2 \cos(\pi t). \tag{6.10}$$

$x(t)$ is plotted in Figure 6.3 for $\beta_1 = \beta_2 = 0.5$.

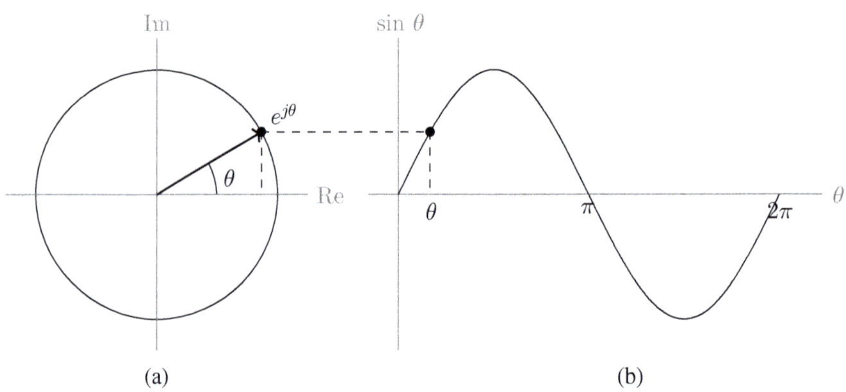

Figure 6.2 A simple rotation on a circle in a complex plane creates trigonometric waveforms of sine and cosine. As θ increases, the point on the circle (a) travels in the counterclockwise direction, at the same time, θ vs. the y-coordinate of that point follows a sine curve, shown on (b).

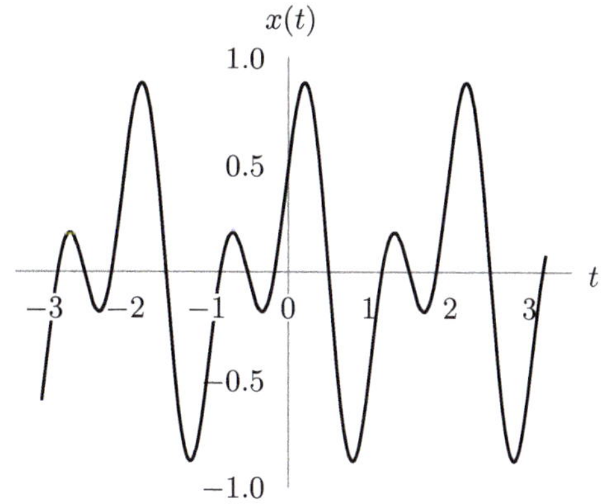

Figure 6.3 Plot of the periodic function, $x(t) = 0.5\sin(2\pi t) + 0.5\cos(\pi t)$.

Solution

Let us use the Euler formula to represent the sines and cosines in terms of complex exponentials.

$$x(t) = \beta_1 \frac{e^{j2\pi t} - e^{-j2\pi t}}{2j} + \beta_2 \frac{e^{j\pi t} + e^{-j\pi t}}{2}. \tag{6.11}$$

Let us put the function $x(t)$ into the following general form:

$$x(t) = \sum_{k=-\infty}^{\infty} a_k e^{jk\omega_0 t} = a_0 + a_1 e^{j\omega_0 t} + a_{-1} e^{-j\omega_0 t} + a_2 e^{2j\omega_0 t} + a_{-2} e^{-2j\omega_0 t} + \cdots, \tag{6.12}$$

where $e^{jk\omega_0 t}$s are infinitely many harmonically related complex exponentials, since the integer value, k, ranges in the interval of $(-\infty, \infty)$.

According to this equation,

- for $k = \pm 1 \rightarrow a_1 = a_{-1} = \frac{\beta_2}{2}$
- for $k = 2 \rightarrow a_2 = \frac{\beta_1}{2j}$,
- for $k = -2 \rightarrow a_{-2} = -\frac{\beta_1}{2j}$,
- otherwise, $a_k = 0$.

Thus, function $x(t)$ is represented by four coefficients of the harmonically related complex exponential functions as:

$$\{a_1, a_{-1}, a_2, a_{-2}\}. \tag{6.13}$$

Note: Coefficients of four harmonically related complex exponentials are sufficient to represent the signal $x(t) = \beta_1 \sin(2\pi t) + \beta_2 \cos(\pi t)$.

Let us set $\beta_1 = \beta_2 = 0.5$. Then, function $x(t)$ becomes

$$x(t) = 0.5\sin(2\pi t) + 0.5\cos(\pi t), \tag{6.14}$$

and the four coefficients of the harmonically related complex exponential becomes,

$$a_1 = a_{-1} = 0.25, \quad a_2 = -a_{-2} = \frac{0.25}{j} = -0.25j. \tag{6.15}$$

Each coefficient a_k shows the contribution of the corresponding harmonic of the complex exponential to make the signal $x(t)$. For this particular example, there are two pairs of nonzero harmonics; a_1, a_{-1} and a_2, a_{-2}. The rest of the coefficients are zero, showing that the function is made of two pairs of harmonics of the complex exponential function.

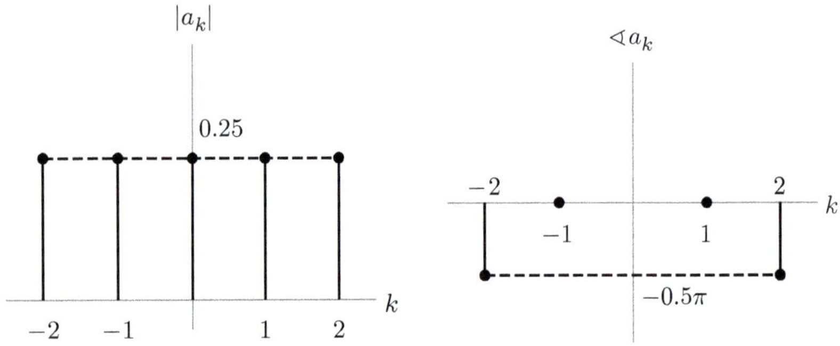

Figure 6.4 Magnitude $|a_k|$ vs. k and phase $\sphericalangle a_k$ vs. k plots. The magnitudes are all the same for all values of k, indicating equal contribution of all of the harmonics. There are two nonzero values in the phase plot, at $k = \pm 2$, which shows that the harmonics line up at $\sphericalangle a_k = -0.5\pi$.

The plot of the coefficients a_k with respect to k provides information about the frequency content of the signals. When the coefficients are complex numbers, we need to generate two plots: one for the magnitude and the other for the phase. For this particular example:

Magnitudes are $|a_1| = |a_2| = |a_{-1}| = |a_{-2}| = 0.25$ and
Phases are $\sphericalangle a_1 = \sphericalangle a_{-1} = 0, \quad \sphericalangle a_2 = -\sphericalangle a_{-2} = -0.5\pi$, which are plotted in Figure 6.4.

The simple example of Exercise 6.1 shows that it is possible to represent combinations of trigonometric functions in terms of the superposition of the harmonically related complex exponentials, using the Euler formula. What if we had a complicated function, which does not consist of any sines and cosines or does not have any known analytical forms?

WATCH Representation of a complicated function by superposition of harmonically related complex exponentials @ https://384book.net/v0602

WATCH Learn more about the Fourier series @ https://384book.net/v0603

6.3 Dirichlet Conditions

Motivating Question: Is it possible to represent any periodical signal as a superposition of infinitely many harmonics of complex exponentials?

When Fourier presented his original paper stating that **any** function can be represented in terms of superposition of harmonically related complex exponentials, he received objections from some of the mathematicians of that time, including his advisor Lagrange. Indeed, Fourier series representation does not exist for all periodic signals. This fact is shown by the famous student of Fourier, named P. G. Lejeune Dirichlet.

Dirichlet established the conditions for the existence of Fourier series representation of a continuous time periodic function.

Dirichlet conditions for the existence of a Fourier series of a **continuous time periodic** functions are summarized here:

Condition 1. The function $x(t)$ must be absolutely integrable in a finite interval. Formally, we should have

$$\int_T |x(t)|dt < \infty. \tag{6.16}$$

Exercise 6.2 Does the following signal satisfy Condition 1?

$$x(t) = \ln(t) \quad \text{for} \quad 0 \le t \le 1,$$

and $x(t)$ is periodic: $x(t) = x(t + T)$ for $T = 1$.

Solution

No, because the absolute integral of this function is not finite:

$$\int_0^1 |\ln(t)|dt \to \infty. \tag{6.17}$$

The plot of this function is shown in Figure 6.5.

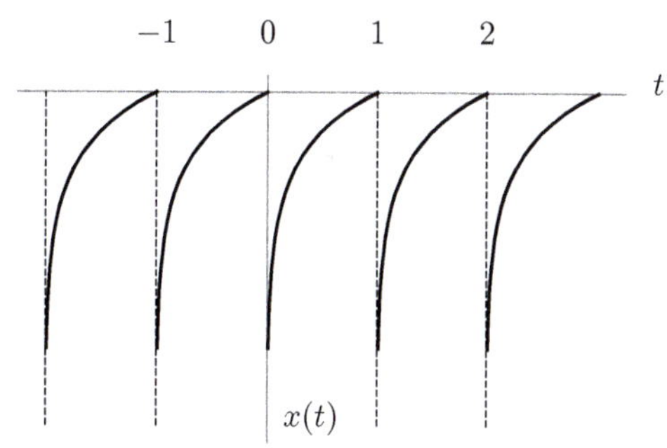

Figure 6.5 The plot of $x(t) = \ln(t)$, for $0 \le t \le 1$ and $x(t) = x(t + 1)$.

Condition 2. The function $x(t)$ must have bounded variation in any finite interval. That is, the number of minima and maxima should be bounded in any finite interval.

Exercise 6.3 Does the following signal satisfy Condition 2?

$$x(t) = \sin\left(\frac{3\pi}{4t}\right) \text{ for } 0 \le t \le 1 \quad \text{and} \quad x(t) = x(t + T) \quad \text{for} \quad T = 1.$$

Solution

No, because the number of minima and maxima of this function in a finite period is ∞. The plot of this function is shown in Figure 6.6.

Condition 3. In a finite interval, there are to be finitely many discontinuities.

Exercise 6.4 Does the following function satisfy Condition 3?

$$x(t) = \begin{cases} 1, & t = \frac{1}{n} \text{ for any positive integer } n \\ 0, & \text{otherwise.} \end{cases} \tag{6.18}$$

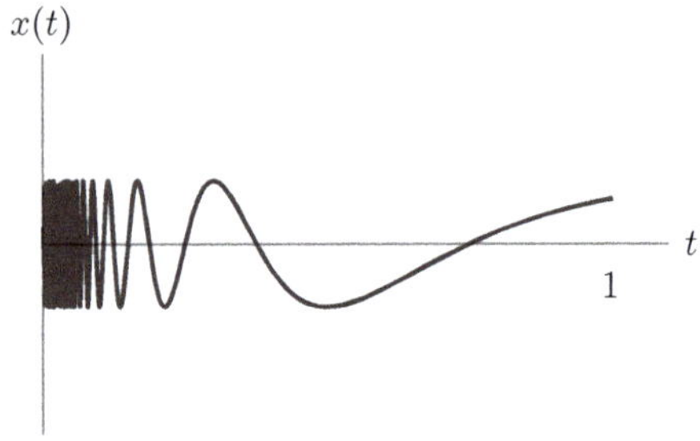

Figure 6.6 A function with infinitely many discontinuities in a finite interval.

Solution
Since there are infinitely many positive integers, this function switches between 0 and 1 infinitely many times. This implies an infinite number of discontinuities. Therefore, this function does not satisfy Condition 3.

Fact: If a continuous time periodic function does not satisfy Dirichlet conditions, then, it is not possible to find a set of coefficients $\{a_k\}$ to represent this function by the superposition of the harmonically related complex exponentials.

6.4 Fourier Theorem

Based upon the aforementioned analyses and conditions we can state the following theorem:

Theorem 6.1 A periodic function, $x(t)$ with fundamental period, T, satisfying the Dirichlet conditions, can be represented as the superposition of harmonically related complex exponentials:

$$\boxed{x(t) = \sum_{k=-\infty}^{\infty} a_k e^{jk\omega_0 t}} \quad \text{Synthesis Equation} \tag{6.19}$$

The coefficients a_k are called the **Fourier series coefficients** or **spectral coefficients** and can be uniquely obtained from the following integral:

$$\boxed{a_k = \frac{1}{T}\int_T x(t)e^{-jk\omega_0 t}dt} \quad \text{Analysis Equation} \tag{6.20}$$

The limits of the integral cover one full period T of the periodic function $x(t)$.

Remark 6.1 When the function $x(t)$ does not satisfy the Dirichlet conditions, it is not possible to find the Fourier series coefficients, a_k, since the integral, which defines the coefficients, is not bounded or does not exist. Although the Dirichlet conditions are rather intuitive, the formal proof of the conditions is beyond the scope of this book.

6.4.1 Proof Sketch for the Fourier Theorem

Multiply both sides of the Fourier series representation equation by $e^{-jn\omega_0 t}$:

$$x(t)e^{-jn\omega_0 t} = \sum_{k=-\infty}^{\infty} a_k e^{jk\omega_0 t} e^{-jn\omega_0 t} = \sum_{-\infty}^{\infty} a_k e^{j(n-k)\omega_0 t}. \tag{6.21}$$

Integrate both sides of the equation in the interval (0, T):

$$\int_0^T x(t)e^{-jn\omega_0 t}dt = \sum_{k=-\infty}^{\infty} a_k \int_0^T e^{j(k-n)\omega_0 t}dt = Ta_k, \tag{6.22}$$

which gives us the synthesis equation of the Fourier theorem. In this representation, each a_k measures the amount of the kth harmonic in the function $x(t)$. The following remark is about the evaluation of the integral of Equation (6.22).

Remark 6.2 Harmonically related complex exponentials are **orthogonal to each other**. In other words, their inner product is

$$<e^{jn\omega_0 t}, e^{jk\omega_0 t}> = \int_0^T e^{jn\omega_0 t}e^{-jk\omega_0 t}dt = \begin{cases} T & \text{for } n = k, \\ 0 & \text{otherwise.} \end{cases} \tag{6.23}$$

As an exercise, you can try to prove Equation (6.23). (*Hint*: Handle $n = k$ and $n \neq k$ cases separately, and use the relationship between the angular frequency and the fundamental period, $\omega_0 = \frac{2\pi}{T}$).

6.4.2 Terminology

The coefficients a_k of the synthesis equation can be considered as measures of the frequency content of the function $x(t)$. Depending on the harmonic values k, the following terminology is commonly used for the Fourier series coefficients:

- Fourier series coefficients, a_k, for $k = \pm 1, 2, 3, \ldots \pm \infty$, measures the contribution of each frequency to build up the signal. They form a frequency spectrum of the signal. For this reason, they are called the **spectral coefficients.**
- The 0th spectral coefficient,

$$a_0 = \frac{1}{T}\int_T x(t)dt, \tag{6.24}$$

  is constant, showing the area under the curve of the function $x(t)$. For this reason, a_0 is called the **average term.**
- The spectral coefficients of the lowest frequency harmonics, a_k and a_{-k}, for $|k| = 1$,

$$a_{\pm 1} = \frac{1}{T}\int_T x(t)e^{\pm j\omega_0 t}dt, \tag{6.25}$$

  is called the **spectral coefficients of the fundamental frequency**.
- The rest of the frequencies of a_k for $|k| \geq 2$ are called the k**th harmonic.**

Fourier series representation of a function, $x(t)$, forms a rigorous framework for the development of digital technology, since it bridges the continuous and discrete time world, as we shall later see in the **Sampling Theorem** (Chapter 11). It has a great impact in many fields of science and engineering, whenever the frequency content of the functions provides us with useful information about the underlying physical phenomenon.

Motivating Question: What do the analysis and synthesis equation tell us about the function $x(t)$?

Let's give a glimpse of the meaning of Fourier series representation, by introducing a new space, called the Hilbert space.

INTERACTIVE

Explore Fourier series representation for continuous time signals @ https://384book.net/i0601

6.5 Frequency Domain and Hilbert Spaces

A Hilbert space is considered a vector space, spanned by functions, rather than vectors, where the distance between the functions is defined by the **inner products**.

Recall a vector space of dimension n, where we represent a vector, $\mathbf{x} \in V^n$, as the linear combination of the basis vectors as follows:

$$\mathbf{x} = \sum_{k=1}^{n} a_k \mathbf{e}_k. \tag{6.26}$$

Recall, also, that a_k's are called the coordinates of the vector $\mathbf{x} = [a_1\, a_2 \ldots .a_n]^T$, with respect to the set of basis vectors, $\{\mathbf{e}_k\}_{k=1}^{n}$.

In Equation (6.26), we may use the standard basis vectors, $\mathbf{e}_k = [0 \ldots . 1 \ldots . 0]^T$ for $k = 1, \ldots, n$, where the kth entry has value 1 and all other entries are zeros. The superscript, T, of the vectors indicates the vector transpose operation.

Remark 6.3 The total of n basis vectors of a vector space V^n, are to be orthogonal to each other, in order to span the entire vector space. In this case, the basis vectors are called linearly independent. It is easy to show that the standard basis vectors, $\mathbf{e}_k$, are orthogonal to each other. In other words, the inner product should satisfy the following condition:

$$<\mathbf{e}_k, \mathbf{e}_j> = \mathbf{e}_k^T \mathbf{e}_j = \begin{cases} 1 & \text{for } j = k, \\ 0 & \text{otherwise.} \end{cases} \tag{6.27}$$

The total of n standard basis vectors, $\mathbf{e}_k, \forall k = 1, \ldots, n$ span the n-dimensional Euclidean vector space.

Motivating Question: What does orthogonality mean?

In geometry, orthogonality is defined as the perpendicularity of two lines. This property is generalized to other mathematical objects, such as sets, graphs, functions, matrices, and vectors. Loosely speaking, orthogonality assures that there are no interference, dependence or correlation among the orthogonal objects. Given a class of mathematical objects, an object cannot be represented in terms of other objects, which are orthogonal to that object.

Let us study the concept of orthogonality in the following simple exercise.

Exercise 6.5 Consider the vector $\mathbf{x} \in R^2$ of Figure 6.7, represented by the coordinates, $\mathbf{x} = [a_1 \quad a_2]$ with respect to the standard basis vector, $\mathbf{e}_1 = [1 \quad 0]$, $\quad \mathbf{e}_2 = [0 \quad 1]$.

a) Find the inner product between $\mathbf{x}$ and $\mathbf{e}_1$, and between $\mathbf{x}$ and $\mathbf{e}_2$.
b) Find the projection of $\mathbf{x}$ on $\mathbf{e}_1$ and $\mathbf{e}_2$.
c) Find the projection of $\mathbf{e}_1$ on $\mathbf{e}_2$.

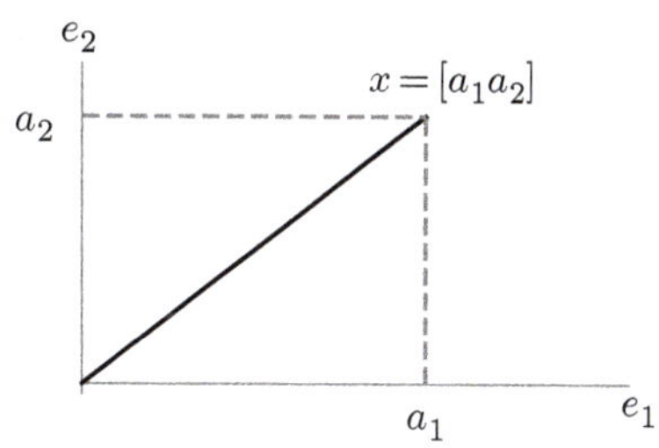

Figure 6.7 A two-dimensional Euclidean vector space, called two-tuples. A vector $\mathbf{x} \in R^2$ is represented by its coordinates a_1, a_2 with respect to the orthogonal basis vectors, $\mathbf{e}_1 = [1\,0]$ and $\mathbf{e}_2 = [0\,1]$.

Solution

a) The coordinates a_1 and a_2 are the inner product of the vector $\mathbf{x}$ with the basis vectors, $\mathbf{e}_1$ and $\mathbf{e}_2$. Mathematically,

$$a_1 = <\mathbf{x}, \mathbf{e}_1> = \mathbf{x}^T \mathbf{e}_1, \tag{6.28}$$

and

$$a_2 = <\mathbf{x}, \mathbf{e}_2> = \mathbf{x}^T \mathbf{e}_2. \tag{6.29}$$

b) The inner products of $<\mathbf{x}, \mathbf{e}_1>$ and $<\mathbf{x}, \mathbf{e}_2>$ are actually the projections of the vector $\mathbf{x}$ on the basis vectors, $\mathbf{e}_1$ and $\mathbf{e}_2$.

Motivating Question: What is the meaning of projection?
The projections of $\mathbf{x}$ on the basis vectors, which are the coordinates of a_1 and a_2 are the measures of the amount of $\mathbf{x}$ in the basis vectors, $\mathbf{e}_1$ and $\mathbf{e}_2$.

c) The projection of $\mathbf{e}_1$ on $\mathbf{e}_2$ is

$$<\mathbf{e}_1, \mathbf{e}_2> = \mathbf{e}_1^T \mathbf{e}_2 = 0.$$

Motivating Question: What is the amount of the basis vector $\mathbf{e}_1$ in $\mathbf{e}_2$?
The vector, $\mathbf{e}_1$ **is not** measured in the vector $\mathbf{e}_2$.
If a mathematical object is orthogonal to other mathematical objects, then this object cannot be observed or measured in the other objects. This property of vectors, in a vector space, is called **linear independence.** If a set of mathematical objects are linearly independent, one of them cannot be represented in terms of the others.

WATCH

Learn more about vector spaces @ https://384book.net/v0604

Let us, now, compare

$$\mathbf{x} = \sum_{k=1}^{n} a_k \mathbf{e}_k, \tag{6.30}$$

to the Fourier series representation of a function,

$$x(t) = \sum_{k=-\infty}^{\infty} a_k e^{jk\omega_0 t}, \tag{6.31}$$

There is a marvelous resemblance between the representations of Equations (6.30) and (6.31). Representation of a vector $\mathbf{x}$ in an n-dimensional Euclidean space looks very similar to the representation of a function $x(t)$ in a space spanned by the harmonically related complex exponentials of infinite dimension. This representation brings us to an entirely new space, called a function space, introduced by **David Hilbert**. Thus, it is named after him, as **Hilbert space**. In this new space, the coordinates of a function are the Fourier series coefficients, $\{a_k\}_{k=-\infty}^{\infty}$, and the basis functions are harmonically related complex exponentials, $\{e^{jk\omega_0 t}\}_{k=-\infty}^{\infty}$, which are orthogonal to each other. Thus, harmonically related complex exponentials span an infinite dimensional Hilbert space. Each vector in a Hilbert space is a function, $x(t)$, with the coordinates, $\{a_k\}_{k=-\infty}^{\infty}$.

Thus, the concept of vector space can be easily extended to Hilbert Space, where the bases of this infinite-dimensional space are functions. In other words, Hilbert space generalizes the Euclidean space. It extends the methods of linear algebra from the finite-dimensional vector spaces to the spaces with infinite number of dimensions.

Remark 6.4 It is possible to define Hilbert spaces spanned by functions other than the harmonically related complex exponentials. These Hilbert spaces, such as wavelet, Hadamard, Haar, sine, and cosine spaces, are beyond the scope of this book.

A loose definition of Hilbert space is given in the following text.

Definition 6.1 A Hilbert space H is a vector space, spanned by a set of orthogonal functions, $\Psi_k(t)$ for integer k, where a function $x(t)$ corresponds to a vector, in this space. The distance between two functions, $x(t), y(t) \in H$ are defined by the inner product on an interval $[a, b]$,

$$<x(t), y(t)> = \int_a^b x(t)y^*(t)dt, \tag{6.32}$$

where [*] indicates the complex conjugate operation.

A signal, which is represented in the time domain as a continuous time periodic function, $x(t)$ can be equivalently represented as a vector in a Hilbert space by its coordinates $\{a_k\}_{k=-\infty}^{\infty}$. This particular Hilbert space, spanned by infinitely many harmonically related complex exponential basis functions is called as the **frequency domain**. Thus, we have two equivalent representations of a continuous time periodic function, in time and frequency domains:

$$x(t) \longleftrightarrow a_k. \tag{6.33}$$

In time domain, a signal is represented by a function, where the domain of the function is time, whereas in the frequency domain a signal is represented by the coordinates $\{a_k\}$ of harmonically related frequencies, called spectral coefficients. Notice that since the function $x(t)$ is continuous, the time variable is a real number, $t \in \mathbb{R}$. On the other hand, the spectral coefficients have integer harmonics, $k \in \mathbb{I}$.

There are beautiful properties of Hilbert spaces, which is beyond the scope of this book.

WATCH

Learn more about the Hilbert space and inner products @ https://384book.net/v0605

Exercise 6.6 Consider the signal represented by the following function:

$$x(t) = 1 + \cos(\omega_0 t). \tag{6.34}$$

Find the coordinates of $x(t)$ in Hilbert space, spanned by the harmonically related complex exponential functions. Plot a_k vs. k.

Solution

The coordinates of a periodic function in Hilbert space are the Fourier series coefficients. The Euler formula provides us with the representation of a periodic function in terms of harmonically related complex exponentials:

$$x(t) = 1 + \frac{1}{2}\left(e^{j\omega_0 t} + e^{-j\omega_0 t}\right). \tag{6.35}$$

Comparing this to the Fourier synthesis equation,

$$x(t) = \sum_{k=-\infty}^{\infty} a_k e^{jk\omega_0 t} = 1 + \frac{1}{2}(e^{j\omega_0 t} + e^{-j\omega_0 t}), \tag{6.36}$$

and equating the coefficients of harmonic exponential functions with the same harmonics k in both sides of the equation, we obtain the Fourier series coefficients as $a_0 = 1$, $a_1 = a_{-1} = \frac{1}{2}$, and $a_k = 0$ for $|k| > 1$. The plot of a_k is given in Figure 6.8.

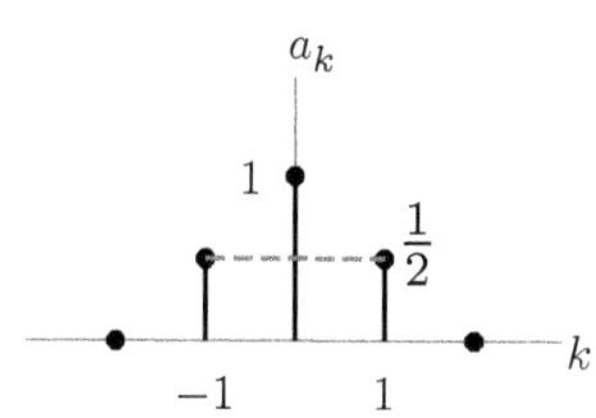

Figure 6.8 The plot of the spectral coefficients a_k vs. k in Exercise 6.6.

Exercise 6.7 Consider the signal represented by the following function:

$$x(t) = 1 + \cos(\omega_0 t) + \sin(\omega_0 t). \tag{6.37}$$

Find the coordinates of $x(t)$ in Hilbert space, where the basis functions are harmonically related complex exponential functions. Plot the coordinates, a_k, vs. k.

Solution

As we did in the previous example, we use the Fourier series synthesis equation together with the Euler formula:

$$x(t) = 1 + \frac{1}{2}(e^{j\omega_0 t} + e^{-j\omega_0 t}) + \frac{1}{2j}(e^{j\omega_0 t} - e^{-j\omega_0 t}) \tag{6.38}$$

$$= 1 + \frac{1}{2}(1 - j)e^{j\omega_0 t} + \frac{1}{2}(1 + j)e^{-j\omega_0 t} = \sum_{k=-\infty}^{\infty} a_k e^{jk\omega_0 t}. \tag{6.39}$$

Equating the coefficients of the exponential functions with the same harmonics, we obtain $a_0 = 1, a_1 = \frac{1}{2}(1 - j), a_{-1} = \frac{1}{2}(1 + j)$ and $a_k = 0$ for $|k| > 1$.

Note: The coordinates of the function in Hilbert space are complex numbers.

Recall that complex numbers can be represented in two different coordinate systems:

1) **In Cartesian Coordinate System:** $a_k = \text{Re}\{a_k\} + j\text{Im}\{a_k\}$,
2) **In Polar Coordinate System:** $a_k = |a_k|e^{\theta_k}$, where the magnitude square of a_k is

$$|a_k|^2 = (\text{Re}\{a_k\})^2 + (\text{Im}\{a_k\})^2, \tag{6.40}$$

and the phase of a_k is

$$\theta_k = \tan^{-1} \frac{\text{Re}\{a_k\}}{\text{Im}\{a_k\}}. \tag{6.41}$$

Let us plot the spectral coefficients a_k vs. k in the polar coordinate system. In this case, we need two plots:

1) **Magnitude spectrum**, which is the plot of the magnitude of Fourier series coefficients $|a_1| = |a_{-1}| = \frac{1}{\sqrt{2}}$, $a_0 = 1$.
2) **Phase spectrum**, which is the plot of the phase of the Fourier series coefficients, $\theta_1 = -\frac{\pi}{4}$, $\theta_{-1} = \frac{\pi}{4}$, and $\theta_0 = 0$.

The magnitude and phase spectrum are given in Figure 6.9.

Exercise 6.8 Consider a periodic signal represented by the following equation:

$$x(t) = \begin{cases} 1, & \text{if } |t| < T_0, \\ 0, & \text{if } T_0 \le |t| \le T - T_0, \end{cases} \tag{6.42}$$

and $x(t) = x(t + T)$. Signals of such form are called **pulse trains** (Figure 6.10). Find and plot the Fourier series coefficients.

Solution

Since the signal $x(t)$ does not consist of trigonometric functions, it is not possible to apply Euler formula to find the coefficients of harmonically related complex exponentials. Thus, we use the Fourier analysis equation to compute the average term,

$$a_0 = \frac{1}{T} \int_{-T_0}^{T_0} dt = \frac{2T_0}{T}, \tag{6.43}$$

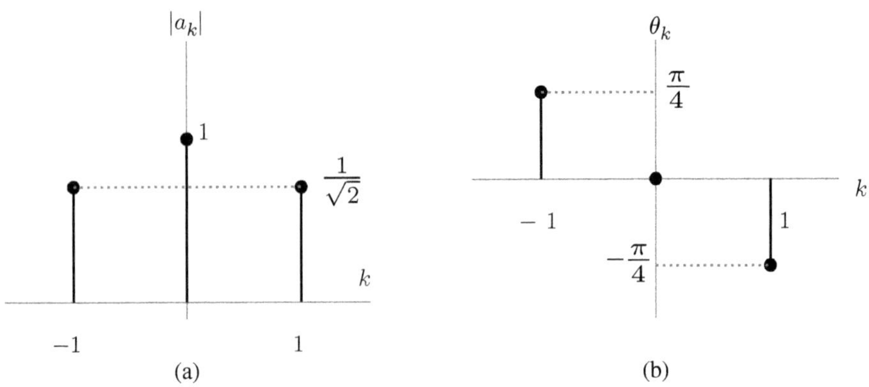

Figure 6.9 Plot of magnitude spectrum (a) and phase spectrum (b) of the Fourier series coefficients of $x(t) = 1 + \cos(\omega_0 t) + \sin(\omega_0 t)$.

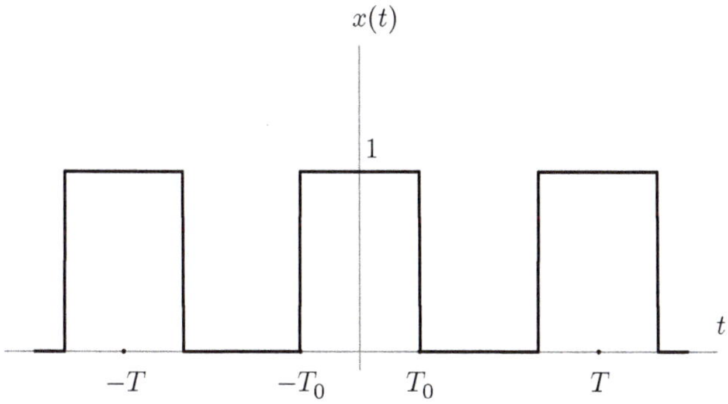

Figure 6.10 A periodic function, called pulse train, which repeats itself at every period, T.

and the spectral coefficient of kth harmonic,

$$\begin{aligned} a_k &= \frac{1}{T}\int_{-T_0}^{T_0} e^{-jk\omega_0 t}dt = \frac{1}{jk\omega_0 T}\left(e^{jk\omega_0 T_0} - e^{-jk\omega_0 T_0}\right) \\ &= \frac{1}{\pi k}\sin(k\omega_0 T_0) = \frac{1}{\pi k}\sin\left(2\pi k\frac{T_0}{T}\right), \end{aligned} \tag{6.44}$$

where the angular frequency is $\omega_0 = \frac{2\pi}{T}$.

Motivating Question: What is the effect of the fundamental period T of $x(t)$ on the spectral coefficients a_k?

In order to answer this question, let us plot the spectral coefficients for three cases of the fundamental period:

Case 1: For $T = 4T_0 \rightarrow a_k = \frac{1}{\pi k}\sin(k\pi/2)$.

Case 2: For $T = 8T_0 \rightarrow a_k = \frac{1}{\pi k}\sin(k\pi/4)$.

Case 3: For $T = 16T_0 \rightarrow a_k = \frac{1}{\pi k}\sin(k\pi/8)$.

Figure 6.11 shows the spectrum of a_k vs. k for $T = 4T_1$, $T = 8T_1$, and $T = 16T_1$. When you compare the spectral coefficients of the function $x(t)$, in the Fourier domain, what do you observe?

Recall that, the spectral coefficient, a_k, for each k shows the amount of the corresponding harmonic frequency in the signal, $x(t)$. For small periods, e.g. $T = 4T_0$, the signal $x(t)$ has relatively less low-frequency components, compared to the other signals. As we increase the period of the signal, the low-frequency components increase, and the rate of change of the spectral coefficients

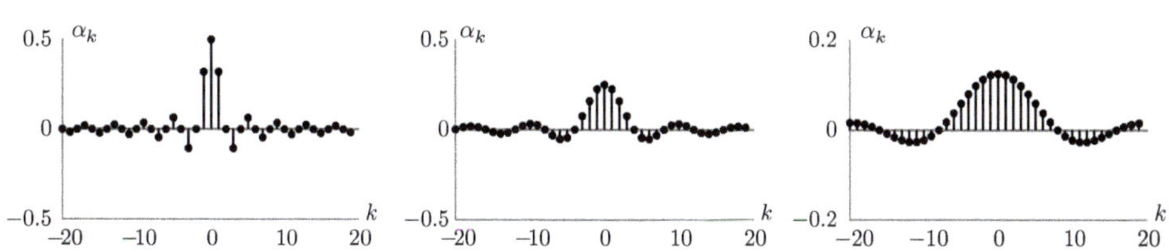

Figure 6.11 Plot of Fourier spectrum a_k vs. k, for $T = 4T_0$ $T = 8T_0$ and $T = 16T_0$, for the signal in Exercise 6.8.

decreases. Investigation of the behavior of the spectral coefficients shows the amount of each frequency component relative to each other, which makes the signal. This is why we call the plot, a_k vs. k as **spectrum**, meaning the band of frequencies, in $x(t)$.

6.6 Response of a Linear Time-Invariant System to the Continuous Time Complex Exponential Input Signal

Until now, we studied various interesting properties of the complex exponential function, $x(t) = e^{j\omega_0 t}$, which can be summarized as follows:

- Complex exponential functions can be written in terms of the trigonometric functions through the Euler formula, $e^{j\omega_0 t} = \cos\omega_0 t + j\sin\omega_0 t$.
- Complex exponentials are periodic functions with period $T = 2\pi/\omega_0$.
- Harmonically related complex exponentials, $\phi_k(t) = e^{jk\omega_0 t}$ represent periodic motions of different speeds on a unit circle, defined in the complex plane. The speed of the motion depends on the harmonics $k\omega_0$.
- Harmonically related complex exponentials are orthogonal to each other, i.e., their inner product is

$$<e^{jn\omega_0}, e^{jk\omega_0}> = \int_0^T e^{jn\omega_0 t} e^{-jk\omega_0 t} dt = \begin{cases} T & \text{for } n = k, \\ 0 & \text{otherwise.} \end{cases} \tag{6.45}$$

- Infinitely many harmonically related complex exponentials span a Hilbert space of functions, called frequency domain.
- A periodic function can be uniquely represented in terms of the superposition of the harmonically related complex exponentials, called Fourier series representation,

$$x(t) = \sum_{k=-\infty}^{\infty} a_k e^{jk\omega_0 t}, \tag{6.46}$$

where

$$a_k = \frac{1}{T}\int_T x(t)e^{-jk\omega_0 t} dt. \tag{6.47}$$

In this section, we observe one more property of the complex exponential function from the systems point of view: When a complex exponential is fed at the input of an LTI system, we obtain a response, which is just the scaled version of the input. The functions satisfying this property are called the **eigenfunction** of an LTI system. We hinted at this behavior when we described the "transfer function" of LTI systems represented by differential and difference equations in Chapter 5.

Definition 6.2 Eigenfunction of a continuous time LTI system, defined on a Hilbert Space, is a nonzero input function $\Phi(t)$, which outputs the scaled input $\Phi(t)$. Mathematically speaking, the eigenfunction, $\Phi(t)$, of an LTI system, represented by an operator, $\mathbf{L}$, satisfies the following equation:

$$\mathbf{L}\Phi(t) = \Lambda\Phi(t), \tag{6.48}$$

where the scalar value, Λ is called the **eigenvalue** of the continuous time LTI system, represented by the operator, $\mathbf{L}$.

Recall the definition of the eigenvalues and eigenvectors of matrices, in linear algebra: Given a matrix, **A**, its eigenvectors, ξ, and eigenvalues, λ, satisfy the following equation, called the characteristic equation:

$$\mathbf{A}\xi = \lambda\xi. \tag{6.49}$$

Characteristic equation states that an $N \times N$ matrix operator, **A**, can be represented by a simple scalar, λ, when it is multiplied by an eigenvector, ξ, of that matrix. In other words, instead of multiplying the matrix A by one of its eigenvectors, we simply multiply it with the corresponding eigenvalue to get the same result of multiplication. The scalar eigenvalue λ replaces the entire N^2 elements of the matrix when it is multiplied with an eigenvector. That is why we call the λ value and the corresponding vector as **eigen,** which is a German word, meaning "**own**," in English. The scalar λ characterizes a matrix with N^2 entries on its own, in a very compact form, when the matrix is multiplied by its own (eigen) vectors.

Recall from linear algebra that the eigenvectors of a matrix are orthogonal to each other and they form a basis of a vector space, where the entries of each vector in this space are the coordinates of the vector with respect to the basis vectors.

6.6.1 Eigenfunctions and Eigenvalues of a Continuous Time LTI Systems

In this section, we are going to show that the complex exponentials are the **eigenfunctions** of a continuous time LTI system, represented by the impulse response, $h(t)$, via convolution integral,

$$y(t) = \int_{-\infty}^{\infty} h(\tau)x(t-\tau)d\tau, \tag{6.50}$$

and we are going to find the corresponding **eigenvalue** of this system.

Let us start by replacing the input by $x(t) = e^{j\omega_0 t}$, in the convolution integral:

$$y(t) = \int_{-\infty}^{\infty} h(\tau)e^{j\omega_0(t-\tau)}d\tau, \tag{6.51}$$

where $\omega_0 = 2\pi/T_0$ is the angular frequency of the periodic input. The output of the LTI system to the input, $x(t) = e^{j\omega_0 t}$, can be written as:

$$y(t) = e^{j\omega_0 t}\int_{-\infty}^{\infty} h(\tau)e^{-j\omega_0\tau}d\tau. \tag{6.52}$$

Note that the integral,

$$H(j\omega_0) = \int_{-\infty}^{\infty} h(\tau)e^{-j\omega_0\tau}d\tau, \tag{6.53}$$

is just a scaling factor of the input, $x(t) = e^{j\omega_0 t}$, of the LTI system.

Therefore, when the input is a complex exponential, $x(t) = e^{j\omega_0 t}$, the corresponding output, $y(t) = H(j\omega_0)e^{j\omega_0 t}$, is just the scaled form of the input.

Definition 6.3 Eigenvalue of a continuous time LTI system for a complex exponential input $x(t) = e^{j\omega_0 t}$ is defined as:

$$H(j\omega_0) = \int_{-\infty}^{\infty} h(\tau)e^{-j\omega_0\tau}d\tau, \tag{6.54}$$

where $\omega_0 = 2\pi/T_0$ is the angular frequency of the periodic input.

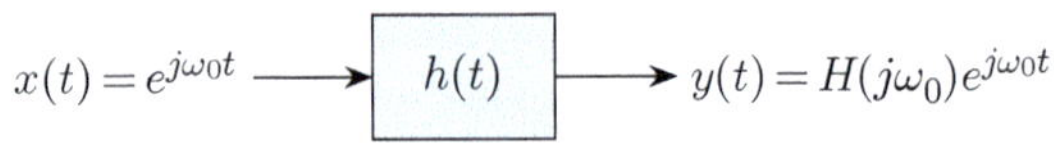

Figure 6.12 An LTI system, represented by $h(t)$, receives a complex exponential, $x(t) = e^{j\omega_0 t}$, as input and outputs the same complex exponential scaled by $H(j\omega_0)$.

Remark 6.5 A LTI system has eigenvalues for each harmonic of the complex exponential. Formally speaking, when the input is an eigenfunction, which corresponds to the kth harmonic of the complex exponential function,

$$x(t) = e^{jk\omega_0 t}, \tag{6.55}$$

the corresponding output is

$$y(t) = H(jk\omega_0)e^{jk\omega_0 t}, \tag{6.56}$$

where $H(jk\omega_0)$ is the kth eigenvalue of the LTI system, defined as:

$$H(jk\omega_0) = \int_{-\infty}^{\infty} h(\tau)e^{-jk\omega_0 \tau}\,d\tau. \tag{6.57}$$

This input output relation is illustrated in Figure 6.12.

Remark 6.6 The eigenfunctions of an LTI system are the harmonically related complex exponentials, which span a Hilbert space. Each periodic function, $x(t)$, in this space can be represented by the superposition of the basis functions, $\Phi_k(t) = e^{jk\omega_0 t}$, as shown in the Fourier theorem. The Fourier series coefficients are the coordinates of this function with respect to the harmonically related complex exponentials.

Remark 6.7 When we set $\lambda = jk\omega_0$, the exponential input of Equation (6.55) becomes $x(t) = e^{\lambda t}$, and the corresponding eigenvalue becomes the transfer function of an LTI system, defined in Section 5.4.3.

Exercise 6.9 Find an eigenvalue and impulse response of the following continuous time system:

$$y(t) = x(t-2). \tag{6.58}$$

Solution
Since this is an LTI system, an eigenfunction of this system is the complex exponential input:

$$x(t) = e^{j\omega_0 t}. \tag{6.59}$$

Inserting this input into the system equation, we get

$$y(t) = e^{j\omega_0(t-2)} = e^{j\omega_0 t}e^{-2j\omega_0} = H(j\omega_0)e^{j\omega_0 t}. \tag{6.60}$$

The multiplicative term in the right-hand side of the Equation (6.60),

$$H(j\omega_0) = e^{-2j\omega_0}, \tag{6.61}$$

is the eigenvalue of the LTI system.

The impulse response can be easily obtained from the convolution equation,

$$y(t) = x(t) * h(t) = x(t-2). \tag{6.62}$$

Thus, the impulse response is $h(t) = \delta(t-2)$.

Indeed, we have

$$H(j\omega_0) = \int_{-\infty}^{\infty} \delta(\tau - 2)e^{-j\omega_0\tau}d\tau = e^{-2j\omega_0}. \tag{6.63}$$

Note that the function $H(jk\omega_0)$ is an eigenvalue of the LTI system, when the input is the eigenfunction $x(t) = e^{jk\omega_0}$, for each integer value of $k = 0, \pm1, \pm2, \pm3 \ldots$. Hence, an LTI system has infinitely many eigenvalues for infinitely many harmonically related complex exponentials.

6.7 Convergence of the Fourier Series and Gibbs Phenomenon

In theory, the Fourier series representation of a function has infinitely many terms under the summation, each of which corresponds to a coordinate a_k multiplied by complex exponential harmonics. For some functions, such as $x(t) = \cos\omega_0 t$ or $x(t) = \sin\omega_0 t$, only finitely many coefficients are nonzero. Thus, we avoid infinite summations.

Unfortunately, most of the signals have infinitely many spectral coefficients. In order to find the Fourier series representation of these types of functions, we use some approximation techniques. A practical way is to truncate the series after a relatively high value of $k = N$ and compute an approximation of the Fourier series representation:

$$x_N(t) = \sum_{k=-N}^{N} a_k e^{jk\omega_0 t}. \tag{6.64}$$

As we increase the number of the terms, N, the function gets better approximated by the series sum. The error between the theoretical and practical computation can be defined as:

$$e_N(t) = |x(t) - x_N(t)|. \tag{6.65}$$

Thus, when we compute the spectral parameters a_k, we need to pay attention to reduce the energy of the error,

$$E_N = \int_{-\infty}^{\infty} |e_N(t)|^2 dt, \tag{6.66}$$

as much as possible.

We expect that as we increase N, the energy of the error gets smaller. However, there is a peculiar behavior of the Fourier series while approximating the functions, which have discontinuities. As we increase the number of terms, the width of the ripples around the discontinuities decreases and converges to a constant value of oscillation around the discontinuities. This behavior is known as the **Gibbs phenomenon**, discovered by Henry Wilbraham and J. Willard Gibbs. An example is illustrated in Figure 6.13.

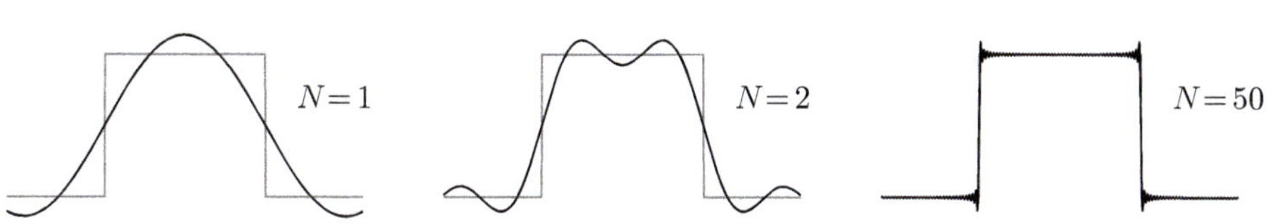

Figure 6.13 Gibbs phenomenon. From left to right, we are approximating a square pulse signal (shown with gray color) using $N = 1$, $N = 2$, and $N = 50$ harmonics. Although the Fourier theorem shows that the series representation perfectly represents any function, satisfying the Dirichlet conditions, even for very large sums, the approximated function oscillates around the discontinuities.

INTERACTIVE

Explore the Gibbs phenomenon @ https://384book.net/i0602

6.8 Properties of Fourier Series for Continuous Time Functions

Fourier series representation of periodic functions in **Hilbert space**, called the **frequency domain** has a wide range of interesting properties. These properties not only link the time domain and frequency domain; they also enable us to solve many problems, which are not mathematically tractable in time domain. In addition, the frequency domain representation of signals enables us to observe properties of signals and systems, which are not possible to observe in the time domain.

The most crucial characteristic of the relationship between the time and frequency domain is that the representation of functions in these two separate domains is one-to-one and onto,

$$x(t) \longleftrightarrow a_k. \tag{6.67}$$

In other words, if a periodic function, $x(t)$, satisfies the Dirichlet conditions, we can uniquely obtain its Fourier series representation by finding the spectral coefficients using the analysis equation of Fourier Theorem:

$$a_k = \frac{1}{T}\int_T x(t)e^{-jk\omega_0 t}dt. \tag{6.68}$$

Equivalently, if the spectral coefficients, $\{a_k\}, \forall k$, are given, we can obtain the time domain representation of the function, $x(t)$, uniquely, using the synthesis equation:

$$x(t) = \sum_{k=-\infty}^{\infty} a_k e^{jk\omega_0 t}. \tag{6.69}$$

Let us briefly outline the relationships among the time and frequency domain representation of functions and their properties.

6.8.1 Linearity Property

Finding the Fourier series representation of a function, $x(t)$, is a linear operation. Mathematically speaking, given the time and frequency domain representations of two signals,

$$x(t) \longleftrightarrow a_k \quad \text{and} \quad y(t) \longleftrightarrow b_k. \tag{6.70}$$

The following superposition property holds in time and frequency domains:

$$Ax(t) + By(t) \longleftrightarrow Aa_k + Bb_k. \tag{6.71}$$

Linearity property follows from the linearity of the integration operation, which defines the analysis and synthesis equations.

6.8.2 Time Shifting Property

A time shift of the function, $x(t)$ by the amount of t_0, is equivalent to the multiplication of its spectral coefficients by the complex exponential, $e^{-jk\omega_0 t_0}$. Mathematically, if

$$x(t) \longleftrightarrow a_k \quad \text{and} \quad y(t) = x(t - t_0) \longleftrightarrow b_k, \tag{6.72}$$

then,

$$b_k = e^{-jk\omega_0 t_0} a_k. \tag{6.73}$$

Time shifting property can be easily shown by inserting the shifted signal into the analysis equation:

$$y(t) \longleftrightarrow b_k = \frac{1}{T}\int_T x(t - t_0)e^{-jk\omega_0 t}dt. \tag{6.74}$$

Let us replace the dummy variable of the integral by $t' = t - t_0 =$. Then, the analysis equation becomes

$$b_k = \frac{1}{T}\int_T x(t')e^{-jk\omega_0(t_0+t')}dt = e^{-jk\omega_0 t_0}\frac{1}{T}\int_T x(t')t_0 e^{-jk\omega_0 t'}dt' = e^{-jk\omega_0 t_0}a_k. \tag{6.75}$$

6.8.3 Time Scale Property

Scaling the time of a function, $x(t)$, does not change its spectral coefficients, a_k, but the fundamental frequency of the spectral coefficient is scaled to $\alpha\omega_0$.

Formally speaking, the Fourier series representation of $x(\alpha t)$ is defined as:

$$x(\alpha t) = \sum_{k=-\infty}^{\infty} a_k e^{jk(\alpha\omega_0)t}. \tag{6.76}$$

This synthesis equation shows that the spectral coefficients of the scaled function $x(\alpha t)$ are the same as the spectral coefficients of the signal $x(t)$. However, the angular frequency is scaled by α. Hence, the fundamental period of $x(\alpha t)$ is $T = \frac{2\pi}{\alpha\omega_0}$.

6.8.4 Time Reversal Property

Reverse of the time of a function $x(t)$ is equivalent to taking the time-scaling parameter $\alpha = -1$. Hence,

$$x(-t) = \sum_{k=-\infty}^{\infty} a_k e^{-jk\omega_0 t}. \tag{6.77}$$

This equation reverses the harmonics of the spectral coefficients. Mathematically speaking, if

$$x(t) \longleftrightarrow a_k \quad \text{and} \quad y(t) = x(-t) \longleftrightarrow b_k, \tag{6.78}$$

then,

$$b_k = a_{-k}. \tag{6.79}$$

6.8.5 Convolution Property

Convolution operation in time domain corresponds to the multiplication operation in the frequency domain.

Given two functions and their corresponding spectral coefficients,

$$x(t) \longleftrightarrow a_k \quad \text{and} \quad y(t) \longleftrightarrow b_k, \tag{6.80}$$

with period T, the convolution of $x(t) * y(t)$ has the following spectral coefficients:

$$\begin{aligned} x(t) * y(t) \longleftrightarrow c_k &= \frac{1}{T}\int_{-\infty}^{\infty} (x(t) * y(t))e^{-jk\omega_0 t}dt \\ &= \frac{1}{T}\int_{-\infty}^{\infty}\int_{-\infty}^{\infty} x(\tau)y(t-\tau)e^{-jk\omega_0 t}d\tau dt. \end{aligned} \tag{6.81}$$

Changing the dummy variable of integral to $t' = t - \tau$, we obtain,

$$c_k = \frac{1}{T}\int_{-\infty}^{\infty} x(\tau)e^{-jk\omega_0 t}d\tau \int_{-\infty}^{\infty} y(t')e^{-jk\omega_0 t'}dt'. \tag{6.82}$$

The aforementioned first and second integrals are equivalent to Ta_k and Tb_k, respectively. Hence, the spectral coefficients of two convoluted periodic signals, $x(t) * y(t)$, is the multiplication of their spectral coefficients, scaled by the fundamental period:

$$x[n] * y[n] \longleftrightarrow Ta_k b_k. \tag{6.83}$$

6.8.6 Multiplication Property

Multiplication of two functions in the time domain is equivalent to the convolution of their spectral coefficients in the frequency domain.

Mathematically speaking, given two functions and their corresponding spectral coefficients,

$$x(t) \longleftrightarrow a_k \quad \text{and} \quad y(t) \longleftrightarrow b_k, \tag{6.84}$$

with period T, multiplication of the signals $x(t)$ and $y(t)$ can be written in terms of Fourier series representation as:

$$x(t)y(t) = \left(\sum_{k=-\infty}^{\infty} a_k e^{-j\frac{2\pi}{T}kt}\right)\left(\sum_{l=-\infty}^{\infty} b_l e^{-j\frac{2\pi}{T}lt}\right). \tag{6.85}$$

Arranging the summations and changing the dummy variable $l = m - k$ gives

$$\begin{aligned} x(t)y(t) &= \sum_{k=-\infty}^{\infty}\sum_{l=-\infty}^{\infty} a_k b_l e^{-j2\pi\frac{k+l}{T}}. \\ &= \sum_{k=-\infty}^{\infty}\sum_{m=-\infty}^{\infty} a_k b_{m-k} e^{-j2\pi\frac{m}{T}}. \end{aligned} \tag{6.86}$$

Therefore, the multiplication of signals $x(t)y(t)$ in time domain corresponds to the convolution of their spectral coefficients, a_k and b_k:

$$z(t) = x(t)y(t) \longleftrightarrow c_k = \sum_{\forall m} a_m b_{k-m}. \tag{6.87}$$

Considering the fact that the spectral coefficients, a_k vs. k and b_k vs. k, are discrete functions, the operation,

$$a_k * b_k = \sum_{\forall m} a_m b_{k-m}, \tag{6.88}$$

is the discrete convolution of a_k and b_k.

INTERACTIVE

An example on the duality of convolution and multiplication @ https://384book.net/i0603

6.8.7 Conjugate Symmetry

When $x(t)$ is a complex function, the spectral coefficients of its complex conjugate, $x^*(t)$, satisfy the conjugate symmetry property. Mathematically speaking,

$$\text{If} \quad x(t) \longleftrightarrow a_k, \quad \text{then} \quad x^*(t) \longleftrightarrow a^*_{-k}. \tag{6.89}$$

This property directly follows the Fourier series synthesis equation.

6.8.8 Parseval's Equality

The energy of a signal in time and frequency domain is preserved. Mathematically speaking,

$$\frac{1}{T}\int_T |x(t)|^2 dt = \sum_{\forall k} |a_k|^2. \tag{6.90}$$

We can show Parseval's equality by inserting the analysis equation into the left-hand side of the Equation (6.90):

$$\begin{aligned}\frac{1}{T}\int_T |x(t)|^2 dt &= \frac{1}{T}\int_T x(t)x^*(t)dt \\ &= \frac{1}{T}\int_T \sum_{k=-\infty}^{\infty} a_k e^{jk\omega_0 t} \sum_{l=-\infty}^{\infty} a_l^* le^{-jl\omega_0 t} dt.\end{aligned} \tag{6.91}$$

Recall that harmonically related complex exponential functions are orthogonal:

$$\int_0^T e^{jk\omega_0 t} e^{-l\omega_0 t} dt = \begin{cases} T & \text{for } l = k, \\ 0 & \text{otherwise.} \end{cases} \tag{6.92}$$

Using this fact and arranging the right-hand side of Equation (6.91), we obtain

$$\frac{1}{T}\int_T |x(t)|^2 dt = \frac{1}{T}\int_T \sum_{k=-\infty}^{\infty}\sum_{l=-\infty}^{\infty} a_k a_l^* e^{(k-l)j\omega_0 t} dt = \sum_{\forall k} |a_k|^2. \tag{6.93}$$

Parseval's equality reveals that the representation of signals in Hilbert space conserves the energy of the time domain.

6.8.9 Differentiation Property

The derivative of a function in the time domain corresponds to the multiplication of its spectral coefficients by $(jk\omega_0)$. Mathematically speaking,

$$\frac{dx(t)}{dt} \longleftrightarrow (jk\omega_0)a_k. \tag{6.94}$$

We can derive the differentiation property by taking the derivative of both sides of the synthesis equation with respect to t:

$$\frac{dx(t)}{dt} = \sum_{k=-\infty}^{\infty} (jk\omega_0)a_k e^{jk\omega_0 t}. \tag{6.95}$$

Notice that derivation operation in time domain corresponds to multiplication operation, in the frequency domain.

There are many interesting properties of the Fourier series representation other than the ones summarized earlier. For further studies, the reader is referred to the book, "*Fourier Analysis*" by Eric Stade (Wiley, 2005). We provide a short list of properties in Table 6.1.

We also provide some popular continuous time periodic functions and their spectral coefficients in Table 6.2. The reader is encouraged to derive the spectral coefficients given in the table, using the analysis equation.

Let us now solve some exercises to demonstrate the power of the properties of the Fourier series representation.

Table 6.1 Summary of the continuous time Fourier series properties.

Periodic signal	Fourier series coefficient
$x(t)$ is periodic with fundamental period T_0	a_k
$y(t)$ is periodic with fundamental period T_0	b_k
Synthesis equation: $\boxed{x(t) = \sum_{k=-\infty}^{\infty} a_k e^{jk\omega_0 t}}$ where $\omega_0 = \frac{2\pi}{T_0}$	Analysis equation: $\boxed{a_k = \frac{1}{T_0}\int_{T_0} x(t)e^{-jk\omega_0 t}dt}$
$Ax(t) + By(t)$	$Aa_k + Bb_k$
$x(t - t_0)$	$a_k e^{-jk\omega_0 t_0}$
$e^{jM\omega_0 t}x(t)$	a_{k-M}
$x^*(t)$	a^*_{-k}
$x(-t)$	a_{-k}
$x(\alpha t), \alpha > 0$ (periodic with fundamental period T_0/α)	a_k
$x(t) * y(t)$	$T_0 a_k b_k$
$x(t)y(t)$	$\sum_{l=-\infty}^{\infty} a_l b_{k-l}$
$\frac{d}{dt}x(t)$	$jk\omega_0 a_k$
$\int_{-\infty}^{t} x(\tau)d\tau$ (bounded and periodic only if $a_0 = 0$)	$\frac{1}{jk\omega_0}a_k$
For real-valued $x(t)$:	$\begin{cases} a_k & = a^*_{-k} \\ \mathrm{Re}\{a_k\} & = \mathrm{Re}\{a_{-k}\} \\ \mathrm{Im}\{a_k\} & = -\mathrm{Im}\{a_{-k}\} \\ \lvert a_k\rvert & = \lvert a_{-k}\rvert \\ \sphericalangle a_k & = -\sphericalangle a_{-k} \end{cases}$
Even part of $x(t)$	$\mathrm{Re}\{a_k\}$
Odd part of $x(t)$	$j\mathrm{Im}\{a_k\}$

$$a_k e^{jk\omega_0 t} + a_{-k}e^{-jk\omega_0 t} = 2\mathrm{Re}\{a_k\}\cos(k\omega_0 t) - 2\mathrm{Im}\{a_k\}\sin(k\omega_0 t)$$

$$\text{Parseval's relation: } \frac{1}{T_0}\int_{T_0} |x(t)|^2 dt = \sum_{k=-\infty}^{\infty} |a_k|^2$$

Exercise 6.10 Find the spectral coefficients of the following impulse train:

$$x(t) = \sum_{k=-\infty}^{\infty} \delta(t - kT), \tag{6.96}$$

where $T = 2\pi/\omega_0$ is the fundamental period.

Solution

Apply the analysis equation to cover one full period of the signal $x(t)$,

$$a_k = \frac{1}{T}\int_{-T/2}^{T/2} \delta(t)e^{-jk\omega_0 t} = \frac{1}{T}. \tag{6.97}$$

The signal and its spectral coefficients are plotted in Figure 6.14.

Table 6.2 Some popular continuous time periodic signals and their spectral coefficients.

Periodic signal	Fourier series coefficient
$x(t) = \sum_{n=-\infty}^{\infty} \delta(t - nT)$	$a_k = \frac{1}{T}$ for all k
$x(t) = 1$	$a_0 = 1, a_k = 0$ for all other k, $\forall T_0 > 0$
$x(t) = e^{j\omega_0 t}$	$a_1 = 1, a_k = 0$ for all other k
$x(t) = \cos(\omega_0 t)$	$a_1 = a_{-1} = \frac{1}{2}, a_k = 0$ for all other k
$x(t) = \sin(\omega_0 t)$	$a_1 = -a_{-1} = \frac{1}{2j}, a_k = 0$ for all other k
$x(t) = \begin{cases} 1, & \lvert t\rvert < T_1 \\ 0, & T_1 < \lvert t\rvert \leq \frac{T_0}{2} \end{cases}$ with fundamental period T_0	$a_k = \frac{\omega_0 T_1}{\pi}\text{sinc}\frac{k\omega_0 T_1}{\pi} = \sin\frac{k\omega_0 T_1}{k\pi}$
$x(t) = \begin{cases} 1, & 0 < t < \pi \\ -1, & -\pi < t < 0 \end{cases}$	$\frac{4}{\pi}\left(\frac{\sin t}{1} + \frac{\sin 3t}{3} + \frac{\sin 5t}{5} + \cdots\right)$
$x(t) = \begin{cases} t, & 0 < t < \pi \\ -t, & -\pi < t < 0 \end{cases}$	$\frac{\pi}{2} - \frac{4}{\pi}\left(\frac{\cos t}{1^2} + \frac{\cos 3t}{3^2} + \frac{\cos 5t}{5^2} + \cdots\right)$
$x(t) = t, -\pi < t < \pi$	$2\left(\frac{\sin t}{1} + \frac{\sin 3t}{3} + \frac{\sin 5t}{5} + \cdots\right)$
$x(t) = t,\ 0 < t < 2\pi$	$\pi - 2\left(\frac{\sin t}{1} + \frac{\sin 3t}{3} + \frac{\sin 5t}{5} + \cdots\right)$
$x(t) = \lvert \sin t\rvert, -\pi < t < \pi$	$\frac{2}{\pi} - \frac{4}{\pi}\left(\frac{\cos 2t}{1 \cdot 3} + \frac{\cos 4t}{3 \cdot 5} + \frac{\cos 6t}{5 \cdot 7} + \cdots\right)$
$x(t) = \begin{cases} 0, & 0 < t < \pi - \alpha \\ 1, & \pi - \alpha < t < \pi + \alpha \\ 0, & \pi + \alpha < t < 2\pi \end{cases}$	$\frac{\alpha}{\pi} - \frac{2}{\pi}\left(\frac{\sin\alpha\cos t}{1} + \frac{\sin 2\alpha\cos 2t}{2} + \frac{\sin 3\alpha\cos 3t}{3} + \cdots\right)$

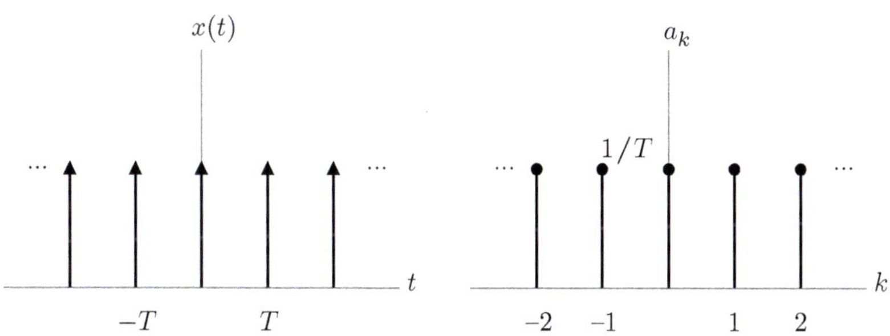

Figure 6.14 A continuous time impulse train, $x(t) = \sum_{k=-\infty}^{\infty} \delta(t - kT)$, has spectral coefficients, a_k, which is a discrete impulse train, scaled by the period T.

Remark 6.8 The Fourier series representation of a continuous time function, $x(t)$, is a discrete function, a_k vs. k. Interestingly, a continuous time impulse train in the time domain is a discrete impulse train in the frequency domain (Figure 6.14). As the period of the impulse train increases in the time domain, the amplitude of the spectral coefficients decreases in the frequency domain.

Exercise 6.11 Find the spectral coefficients of the derivative of the square wave given as follows:

$$g(t) = \begin{cases} 1, & \text{if } |t| < T_1 \\ 0, & \text{if } T_1 \le |t| \le T - T_1, \end{cases} \tag{6.98}$$

where $g(t) = g(t + T)$.

Solution

The plot of $g(t)$ and its derivative are given in Figure 6.15. The derivative $q(t)$ of the square wave is an impulse train with alternating sign, at the discontinuities of $g(t)$, as follows:

$$q(t) = \sum_{k=-\infty}^{\infty} \delta(t - kT + T_1) - \sum_{k=-\infty}^{\infty} \delta(t - kT - T_1). \tag{6.99}$$

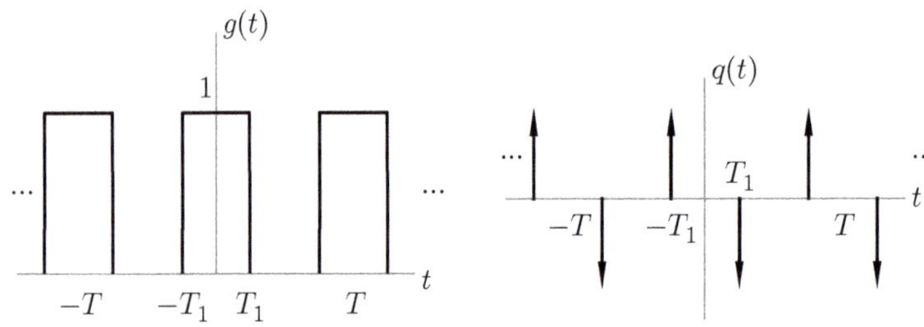

Figure 6.15 Square wave of width $2T_1$ and period T and its derivative.

From the previous example, we know that the spectral coefficients of

$$x(t) = \sum_{k=-\infty}^{\infty} \delta(t - kT), \tag{6.100}$$

is $a_k = \frac{1}{T}$. Using the time shift and linearity property, we obtain the spectral coefficients of $q(t)$, as follows:

$$q(t) \longleftrightarrow b_k = e^{jk\omega_0 T_1} a_k - e^{-jk\omega_0 T_1} a_k. \tag{6.101}$$

$$b_k = \frac{2j}{T} \sin(k\omega_0 T_1). \tag{6.102}$$

In fact, if we use the differentiation property, we can obtain the spectral coefficients of $g(t)$ from that of $q(t)$. Let $g(t) \longleftrightarrow c_k$. Using the differentiation property, we have $b_k = jk\omega_0 c_k$, where the spectral coefficients c_k are

$$c_k = \begin{cases} \dfrac{\sin(k\omega_0 T_1)}{k\pi} & k \ne 0, \\ \dfrac{2T_1}{T} & k = 0. \end{cases} \tag{6.103}$$

6.9 Trigonometric Fourier Series for Continuous Time Functions

Instead of using the harmonically related complex exponentials as the basis functions of Hilbert space, we can define a Fourier series representation of a periodic signal, $x(t)$, using sine and cosine basis functions. This equivalent representation is called the trigonometric Fourier series and can be obtained by using the Euler formula,

$$e^{j\omega_0 t} = \cos\omega_0 t + j\sin\omega_0 t. \tag{6.104}$$

Theorem 6.2 A continuous time periodic function, with period T, satisfying the Dirichlet conditions can be represented by the following trigonometric Fourier series:

$$x(t) = a_0 + \sum_{k=1}^{\infty}(B_k\cos(k\omega_0 t) + C_k\sin(k\omega_0 t)), \tag{6.105}$$

where the average term is

$$a_0 = \frac{1}{T}\int_T x(t)dt, \tag{6.106}$$

and the trigonometric coefficients are

$$B_k = \frac{2}{T}\int_T x(t)\cos(k\omega_0 t)dt, \tag{6.107}$$

and

$$C_k = \frac{2}{T}\int_T x(t)\sin(k\omega_0 t)dt. \tag{6.108}$$

The relationship between the spectral coefficients and trigonometric coefficients is given by

$$a_k = \frac{1}{2}(B_k + jC_k) \quad \text{and} \quad a_{-k} = \frac{1}{2}(B_k - jC_k), \quad \forall k \geq 1. \tag{6.109}$$

Proof: In the trigonometric Fourier series, replace

$$\sin k\omega_0 t = \frac{e^{jk\omega_0 t} - e^{-j2k\omega_0 t}}{2j} \quad \text{and} \quad \cos k\omega_0 t = \frac{e^{jk\omega_0 t} + e^{-j2k\omega_0 t}}{2}, \tag{6.110}$$

to obtain

$$x(t) = a_0 + \sum_{k=1}^{\infty}\left(\frac{B_k}{2}(e^{jk\omega_0 t} + e^{-jk\omega_0 t}) + \frac{C_k}{2j}(e^{jk\omega_0 t} - e^{-j2k\omega_0 t})\right). \tag{6.111}$$

Then, insert $a_k = \frac{1}{2}(B_k + jC_k)$ and $a_{-k} = \frac{1}{2}(B_k - jC_k)$, in the synthesis equation,

$$x(t) = \sum_{k=-\infty}^{\infty} a_k e^{jk\omega_0 t} = a_0 + \frac{1}{2}\sum_{k=1}^{\infty}(B_k + jC_k)e^{jk\omega_0 t} + \frac{1}{2}\sum_{k=-\infty}^{-1}(B_k - jC_k)e^{jk\omega_0 t}, \tag{6.112}$$

to show that

$$x(t) = \sum_{k=0}^{\infty} a_k e^{jk\omega_0 t} = a_0 + \sum_{k=1}^{\infty}(B_k\cos(k\omega_0 t) + C_k\sin(k\omega_0 t)). \tag{6.113}$$

Exercise 6.12 Find trigonometric Fourier series representation of $x(t) = e^{-t/2}$, $0 < t < \pi$, where $T = \pi$, $\omega_0 = \frac{2\pi}{T} = 2$. The plot of this signal is given in Figure 6.16.

Solution

The average term is

$$a_0 = \frac{1}{\pi}\int_0^{\pi} e^{-t/2}dt = \frac{2}{\pi} - \frac{2}{\pi e^{\pi/2}} \approx 0.504. \tag{6.114}$$

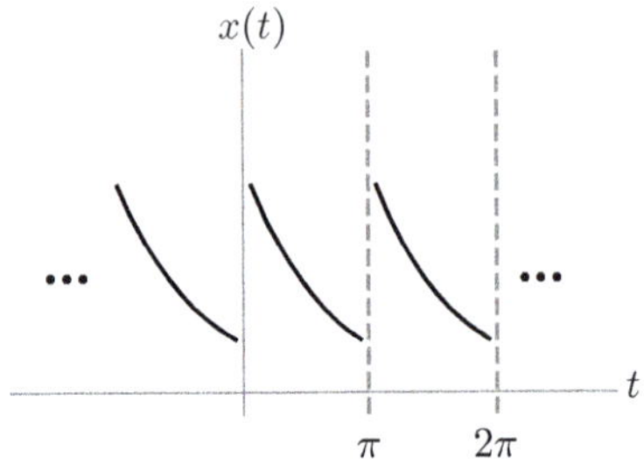

Figure 6.16 Plot of $x(t) = e^{-t/2}$ with period $T = \pi$.

The trigonometric coefficients are

$$B_k = \frac{2}{\pi}\int_0^{\pi} e^{-t/2}\cos(2kt)dt = \frac{2 \times 0.504}{1 + 16k^2}, \tag{6.115}$$

$$C_k = \frac{2}{\pi}\int_0^{\pi} e^{-t/2}\sin(2kt)dt = \frac{8k}{1 + 16k^2}0.504. \tag{6.116}$$

Figure 6.17 presents the plots of B_k and C_k.

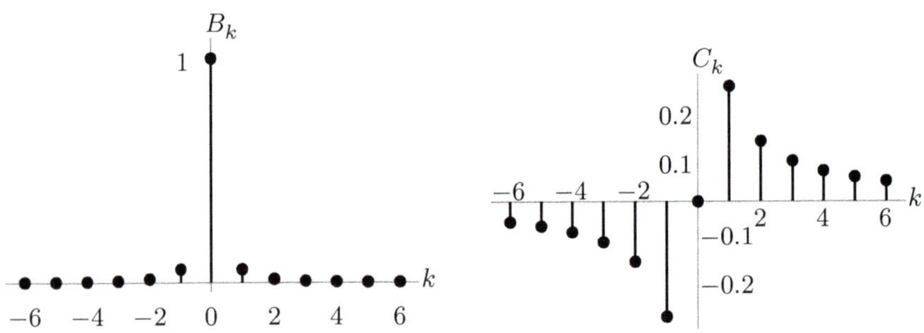

Figure 6.17 Plot of trigonometric coefficients, B_k and C_k, for $x(t) = e^{-t/2}$, $0 < t < \pi$ with period $T = \pi$.

INTERACTIVE

Explore trigonometric waveforms @ https://384book.net/i0604

6.10 Trigonometric Fourier Series for Continuous Time Even and Odd Functions

The trigonometric Fourier series has an interesting property. It decomposes the signal into its even and odd parts.

The basis functions of the trigonometric Fourier series consist of $\cos k\omega_0 t$, which is an even function, and $\sin \omega_0 t$, which is an odd function. When $x(t)$ is an even function, all of the C_k's become zeros. Similarly, when $x(t)$ is odd, then B_k's are all zeros. Thus, the trigonometric Fourier series nicely decomposes the function, $x(t)$, into its even and odd parts,

$$x(t) = a_0 + \sum_{k=1}^{\infty}(\underbrace{B_k\cos(k\omega_0 t)}_{\text{even part}} + \underbrace{C_k\sin(k\omega_0 t)}_{\text{odd part}}), \tag{6.117}$$

where is B_k represents the trigonometric coefficients of the even part and C_k represents the trigonometric coefficients of the odd part of the function $x(t)$.

Exercise 6.13 Find the trigonometric Fourier series representation of the following square wave signal:

$$x(t) = \begin{cases} 1, & \text{if } |t| < T_0 \\ 0, & \text{if } T_0 \leq |t| \leq T - T_0, \end{cases} \tag{6.118}$$

where $x(t) = x(t + T)$ is periodic, as shown in Figure 6.18. Find and plot the Fourier series coefficients.

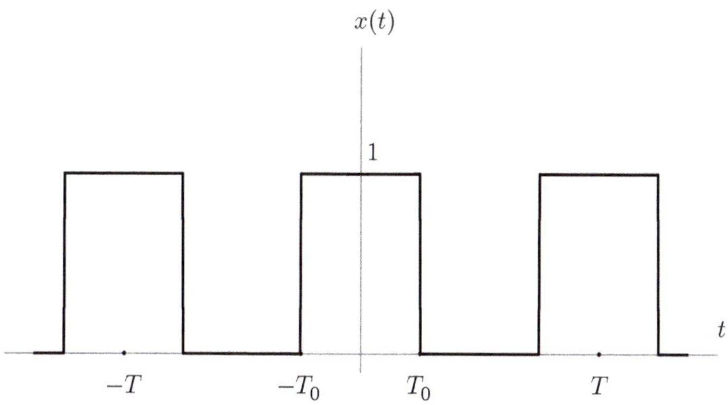

Figure 6.18 A periodic function, called square wave, which repeats itself at every period, T.

Solution

Note that this function is even. Thus, the spectral coefficients corresponding to the odd part, $C_k = 0$ for all k.

The spectral coefficients corresponding to the even part can be computed from the following equation:

$$B_k = \frac{2}{T}\int_{-T_0}^{T_0} \cos(k\omega_0 t)dt = \frac{4\sin(k\omega_0 T_0)}{k\omega_0 T}. \tag{6.119}$$

Exercise 6.14 Find the trigonometric Fourier series representation of the triangle wave given as follows:

$$x(t) = \begin{cases} -t + 1, & \text{if } 0 < t < 2, \\ t + 1, & \text{if} -2 \leq t \leq 0, \end{cases} \tag{6.120}$$

where $x(t) = x(t + T)$ is periodic, with $T = 4$. Find the Fourier series coefficients.

Solution

This is another even function. Hence, the odd part of the trigonometric Fourier series is zero and the even part is

$$B_k = \frac{2}{T}\int_{-T/2}^{T/2} x(t)\cos(k\omega_0 t)dt = \frac{1}{2}\int_{-2}^{0} (t + 1)\cos(k\omega_0 t)dt + \int_{0}^{2} (-t + 1)\cos(k\omega_0 t)dt. \tag{6.121}$$

Using the integration by parts and by replacing the angular frequency $\omega_0 = 2\pi/T = \pi/2$, we obtain,

$$B_k = \frac{2\sin\pi k}{\pi k} + \frac{8\sin(\frac{\pi k}{2})^2 - \pi k \sin\pi k}{\pi^2 k^2} = \frac{8\sin(\frac{\pi k}{2})^2}{\pi^2 k^2}. \tag{6.122}$$

The coefficients, B_k can be further simplified considering the even and odd values of k as follows:

$$B_k = \begin{cases} \frac{4}{\pi^2 k^2}(1-(-1)^k), & \text{for } k \text{ is even,} \\ 0, & \text{for } k \text{ is odd.} \end{cases} \tag{6.123}$$

The average value, a_0, is 0. Hence, the trigonometric Fourier series representation of this function is

$$x(t) = \sum_{k=1}^{\infty} \frac{8\sin\left(\frac{\pi k}{2}\right)^2}{\pi^2 k^2} \cos\left(k\frac{\pi}{2}t\right). \tag{6.124}$$

Exercise 6.15 Find the trigonometric Fourier series representation of the sawtooth wave given as follows:

$$x(t) = t \quad \text{for} \quad -1 < t < 1, \tag{6.125}$$

where $x(t) = x(t+T)$ is periodic, with $T = 2$. Find the Fourier representation of this function.

Solution

This is an odd function. Hence, the even part of the trigonometric Fourier series is zero and the odd part is

$$C_k = \frac{2}{T}\int_{-T/2}^{T/2} x(t)\sin(k\omega_0 t)dt = \int_{-1}^{1} t\sin(k\omega_0 t)dt. \tag{6.126}$$

Using the integration by parts, we obtain

$$C_k = \frac{2(\sin\pi k - \pi k\cos\pi k)}{\pi^2 k^2} = -\frac{2}{\pi k}(-1)^k. \tag{6.127}$$

The average value is $a_0 = 0$ and the angular frequency is $\omega_0 = 2\pi/2 = \pi$. Hence, the trigonometric Fourier series representation of this function is

$$x(t) = -\frac{2}{\pi}\sum_{k=1}^{\infty}\frac{(-1)^k}{k}\sin(k\pi t). \tag{6.128}$$

Exercise 6.16 Consider the following continuous time LTI system, represented by its impulse response $h(t)$.

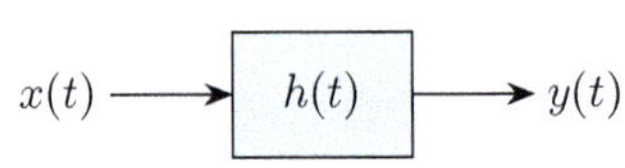

a) Find the impulse response of this system, if the unit step response is $s(t) = e^{-2t}u(t)$.
b) Suppose the input to this system is $x(t) = \cos(\pi t) + \sin(2\pi t)$. Find the spectral coefficients of $x(t)$.
c) Find and plot the spectral coefficients of the output $y(t)$ for the input given in part b).

Solution

a) Impulse response is simply the derivative of the unit step response,

$$h(t) = -2e^{-2t}u(t). \tag{6.129}$$

b) We can split $x(t)$ as $x(t) = x_1(t) + x_2(t)$ and find the spectral coefficients of each component:
- For $x_1(t)$, the angular frequency is $\omega_0 = \pi$ and the period is $T = 2$. The spectral coefficients are $a_1 = a_{-1} = 1/2$.
- For $x_2(t)$, the angular frequency is $\omega_0 = 2\pi$ and the period is $T = 1$. The spectral coefficients are $a_2 = 1/(2j), a_{-2} = -1/(2j)$.

Using the linearity property of Fourier series, we obtain the spectral coefficients of $x(t)$, as follows:

$$a_1 = a_{-1} = \frac{1}{2}, \quad a_2 = -a_{-2} = \frac{1}{2j}. \tag{6.130}$$

c) Recall that when the input is the eigenfunction of the LTI system, the corresponding output is

$$y(t) = H(j\omega_0)e^{j\omega_0 t}, \tag{6.131}$$

where the eigenvalue of the system is

$$H(j\omega_0) = \int_{-\infty}^{\infty} h(t)e^{-j\omega_0 t}dt. \tag{6.132}$$

For the impulse response we obtain in part a), $h(t) = -2e^{-2t}u(t)$, the eigenvalue of the system is obtained as follows:

$$H(j\omega_0) = -2\int_{0}^{\infty} e^{-(2+j\omega_0)t}dt = -\frac{2}{2+j\omega_0}. \tag{6.133}$$

Recall, also, that output $y(t)$ can be represented by Fourier series and the convolution integral as follows:

$$y(t) = \sum_{k=-\infty}^{\infty} b_k e^{jk\omega_0 t} = \int_{k=-\infty}^{\infty} x(\tau)h(t-\tau)d\tau. \tag{6.134}$$

Replace $x(\tau)$ by its Fourier series representation, in Equation (6.134),

$$x(\tau) = \sum_{k=-\infty}^{\infty} a_k e^{jk\omega_0 \tau}, \tag{6.135}$$

and arrange to obtain the spectral coefficients of $y(t)$ as follows:

$$b_k = a_k H(jk\omega_0). \tag{6.136}$$

Since the fundamental frequency of the input signal is $\omega_0 = \pi$, we obtain the spectral coefficients of the output as follows:

$$b_k = -a_k\frac{2}{2+jk\pi}. \tag{6.137}$$

Insert the values of the spectral coefficients, a_k, found in the previous part into Equation (6.137) to obtain

$$b_1 = -\frac{1}{2+j\pi}, \ b_{-1} = -\frac{1}{2-j\pi}, \ b_2 = \frac{1}{2(\pi-j)}, \ b_{-2} = \frac{1}{2(\pi+j)}. \tag{6.138}$$

6.11 Chapter Summary

Harmony is an essential concept in natural and human-made systems. A very powerful mathematical object to represent harmony is **harmonically related complex exponential function,** $\Phi_k(t) = e^{jk\omega_0 t}$, for continuous time signals, where each k defines a harmonic, for $k = 0, 1, \dots, \infty$.

Harmonically related exponentials bear very interesting properties. First of all, they are periodic functions with angular frequencies, $k\omega_0$. As k increases, they represent periodic motion on a unit circle of the complex plane with higher speeds. Secondly, they are orthogonal to each other. Thus, they span an infinite-dimensional function space, called **Hilbert space,** where each periodic function is uniquely represented by a set of coordinates $\{a_k\}$, called the spectral coefficients, provided that the function satisfies the Dirichlet conditions. The representation of a signal in Hilbert space is called **Fourier series**, named after J. B. Fourier, a revolutionary French politician and mathematician. Fourier series representation enables us to decompose a periodic signal into its harmonically related frequency components. Since a periodic function is represented in terms of its frequency content, we call this specific Hilbert space as **frequency domain.**

Finally, harmonically related complex exponentials are the **eigenfunctions of an LTI system.** In other words, when we feed harmonics of the complex exponential at the input of an LTI system, the output is just the scaled version of the input. This scale is called the **eigenvalue** or **the transfer function of the system** and it uniquely describes an LTI system in the frequency domain.

Problems

6.1 Represent the following signals in terms of the superposition of complex exponentials:
a) $x(t) = 1 + \sin(\pi t)$.
b) $x(t) = \cos(\pi t + \frac{\pi}{2})$.
c) $x(t) = 1 + \sin(\pi t) + \cos(\frac{\pi}{10} t)$.

6.2 Does the following function satisfy Dirichlet conditions:

$$x(t) = \begin{cases} |t| & \text{for } -1 \leq t \leq 1 \\ 0 & \text{o.w.} \end{cases}$$

6.3 Let $x(t)$ be a continuous time square wave signal with a fundamental period $T = 6$ seconds, represented by the following analytical expression in one full period:

$$x(t) = \begin{cases} 1 & |t| < 2 \\ 0 & 2 \leq |t| \leq 4. \end{cases}$$

a) Find and plot the Fourier series coefficients for $x(t)$.
b) Find and plot the Fourier series coefficients for $\frac{dx}{dt}$.
c) Find the Trigonometric Fourier series representation of $x(t)$.

6.4 Show that the inner product between two harmonically related complex exponential functions satisfies the following equation:

$$<e^{jn\omega_0 t}, e^{jk\omega_0 t}> = \int_0^T e^{jn\omega_0 t} e^{-jk\omega_0 t} dt = \begin{cases} T & \text{for } n = k, \\ 0 & n \neq k. \end{cases}$$

6.5 Show that the functions $x_1 = e^{j\omega_0 t}$ and $x_2 = e^{2j\omega_0 t}$ are orthogonal to each other.

6.6 Let $x(t)$ be a continuous time signal, represented by the following function:

$$x(t) = \cos\left(\frac{\pi t}{3}\right) + 2\cos\left(\pi t + \frac{\pi}{2}\right).$$

a) Find the fundamental period of this signal.
b) Find and plot the Fourier series coefficients for $x(t)$.
c) Find the trigonometric Fourier series representation of $x(t)$.

6.7 Let $x(t)$ be a continuous time signal represented by the following function:

$$x(t) = 1 + \sin\left(\frac{\pi t}{2}\right) + 4\sin(\pi t).$$

a) Find the fundamental frequency ω_0 of this signal.
b) Find and plot the nonzero Fourier series coefficients for $x(t)$.
c) Find the trigonometric Fourier series representation of $x(t)$.

6.8 The eigenvalues of a continuous time LTI system are given by the following equation:

$$H(jk\omega_0) = \frac{\cos(2k\omega_0)}{kw_0}.$$

Let us define the input signal $x(t)$ by the following function in one full period;

$$x(t) = \begin{cases} 1 & 0 \leq t < 3 \\ -1 & 3 \leq t < 6, \end{cases}$$

where the fundamental period is $T = 6$.
a) Find and plot the Fourier series coefficients of the input signal $x(t)$.
b) Find and plot the Fourier series coefficients of the output $y(t)$.
c) Find and plot the output signal $y(t)$, approximately.

6.9 The eigenvalues of a continuous time LTI system is represented by the following equation:

$$H(jk\omega_0) = \begin{cases} 1 & |\omega| < 50 \\ 0 & |\omega| \geq 50. \end{cases}$$

Suppose that a periodic input signal $x(t) \longleftrightarrow a_k$ with fundamental period $\mathrm{T} = \frac{\pi}{3}$ is fed to the LTI system. If the output is the same as the input, i.e., $y(t) = x(t)$, what are the Fourier series coefficients of the input and output?

6.10 Find the eigenvalues of the LTI systems described by the following input–output pairs:
a) $x(t) = e^{jt}, \quad y(t) = je^{5jt}$
b) $x(t) = e^{j\pi t}, \quad y(t) = e^{j\pi t - 2\pi}.$
c) $x(t) = e^{j3t}, \quad y(t) = \dfrac{\cos 3t}{j} + \sin(3t).$

6.11 Find the eigenvalue of the following differential equation, when the input is the eigen function, $x(t) = e^{jk\omega_0 t}$:

$$\ddot{y}(t) + 6\dot{y}(t) + 9y(t) = x(t).$$

6.12 Find the spectral coefficients of the signal given as follows.

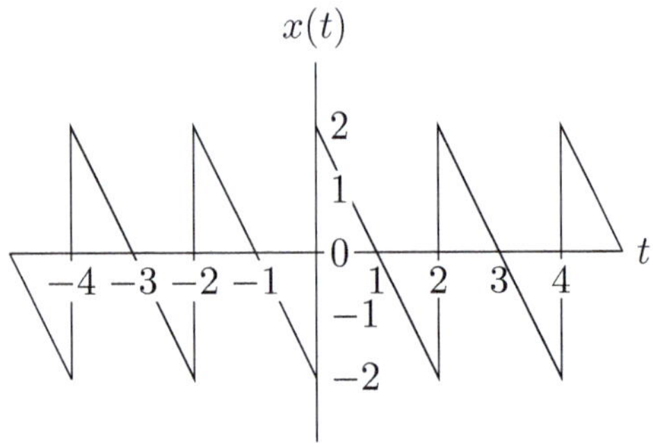

6.13 Find the spectral coefficients of the function given as follows.

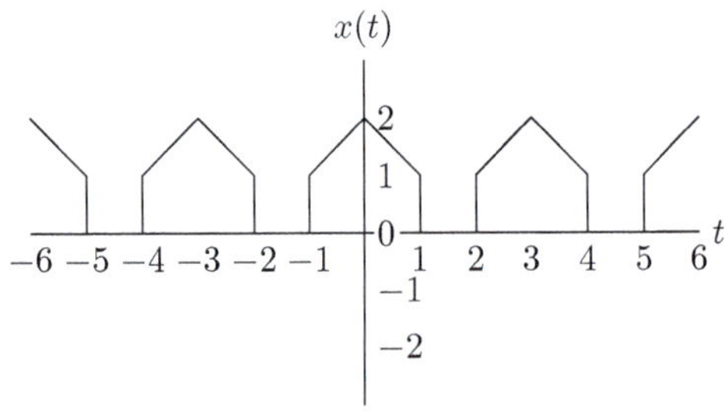

6.14 Find the spectral coefficients of the function given as follows.

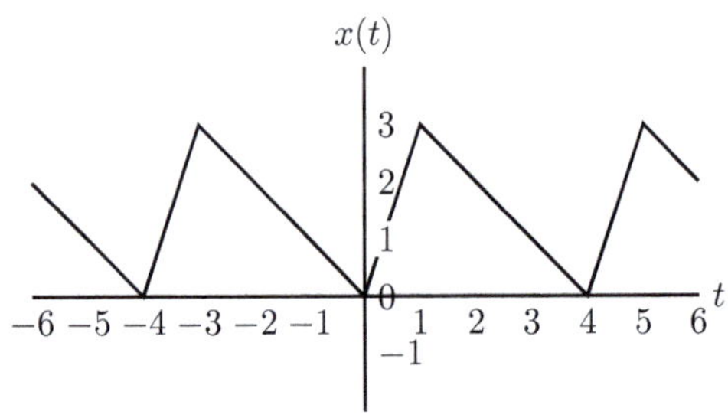

6.15 Determine the periodic continuous time signal $x(t)$ with a period $T = 4$ whose Fourier series coefficients of $x(t)$ are as follows:

$$\alpha_k = \begin{cases} k \cdot j & k \leq 2 \\ 0 & \text{otherwise.} \end{cases}$$

6.16 Determine the periodic continuous time signal $x(t)$ with a period $T = 6$ whose Fourier series coefficients of $x(t)$ are as follows:

$$\alpha_k = \begin{cases} k \cdot j & -2 \leq k \leq 2 \\ 0 & \text{otherwise.} \end{cases}$$

6.17 Determine the periodic continuous time signal $x(t)$ with a period $T = 4$ whose Fourier series coefficients of $x(t)$ are as follows:

$$\alpha_k = \begin{cases} -1 & k \text{ is even} \\ 1 & k \text{ is odd.} \end{cases}$$

6.18 Consider a continuous time periodic signal in one full period;

$$x(t) = \begin{cases} 2t & 0 \leq x \leq 2 \\ 4 - t & 2 < x \leq 4, \end{cases}$$

where the fundamental period is $T = 4$.
and Fourier series coefficients a_k.

a) Find the spectral coefficients of this function.
b) Find the spectral coefficients of $\frac{dx}{dt}$ using differentiation property.

6.19 Consider the following periodic continuous time signals:

$$x_1(t) = \cos\left(\frac{\pi t}{2}\right)$$

$$x_2(t) = \sin\left(\frac{\pi t}{2}\right)$$

$$x_3(t) = x_1(t)x_2(t).$$

a) Find the spectral coefficients of $x_1(t)$, $x_2(t)$, and $x_3(t)$.
b) Find the spectral coefficients of $x_3(t)$ using the multiplication property and compare your results with what you found in part (a).

6.20 Consider a continuous time LTI system whose unit step response is

$$s(t) = e^{-3t}u(t).$$

a) Find the impulse response $h(t)$ of the system.
b) Find and plot the spectral coefficients of the input $x(t) = 2\cos(2\pi t) - \sin(\pi t)$.
c) Find and plot the spectral coefficients of the output of this system, when the input is $x(t) = 2\cos(2\pi t) - \sin(\pi t)$.

7

Fourier Series Representation of Discrete Time Periodic Signals

"In the middle of difficulty lies opportunity."

Albert Einstein

Remember that Fourier series are useful mathematical representations of continuous time periodic signals, which enable us to decompose them into their frequency components. Loosely speaking, Fourier series provide the "amount" of each harmonic frequency in a signal. These amounts are called **spectral coefficients** of the signal.

WATCH Learn more about the Fourier series and the frequency spectrum @ https://384book.net/v0701

In this chapter, we extend the methods of continuous time Fourier analysis methods into discrete time signals. We shall represent a discrete time periodic function in terms of weighted summation of harmonically related discrete time complex exponentials, $\Phi_k[n] = e^{jk\omega_0 n}$, where the weights are the spectral coefficients, $\{a_k\}$.

7.1 Fourier Series Theorem for Discrete Time Functions

Although the representation of discrete time periodic signals in frequency domain resembles to that of the continuous time signals, there are three major differences between them:

1) Since the signal $x[n]$ is discrete, the integral operations of the analysis equation, which uniquely determines the spectral coefficients, are replaced by the **summation operation**. Thus, we do not bother to take complicated integrals.
2) Furthermore, the limits of the summation of the analysis and synthesis equation are **finite**. Thus, we do not have a convergence problem, nor do we need Dirichlet conditions.
3) The spectral coefficients of discrete time Fourier series of a periodic function are always **periodic**, with the fundamental period of the signal.

Let us formally introduce the Fourier series representation of discrete time periodic signals and investigate the aforementioned facts.

Signals and Systems: Theory and Practical Explorations with Python,
First Edition. Fatoş Tunay Yarman Vural and Emre Akbaş.

Companion website: www.wiley.com/go/signalsandsystems

Theorem 7.1 A discrete time **periodic signal**, $x[n]$, with the fundamental period N, can be decomposed into harmonically related complex exponentials, $e^{jk\omega_0 n}$, weighted by a set of coefficients, called the spectral coefficients, a_k, as follows:

$$\boxed{x[n] = \sum_{k=\langle N \rangle} a_k e^{jk\omega_0 n}} \quad \text{Synthesis equation,} \tag{7.1}$$

where the **spectral coefficients** are uniquely obtained from,

$$\boxed{a_k = \frac{1}{N} \sum_{n=\langle N \rangle} x[n] e^{-jk\omega_0 n}} \quad \text{Analysis equation} \tag{7.2}$$

In synthesis and analysis equations, the limit of the summations, $\langle N \rangle$, indicates N successive integers to cover one full range of fundamental period.

Furthermore, **the spectral coefficients**, $\{a_k\}$, are periodic with period N. In other words, $\{a_k\}$ repeats with $a_k = a_{k+N}, \quad \forall k \in (-\infty, \infty)$.

Since both the spectral coefficients, $\{a_k\}$, and the harmonically related complex exponentials, $\Phi_k[n] = e^{jk\omega_0 n}$, are periodic, with the fundamental period of the signal $x[n]$, the summations of synthesis and analysis equations are evaluated over one full range of the fundamental period, N.

Proof Sketch: Let us start by showing that the spectral coefficients, $\{a_k\}$ are periodic, with the fundamental period of the signal, $x[n]$. Recall that harmonically related complex exponentials are periodic, that is, for integer value $N = 2\pi/\omega_0$, we have $\Phi_k[n] = \Phi_{k+N}[n]$, which is easy to show:

$$\Phi_{k+N}[n] = e^{j(k+N)\omega_0 n} = e^{jk\omega_0 n} e^{jN\omega_0 n} = e^{jk\omega_0 n} = \Phi_k[n], \tag{7.3}$$

since $e^{jN\omega_0 n} = 1$ due to the fact that $e^{jN\omega_0 n} = \cos(2\pi n) + j\sin(2\pi n) = 1$.

The analysis equation,

$$a_k = \frac{1}{N} \sum_{n=\langle N \rangle} x[n] e^{-jk\omega_0 n}, \tag{7.4}$$

is simply a superposition of the harmonically related complex exponentials, weighted by the periodic function, $x[n]$, where the fundamental period is N. Since both the signal $x[n]$ and the complex exponentials $\Phi_k[n]$ are periodic, summation of the multiplications of these functions is also periodic, with the fundamental period N. Therefore, the spectral coefficients, obtained from the analysis equation, repeat themselves after the fundamental period N:

$$\boxed{a_k = a_{k+N}.} \tag{7.5}$$

Next, let us show that the coefficients a_k obtained from the analysis equation satisfy the synthesis equation.

In order to show that the synthesis equation can be obtained from the analysis equation, we multiply both sides of the synthesis equation by $\Phi_r[n] = e^{-jr\omega_0 n}$ and sum them over one period $n = \langle N \rangle$, to obtain the following equation:

$$\sum_{n=\langle N \rangle} x[n] e^{-jr\omega_0 n} = \sum_{n=\langle N \rangle} \sum_{k=\langle N \rangle} a_k e^{jk\omega_0 n} e^{-jr\omega_0 n}. \tag{7.6}$$

We arrange the summations to get

$$\sum_{n=\langle N \rangle} x[n] e^{-jr\omega_0 n} = \sum_{k=\langle N \rangle} a_k \sum_{n=\langle N \rangle} e^{j(k-r)\omega_0 n}. \tag{7.7}$$

The second summation on the right-hand side is the inner product of two harmonically related discrete time complex exponentials. We visited the continuous time case of this inner product in Equation (6.23). The discrete time case is similar:

$$\langle e^{jk\omega_0 n}, e^{jr\omega_0 n}\rangle = \sum_{n=\langle N\rangle} e^{j(k-r)\omega_0 n} = \begin{cases} N, & \text{if } r = k, \\ 0, & \text{otherwise.} \end{cases} \tag{7.8}$$

The $r = k$ case is trivial. For the $r \neq k$ case, let $c = e^{j(k-r)\omega_0 n}$ and S be the sum:

$$S = c + c^2 + c^3 + \cdots + c^N. \tag{7.9}$$

Multiplying both sides of Equation (7.9) by c, we obtain

$$cS = c^2 + c^3 + c^4 + \cdots + c^{N+1}. \tag{7.10}$$

Subtracting Equation (7.10) from (7.9) and with some arrangement, we get

$$S = \frac{c(1 - c^N)}{1 - c} = \frac{e^{j(k-r)\omega_0 n}(1 - e^{j(k-r)\omega_0 nN})}{1 - e^{j(k-r)\omega_0 n}}. \tag{7.11}$$

Since $\omega_0 = 2\pi/N$ and r, k, n are integers, $e^{j(k-r)\omega_0 nN}$ is equal to $e^{j2\pi} = 1$, which yields $S = 0$. Using this result in Equation (7.7), we obtain

$$\sum_{n=\langle N\rangle} x[n]e^{-jk\omega_0 n} = a_k N, \tag{7.12}$$

which directly implies the analysis equation,

$$a_k = \frac{1}{N}\sum_{n=\langle N\rangle} x[n]e^{-jk\omega_0 n}. \tag{7.13}$$

Remark 7.1 Although the spectral coefficients $\{a_k\}$ are generated for only one full period, $\langle N\rangle$, we need to keep in mind that they repeat themselves at every period N for all $k \in (-\infty, \infty)$:

$$a_k = a_{k\pm N}. \tag{7.14}$$

Since the summations of analysis and synthesis equations have finite limits, there is no concern about the existence of spectral coefficients. This relaxes the convergence constraints imposed by Dirichlet conditions. Furthermore, periodicity of the spectral coefficients brings a substantial computational efficiency in representing discrete time periodic signals in the frequency domain. This is not the case for the continuous time signals, as their spectral coefficients are the **integral** of product of two periodic functions, $x(t)$ and $e^{jk\omega_0 t}$, which may not result in a periodic function.

INTERACTIVE

Explore Fourier series representation for discrete time signals @ https://384book.net/i0701

7.2 Discrete Time Fourier Series Representation in Hilbert Space

In Chapter 6, we defined a continuous time periodic function as a vector, represented by the coordinates, $\{a_k\}$, corresponding to the spectral coefficients, in an infinite dimensional space, called

Hilbert space. Recall that a Hilbert space H is a vector space, where the vectors are functions and the distance between the functions is defined by the inner product.

An extension of the representation of continuous time periodic functions to that of the discrete time functions is possible by defining a Hilbert space, spanned by the discrete time harmonically related complex exponential functions. In this case, the inner product between two discrete time periodic functions, $x[n]$ and $y[n]$ is defined as:

$$\langle x[n], y[n]\rangle = \sum_{n=\langle N\rangle} x[n]y^*[n], \tag{7.15}$$

where $(*)$ indicates the complex conjugate operation. This time, the limit of the summation is finite. Thus, we can represent one period of discrete time functions, $x[n], y[n] \in H$, by finite length vectors, $\mathbf{x}$ and $\mathbf{y}$:

$$\mathbf{x} = [x[0] \quad x[1] \quad \dots \quad x[N-1]]^T \text{ and} \tag{7.16}$$

$$\mathbf{y} = [y[0] \quad y[1] \quad \dots \quad y[N-1]]^T, \tag{7.17}$$

where T represents the vector transpose operation. The entries of the vector $\mathbf{x}$ and $\mathbf{y}$, cover one full period of the function $x[n]$ and $y[n]$, respectively. This representation enables us to show that the inner product, defined for a Hilbert space H, is reduced to that of a classical vector space as follows:

$$\langle x[n], y[n]\rangle = \sum_{n=\langle N\rangle} x[n]y^*[n] = \mathbf{x}^T\mathbf{y}^*. \tag{7.18}$$

Furthermore, we can define an N-dimensional basis vector, $\mathbf{e}^{jn\omega_0}$, which only covers one full period, $N = 2\pi/\omega_0$, of the complex exponential function $\phi_k[n] = e^{jk\omega_0 n}$ as follows:

$$\mathbf{e}^{jn\omega_0} = [1 \quad e^{jn\omega_0} \quad e^{2jn\omega_0} \quad \dots \quad e^{j(N-1)n\omega_0}]^T. \tag{7.19}$$

Finally, we can define the spectral coefficient vector over one period of N,

$$\mathbf{a} = [a_0 \quad a_1 \quad \dots \quad a_{N-1}]^T, \tag{7.20}$$

so that we represent the Fourier series of a discrete time function as:

$$\mathbf{x} = \mathbf{a}^T\mathbf{e}^{j\omega_0 n}. \tag{7.21}$$

Using the definition of the inner product given in Equation (7.15), we can show that discrete time harmonically related complex exponential functions are orthogonal to each other. Mathematically, the inner product of two complex exponential functions with different harmonics is

$$\langle e^{jk\omega_0}, e^{jr\omega_0}\rangle = (\mathbf{e}^{jk\omega_0})^T(\mathbf{e}^{jr\omega_0})^* = \sum_{n=\langle N\rangle} e^{j(k-r)(2\pi/N)} = \begin{cases} N, & \text{if } r = k, \\ 0, & \text{otherwise.} \end{cases} \tag{7.22}$$

Hence, harmonically related discrete time exponentials form a basis and they span the Hilbert space of discrete time periodic functions. The spectral coefficients, each of which shows the amount of a particular harmonic frequency, are the coordinates of a discrete time function in this Hilbert space.

Exercise 7.1 Consider the discrete time function, $x[n] = \sin 0.1\pi n$.

a) Plot this signal. Is this a periodic function? If yes, what is the period?

b) Find and plot the spectral coefficients, $\{a_k\}$, which are the coordinates of this function in Hilbert space.
c) Comment on the frequency content of this signal, analyzing the magnitude and phase spectrum.

Solution

a) This signal is periodic, provided that there exists an integer N, satisfying the following equation:

$$x[n] = x[n + N]. \tag{7.23}$$

The fundamental period of the signal is the smallest integer N, which is related to the angular frequency, as follows:

$$\omega_0 = 0.1\pi = \frac{2\pi}{N}m.$$

The smallest integer value for the fundamental period N is obtained for $m = 1$, as $N = \frac{2}{0.1} = 20$. Hence, this signal is periodic! The plot of this signal is given in Figure 7.1.

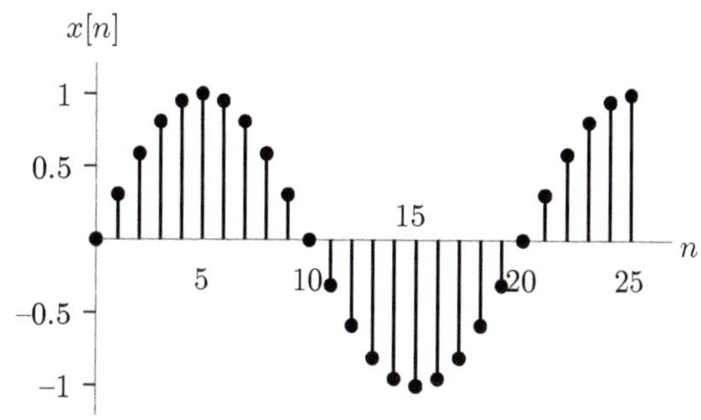

Figure 7.1 The plot of $x[n] = \sin 0.1\pi n$. Its fundamental period is $N = 20$.

b) The spectral coefficients, $\{a_k\}$, can be easily obtained by applying the Euler formula:

$$x[n] = \frac{1}{2j}\left(e^{j0.1\pi n} - e^{-j0.1\pi n}\right). \tag{7.24}$$

$$a_1 = \frac{1}{2j}, \quad a_{-1} = -\frac{1}{2j}. \tag{7.25}$$

Since the spectral coefficients are periodic with the fundamental period of the signal ($N = 20$), we have

$$a_1 = a_{21} = a_{41} = \cdots a_{kN+1}, \quad \forall k. \tag{7.26}$$

and

$$a_{-1} = a_{19} = a_{39} = \cdots a_{kN-1}, \quad \forall k. \tag{7.27}$$

Recall that when the spectral coefficients are complex numbers, we need two plots:
1) **Magnitude spectrum:** $|a_1| = |a_{-1}| = \frac{1}{2}$
2) **Phase spectrum:** $\sphericalangle a_1 = -\frac{\pi}{2}, \quad \sphericalangle a_2 = \frac{\pi}{2}$

Figure 7.2 provides the plots for these spectra.

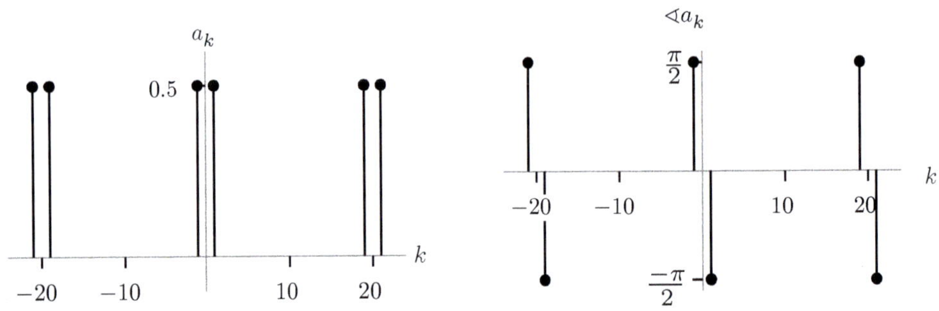

Figure 7.2 Magnitude and phase spectrum of $x[n] = \sin 0.1\pi n$. Both $|a_k|$ vs. k and the $\sphericalangle a_k$ vs. k are discrete periodic functions, which repeat themselves at every $N = 20$.

c) Analyzing the magnitude spectrum (Figure 7.2) for one period, we observe that, the signal consists of the fundamental frequency corresponding to $k\omega_0$ for $k = \pm 1$. Similarly, analyzing the phase spectrum, we observe that there is a phase shift of the signal by the amount of $\pm\pi/2$ at the fundamental frequency corresponding to $k\omega_0$ for $k = \pm 1$.

Exercise 7.2 Consider the following signal:

$$x[n] = 1 + \sin\left(\frac{\pi}{5}n\right) + \cos\left(\frac{2\pi}{5}n + \frac{\pi}{2}\right). \tag{7.28}$$

a) Find the coordinates of this function in a Hilbert space spanned by discrete complex exponentials, $\Phi_k[n] = e^{jk\omega_0 n}, \forall k, \forall n$.
b) Plot the spectral coefficients using the Cartesian coordinates.
c) Plot the spectral coefficients in polar coordinates.
d) Comment on the frequency content of this signal, compared to that of the previous example.

Solution

a) Spectral coefficients of a periodic signal correspond to the coordinates in Hilbert space. Using the Euler formula for cosines and sines, we can directly find the spectral coefficients as follows:

$$x[n] = 1 + \frac{1}{2j}[e^{j(\pi/5)n} - e^{-j(\pi/5)n}] + \frac{1}{2}[e^{j(2\pi n/5+\pi/2)} + e^{-j(2\pi n/5+\pi/2)}]. \tag{7.29}$$

Arranging the terms, we obtain,

$$x[n] = 1 + \frac{1}{2j}e^{j(\pi/5)n} - \frac{1}{2j}e^{-j(\pi/5)n} + \left(\frac{1}{2}e^{j\pi/2}\right)e^{j(2\pi/5)n} + \left(\frac{1}{2}e^{-j\pi/2}\right)e^{-j(2\pi/5)n}. \tag{7.30}$$

Thus, the Fourier series coefficients of this function, which are the coordinates in Hilbert space are as follows:

$$a_0 = 1,$$
$$a_1 = \frac{1}{2j} = -\frac{1}{2}j,$$
$$a_{-1} = -\frac{1}{2j} = \frac{1}{2}j, \tag{7.31}$$

$$a_2 = \frac{e^{j\pi/2}}{2} = \frac{1}{2}j,$$
$$a_{-2} = \frac{e^{-j\pi/2}}{2} = -\frac{1}{2}j,$$

with $a_k = 0$ for other values of k in the interval of summation in the synthesis equation.

For this signal, $\omega_0 = \pi/5$, which implies that the period is $N = 10$ ($\omega_0 = 2\pi/N$). Thus, the Fourier coefficients are periodic with period $N = 10$. In other words, $a_k = a_{k\pm N}$.

b) In the Cartesian coordinate system, we plot the real and imaginary part of the spectral coefficients. For real parts of the spectral coefficients, we plot

$$\text{Re}\left\{a_k\right\} = \text{Re}\left\{a_{k+mN}\right\} = \begin{cases} 1, & \text{for} \quad k = 0, \forall m, \\ 0, & \text{otherwise.} \end{cases} \tag{7.32}$$

For the imaginary parts of the spectral coefficients, we plot

$$\text{Im}\left\{a_k\right\} = \text{Im}\left\{a_{k+mN}\right\} = \begin{cases} \frac{1}{2}, & \text{for} \quad k = -1, 2, \forall m, \\ -\frac{1}{2}, & \text{for} \quad k = 1, -2, \forall m, \\ 0 & \text{otherwise.} \end{cases} \tag{7.33}$$

Figure 7.3 (top row) shows the plots of the real part and imaginary part of the spectral coefficients.

c) In the polar coordinate system, we plot the magnitude of the spectral coefficients.

$$|a_k| = |a_{k+mN}| = \begin{cases} \frac{1}{2}, & \text{for} \quad k = \pm 1, \pm 2, \forall m, \\ 0, & \text{otherwise,} \end{cases} \tag{7.34}$$

and the phase of the spectral coefficients,

$$\sphericalangle a_k = \sphericalangle a_{k+mN} = \begin{cases} -\text{sign}(k)\frac{\pi}{2}, & \text{for} \quad k = \pm 1, \forall m, \\ \text{sign}(k)\frac{\pi}{2}, & \text{for} \quad k = \pm 2, \forall m, \\ 0, & \text{otherwise.} \end{cases} \tag{7.35}$$

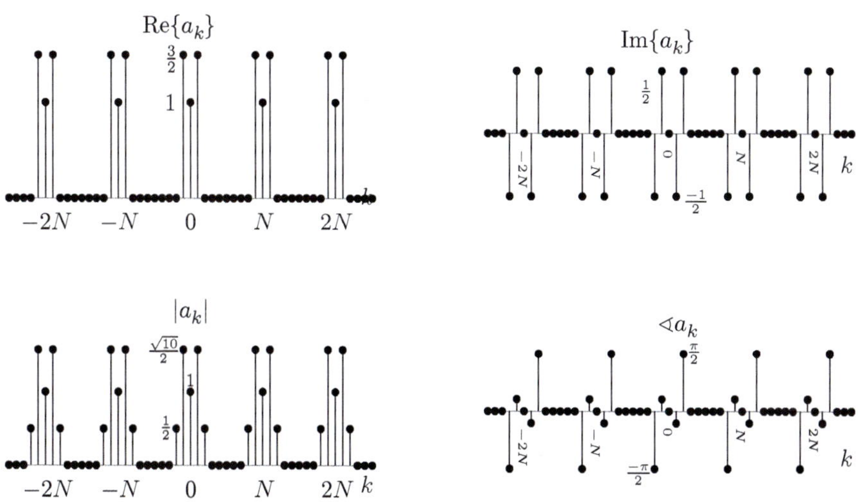

Figure 7.3 Plot of the real and imaginary parts (top row) and the magnitude and phase of spectral coefficients, for $x[n] = 1 + \sin\left(\frac{2\pi}{10}\right)n + 3\cos\left(\frac{2\pi}{10}\right)n + \cos\left(\frac{4\pi}{10}n + \frac{\pi}{2}\right)$ (bottom row).

Figure 7.3 (bottom row) shows the plots of the magnitude and phase of the spectral coefficients for $N = 10$.

d) Analyzing the magnitude and phase spectrum of this signal and comparing it to that of the previous example, we observe that the second signal has additional harmonics for $k = \pm 2$, indicating a "richer" signal in terms of the frequency content, compared to the first one.

The above two examples are relatively easy to be represented in the frequency domain, since the application of Euler formula directly yields the spectral coefficients. The frequency content of these signals can be observed in both time and frequency domains. However, signals, such as speech and music, which consist of a large variety of harmonics, cannot be analyzed in the time domain. On the other hand, the number of spectral coefficients provides us a measurable value for each harmonic frequency, contained in the signal.

Let us investigate the frequency content of the following signal given.

Exercise 7.3 Find the coordinates of the discrete time periodic square wave shown in Figure 7.4, in a Hilbert space spanned by discrete complex exponentials, $\Phi_k[n] = e^{jk\omega_0 n}$ and investigate the frequency content of this signal.

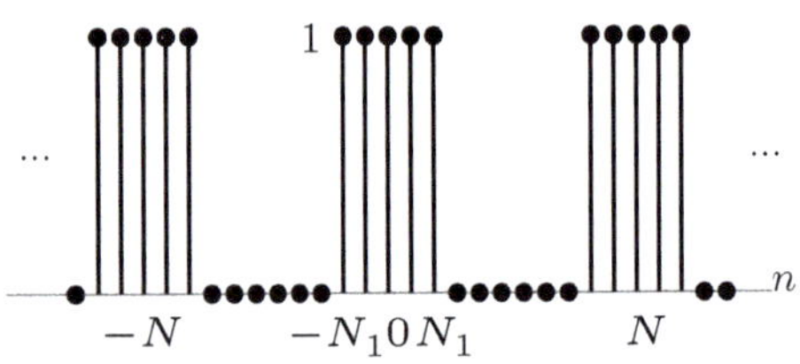

Figure 7.4 Discrete time periodic square wave with a fundamental period N, for $N = 11$ and $N_1 = 2$.

Solution

Analyzing the frequency content of this signal in time domain is not possible. All we observe is a square wave of fundamental period, N. Let us investigate the signal in the frequency domain by finding the coordinates from the analysis equation for the nonzero values of $x[n]$ over one full period, N,

$$a_k = \frac{1}{N} \sum_{n=-N_1}^{N_1} e^{-jk(2\pi/N)n}. \tag{7.36}$$

Changing the dummy variable of sum as, $m = n + N_1$, we observe that this equation becomes

$$\begin{aligned} a_k &= \frac{1}{N} \sum_{m=0}^{2N_1} e^{-jk(2\pi/N)(m-N_1)} \\ &= \frac{1}{N} e^{jk(2\pi/N)N_1} \sum_{m=0}^{2N_1} e^{-jk(2\pi/N)m}. \end{aligned} \tag{7.37}$$

The summation is a finite geometric series, which has the following closed form:

$$\begin{aligned} a_k &= \frac{1}{N} e^{jk(2\pi/N)N_1} \left(\frac{1 - e^{-jk2\pi(2N_1+1)/N}}{1 - e^{-jk(2\pi/N)}} \right) \\ &= \frac{1}{N} \frac{e^{-jk(2\pi/2N)}[e^{jk2\pi(N_1+1/2)/N} - e^{-jk2\pi(N_1+1/2)/N}]}{e^{-jk(2\pi/2N)}[e^{jk(2\pi/2N)} - e^{-jk(2\pi/N)}]} \\ &= \frac{1}{N} \frac{\sin[2\pi k(N_1 + 1/2)/N]}{\sin(\pi k/N)}, \quad \text{for} \quad k \neq 0, \pm N, \pm 2N, \dots \end{aligned} \tag{7.38}$$

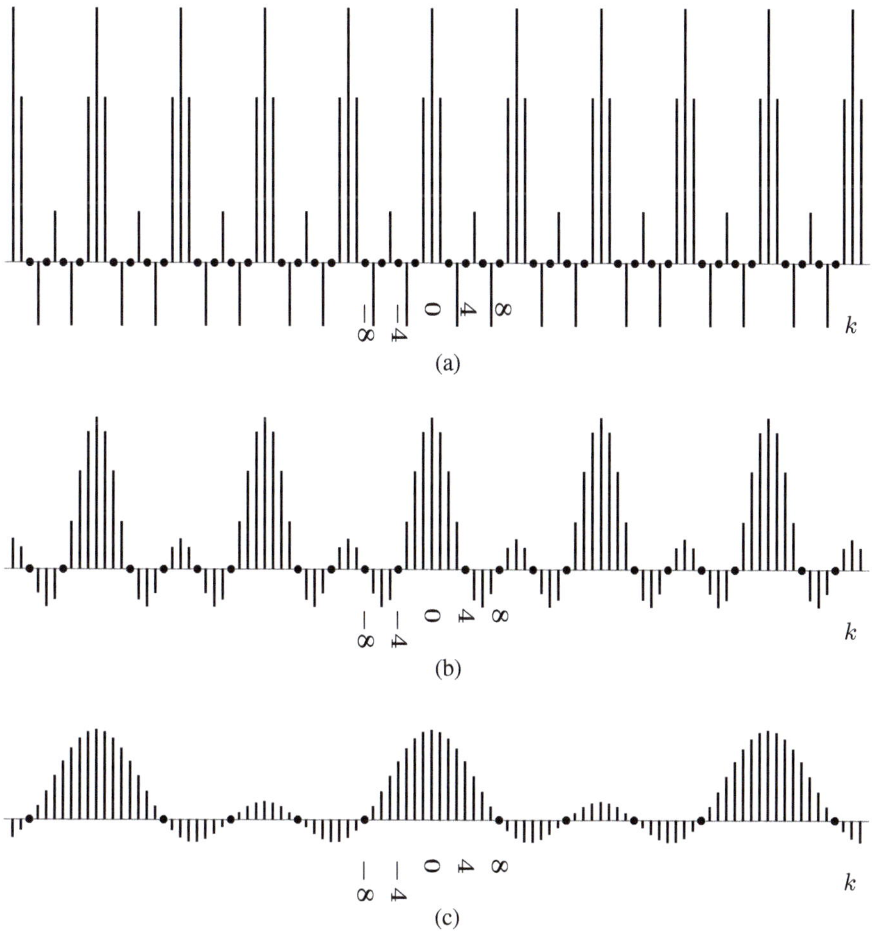

Figure 7.5 Fourier series coefficients for the periodic square wave of Example 7.4; plots of Na_k for $2N1 + 1 = 5$ and (a) $N = 10$; (b) $N = 20$; and (c) $N = 40$.

and

$$a_k = \frac{2N_1 + 1}{N}, \quad \text{for} \quad k = 0, \pm N, \pm 2N, \ldots.$$

The coefficients a_k for $2N_1 + 1 = 5$ are sketched for $N = 10, 20$, and 40 in Figure 7.5a, b, and c, respectively.

Analysis of Figure 7.5 reveals that as we increase the period from $N = 10$ to 40, we obtain a smoother periodic discrete representation of the coefficients, a_k vs. k.

7.3 Properties of Discrete Time Fourier Series

Most of the properties of discrete time Fourier series are similar to those of the continuous time Fourier series, such as linearity, time reverse, conjugate symmetry, and frequency shifting. For this reason, we suffice to provide them in Table 7.1. The properties, listed in this table, can be easily proved by the direct application of the analysis and synthesis equation of discrete time Fourier series.

Table 7.1 Summary of the properties of discrete time Fourier series.

Periodic signal	**Fourier series coefficient**
$x[n]$ is periodic with fundamental period N	a_k
$y[n]$ is periodic with fundamental period N	b_k
Synthesis equation: $x[n] = \sum_{k=\langle N\rangle} a_k e^{jk\omega_0 n}$ where $\omega_0 = \frac{2\pi}{N}$	Analysis equation: $a_k = \frac{1}{N}\sum_{n=\langle N\rangle} x[n]e^{-jk\omega_0 n}$
$Ax[n] + By[n]$	$Aa_k + Bb_k$
$x[n-n_0]$	$a_k e^{-jk\omega_0 n_0}$
$e^{jM\omega_0 n}x[n]$	a_{k-M}
x_n^*	a^*_{-k}
$x[-n]$	a_{-k}
$x_{(m)}[n] = \begin{cases} x[n/m], & \text{if } n \text{ is a multiple of } m \\ 0, & \text{otherwise} \end{cases}$	$\frac{1}{m}a_k$, period mN
$x[n] * y[n]$	$Na_k b_k$
$x[n]y[n]$	$\sum_{l=\langle N\rangle} a_l b_{k-l}$
$x[n] - x[n-1]$	$(1 - e^{-j\omega_0 k})a_k$
$\sum_{k=-\infty}^{\infty} x[k]$, bounded and periodic only if $a_0 = 0$	$\frac{1}{1 - e^{-jk\omega_0}}a_k$
For real-valued $x[n]$:	$\begin{cases} a_k = a^*_{-k} \\ \text{Re}\{a_k\} = \text{Re}\{a_{-k}\} \\ \text{Im}\{a_k\} = -\text{Im}\{a_{-k}\} \\ \lvert a_k\rvert = \lvert a_{-k}\rvert \\ \sphericalangle a_k = -\sphericalangle a_{-k} \end{cases}$
Even part of $x[n]$	$\text{Re}\{a_k\}$
Odd part of $x[n]$	$j\text{Im}\{a_k\}$
Parseval's relation: $\frac{1}{N}\sum_{n=\langle N\rangle} \lvert x[n]\rvert^2 = \sum_{k=\langle N\rangle} \lvert a_k\rvert^2$	

We, also, provide some popular discrete time signals and their spectral coefficients, in Table 7.2. It is highly recommended to the readers to derive the spectral coefficients from the functions, $x[n]$ of Table 7.2.

Recall that time and frequency domain representations of the signals are one-to-one and onto. In other words, given the signal $x[n]$ it is possible to compute the spectral coefficients $\{a_k\}$, uniquely, from the synthesis equation. Equivalently, given the spectral coefficients $\{a_k\}$, it is possible recover the discrete time function, $x[n]$, uniquely, from the analysis equation. This property is shown as follows:

$$x[n] \leftrightarrow \{a_k\}. \tag{7.39}$$

Table 7.2 Some popular discrete time periodic signals and their spectral coefficients.

Periodic signal $x[n]$ with fundamental period N	Spectral coefficients a_k
$\sum_{k=-\infty}^{\infty} \delta(n-kN)$	$a_k = \frac{1}{N}$, for all k
1	$a_k = \begin{cases} 1, & k = 0, \pm N, \pm 2N, \dots \\ 0, & \text{otherwise} \end{cases}$
In the following, $\omega_0 = \frac{2\pi m}{N}$ and m, N are integers; otherwise, the signal is not periodic.	
$e^{j\omega_0 n}$	$a_k = \begin{cases} 1, & k = \pm m, \pm m \pm N, \pm m \pm 2N, \dots \\ 0, & \text{otherwise} \end{cases}$
$\cos \omega_0 n$	$a_k = \begin{cases} \frac{1}{2}, & k = \pm m, \pm m \pm N, \pm m \pm 2N, \dots \\ 0, & \text{otherwise} \end{cases}$
$\sin \omega_0 n$	$a_k = \begin{cases} \frac{1}{2j}, & k = \pm m, \pm m \pm N, \pm m \pm 2N, \dots \\ -\frac{1}{2j}, & k = -m, -m \pm N, -m \pm 2N, \dots \end{cases}$
$x[n] = \begin{cases} 1, & \lvert n \rvert \le N_1 \\ 0, & N_1 < \lvert n \rvert \le \frac{N}{2} \end{cases}$	$a_k = \begin{cases} \frac{\sin \frac{2\pi k}{N}(N_1 + \frac{1}{2})}{N \sin \frac{\pi k}{N}}, & k \neq 0, \pm N, \pm 2N, \dots \\ \frac{2N_1 + 1}{N}, & k = 0, \pm N, \pm 2N, \dots \end{cases}$

Let us study the following examples to see the application of the properties on solving the problems.

Exercise 7.4 Find the discrete time signal $x[n]$, which has the following spectral coefficients:

$$a_k = \cos\left(k\frac{\pi}{4}\right) + \sin\left(3k\frac{\pi}{4}\right). \tag{7.40}$$

Solution

Using the linearity property, we can split a_k into two terms, as:

$$a_k = \underbrace{a_k^{(1)}}_{\cos(k\frac{\pi}{4})} + \underbrace{a_k^{(2)}}_{\sin(3k\frac{\pi}{4})}, \tag{7.41}$$

each of which represents the signal $x_1[n]$ and the signal $x_2[n]$, with the corresponding spectral coefficients:

$$x_1[n] \leftrightarrow a_k^{(1)} \quad \text{and} \quad x_2[n] \leftrightarrow a_k^{(2)}.$$

Then, linearity property implies that,

$$x[n] = x_1[n] + x_2[n] \leftrightarrow a_k = a_k^{(1)} + a_k^{(2)}.$$

Let us first find the signals $x_1[n]$ and $x_2[n]$, represented by the spectral coefficients of $a_k^{(1)}$ and $a_k^{(2)}$, then add them to find the signal $x[n]$, as follows:

The spectral coefficients $a_k^{(1)}$ and $a_k^{(2)}$, can be written in terms of complex exponential functions as:

$$a_k^{(1)} = \cos\left(k\frac{\pi}{4}\right) = \frac{e^{jk\frac{\pi}{4}} + e^{-jk\frac{\pi}{4}}}{2} \tag{7.42}$$

and

$$a_k^{(2)} = \sin\left(3k\frac{\pi}{4}\right) = \frac{e^{j3k\frac{\pi}{4}} - e^{-j3k\frac{\pi}{4}}}{2j}, \tag{7.43}$$

respectively.

The angular frequency of $a_k^{(1)}$ and $a_k^{(1)}$ is $\omega_0 = \frac{\pi}{4}$. The fundamental period is $N = \frac{2\pi}{\omega_0} = 8$. Thus, the fundamental periods of $x[n]$, as well as, $x_1[n]$ and $x_2[n]$ are all $N = 8$. In other words, $a_k = a_{k\pm 8}$ and $x[n] = x[n \pm 8]$.

The analysis equations for $a_k^{(1)}$ and $a_k^{(2)}$ can be written as:

$$a_k^{(1)} = \frac{1}{8}\sum_{n=-3}^{4} x_1[n]e^{-jkn\frac{\pi}{4}} = \frac{1}{2}e^{jk\frac{\pi}{4}} + \frac{1}{2}e^{-jk\frac{\pi}{4}}, \tag{7.44}$$

and

$$a_k^{(2)} = \frac{1}{8}\sum_{n=-3}^{4} x_2[n]e^{-jkn\frac{\pi}{4}} = \frac{1}{2j}e^{j3k\frac{\pi}{4}} - \frac{1}{2j}e^{-j3k\frac{\pi}{4}}. \tag{7.45}$$

Comparing the left-hand side and the right-hand side of the Equations (7.44) and (7.45), gives the signal, in $-3 \leq n \leq 4$ as follows:

$x_1[1] = x_1[-1] = 4$ and $x_1[n] = 0$ for $k \neq \pm 1$, in the duration $-3 \leq n \leq 4$,
$x_2[3] = -x_2[-3] = 4j$ and $x[n] = 0$ for $n \neq \pm 3$, in the duration $-3 \leq n \leq 4$.

Hence, in one full period $-3 \leq n \leq 4$,

$$x_1[n] = 4\delta[n-1] + 4\delta[n+1]$$

and

$$x_2[n] = 4j\delta[n-3] - 4j\delta[n+3].$$

Finally, in the period of $-3 \leq n \leq 4$, the signal $x[n]$ is written as follows:

$$x[n] = x_1[n] + x_2[n] = 4\delta[n-1] + \delta[n+1] + 4j(\delta[n-3] + \delta[n+3]). \tag{7.46}$$

The periodicity, $x[n] = x[n+N]$ implies that

$$x[-3] = x[-3+8] = x[5]$$

and

$$x[-1] = x[-1+8] = x[7].$$

Hence, $x[n]$ can also be written in the interval of $0 \leq n \leq 7$, as follows:

$$x[n] = 4\delta[n-1] - 4j\delta[n-3] + 4j\delta[n-5] + 4\delta[n-7] \quad \text{for} \quad 0 \leq n \leq 7. \tag{7.47}$$

Note that, $x[n]$ is periodic, where $x[n] = x[n \pm 8]$ for all $-\infty < n < \infty$.

Exercise 7.5 Find the fundamental period and the spectral coefficients of the following discrete time function:

$$x[n] = \cos\left(\frac{6\pi}{13}n + \frac{\pi}{6}\right). \tag{7.48}$$

Solution

The angular frequency of this function is $\omega_0 = \frac{6\pi}{13}$. Recall that $\omega_0 = \frac{2\pi}{N} \cdot m = \frac{2\pi}{13} \cdot 3$. Hence, the fundamental period is $N = 13$.

From Table 7.2, we see that the spectral coefficients of $x'[n] = \cos\omega_0 n$ for $\omega_0 = \frac{2\pi}{N} \cdot m$ is

$$a'_k = \begin{cases} \frac{1}{2}, & \text{for} \quad k = \pm m, \pm m \pm N, \pm m \pm 2N, \dots, \\ 0, & \text{o.w.} \end{cases} \tag{7.49}$$

In this example, $m = 3$ and $N = 13$. Thus, the spectral coefficients for $\cos\omega_0 n$ for $\omega_0 = \frac{2\pi}{N} \cdot m$ is

$$a'_k = \begin{cases} \frac{1}{2}, & \text{for} \quad k = \pm 3, (\pm 3 \pm 13), (\pm 3 \pm 26), \dots, \\ 0, & \text{o.w.} \end{cases} \tag{7.50}$$

From Table 7.1, we see that a time-shift gives a multiplicative exponential factor in the frequency domain:

$$x[n] = x'[n - n_0] \leftrightarrow a'_k \cdot e^{jk\omega_0 n_0}.$$

In this exercise, $\omega_0 = \frac{6\pi}{13}$. In order to find the amount of shift n_0, we factorize ω_0, as follows:

$$x[n] = \cos\left(\frac{6\pi}{13}n + \frac{\pi}{6}\right) = \cos\left(\frac{6\pi}{13}n + \frac{13}{36}\right). \tag{7.51}$$

Hence, $n_0 = -13/36$. Replacing the values of n_0 and ω_0, we obtain

$$a_k = \begin{cases} \frac{1}{2} \cdot e^{jk\frac{\pi}{6}}, & \text{for} \quad k = \pm 3, (\pm 3 \pm 13), (\pm 3 \pm 26), \dots, \\ 0, & \text{o.w.} \end{cases} \tag{7.52}$$

The spectral coefficients are complex numbers. In this case, we need two plots for the magnitude spectrum,

$$|a_k| = \begin{cases} \frac{1}{2}, & \text{for} \quad k = \pm 3, (\pm 3 \pm 13), (\pm 3 \pm 26), \dots, \\ 0, & \text{o.w.,} \end{cases} \tag{7.53}$$

and the phase spectrum,

$$\angle a_k = \begin{cases} k\frac{\pi}{6}, & \text{for} \quad k = \pm 3, (\pm 3 \pm 13), (\pm 3 \pm 26), \dots, \\ 0, & \text{o.w.} \end{cases} \tag{7.54}$$

There are two major differences in the properties of the continuous time and discrete time Fourier series representations:

1) Instead of the differentiation operation of continuous time, we have the difference operation in discrete time.
2) Since both the signal and its corresponding spectral coefficients are periodic, convolution property in time domain requires **circular convolution** of the signals. Similarly, the multiplication property requires **circular convolution** of the spectral coefficients in the frequency domain.

In the rest of Section 7.3, we focus on three properties: the difference, convolution, and multiplication properties as follows.

7.3.1 Difference Property

Given a discrete time periodic function and the corresponding spectral coefficients,

$$x[n] \leftrightarrow a_k, \tag{7.55}$$

the delay operation in time domain corresponds to the multiplication operation in the frequency domain, as follows:

$$x[n-n_0] = \sum_{k=-\infty}^{\infty} a_k e^{jk\omega_0(n-n_0)} = \sum_{k=-\infty}^{\infty} (a_k e^{-j\omega_0 n_0}) e^{jk\omega_0 n} = \sum_{k=-\infty}^{\infty} a'_k e^{jk\omega_0 n}. \tag{7.56}$$

Thus, the spectral coefficients of $x[n-n_0]$ is $a'_k = a_k e^{-j\omega_0 n_0}$:

$$x[n-n_0] \leftrightarrow a_k e^{-jkn_0\frac{2\pi}{N}}. \tag{7.57}$$

When the input–output pairs of an linear time-invariant (LTI) system are periodic, we can use the Fourier series representation to find the spectral coefficients of the output from the spectral coefficients of the input. Difference property converts difference equations into algebraic equations in terms of the spectral coefficients of the input and output. Therefore, the spectral coefficients of the output of an LTI system are obtained from the spectral coefficients of the input, without solving the difference equation by recursive methods.

Exercise 7.6 Consider the discrete time LTI system represented by the following difference equation:

$$y[n] = x[n] - x[n-1], \tag{7.58}$$

a) Find the spectral coefficients of the output of this system, when the spectral coefficients of the input are $\{a_k\}$.
b) Find the spectral coefficients of the output, when the input is $x[n] = \sin(0.1\pi n)$.

Solution

a) Using the difference property, for $x[n] \leftrightarrow a_k$, we obtain,

$$x[n-1] \leftrightarrow a_k e^{-jkn\frac{2\pi}{N}}. \tag{7.59}$$

Since Fourier series representation is linear, and one-to-one and onto, we can replace the corresponding spectral coefficients into the difference equation to obtain its frequency domain representation, which gives the spectral coefficients of the output, in terms of the spectral coefficients of the input,

$$y[n] = x[n] - x[n-1] \leftrightarrow b_k = (1 - e^{-jk\frac{2\pi}{N}})a_k. \tag{7.60}$$

b) Recall that the nonzero spectral coefficients of the input signal, $x[n] = \sin(0.1\pi n)$, is obtained in the previous example as:

$$a_1 = \frac{1}{2j}, \quad a_{-1} = -\frac{1}{2j}, \tag{7.61}$$

where a_k is periodic with $a_k = a_{k\pm N}, \quad \forall k$.

Considering the fact that the fundamental period of the input signal is $N = 20$, the nonzero spectral coefficients of the output is

$$b_1 = \frac{1}{2j}(1 - e^{-j\frac{\pi}{10}}) \quad b_{-1} = \frac{-1}{2j}(1 - e^{j\frac{\pi}{10}}), \tag{7.62}$$

where b_k is periodic, so that $b_k = b_{k\pm 20}, \quad \forall k$.

Exercise 7.7 Consider the discrete time LTI system represented by the following difference equation:

$$y[n] + 0.5y[n-1] = x[n]. \tag{7.63}$$

a) Find the spectral coefficients b_k of the output of this system, when the spectral coefficients of the input are $\{a_k\}$.
b) Find the spectral coefficients, b_k, of the output, when the input is an impulse train, $x[n] = \sum_{k=-\infty}^{\infty} \delta[n-2k]$.

Solution

a) Using the difference property, for $y[n] \leftrightarrow b_k$, we obtain,

$$y[n-1] \leftrightarrow b_k e^{-jkn\frac{2\pi}{N}}. \tag{7.64}$$

Using the linearity property, we can represent both sides of Equation (7.63) by the spectral coefficients:

$$y[n] - 0.5y[n-1] = x[n] \leftrightarrow b_k(1 - 0.5e^{-jkn\frac{2\pi}{N}}) = a_k. \tag{7.65}$$

Leaving the spectral coefficients of b_k in the right-hand side of the equation, we obtain:

$$b_k = \frac{a_k.}{1 - 0.5e^{-jkn\frac{2\pi}{N}}}. \tag{7.66}$$

b) From Table 7.2, we see that,

$$\sum_{-\infty}^{\infty} \delta[n-2k] \leftrightarrow a_k = 1/2. \tag{7.67}$$

where a_k is periodic with $a_k = a_{k\pm 2}, \quad \forall k$.

Considering the fact that the fundamental period of the input signal is $N = 2$,

$$b_k = \frac{1}{2 - e^{-jkn\pi}}, \tag{7.68}$$

where b_k is periodic, so that $b_k = b_{k\pm 2}, \quad \forall k$.

7.3.2 Convolution Property

Convolution operation in the time domain corresponds to the multiplication operation in the frequency domain. Given two functions and their corresponding spectral coefficients,

$$x[n] \leftrightarrow a_k \quad \text{and} \quad y[n] \leftrightarrow b_k, \tag{7.69}$$

with period N, the spectral coefficients of the convolution $x[n] * y[n]$ has the following spectral coefficient, c_k:

$$x[n] * y[n] \leftrightarrow c_k = \frac{1}{N}\sum_{n=0}^{N-1}(x[n] * y[n])e^{-jk\omega_0 n} = \frac{1}{N}\sum_{n=0}^{N-1}\sum_{l=0}^{N-1} x[l]\ y[n-l]e^{-jk\omega_0 n}. \tag{7.70}$$

Changing the dummy variable of summation to $m = n - l$, we obtain,

$$c_k = \frac{1}{N}\sum_{l=0}^{N-1} x[l]e^{-jk\omega_0 l}\sum_{m=0}^{N-1} y[m]e^{-jk\omega_0 m}. \tag{7.71}$$

In Equation (7.71), the first sum is

$$Na_k = \sum_{l=0}^{N-1} x[l]e^{-jk\omega_0 n}, \tag{7.72}$$

and the second sum is

$$Nb_k = \sum_{m=0}^{N-1} y[m]e^{jk\omega_0 m}. \tag{7.73}$$

Hence,

$$x[n] * y[n] \leftrightarrow Na_k b_k. \tag{7.74}$$

Remark 7.2 In Equations (7.70), the limits of the convolution summation do not range in $(-\infty, \infty)$, as in the classical definition of convolution. Instead, we cover one full period,

$$x[n] * y[n] = \sum_{l=0}^{N-1} x[l]\ y[n-l]. \tag{7.75}$$

This is a special case of convolution operation used for periodic signals, called circular convolution, where we slide one of the functions on the other one, as we multiply the overlaps, until we cover one period, N.

Remark 7.3 We keep in mind that both $x[n \pm N] \leftrightarrow a_{k\pm N}$ and $y[n \pm N] \leftrightarrow b_{k\pm N}$ are periodic, for $-\infty < k < \infty$ and $-\infty < n < \infty$.

Exercise 7.8 Consider the following discrete time signal,

$$x[n] = \sin 0.1\pi n + \sin 0.1\pi(n-1). \tag{7.76}$$

a) What is the fundamental period of $x[n]$?
b) Find the spectral coefficients a_k of the signal $x[n]$.

Solution

a) The fundamental period is
$N = m\frac{2\pi}{\omega_0} = m\frac{2\pi}{0.1\pi} = 20.$
b) Define, $x'[n] = \sin 0.1\pi n$. Then, the corresponding spectral coefficients in one full period is are

$$a'_1 = \frac{1}{2j}, \quad a'_{-1} = -\frac{1}{2j}.$$

Since a_k is periodic, for all $-\infty < k < \infty$,

$$a_{k\pm 20} = a_k.$$

Using the linearity and time shift properties, we get,

$$x[n] = x'[n] + x'[n-1] \leftrightarrow (1 + e^{-jk\frac{2\pi}{N}})a_k. \tag{7.77}$$

Remark 7.4 In the aforementioned example, we can use the linearity property easily, since both terms have the same fundamental period.

Motivating Question: What if the fundamental period of two terms were different?
The following exercise answers this question.

Exercise 7.9 Consider a periodic signal,

$$y[n] = x_1[n] + x_2[n],$$

where $x_1[n]$ and $x_2[n]$ has the fundamental period N_1 and N_2, respectively. What is the fundamental period of the signal $y[n]$?

Solution

The fundamental period of $y[n]$ is the greatest common divisor of $N_1 \cdot N_2$. Indeed,

$$y[n] = x_1[n] + x_2[n] = x_1[n + N_1N_2] + x_2[n + N_1N_2] = y[n + N_1N_2] \tag{7.78}$$

is satisfied for N_1N_2. Hence, it is also satisfied for the greatest common divisor of N_1N_2.

Exercise 7.10 Find the spectral coefficients a_k of the sequence $x[n]$ shown in Figure 7.6a.

This sequence has a fundamental period of $N = 5$. We observe that $x[n]$ may be viewed as the sum of the square wave $x_1[n]$ in Figure 7.6b and the de sequence $x_2[n]$ in Figure 7.6c.

Solution

Denoting the Fourier series coefficients representations by

$$x_1[n] \leftrightarrow b_k \tag{7.79}$$

and

$$x_2[n] \leftrightarrow c_k, \tag{7.80}$$

we can use the linearity property of Table 7.2 to obtain the spectral coefficients of the signal $x[n]$ as follows:

$$a_k = b_k + c_k. \tag{7.81}$$

Using the result of Exercise 5.3 for $N = 5$ and $N_1 = 1$, Fourier series coefficients b_k corresponding to $x_1[n]$ can be expressed as:

$$b_k = \begin{cases} \frac{1}{5}\frac{\sin(3\pi k/5)}{\sin(\pi k/5)}, & \text{for } k \neq 0, \pm 5, \pm 10, \dots, \\ \frac{3}{5}, & \text{for } k = 0, \pm 5, \pm 10, \dots, \end{cases} \tag{7.82}$$

The sequence $x_2[n]$ has only a constant value, which is captured by its zeroth Fourier series coefficient:

$$c_0 = \frac{1}{5}\sum_{n=0}^{4} x_2[n] = 1. \tag{7.83}$$

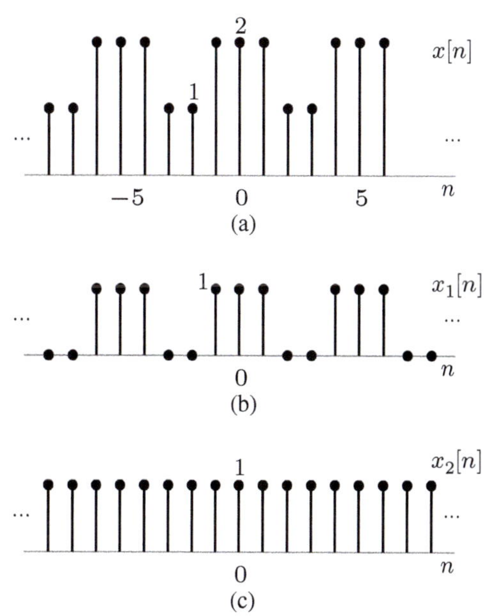

Figure 7.6 (a) Periodic signal $x[n]$ and its representation as a sum of (b) the square wave $x_1[n]$ and (c) the constant signal $x_2[n]$.

Since the discrete time Fourier series coefficients are periodic, it follows that $c_k = 1$ whenever k is an integer multiple of 5. The remaining coefficients of $x_2[n]$ must be zero, because; $x_2[n]$ contains only a dc component. We can now substitute the expressions for b_k and c_k into $a_k = b_k + c_k$ to obtain

$$a_k = \begin{cases} b_k = \frac{1}{5}\frac{\sin(3\pi k/5)}{\sin(\pi k/5)}, & \text{for } k \neq 0, \pm 5, \pm 10, \dots, \\ \frac{8}{5}, & \text{for } k = 0, \pm 5, \pm 10, \dots. \end{cases} \tag{7.84}$$

Exercise 7.11 Find the signal $x[n]$, described by the following properties:

1) $x[n]$ is periodic with period $N = 6$.
2) $\sum_{n=0}^{5} x[n] = 2$
3) $\sum_{n=2}^{7} (-1)^n x[n] = 1$
4) $x[n]$ has the minimum power per period among the set of signals satisfying the preceding three conditions.

Solution

We denote the Fourier series coefficients of the signal, $x[n]$, as follows:

$$x[n] \leftrightarrow a_k. \tag{7.85}$$

From Fact 2, we conclude that $a_0 = 1/3$. Noting that $(-1)^n = e^{-j\pi n} = e^{-j(2\pi/6)3n}$, we see from Fact 3 that $a_3 = 1/6$. From Parseval's relation (see Table 7.2), the average power in $x[n]$ is

$$P = \sum_{k=0}^{5} |a_k|^2. \tag{7.86}$$

Since each nonzero coefficient contributes a positive amount to P, and since the values of a_0 and a_3 are prespecified, the value of P is minimized by choosing $a_1 = a_2 = a_4 = a_5 = 0$. It then follows that

$$x[n] = a_0 + a_3 e^{j\pi n} = (1/3) + (1/6)(-1)^n, \tag{7.87}$$

which is sketched in Figure 7.7.

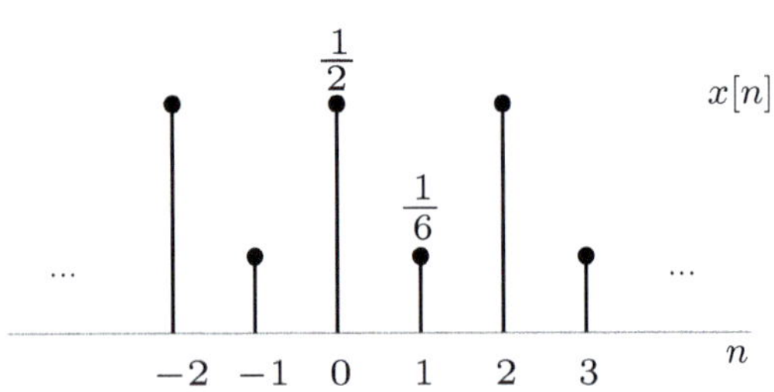

Figure 7.7 Sequence $x[n]$ that is consistent with the properties specified in the example.

7.3.3 Multiplication Property

Multiplication operation in the time domain corresponds to the convolution operation in the frequency domain.

Given two functions and their corresponding spectral coefficients,

$$x[n] \leftrightarrow a_k \quad \text{and} \quad y[n] \leftrightarrow b_k, \tag{7.88}$$

with period N, multiplication of the signals $x[n]$ and $y[n]$ can be written in terms of Fourier series representation, as follows:

$$x[n] \cdot y[n] = \left(\sum_{k=0}^{N-1} a_k e^{jk\frac{2\pi}{N}kn} \right) \cdot \left(\sum_{l=0}^{N-1} b_l e^{-j\frac{2\pi}{N}ln} \right). \tag{7.89}$$

Arranging the summations and changing the dummy variable $l = m - k$ gives,

$$x[n] \cdot y[n] = \sum_{k=0}^{N-1}\sum_{l=0}^{N-1} a_k b_l e^{-jn2\pi\frac{k+l}{N}} = \sum_{k=0}^{N-1}\sum_{m=0}^{N-1} a_k b_{m-k} e^{-jn2\pi\frac{m}{N}}. \tag{7.90}$$

Therefore, the multiplication of two discrete time signals in the time domain corresponds to the convolution of the spectral coefficients in the frequency domain:

$$x[n] \cdot y[n] \leftrightarrow \sum_{k=0}^{N-1} a_k b_{n-k}. \tag{7.91}$$

In the Fourier domain, we perform circular convolution, where the limits of the summation cover only one full period, N.

Remark 7.5 We keep in mind that both $x[n] \leftrightarrow a_k$ and $y[n \pm N] \leftrightarrow b_{k \pm N}$ are periodic, with $-\infty < k < \infty$ and $-\infty < n < \infty$.

Exercise 7.12 Given two periodic signals, with the fundamental period $N = 7$ and the corresponding Fourier series representation,

$$x[n] \leftrightarrow a_k \quad \text{and} \quad y[n] \leftrightarrow b_k, \tag{7.92}$$

a) Find the spectral coefficients of the following signal, in terms of a_k and b_k:

$$w[n] = \sum_{\forall r} x[r] \cdot y[n - r]. \tag{7.93}$$

b) Suppose, now, that $x[n] = y[n]$ and the spectral coefficients of

$$w[n] = \sum_{\forall r} x[r] \cdot x[n - r] \leftrightarrow c_k \tag{7.94}$$

is as follows:

$$c_k = \frac{1}{7} \frac{\sin^2(3\pi k/7)}{\sin^2(\pi k/7)}. \tag{7.95}$$

Find and plot the signal, $x[n]$.

c) Find and plot $w[n]$.

Solution

a) The signal $w[n]$ is also periodic with $N = 7$. Therefore, the limits of the summation should cover only one full period of N consecutive values of r, which indicates a circular convolution operation.

From the Convolution property, we know that

$$x[n] * y[n] \leftrightarrow N a_k b_k. \tag{7.96}$$

Since $w[n] = x[n] * y[n]$, Fourier series coefficients of $w[n]$ for $N = 7$ is

$$c_k = Na_k b_k = 7a_k b_k. \tag{7.97}$$

b) When $x[n] = y[n]$

Then, the convolution of the signal $x[n]$ by itself has the following spectral coefficients:

$$x[n] * x[n] \leftrightarrow Na_k^2. \tag{7.98}$$

In this case, the spectral coefficients of $w[n]$ is the $c_k = 1/7a_k^2$,

$$c_k = \frac{1}{7}\frac{\sin(3\pi k/7)}{\sin(\pi k/7)}\frac{\sin(3\pi k/7)}{\sin(\pi k/7)} = \frac{1}{7}a_k a_k. \tag{7.99}$$

From Table 7.2, we can see that the spectral coefficients

$$a_k = \frac{\sin(3\pi k/7)}{\sin(\pi k/7)}, \tag{7.100}$$

belongs to the square wave signal, with $N_1 = 1$ and $N = 7$ (see Figure 7.8a).

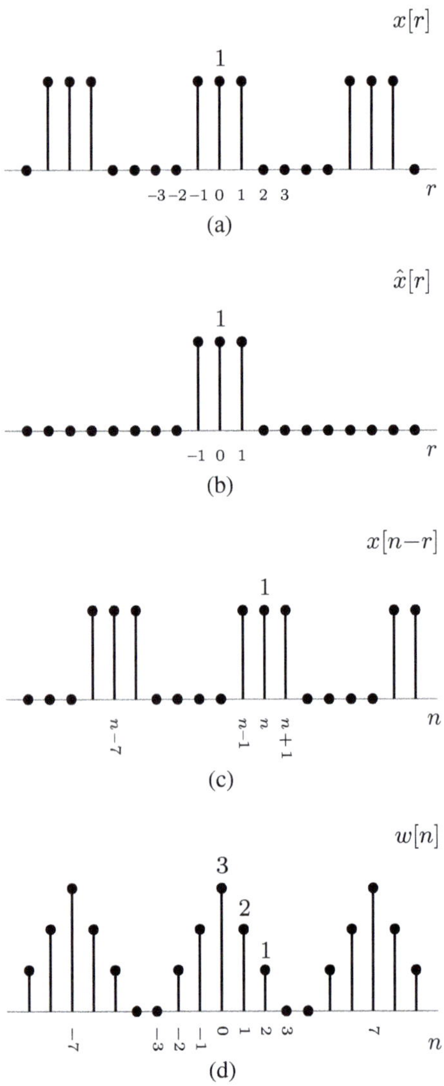

Figure 7.8 (a) The square-wave sequence $x[r]$ in Example 7.12; (b) the sequence $\hat{x}[r]$ equal to $x[r]$ for $-3 \leq r \leq 3$ and zero otherwise; (c) the sequence $x[n-r]$; (d) the sequence $w[n]$ equal to the periodic convolution of $x[n]$ with itself and to the aperiodic convolution of $\hat{x}[n]$ with $x[n]$.

c) Using the periodic convolution property, we see that

$$w[n] = \sum_{r=\langle 7\rangle} x[r] \cdot x[n-r] = \sum_{r=-3}^{3} x[r] \cdot x[n-r], \tag{7.101}$$

where, in the last equality, we have chosen to sum over the interval $-3 \leq r \leq 3$. Except for the fact that the sum is limited to a finite interval, the product-and-sum method for evaluating convolution is applicable here. In fact, we can convert this equation to an ordinary convolution by defining a signal $\hat{x}[n]$ that equals $x[n]$ for $-3 \leq n \leq 3$ and is zero otherwise. Then, from this equation,

$$w[n] = \sum_{r=-3}^{3} \hat{x}[r] \cdot x[n-r] = \sum_{r=-\infty}^{+\infty} \hat{x}[r] \cdot x[n-r]. \tag{7.102}$$

That is, $w[n]$ is the aperiodic convolution of the sequences $\hat{x}[n]$ and $x[n]$.

The sequences $x[r], \hat{x}[r]$, and $x[n-r]$ are sketched in Figure 7.8a, b, and c. From the figure, we can immediately calculate $w[n]$. In particular, we see that $w[0] = 3$; $w[-1] = w[1] = 2$; $w[-2] = w[2] = 1$; and $w[-3] = w[3] = 0$. Since $w[n]$ is periodic with period 7, we can sketch $w[n]$, as shown in Figure 7.8d.

7.4 Discrete Time LTI Systems with Periodic Input and Output Pairs

Consider an LTI system, represented by the impulse response $h[n]$ and equivalently by the following difference equation:

$$\sum_{k=0}^{N} a_k y[n-k] = \sum_{k=0}^{M} b_k x[n-k]. \tag{7.103}$$

Suppose that a periodic input $x[n]$ generates a periodic output $y[n]$ (Figure 7.9).

Motivating Question: What is the relationship between the spectral coefficients of the input, $\{a_k\}$ and spectral coefficients of the output, $\{b_k\}$?

This is a crucial question, because, if we can establish a relationship between the spectral coefficients of the input and output, then, we can just feed the spectral coefficients of the input to receive the spectral coefficients of the output. Then, based on the spectral coefficients of the output, we can construct the output signal. This approach saves a great deal of cost in implementing the LTI systems.

In order to find the relationship between the spectral coefficients of the input and that of the output, we use the harmonically related complex exponential eigen-functions and the linearity property. Then, we find the corresponding eigenvalues of the LTI system, as explained in Section 7.4.1.

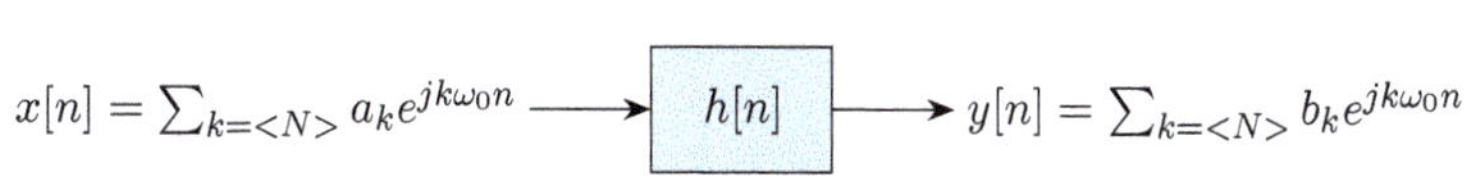

Figure 7.9 An LTI system, where the periodic input output pairs are represented by Fourier series.

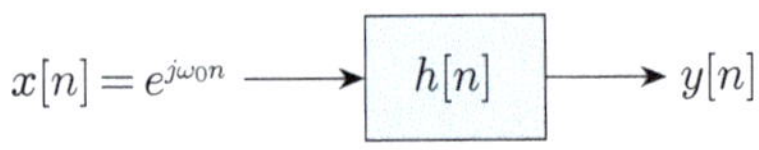

Figure 7.10 A discrete time LTI system, represented by the impulse response, $h[n]$, is fed by the eigenfunction, $x[n] = e^{j\omega_0 n}$. The corresponding output is $y[n] = e^{j\omega_0 n}H(e^{j\omega_0})$.

7.4.1 Eigenfunctions, Eigenvalues, and Transfer Functions of a Discrete Time LTI Systems

Suppose that the input to an LTI system, shown in Figure 7.10, is a discrete time complex exponential,

$$x[n] = e^{j\omega_0 n}, \tag{7.104}$$

then, convolution summation for this input is

$$y[n] = x[n] * h[n] = \sum_{k=-\infty}^{\infty} h[k]e^{j\omega_0(n-k)} = e^{j\omega_0 n}\sum_{k=-\infty}^{\infty} h[k]e^{-j\omega_0 k} = e^{j\omega_0 n}H(e^{j\omega_0}), \tag{7.105}$$

where

$$H(e^{j\omega_0}) = \sum_{k=-\infty}^{\infty} h[k]e^{-j\omega_0 k}. \tag{7.106}$$

The function $H(e^{j\omega_0})$ is the superposition of the harmonically related complex exponentials, where the weights are the range of the impulse response for each $k = 0, \pm1, \pm2, \pm3 \dots$. The purely imaginary exponent of the exponential function is scaled by the angular frequency ω_0. Note that since the impulse response is not a periodic function, the limits of the summation range in $(-\infty, \infty)$. This equation is not to be confused with the Fourier series representation of periodic signals.

As in the continuous case, we can define the eigenvalue of a discrete time system, when the input is a discrete time complex exponential eigenfunction, $x[n] = e^{j\omega_0 n}$, as follows.

Definition 7.1 Definition: Eigenvalue of a discrete time LTI system for a complex exponential eigenfunction $x[n] = e^{j\omega_0 n}$ is defined as:

$$H(e^{j\omega_0}) = \sum_{k=\infty}^{\infty} h[k]e^{-j\omega_0 k}. \tag{7.107}$$

Thus, when the input is an eigenfunction, $x[n] = e^{j\omega_0 n}$, the corresponding output is the scaled version of the eigen-function, where the scaling factor is the eigenvalue of the LTI system (Figure 7.10),

$$y[n] = e^{j\omega_0 n}H(e^{j\omega_0}). \tag{7.108}$$

In general, any exponential input $e^{\lambda n}$ is directly passed to the output with a scaling factor,

$$H(e^{\lambda}) = \sum_{k=-\infty}^{\infty} h[k]e^{-\lambda k}, \tag{7.109}$$

which is called **transfer function**.

The output to the exponential input $x[n] = e^{\lambda n}$ is then written as:

$$y[n] = e^{\lambda n}H(e^{\lambda}).$$

Inserting the output $y[n]$ and the exponential input $x[n] = e^{\lambda n}$, in the difference equation, we obtain,

$$\sum_{k=0}^{N} a_k H(e^{\lambda})e^{n-k} = \sum_{k=0}^{M} b_k e^{n-k}. \tag{7.110}$$

Arranging Equation (7.111), we obtain the transfer function for discrete time LTI systems in terms of the parameters of the difference equation, as follows:

$$H(e^{\lambda}) = \frac{\sum_{k=0}^{M} b_k e^{\lambda k}}{\sum_{k=0}^{N} a_k e^{\lambda k}}. \tag{7.111}$$

Hence, when the input of a discrete time LTI system is an exponential function, the corresponding output is just the scaled version of the input, where the scaling factor $H(e^{\lambda})$ is the **transfer function**.

7.4.2 Relationship Between the Fourier Series of Periodic Input and Output Pairs of Discrete Time LTI Systems

Suppose, now, that the input to an LTI system is the kth harmonic of a discrete time complex exponential function,

$$x[n] = e^{jk\omega_0 n}, \tag{7.112}$$

then, the corresponding output becomes

$$y[n] = x[n] * h[n] = \sum_{l=-\infty}^{\infty} h[l]e^{jk\omega_0(n-l)} = e^{jk\omega_0 n}H(e^{jk\omega_0}), \tag{7.113}$$

where

$$H(e^{jk\omega_0}) = \sum_{l=-\infty}^{\infty} h[l]e^{-jk\omega_0 l}. \tag{7.114}$$

In general, we can superpose harmonically related complex exponentials to represent the input signal by the Fourier series,

$$x[n] = \sum_{k=\langle N\rangle} a_k e^{jk\omega_0 n}, \tag{7.115}$$

to obtain the Fourier series representation of the output,

$$y[n] = \sum_{k=\langle N\rangle} b_k e^{jk\omega_0 n} = \sum_{k=\langle N\rangle} a_k H(e^{jk\omega_0})e^{jk\omega_0 n}, \tag{7.116}$$

Therefore, the relationship between the spectral coefficients of the input and output pairs of a discrete time LTI system is

$$\boxed{b_k = a_k H(e^{jk\omega_0}),} \tag{7.117}$$

where the eigenvalue for the kth harmonic eigenfunction is

$$\boxed{H(e^{jk\omega_0}) = \sum_{n=-\infty}^{\infty} h[n]e^{-jk\omega_0 n}.} \tag{7.118}$$

Exercise 7.13 Given the following discrete time LTI system:

$$y[n] = x[n-1] - x[n-2]. \tag{7.119}$$

a) Find the impulse response of this system.
b) Find the eigenvalue of this system corresponding to the kth harmonic eigen-function.

Solution

a) The impulse response of this LTI system is easily obtained by replacing the input by the impulse function,

$$h[n] = \delta[n-1] - \delta[n-2]. \tag{7.120}$$

b) The eigenvalue of this system corresponding to the kth harmonic eigenfunction, $x[n] = e^{jk\omega_0 n}$, can be obtained by replacing $z = e^{jk\omega_0}$, as follows:

$$H(e^{jk\omega_0}) = e^{-jk\omega_0} - e^{-2kj\omega_0}. \tag{7.121}$$

Exercise 7.14 Consider an LTI system represented by the impulse response, $h[n] = \alpha^n u[n], -1 < \alpha < 1$,

Find the Fourier series representation of the output, when the input is $x[n] = \cos \omega_0 n$, where the fundamental period is $N = 4$.

Solution

We know that the spectral coefficients a_k of the input and the spectral coefficients b_k of the output are related by the eigenvalue of the LTI system, as follows:

$$b_k = a_k H(e^{jk\omega_0}). \tag{7.122}$$

Let us first find the spectral coefficients of the input:

$$x[n] \leftrightarrow a_1 = a_{-1} = \frac{1}{2}. \tag{7.123}$$

Since the period is $N = 4$, the spectral coefficients will repeat at every N, $a_k = a_{k\pm 4}$ for all k. Next, let us find the eigenvalue of the system for the eigenfunction input, $x[n] = e^{j\omega_0 n}$, as follows:

$$H(e^{j\omega_0}) = \sum_{k=-\infty}^{\infty} h[k]e^{-j\omega_0 k} = \sum_{k=0}^{\infty} \alpha^k e^{-j\omega_0 k} = \frac{1}{1-\alpha e^{-j\omega_0}}. \tag{7.124}$$

Therefore, kth harmonics of the spectral coefficient of the output is

$$b_k = a_k H(e^{jk\omega_0}) = \frac{a_k}{1-\alpha e^{-jk\omega_0}}. \tag{7.125}$$

Considering the nonzero spectral coefficients of the input for $k = \pm 1$ and inserting the value of the fundamental period $N = 4$ in angular frequency, $\omega_0 = \frac{2\pi}{N} = \frac{\pi}{2}$, the spectral coefficients of the output becomes

$$b_k = \begin{cases} \frac{0.5}{1-\alpha e^{-jk\frac{\pi}{2}}}, & \text{for} \quad k = \pm 1, (\pm 1 \pm N), (\pm 1 \pm 2N), (\pm 1 \pm 3N), \ldots, \\ 0, & \text{otherwise.} \end{cases} \tag{7.126}$$

Finally, the Fourier series representation of the output, then, becomes

$$y[n] = \sum_{-\infty}^{\infty} a_k H(e^{jk\omega_0}) e^{jk\omega_0 n} = \frac{0.5}{1-\alpha e^{jk\frac{\pi}{2}}} + \frac{0.5}{1-\alpha e^{-jk\frac{\pi}{2}}}. \tag{7.127}$$

Remark 7.6 The eigenvalues of a discrete time LTI system not only relate the harmonical complex exponential inputs and the corresponding outputs of the system,

$$L\Phi_k[n] = H(e^{jk\omega_0 n})\Phi_k[n], \tag{7.128}$$

but it, also, relates the spectral coefficients of the periodic input and output signals,

$$b_k = a_k H(e^{jk\omega_0}). \tag{7.129}$$

This is due to the beauty of linearity, and time invariance, together with the harmony of the exponential functions.

7.5 Chapter Summary

Is it possible to extend the Fourier series theorems to discrete time periodic signals and systems? If yes, what type of Hilbert space can be defined to represent spectral coefficients of a discrete time periodic function, in a function space? What are the basis functions, which span this space? Is this an infinite dimensional space as in the continuous time signals or is it finite dimensional? What is the dimension of this space?

In this chapter, we try to answer the above questions. We define discrete time Fourier series for discrete time periodic functions and represent them in a Hilbert space spanned by the orthogonal harmonically related complex exponential functions. The discrete nature of signal, $x[n]$, with period N, enables us to replace the integral operation with the **summation operation** to compute the spectral coefficients, $\{a_k\}$, of the signal. Considering the fact that harmonically related complex exponentials are periodic functions and superposition of the periodic functions are also periodic, we end up with periodic and discrete spectral coefficients, $a_k = a_{k+N}$, which corresponds to the coordinates of the discrete time periodic functions in Hilbert space, spanned by N harmonically related and orthogonal discrete time complex exponentials. Thus, discrete time functions can be represented in **finite-dimensional Hilbert spaces**. The period of the spectral coefficients is the same as the period of the signal in the time domain, which defines the dimension of the Hilbert space.

The relationship between the input and output signals can be uniquely identified by the spectral coefficients of the input and output functions. The spectral coefficients of the output are just the scaled version of the spectral coefficients of the input, where the scale factor is the eigenvalue of the discrete time LTI systems, corresponding to the complex exponential eigenfunctions. This scaling factor is called the **transfer function** and it uniquely represents the discrete time LTI system.

In summary, we represent a discrete time periodic signal in an N-dimensional Hilbert space spanned by harmonically related discrete complex exponentials, where the coordinate of each function is also periodic with the same period of the discrete time function. Furthermore, a discrete time LTI system is uniquely represented by the eigenvalue of the system corresponding to the complex exponential eigenfunction, given at the input.

Problems

7.1 Consider the discrete time signal given as follows:

$$x[n] = 1 + \sin\left(\frac{2\pi}{3}\right)n + 3\cos\left(\frac{\pi}{5}\right)n.$$

a) Find the fundamental period of this signal.
b) Find the coordinates of this function, in a Hilbert space spanned by discrete complex exponentials, $\Phi_k[n] = e^{jk\omega_0 n}, \forall k, \forall n$.
c) Plot the spectral coefficients in polar coordinates.
d) Plot the spectral coefficients in Cartesian coordinates.

7.2 Consider the discrete time periodic signal given in one full period,

$$x[n] = \begin{cases} 1, & \text{for } 0 \leq n \leq 4, \\ 0, & 5 \leq n \leq 9. \end{cases}$$

a) Find the coordinates of this function, in a Hilbert space spanned by discrete complex exponentials, $\Phi_k[n] = e^{jk\omega_0 n}, \forall k, \forall n$.
b) Plot the spectral coefficients of this function.

7.3 Find the discrete time signal $x[n]$, which has the following spectral coefficients:

$$a_k = \cos\left(k\frac{\pi}{3}\right) + \cos\left(k\frac{\pi}{4}\right).$$

7.4 Consider the following discrete time periodic signal:

$$x[n] = \sin\left(\frac{6\pi}{13}n + \frac{\pi}{2}\right)$$

a) Find the fundamental period of this signal.
b) Find and plot the spectral coefficients in the polar coordinate system.

7.5 Consider the following discrete time periodic function:

$$x[n] = \sum_{m=-\infty}^{+\infty} \{2\delta[n-m] + \delta[n-2m-1]\}.$$

a) Plot this function and find the fundamental period.
b) Calculate the values of spectral coefficients of $x[n]$ over a period using the **analysis equation**.

7.6 A discrete time, real and **odd** signal $x[n]$ with period $N = 4$, has the following Fourier series coefficients a_k:

$$a_9 = 2j, \quad a_{14} = j, \quad a_{19} = 4j.$$

Find a_0, a_1, a_{-6}, and a_{-3}.

7.7 A discrete time real and periodic signal $x[n]$ has the fundamental period $N = 5$. The nonzero spectral coefficients of $x[n]$ are

$$a_0 = 4, \ a_2 = a_{-2} = 4e^{-j\pi/6}, \quad \text{and} \quad a_4 = a_{-4} = 2e^{j\pi/3}.$$

Find $x[n]$.

7.8 Consider the discrete time periodic signal $x[n]$ with $N = 8$ given as follows:

$$x[n] = 2 - \sin\frac{\pi n}{4} \quad \text{for } 0 \leq n \leq 7.$$

a) Find Fourier series coefficients a_k of $x[n]$.
b) Plot the magnitude and phase diagram for a_k.

7.9 A discrete time periodic signal $x[n]$ has the fundamental period $N = 16$. Fourier series coefficients a_k and the signal $x[n]$ satisfies the following properties:

$$a_k = -a_{k-8}$$

$$x[2n+1] = (-1)^{n+1}.$$

a) Find and plot $x[n]$.
b) Find and plot the magnitude and phase of the Fourier series coefficients a_k.

7.10 The signal $x[n]$ is a real-valued discrete time period signal whose fundamental period is N. The complex spectral coefficients of $x[n]$ have the following form:

$$a_k = b_k - jc_k,$$

where b_k and c_k are real-valued sequences.
a) Prove that $a_k^* = a_{-k}$.
b) Find the relation between b_k and b_{-k}.
c) Find the relation between c_k and c_{-k}.

7.11 Fourier series representation of a discrete time periodic signal is as follows:

$$x[n] = \sum_{k=\langle N\rangle} a_k e^{jk(2\pi/N)n}.$$

Find Fourier series representations of the following signals in terms of a_k.
a) $x[n] - x\left[n - \frac{N}{2}\right]$, assume that N is even.
b) $x[n] + x\left[n + \frac{N}{2}\right]$, assume that N is even.
(Hint: This signal is periodic with fundamental period $N/2$)
c) $(-1)^n x[n]$, assume that N is even.
d) $(-1)^n x[n]$, assume that N is even.
(Hint: This signal is periodic with fundamental period $2N$)
e) $\dfrac{(1-(-1)^{n+1})}{2} x[n]$

7.12 Consider the following discrete time signals,

$$x[n] = 1 + \cos\left(\frac{\pi n}{3}\right), \quad y[n] = \sin\left(\frac{\pi n}{3} + \frac{\pi}{4}\right).$$

a) Find the Fourier series coefficients of $x[n]$.
b) Find the Fourier series coefficients of $y[n]$.
c) Use convolution property to calculate the Fourier Series coefficients of $z[n] = x[n] * y[n]$.

7.13 The discrete time periodic signals $x[n]$ and $y[n]$ have the same fundamental period N. Let

$$g[n] = \sum_{k=-\infty}^{+\infty} x[k]y[n-k].$$

be their periodic convolution. What is the fundamental period of $g[n]$? Verify your answer.

7.14 The discrete time periodic signals $x[n]$ and $y[n]$ has the fundamental period $N = 4$. The corresponding Fourier series coefficients are a_k and b_k, respectively. Mathematically,

$$x[n] \longleftrightarrow a_k \quad \text{and} \quad y[n] \longleftrightarrow b_k.$$

Moreover, it is given that

$$2a_0 = 2a_3 = a_1 = a_2 = 1 \quad \text{and} \quad b_0 = b_1 = b_2 = b_3 = 4$$

and

$$g[n] = x_1[n]x_2[n] \longleftrightarrow c_k.$$

a) What is the period of $g[n]$? Verify your answer.
b) Use multiplication property to find c_k.

7.15 The impulse train

$$x[n] = \sum_{n=-\infty}^{+\infty} 2\delta[n-k]$$

is fed into a LTI system. The corresponding output of the system is found as

$$y[n] = \cos\left(\frac{5\pi}{2}n + \frac{9\pi}{4}\right).$$

Find the eigenvalues of $H(e^{j5k\pi/2})$ at $k = 0, 1, 2,$ and 3.

Assume that the aforementioned LTI system is applied to the following periodic input signals. Find the corresponding output for each input signal.

a) $x_1[n] = (-1)^{n+2}$
b) $x_2[n] = 1 + \cos\left(\frac{\pi}{2}n + \frac{\pi}{2}\right)$
c) $x_3[n] = \sum_{k=-\infty}^{+\infty} 2^{4k-n}u[n-k].$

7.16 Consider a causal discrete time LTI system whose difference equation is given here:

$$y[n] - \frac{1}{4}y[n-1] = x[n-1].$$

a) Plot the block diagram of the system.
b) Determine the Fourier series representation of the output $y[n]$ if the input $x[n] = \cos\left(\frac{\pi}{4}n\right) + \cos\left(\frac{\pi}{2}n\right)$ is fed into the system.

7.17 Consider a discrete time LTI system whose impulse response $h[n]$ is

$$h[n] = \begin{cases} 2, & 0 \le n \le 3 \\ -1, & -3 \le n \le -2, \\ 0, & \text{otherwise.} \end{cases}$$

Calculate the Fourier series coefficients of the output $y[n]$ if the input to the system is $x[n]$ is

$$x[n] = \sum_{n=-\infty}^{+\infty} \delta[n-6k].$$

7.18 Consider the discrete time LTI system represented by the following difference equation:

$$y[n] - 2y[n-1] + y[n-2] = x[n]. \tag{7.130}$$

a) Find the spectral coefficients b_k of the output of this system, when the spectral coefficients of the input are $\{a_k\}$.
b) Find the spectral coefficients, b_k, of the output, when the input is, $x[n] = \cos\frac{3\pi}{2}n.$

7.19 Consider an LTI system represented by the impulse response,

$$h[n] = a^{|n|}, \qquad |a| < 1. \tag{7.131}$$

Find the Fourier series representation of the output, when the input is $x[n] = \cos\omega_0 n$, where the fundamental period is $N = 3$.

7.20 In this programming task, we try to approximate two different periodic functions by using their Fourier Series representations.

a) Firstly, write a function that computes the first $n + 1$ Fourier Series coefficients of a given signal. Your function takes the given signal, the period of the signal, and the number of coefficients as input. You will need to compute the DC component and the coefficients of n harmonic. (For safety you can compute one DC coefficient, n coefficients for cosine components, and n coefficients for sine components.)

b) Write a function to generate the approximate function by using Fourier Series coefficients.

c) Generate the following square wave function by dividing $[-0.5, 0.5]$ range into 1000 points.

$$s[n] = \begin{cases} -1, & \text{if } -0.5 < n < 0, \\ 1, & \text{if } 0 < n < 0.5. \end{cases}$$

You can assume that this function is periodic and aforementioned definition belongs to one cycle of the signal. Compute n Fourier Series coefficients of the given function by using the function you implemented in the first part. Then, generate the approximate function by using the function you implemented in the second part. Plot both the original function and the approximated function on the same plot by setting $n = [1, 5, 10, 50, 100]$. (You can use plt.plot() function for better visualization.)

d) Generate the following sawtooth function by dividing $[-0.5, 0.5]$ range into 1000 points. (You can use **scipy.signal.sawtooth()** function or you can implement it by hand.)

$$s[n] = \begin{cases} 1 + 2n, & \text{if } -0.5 < n < 0, \\ -1 + 2n, & \text{if } 0 < n < 0.5. \end{cases}$$

Apply the procedure in the third part to the new signal. What is the effect of increasing n?

You should write your code in `Python` and no library is allowed other than `matplotlib.pyplot`, `numpy` and `scipy.signal.sawtooth()`.

8

Continuous Time Fourier Transform and Its Extension to Laplace Transform

"Primary causes are unknown to us; but are subject to simple and constant laws, which may be discovered by observation, the study of them being the object of natural philosophy."

Jean Baptiste Joseph Fourier

A visual introduction to Fourier Transform @ https://384book.net/v0801

In Chapter 6, we decomposed a continuous time periodic signal into its frequency harmonics by using Fourier series representation. The spectral coefficients of Fourier series give us the amount of each frequency component in a periodic signal.

We have seen that Fourier series representation enables us to observe and analyze the frequency content of the periodic signals. In this representation, it is possible to attenuate the *unwanted* frequency components or to boost some *important* frequencies of signals by changing the spectral coefficients of the signal. Unfortunately, Fourier series analysis is applicable **only** to the **periodic functions**.

Motivating Question: What if the function is not periodic? Can we extend the Fourier series representation to aperiodic functions?

The answer is **yes**!

The generalized form of Fourier series, which enables us to represent both periodic and aperiodic functions in terms of their frequency content, is called the Fourier transform.

In this chapter, we shall extend the Fourier series representation of periodic functions to aperiodic functions to define **Fourier transforms**. We shall study the properties of Fourier transform. We shall see that Fourier transforms are very important tools for analyzing natural systems. They are also extremely useful to design and implement a wide range of human-made signals and systems. However, it is not possible to find a finite Fourier transform of all functions. In order to utilize the transform domain techniques for such functions, we shall generalize the continuous time Fourier transform to **Laplace transform** by extending the exponential basis functions with a purely imaginary exponent to exponential functions with complex exponent.

Signals and Systems: Theory and Practical Explorations with Python,
First Edition. Fatoş Tunay Yarman Vural and Emre Akbaş.

Companion website: www.wiley.com/go/signalsandsystems

8.1 Fourier Series Extension to Aperiodic Functions

The idea of extending the Fourier analysis to aperiodic functions is simple. We can assume an aperiodic continuous time function, $x(t)$, as a function with infinite period. In other words, we assume that the fundamental period of the function, $x(t)$, approaches infinity, $T \to \infty$.

Recall that a function is periodic if $\exists T$, such that

$$x(t) = x(t + T). \tag{8.1}$$

Note: The period, T, can be as long as we need. When a signal has an infinite period, we assume that it repeats itself at every $T \to \infty$.

Interestingly, as $T \to \infty$, the sum operation of the Fourier series synthesis equation converges to an integral operation.

Let us see how.

Formally speaking, consider a continuous time aperiodic function, $x(t)$,

$$x(t) = \begin{cases} \tilde{x}(t) & -T_1 < t < T_2 \\ 0 & \text{otherwise}, \end{cases} \tag{8.2}$$

where $\tilde{x}(t)$ is a periodic function, generated by repeating the aperiodic function, $x(t)$. For the time being, let us assume that the nonzero range in the interval, $(-T_1, T_2)$ is finite, in Equation (8.2). Then, the fundamental period of

$$\tilde{x}(t) = \tilde{x}(t + T), \tag{8.3}$$

should be $T \geq (T_1 + T_2)$, as shown in Figure 8.1. Since $(-T_1, T_2)$ is a finite interval, we can repeat the function, $x(t)$ to, generate a periodic function, $\tilde{x}(t)$, with finite fundamental period, T.

Note: The center part of the periodic function, $\tilde{x}(t)$ is the aperiodic function, $x(t)$, which is nonzero in a finite interval, (T_1, T_2).

If we stretch the aperiodic function, $x(t)$, such that $T_1 \to \infty$ and $T_2 \to \infty$, then, we obtain a function with infinite period. We need to use our imagination to think about a signal that repeats itself at every period, $T \to \infty$. In summary, any aperiodic function can be considered as a periodic function with an infinite fundamental period.

Now, we can extend the Fourier series theorem to the signals of infinite period as follows.

Recall that the Fourier series representation of a periodic signal was given by the following synthesis and analysis equations:

$$\tilde{x}(t) = \sum_{k=-\infty}^{\infty} a_k e^{jk\omega_0 t} \quad \text{and} \quad a_k = \frac{1}{T}\int_T \tilde{x}(t)e^{-jk\omega_0 t}dt, \tag{8.4}$$

respectively.

Consider a signal, $x(t) = 0$ for $t < -T_1$ and $t > T_2$, which can be defined in one full period of a periodic function $\tilde{x}(t)$. Then,

$$Ta_k = \int_T \tilde{x}(t)e^{-jk\omega_0 t}dt = \int_{-T_1}^{T_2} x(t)e^{-jk\omega_0 t}dt. \tag{8.5}$$

Let us define a complex-variable function, as follows:

$$X(jk\omega_0) \triangleq Ta_k. \tag{8.6}$$

$X(jk\omega_0)$ is proportional to the spectral coefficients of the periodic signal, $\tilde{x}(t)$.

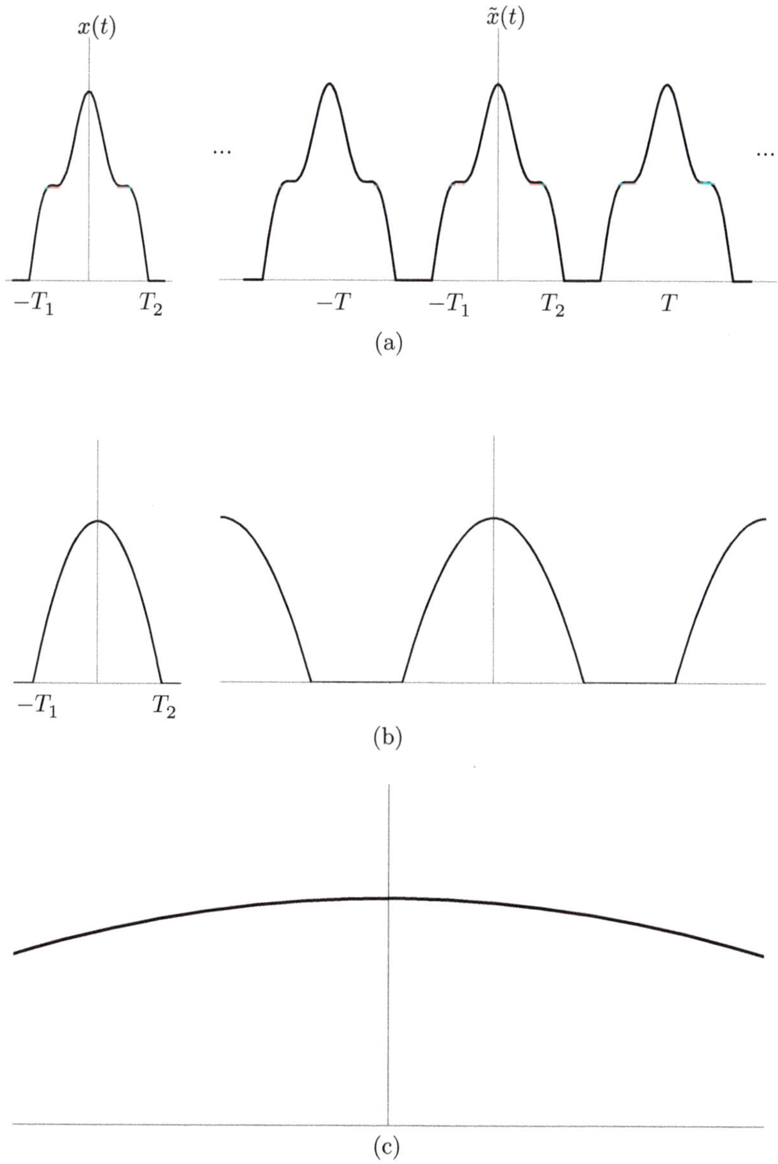

Figure 8.1 (a) Given an aperiodic function $x(t)$, which is nonzero in a finite interval $-T_1 < t < T_2$, we generate a periodic signal, $\tilde{x}(t)$ by repeating $x(t)$ with period, $T \geq (T_1 + T_2)$. (b) We can enlarge the nonzero interval of $x(t)$ and the period, T of $\tilde{x}(t)$ as much as we like. (c) We stretch the nonzero interval, $(T_1 + T_2)$, of $x(t)$ to ∞.

Then, we can replace a_k by $X(jk\omega_0)/T$ and $T = 2\pi/\omega_0$ in the Fourier series synthesis equation to obtain,

$$\begin{aligned}\tilde{x}(t) &= \frac{1}{T}\sum_{k=-\infty}^{\infty} X(jk\omega_0)e^{jk\omega_0 t} \\ &= \frac{1}{2\pi}\sum_{k=-\infty}^{\infty} X(jk\omega_0)e^{jk\omega_0 t}\omega_0. \end{aligned} \tag{8.7}$$

Finally, we stretch the nonzero interval of $x(t)$ and the fundamental period of $\tilde{x}(t)$ to infinity,

$$(-T_1, T_2) \to \infty \quad \text{and} \quad T \to \infty, \tag{8.8}$$

and take the limit,

$$\lim_{T\to\infty} \tilde{x}(t) = x(t) = \lim_{\omega_0 \to 0} \frac{1}{2\pi} \sum X(jk\omega_0) e^{jk\omega_0 t} \omega_0. \tag{8.9}$$

As the fundamental period, $T \to \infty$, the angular frequency, $\omega_0 \to 0$, in the limit, harmonically related discrete frequencies converge to continuous frequency,

$$\lim_{\omega_0 \to 0} k\omega_0 \to d\omega, \tag{8.10}$$

which implies,

$$X(jk\omega_0) \to X(j\omega). \tag{8.11}$$

Then, the function, $x(t) = \lim_{T\to\infty} \tilde{x}(t)$ can be represented by,

$$x(t) = \frac{1}{2\pi} \int_{-\infty}^{\infty} X(j\omega) e^{j\omega t} d\omega. \tag{8.12}$$

This equation represents an aperiodic function, $x(t)$, in terms of the weighted integral of complex exponential function, where the weight $X(j\omega)$ is a complex function of a continuous frequency variable, ω, in the frequency domain.

Considering the fact that

$$X(jk\omega_0) \triangleq Ta_k, \tag{8.13}$$

we can uniquely obtain the weight function, $X(j\omega)$ from the function, $x(t)$ by taking the limit of $X(jk\omega_0)$, as $\omega_0 \to 0$, as follows:

$$X(j\omega) \triangleq \lim_{\omega_0 \to 0} X(jk\omega_0) \triangleq \lim_{\omega_0 \to 0} Ta_k = \int_{-\infty}^{\infty} x(t) e^{-j\omega t} dt. \tag{8.14}$$

The complex weight function, $X(j\omega)$, which is called the **Fourier transform** of the time domain function, $x(t)$, generalizes the Fourier series representation of periodic functions to aperiodic functions. In the aforementioned rough formalism, the idea of representing periodic functions by the weighted **summation** of complex exponentials is brilliantly extended to representing aperiodic functions by weighted **integral** of complex exponentials.

8.2 Existence and Convergence of the Fourier Transforms: Dirichlet Conditions

The validity of the extension of the Fourier series of periodic signals to aperiodic signals relies upon a very major assumption: We need to be able to uniquely obtain the frequency domain function, $X(j\omega)$, from the time domain function, $x(t)$ by following integral:

$$X(j\omega) = \int_{-\infty}^{\infty} x(t) e^{-j\omega t} dt. \tag{8.15}$$

This integral exists when the function, $x(t)$ satisfies the Dirichlet conditions, which can be summarized as follows:

1) The function $x(t)$ should have finite energy,

$$\int_{-\infty}^{\infty} |x(t)|^2 dt < \infty. \tag{8.16}$$

2) The function $x(t)$ should have a finite number of maxima and minima in a finite interval.
3) The function $x(t)$ should have a finite number of discontinuities in a finite interval and all the discontinuities are to be finite.

The Dirichlet conditions assure that the Fourier transform function, $X(j\omega)$, exists. In other words, the Fourier transform is finite,

$$X(j\omega) < \infty. \tag{8.17}$$

Dirichlet conditions, also, assure the convergence of the periodic function $\tilde{x}(t)$ to the aperiodic function $x(t)$, as the period $T = \frac{2\pi}{\omega_0} \rightarrow \infty$, where

$$\tilde{x}(t) = \frac{1}{T} \sum_{k=-\infty}^{\infty} X(jk\omega_0)e^{jk\omega_0 t}. \tag{8.18}$$

Formally speaking, the absolute integral of the error $e(t)$ between the function $x(t)$ and $\tilde{x}(t)$,

$$e(t) = x(t) - \tilde{x}(t), \tag{8.19}$$

converges to 0,

$$\lim_{T\rightarrow\infty} \int_{-\infty}^{\infty} |e(t)|^2 dt \rightarrow 0. \tag{8.20}$$

The satisfaction of the Dirichlet conditions is rather intuitive for the existence and convergence of the Fourier transform. Let us get a feeling about the necessity and sufficiency of the Dirichlet conditions for existence and convergence of the Fourier transform by a simple example given in the following text, leaving the formal proofs to the interested readers.

Exercise 8.1 Does the following function satisfy Dirichlet conditions?

$$x(t) = \frac{1}{4-t^2}. \tag{8.21}$$

Solution

No! This function violates the first Dirichlet condition, mentioned earlier. It has an infinite energy,

$$\int_{-\infty}^{\infty} |\frac{1}{4-t^2}|^2 dt \rightarrow \infty. \tag{8.22}$$

Indeed, the integral to obtain the weight function, $X(j\omega)$, does not exist,

$$X(j\omega) = \int_{-\infty}^{\infty} x(t)e^{-j\omega t} dt = \int_{-\infty}^{\infty} \frac{1}{4-t^2} e^{-j\omega t} dt \rightarrow \infty. \tag{8.23}$$

8.3 Fourier Transforms

The extension of the Fourier series representation of periodic functions to aperiodic functions enables us to define Fourier transform, as follows.

Any continuous time function, $x(t)$, satisfying Dirichlet conditions, can be uniquely represented by the weighted integral of complex exponentials, called **synthesis equation** as follows:

$$\boxed{x(t) = \frac{1}{2\pi} \int_{-\infty}^{\infty} X(j\omega)e^{j\omega t} d\omega}. \tag{8.24}$$

The weight, $X(j\omega)$ of the synthesis equation can be uniquely obtained from the weighted integral of complex conjugate exponentials, called **analysis equation** as follows:

$$\boxed{X(j\omega) = \int_{-\infty}^{\infty} x(t)e^{-j\omega t}dt}, \tag{8.25}$$

where the complex function, $X(j\omega)$ is called the **Fourier transform** of $x(t)$. The weight of the complex exponential in the analysis equation is the function $x(t)$ itself.

Motivating Question: What do analysis and synthesis equations tell us?

Synthesis equation recovers a time domain phenomenon, represented by a function, $x(t)$ from its frequency content, where the amount of each frequency ω is measured by $X(j\omega)$. Compared to the discrete spectral coefficients of Fourier series representation, ω is a continuous variable. Therefore, we can continuously measure the frequency content of a time domain function, $x(t)$ by its Fourier transform, $X(j\omega)$.

Note: The domains of $x(t)$ and $X(j\omega)$ are different. While the domain, ω, of $X(j\omega)$ is **frequency domain**, the domain, t, of $x(t)$ is **time domain**. The argument of the Fourier transform X is not only the frequency variable, ω, but $j\omega$ to remind us that the function $X(j\omega)$ is a **complex function**.

The analysis equation is even more interesting: It tells us that if we observe a physical phenomenon in the time domain, we can uniquely obtain its representation in the frequency domain, where time completely disappears. A physical phenomenon, which is represented by a function of time, can be uniquely represented by a function of frequency. In the frequency domain, a physical phenomenon, such as music, speech, and heartbeats, is independent of time, but it is represented by its frequency variations.

Time domain and frequency domain representations are one-to-one and onto. Loosely speaking, if $x(t)$ exists in the time domain, then $X(j\omega)$ exists in the frequency domain and vice versa. This fact is mathematically formalized as follows:

$$x(t) \leftrightarrow X(j\omega). \tag{8.26}$$

The time domain and the frequency domain representations show two different aspects of the same physical phenomenon, which is represented in two different spaces.

8.4 Comparison of Fourier Series and Fourier Transform

Let us now conceptually compare the Fourier series of a periodic signal and Fourier transform of an aperiodic signal. Recall that Fourier series representation of a continuous time periodic signal,

$$x(t) \leftrightarrow a_k, \tag{8.27}$$

and the Fourier transform of a continuous time aperiodic signal

$$x(t) \leftrightarrow X(j\omega), \tag{8.28}$$

are one-to-one and onto, provided that they satisfy the Dirichlet conditions. Recall, also, that the Fourier transform of an aperiodic function is obtained by assuming that an aperiodic function can be considered as a periodic function of infinite period. In this sense, the Fourier transform can be considered as the generalized form of the Fourier series, which is applicable to both periodic and aperiodic signals.

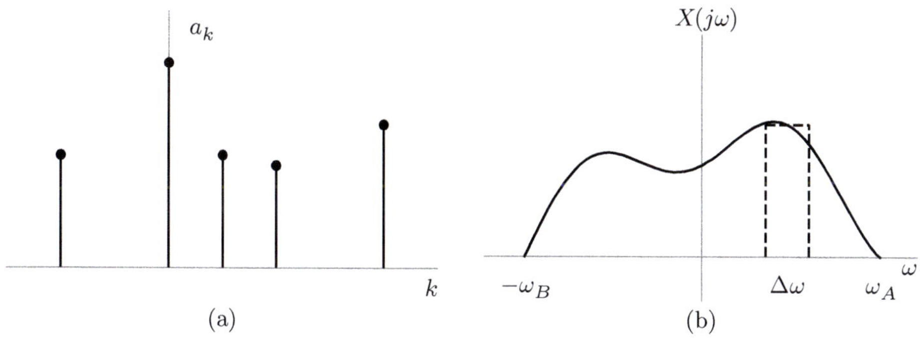

Figure 8.2 (a) The discrete spectral coefficients, a_k of a periodic function, $x(t)$. At each integer value of k, spectral coefficients $\{a_k\}$, measure the amount of the harmonic frequency, $k\omega_0$, which is the integer multiple of angular frequency, ω_0. (b) The Fourier transform of an aperiodic function. As we can observe, $X(j\omega)$ is a continuous function of the frequency variable ω. This time we can measure the amount of a frequency in an interval, which is the area of the rectangle of $X(j\omega) \cdot \Delta\omega$. Note that, in this illustration, the Fourier transform, $X(j\omega)$ $X(j\omega)$ is zero outside of $[-\omega_B, \omega_A]$. This particular class of signals, called band-limited, has a special importance to develop digital technologies.

Motivating Question: Do the Fourier series and transform resemble each other? What are the similarities and distinctions between the two representations in the frequency domain?

The major similarity is that spectral coefficients, a_k of Fourier series and the Fourier transform, $X(j\omega)$ indicate the frequency content of the signal. In both cases, the signal is represented as a function of its frequencies.

The major distinction between the Fourier series and Fourier transform is that the spectral coefficients, $\{a_k\}$, form a discrete function of k, whereas $X(j\omega)$ is a continuous function of frequency variable, ω. While the Fourier series representation provides us the amount of each **harmonic frequency**, $k\omega_0$, which contributes to a periodic signal, Fourier transform, provides us the amount of the **frequency interval** in a continuous band of frequencies, which makes up the signal (Figure 8.2). Thus, instead of the harmonics at integer values of k, we have frequency intervals of the Fourier transform function.

8.5 Frequency Content of Fourier Transform

The functions in the frequency domain bear some properties, which cannot be observed and quantified in the time domain. In order to quantify these properties, we need to introduce new concepts and their formal definitions. In particular, the behavior of a function is to be quantified in terms of its frequency content. An important concept is the **bandwidth** of a function in frequency domain, as defined in the following text.

Definition: Bandwidth of a Function in Frequency domain: The frequency interval, where the Fourier transform $X(j\omega) \neq 0$ is called **bandwidth**.

When the time domain function is real valued its Fourier transform is conjugate symmetric. In other words, if $x(t)$ is real, then, $X(j\omega) = X^*(-j\omega)$. For conjugate symmetric functions, the cutoff frequencies at the lower and upper bounds of the Fourier transform are $-\omega_c$ and ω_c. Hence, the bandwidth of conjugate symmetric Fourier transforms is $\omega_{b\omega} = 2\omega_c$.

Definition: Bandlimited Functions

A function is called **bandlimited** if its Fourier transform has zero values outside a finite frequency interval ω_c. Mathematically speaking,

$$X(j\omega) = 0 \quad \text{for } |\omega| > \omega_c. \tag{8.29}$$

A bandlimited signal, such as speech or music, consists of only a finite band of frequencies. These signals can be easily created, stored, transmitted, and/or processed by digital technologies for many real-world applications.

The bandwidth of a signal determines the frequency content of the signal. The larger the bandwidth is the more frequency it carries.

Definition 8.1 ***(Cutoff Frequency of a Function)*** The cutoff frequency ω_c, is the frequency at the upper and lower limits of a bandlimited function, where beyond these frequencies, the function has 0 values.

In this definition, the cutoff frequency is the angular frequency $\omega_c = \frac{2\pi}{T} = 2\pi f_c$, which is measured by radians. The period T is measured by seconds and the frequency f_c is measured by Hertz (cycle/s).

WATCH

Listen to Maria Callas, Luciano Pavarotti, and others for examples of different human voice bandwidths @ https://384book.net/v0802

In the following examples, we investigate the behavior of signals in the frequency domain.

Exercise 8.2 Consider the following signal:

$$x(t) = e^{-a|t|}, \quad a > 0. \tag{8.30}$$

a) Sketch this signal and give a real-life example that can be approximated by this signal.
b) Find and plot the Fourier transform of this signal.
c) Is this a bandlimited signal? Comment on the frequency content of this signal.

Solution

a) This is an even signal, as shown in Figure 8.3. This type of function can be used to approximate the expected amount of money we have left after optimal gambling in a casino with nonfavorable odds.
b) The Fourier transform of the signal is obtained by using the synthesis equation,

$$\begin{aligned} X(j\omega) &= \int_{-\infty}^{+\infty} e^{-a|t|} e^{-j\omega t} = \int_{-\infty}^{0} e^{at} e^{-j\omega t} + \int_{0}^{+\infty} e^{-at} e^{-j\omega t} \\ &= \frac{1}{a - j\omega} + \frac{1}{a + j\omega} \\ &= \frac{2a}{a^2 + \omega^2}. \end{aligned} \tag{8.31}$$

Fourier transform, $X(j\omega)$ does not have an imaginary part. Thus, it is a real function, as illustrated in Figure 8.4.

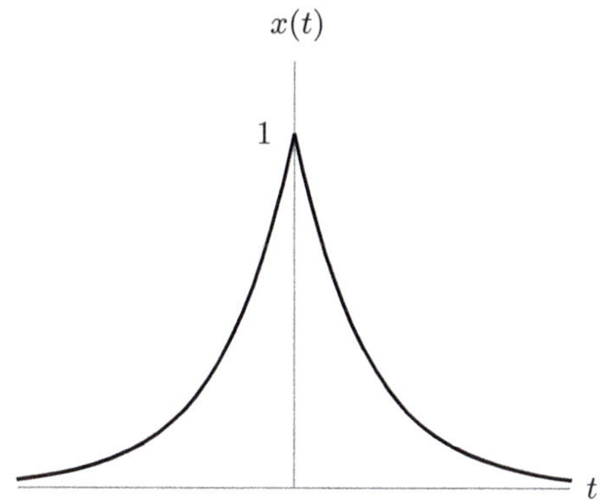

Figure 8.3 Sketch of the signal, $x(t) = e^{-a|t|}$.

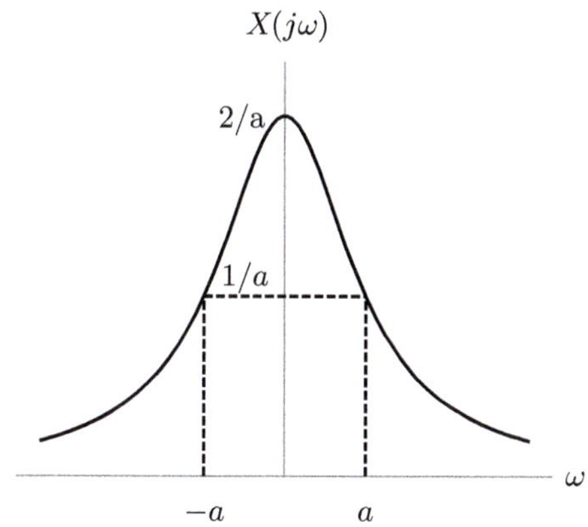

Figure 8.4 Fourier transform of the signal, $x(t)$, as depicted in Figure 8.3.

c) This is **not** a bandlimited signal because $X(j\omega) \neq 0$ for $-\infty < \omega < \infty$.

Recall that the frequency content of signals can be investigated from the magnitude and phase spectrum of the Fourier transform. Since the Fourier transform of this signal is a real function of frequency, the magnitude, $|X(j\omega)| = X(j\omega)$ and $\measuredangle X(j\omega) = 0$. Although the signal consists of all the frequencies, we observe that low-frequency content dominates the signal, allowing relatively less proportion of high frequencies, outside the interval, $|\omega| > a$.

Exercise 8.3 Consider the following impulse function:

$$x(t) = \delta(t - t_0). \tag{8.32}$$

a) Give a real-life example, which can be approximated by this signal.
b) Find the Fourier transform for $t_0 \neq 0$.
c) Find the Fourier transform for $t_0 = 0$.
d) Is this a bandlimited signal?

Solution

a) This function approximates a physical phenomenon, such as a lightening pulse at $t = t_0$, which acts for a short duration with a very high voltage.
b) Let us investigate the frequency domain representation of the impulse function. For this purpose, we compute the Fourier transform by using the analysis equation:

$$X(j\omega) = \int_{-\infty}^{\infty} \delta(t - t_0)e^{-j\omega t}dt = e^{-j\omega t_0}. \tag{8.33}$$

c) If the lightning occurs at $t_0 = 0$, then, $\quad X(j\omega) = 1$.

For this particular example, the Fourier transform, $X(j\omega)$ is a real function. The magnitude is the same as the function itself, $|X(j\omega)| = 1$, for all frequencies and the phase, $\measuredangle X(j\omega) = 0$. This result indicates that a lightning consists of all the frequencies with equal amounts. Formally speaking,

$$x(t) = \delta(t) \leftrightarrow X(j\omega) = 1. \tag{8.34}$$

d) This is **not** a bandlimited signal because $X(j\omega) = 1 \neq 0$ or

$$X(j\omega) = e^{-j\omega t_0} \neq 0, \quad \text{for } -\infty < \omega < \infty. \tag{8.35}$$

Exercise 8.4 Consider the rectangular pulse signal, given as follows:

$$x(t) = \begin{cases} 1, & |t| < T_1 \\ 0, & |t| > T_1. \end{cases} \tag{8.36}$$

as shown in Figure 8.5a.

a) Give a real-life example, which can be approximated by this signal.
b) Find and plot the Fourier transform of this signal.
c) Is this a bandlimited signal? Comment on the frequency content of this signal.

Solution

a) Rectangular pulse signal can be approximately generated by opening and closing a switch of an electrical circuit,
b) The analysis equation gives us the Fourier transform of $x(t)$,

$$X(j\omega) = \int_{-T_1}^{T_1} e^{-j\omega t} dt = 2\frac{\sin \omega T_1}{\omega}, \tag{8.37}$$

as sketched in Figure 8.5b.

c) This is **not** a bandlimited signal because $X(j\omega) \neq 0$, for $-\infty < \omega < \infty$.

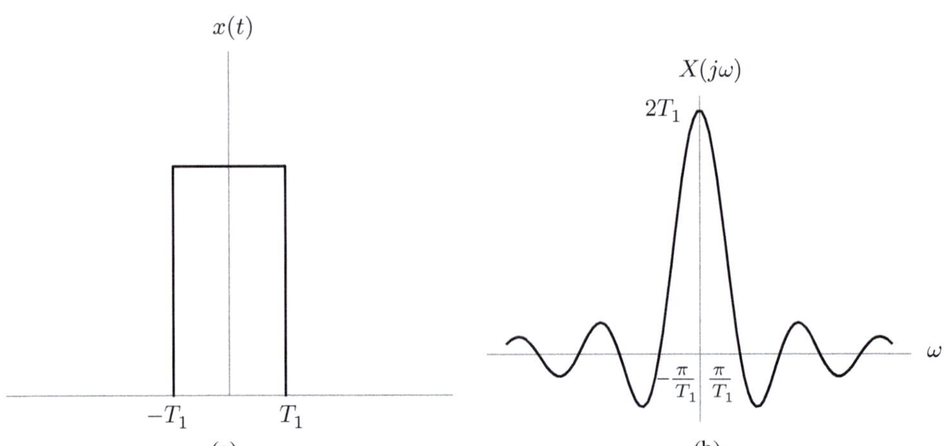

Figure 8.5 (a) The rectangular pulse signal of the example and (b) its Fourier transform.

The analysis of Figure 8.5 reveals that the time domain signal has a limited duration. The frequency content of the signal alternates and attenuates, as $\omega \to \infty$. The high-frequency components of the signal correspond to the discontinuities of the time domain function at $-T_1$ and T_1. Notice that as $-T_1$ and T_1 approaches 0, the Fourier transform $X(j\omega)$ gets flatter.

Exercise 8.5 Consider the signal $x(t)$ whose Fourier transform is given below,

$$X(j\omega) = \begin{cases} 1, & |\omega| < W \\ 0, & |\omega| > W. \end{cases} \tag{8.38}$$

a) Find the signal $x(t)$.
b) Is this a bandlimited signal? If yes, what is the bandwidth? Comment on the frequency content of this signal.

Solution

a) This transform is illustrated in Figure 8.6a. Using the synthesis equation, we can determine the signal, $x(t)$, in time domain,

$$x(t) = \frac{1}{2\pi} \int_{-W}^{W} e^{j\omega t} d\omega = \frac{\sin Wt}{\pi t}, \tag{8.39}$$

which is depicted in Figure 8.6b.

b) This is a bandlimited signal because $X(j\omega) = 0$, for $|\omega| > W$. The bandwidth of this signal is $2W$.

The analysis of Figure 8.6 reveals that the frequency domain signal consists of low frequencies for $|\omega| > W$. The time domain counterpart, $x(t)$ alternates and attenuates, as $t \to \pm\infty$. Notice that as W approaches 0, the time domain function $x(t)$ gets flatter.

Comparison of Figures 8.5 and 8.6 shows the beautiful duality between the analytical shape of the functions in time and frequency domains.

Sine Cardinal (Sinc) Function: The functions in the time and frequency domains appeared in the aforementioned examples,

$$x(t) = \frac{\sin Wt}{\pi t},$$

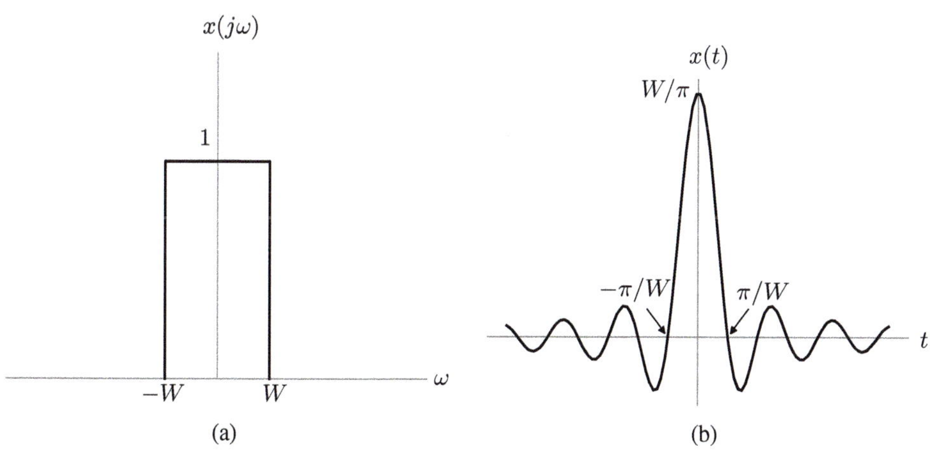

Figure 8.6 Fourier transform pair of Exercise 8.5: (a) Fourier transform, $X(j\omega)$ and (b) the corresponding time domain function, $x(t)$.

and

$$X(j\omega) = \frac{\sin \omega T_1}{\omega},$$

are called **sine cardinal**, or in short, **sinc functions**. The general form of the normalized sinc function is

$$x(t) = \frac{\sin \pi t}{\pi t}.$$

This even function is maximum at $t = 0$ in the time domain and $\omega = 0$ in the frequency domain. It keeps attenuating as $t \to \pm\infty$. Replacing $x(t)$ by $h(t)$, and $X(j\omega)$ by $H(j\omega)$. Later in Chapter 10, we call this LTI system as the ideal low-pass filter,

$$h(t) = \frac{\sin \pi\omega_c t}{\pi t} \leftrightarrow H(j\omega) = \begin{cases} 1, & |\omega| < \omega_c \\ 0, & |\omega| > \omega_c. \end{cases} \tag{8.40}$$

The sinc function is very important in establishing the relationship between the continuous time and discrete time worlds, which lies in the foundation of the entire digital technology, as we shall see in the sampling theorems of Chapters 11 and 12.

8.6 Representation of LTI Systems in Frequency Domain by Frequency Response

In Chapter 6, we have seen that the response of an LTI system to the complex exponential input with the angular frequency ω_0,

$$x(t) = e^{j\omega_0 t}, \tag{8.41}$$

is,

$$y(t) = H(j\omega_0)e^{j\omega_0 t}, \tag{8.42}$$

where the eigenfunction $e^{j\omega_0 t}$ is scaled by the eigenvalue of an LTI system, defined as:

$$H(j\omega_0) = \int_{-\infty}^{\infty} h(t)e^{-j\omega_0 t}dt. \tag{8.43}$$

Note that, for $\omega_0 \to \omega$, the eigenvalue, $H(j\omega_0)$ converges to the Fourier transform of the impulse response. Considering the fact that impulse response uniquely represents an LTI system in time domain, its Fourier transform is very crucial to represent the LTI system in frequency domain. Hence, it requires a special name, called frequency response, as defined in the following text.

Definition: Frequency response of a continuous time LTI system is defined as the Fourier transform of the impulse response. Mathematically, frequency response of a continuous time LTI system is

$$H(j\omega) = \int_{-\infty}^{\infty} h(t)e^{-j\omega t}dt. \tag{8.44}$$

Therefore, impulse response and frequency response,

$$h(t) \leftrightarrow H(j\omega), \tag{8.45}$$

are one-to-one and onto representation of the same LTI system in two different domains, namely, in time and frequency domains. While the time-dependent properties of the LTI system are investigated by its impulse response or by the corresponding differential equation, the frequency-shaping properties of the LTI system are investigated by analyzing the frequency response, which is the Fourier Transform of the impulse response.

Note that, the eigenvalues $H(jk\omega_0)$ of a continuous time LTI system for each harmonic frequency $k\omega_0$ for all integer values of k are specific instances of the frequency response $H(j\omega)$. In other words, the eigenvalues are the values of the frequency response at $\omega = k\omega_0, \forall k$.

Recall that Fourier transform is a complex-valued function. In polar coordinate system, Fourier transform of the impulse response, namely the frequency response is represented by:

$$H(j\omega) = |H(j\omega)|e^{j\measuredangle H(j\omega)}, \tag{8.46}$$

where the real-valued functions $|H(j\omega)|$ and $\measuredangle H(j\omega)$ are called the magnitude and phase spectrum, respectively. Analysis of Fourier transform of a function requires the analysis of magnitude and phase spectrums.

Motivating Question: What does the **magnitude and phase spectrum** of the Fourier transform of the impulse response indicate?

Generally speaking, the magnitude and phase spectrum of the frequency response $H(j\omega)$ indicate the frequency content of the impulse response function, $h(t)$.

- Magnitude spectrum of the Fourier transform, $|H(j\omega)|$, determines the relative presence of frequencies in the impulse response. The magnitude spectrum of the Fourier transform is simply how much each frequency component is contributing to form this function.
- Phase spectrum of the Fourier transform, $\measuredangle H(j\omega)$ determines how the frequencies line up relative to one another. The phase spectrum of the Fourier transform is simply how much the frequency response delays a frequency component of an input signal.

In the following example, we investigate the behavior of an LTI system in various one-to-one and onto representations, namely as impulse and frequency responses, and differential equations.

Exercise 8.6 Consider a continuous time LTI system, represented by the following impulse response:

$$h(t) = e^{-at}u(t), \quad a > 0. \tag{8.47}$$

a) Find the differential equation that represents this system.
b) What type of physical phenomenon does this system represent?
c) Find the Fourier transform of the impulse response and comment on the frequency content of this system.
   Assume that the system is initially at rest at $t = 0$.

Solution

We have seen this impulse response in Chapter 4, where the plot in time domain is shown in Figure 8.7.

a) This impulse response represents a first-order differential equation, given below;

$$\frac{dy(t)}{dt} + ay(t) = x(t), \tag{8.48}$$

   assuming that the system is initially at rest, with $x(t) = 0$ and $y(t) = 0$ for $t \le 0$.
b) For example, this system may approximately represent the velocity decay of a car, running with a unit speed. At time $t = 0$, if we stop pushing the gas pedal, the velocity will decay and approach 0. The decay rate, a, depends on the environmental conditions and the properties of the car.

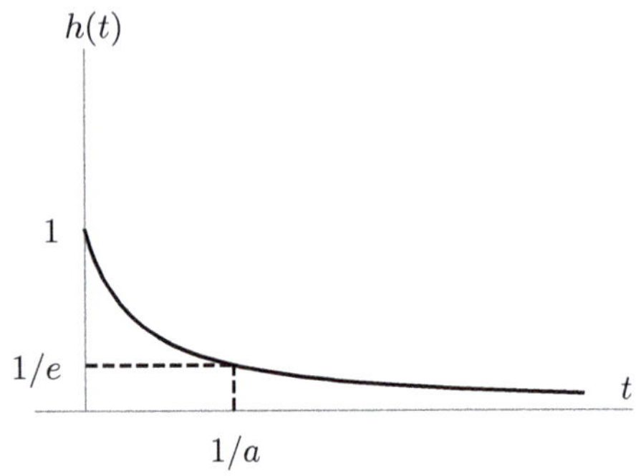

Figure 8.7 The exponential function is obtained at the output of an LTI system represented by a first-order homogeneous differential equation, $\frac{dy(t)}{dt} + ay(t) = x(t)$.

c) Let us investigate the structure of the phenomenon represented by $h(t)$ in frequency domain, using the analysis equation:

$$H(j\omega) = \int_{-\infty}^{\infty} h(t)e^{-j\omega t}dt = \int_{0}^{\infty} e^{-at}e^{-j\omega t}dt = \frac{1}{a + j\omega}. \tag{8.49}$$

Since $H(j\omega)$ is a complex function, we need to find the magnitude and the phase of this function:

$$|H(j\omega)| = \frac{1}{\sqrt{a^2 + \omega^2}}$$
$$\measuredangle H(j\omega) = -\tan^{-1}\left(\frac{\omega}{a}\right). \tag{8.50}$$

The magnitude and phase plot of $H(j\omega)$ are shown in Figure 8.8.

The magnitude spectrum is relatively high for $|\omega| \leq a$, compared to the frequencies outside of this interval. It converges to zero as the absolute value of the angular frequency, $|\omega|$ is increased. Thus, this LTI system passes the low-frequency components and suppresses the high-frequency

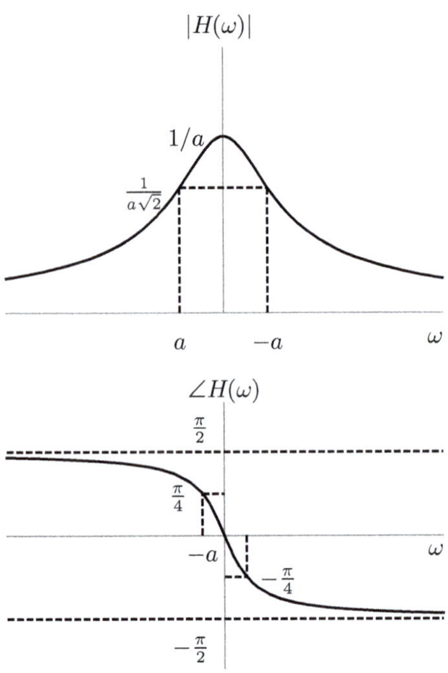

Figure 8.8 Magnitude and phase spectrum of the frequency response, $H(j\omega) = \frac{1}{a+j\omega}$.

components of the input signal (Figure 8.8). The phase spectrum shows that the LTI system delays the high frequencies more compared to the low frequencies of an input signal.

The decay rate a of the impulse response defines the bandwidth of the frequency response. The higher the decay rate gets, the larger the bandwidth we define in the magnitude spectrum. Hence, this LTI system passes more frequencies as a increases.

8.7 Relationship Between the Fourier Series and Fourier Transform of Periodic Functions

Suppose that we are given a continuous time periodic function, $x(t)$, with the corresponding spectral coefficients,

$$x(t) \leftrightarrow a_k. \tag{8.51}$$

Suppose, also, that the Fourier transform of the function, $x(t)$, is

$$x(t) \leftrightarrow X(j\omega). \tag{8.52}$$

Motivating Question: What is the relationship between the spectral coefficients, a_k and Fourier transform $X(j\omega)$ of $x(t)$?

In order to find an answer to this question, let us solve the following exercise.

Exercise 8.7 Consider the following impulse train in frequency domain:

$$X(j\omega) = \sum_{k=-\infty}^{\infty} 2\pi a_k \delta(\omega - k\omega_0). \tag{8.53}$$

Find the inverse Fourier transform, $x(t)$ of the frequency domain signal, $X(j\omega)$.

Solution

The frequency domain function, $X(j\omega)$, is a periodic function, where the fundamental period is $\omega = \omega_0$. Let us find the inverse Fourier transform, $x(t)$ by using the synthesis equation.

Inserting the right-hand side of the Fourier transform, $X(j\omega)$, into the synthesis equation, we obtain,

$$x(t) = \frac{1}{2\pi} \int_{-\infty}^{\infty} \sum_{k=-\infty}^{\infty} 2\pi a_k \delta(\omega - k\omega_0) e^{j\omega t} d\omega. \tag{8.54}$$

Note: Finding the inverse Fourier transform involves taking the weighted integral of a complex function, $X(j\omega)$. This may not be an easy task, which may require contour integration. In order to avoid taking complex integrals most of the time we employ lookup tables. However, for this particular function, we can arrange Equation (8.54) and obtain,

$$x(t) = \frac{1}{2\pi} \int_{-\infty}^{\infty} \sum_{k=-\infty}^{\infty} 2\pi a_k \delta(\omega - k\omega_0) e^{j\omega t} d\omega = \sum_{k=-\infty}^{\infty} a_k \int_{-\infty}^{\infty} \delta(\omega - k\omega_0) e^{j\omega t} d\omega. \tag{8.55}$$

The integral in the right-hand side of Equation (8.55) is easy to take:

$$\int_{-\infty}^{\infty} \delta(\omega - k\omega_0) e^{j\omega t} d\omega = e^{jk\omega_0 t}. \tag{8.56}$$

Replacing the result of the integral in Equation (8.56) into Equation (8.55), we obtain,

$$x(t) = \frac{1}{2\pi} \int_{-\infty}^{\infty} \sum_{k=-\infty}^{\infty} 2\pi a_k \delta(\omega - k\omega_0) e^{j\omega t} d\omega = \sum_{k=-\infty}^{\infty} a_k e^{jk\omega_0 t}. \tag{8.57}$$

Interestingly, the right-hand side of this equation is the Fourier series representation of the function, $x(t)$, which is periodic with fundamental period, $T = \frac{2\pi}{\omega_0}$.

Comparing the Fourier transform and Fourier series of $x(t)$,

$$x(t) = \sum_{k=-\infty}^{\infty} 2\pi a_k e^{jk\omega_0 t} \leftrightarrow X(j\omega) = \sum_{k=-\infty}^{\infty} 2\pi a_k \delta(\omega - k\omega_0), \tag{8.58}$$

provides us with the relationship between Fourier transform and spectral coefficients of a periodic signal, $x(t)$, as follows:

$$\boxed{X(j\omega) = 2\pi \sum_{k=-\infty}^{\infty} a_k \delta(\omega - k\omega_0)}. \tag{8.59}$$

The relationship between the Fourier series coefficients of a periodic signal and its Fourier transform shows that the Fourier transform of a periodic signal is the weighted sum of shifted impulses, where the weights are the spectral coefficients, a_k. Let us not forget the constant multiplicative factor 2π.

Exercise 8.8 Consider the following periodic signal:

$$x(t) = \sin(\omega_0 t). \tag{8.60}$$

a) Find and plot the Fourier series representation of $x(t)$.
b) Find and plot the Fourier transform of $x(t)$.
c) Is this a bandlimited signal? If yes, find the bandwidth. Comment on the frequency content of this signal.

Solution

a) Recall that there are only two nonzero spectral coefficients of this function,

$$a_1 = -a_{-1} = \frac{1}{2j}.$$

b) We can easily compute the Fourier transform of $x(t)$ by inserting the spectral coefficients into Equation (8.59), as follows:

$$X(j\omega) = \sum_{k=-\infty}^{\infty} 2\pi a_k \delta(\omega - k\omega_0) = \frac{\pi}{j}(\delta(\omega - \omega_0) - \delta(\omega + \omega_0)). \tag{8.61}$$

The plots of spectral coefficients, a_k vs. k and Fourier transform, $X(j\omega)$ vs. ω is shown in Figure 8.9. Note that both the spectral coefficients and the Fourier transform consist of two impulse functions. While the spectral coefficients consist of two discrete shifted impulse functions, the Fourier transform consists of two shifted continuous impulses. The amount of shift of the impulse functions in spectral coefficients is integer-valued. On the other hand, the amount of shift of the continuous impulse functions in the Fourier transform is irrational number of angular frequency, which is $|\omega_0| = \frac{2\pi}{T}$.

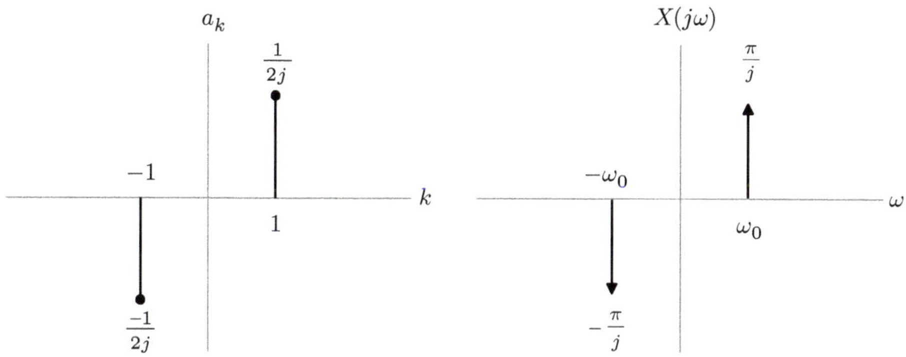

Figure 8.9 Comparison of the spectral coefficients and Fourier transform of the periodic signal $x(t) = \sin(\omega_0 t)$. While the spectral coefficients are discrete impulses, the Fourier transform is continuous impulses.

c) This is a bandlimited signal, where the bandwidth is $2\omega_0$.
   Interestingly, Fourier transform of the sinusoidal signal, $x(t) = \sin \omega_0 t$, which ranges $-\infty < t < \infty$, generates a pair of impulse functions, providing a very compact representation in the frequency domain. This behavior of the Fourier transform enables us to compress time domain signals by storing them compactly, in the frequency domain.

Exercise 8.9 Suppose that the impulse train function in the time domain is given by the following equation:

$$x(t) = \sum_{k=-\infty}^{\infty} \delta(t - kT). \tag{8.62}$$

a) Find the Fourier series representation of the impulse train, $x(t)$.
b) Find the Fourier transform of the impulse train, $x(t)$.
c) Compare the functions in time domain $x(t)$, its spectral coefficients a_k and its Fourier transform $X(j\omega)$.

Solution

a) Fourier series representation of $x(t)$ is

$$a_k = \frac{1}{T}\int_T \delta(t - kT)e^{jk\omega_0 t}dt = e^{-jk\omega_0 T} = \frac{1}{T}. \tag{8.63}$$

b) By using the Fourier series coefficients, we can find the Fourier transform $X(j\omega)$, as follows:

$$X(j\omega) = \sum_{k=-\infty}^{\infty} 2\pi a_k \delta(\omega - k\omega_0) = \frac{2\pi}{T}\sum_{k=-\infty}^{\infty} \delta\left(\omega - k\frac{2\pi}{T}\right). \tag{8.64}$$

c) Interestingly, the impulse train function in the time domain has an impulse train function in the frequency domain. While the Fourier transform of a continuous time impulse train is a continuous frequency impulse train, its Fourier series representation is a discrete impulse train (see Figure 8.10). Thus, the impulse train preserves its analytical form in both frequency and time domains. In other words, the periodic function impulse train is always impulse train in both time and frequency domains, where
   - the fundamental period of the time domain function, $x(t)$, is $T = \frac{2\pi}{\omega_0}$,

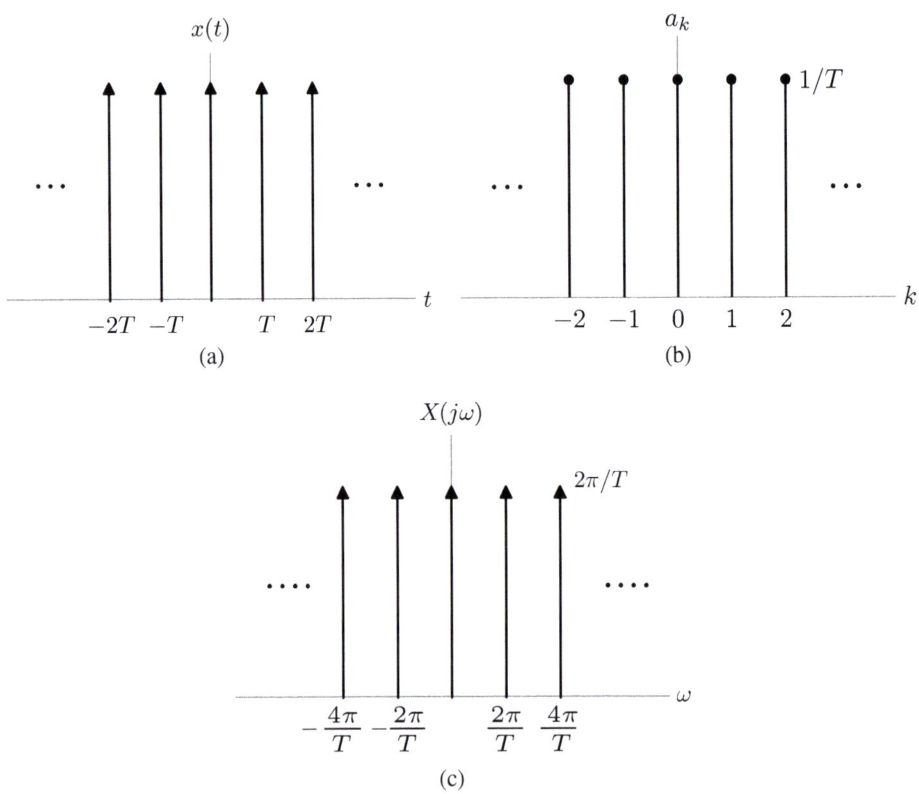

Figure 8.10 Impulse train in time domain (a), its spectral coefficients (b), and its Fourier transform (c).

- the fundamental period of the Fourier transform, $X(j\omega)$ in the frequency domain is $\omega_0 = \frac{2\pi}{T}$,
- the fundamental period of the spectral coefficients, a_k, is $k = 1$.

Note: The smaller period in the time domain results in a larger angular frequency $\omega_0 = \frac{2\pi}{T}$, in the frequency domain.

This conservative behavior of the impulse train function, which preserves the analytical form in both domains, breaks the thick wall between the time and frequency domains and opens a wide window to the digital era, as we shall see in the Sampling theorem of Chapters 11 and 12.

8.8 Properties of Fourier Transform: For Continuous Time Signals and Systems

WATCH

The sound of sine waves and their Fourier transforms @ https://384book.net/v0803

So far, we have seen that Fourier transforms map a time domain function into a new domain, called frequency domain. In this domain, the time variable disappears and the function is represented in terms of a continuous variable of frequencies by the following analysis equation:

$$X(j\omega) = \int_{-\infty}^{\infty} x(t)e^{-j\omega t}dt, \tag{8.65}$$

where the time domain function can be uniquely obtained from its frequency domain representation by the synthesis equation,

$$x(t) = \frac{1}{2\pi}\int_{-\infty}^{\infty} X(j\omega)e^{j\omega t}d\omega. \tag{8.66}$$

Recall that the complex exponential function, in Equations (8.65) and (8.66) represents a **waveform** of frequency of ω by Euler formula,

$$e^{j\omega t} = \cos\omega t + j\sin\omega t. \tag{8.67}$$

Loosely speaking, the Fourier analysis and synthesis equations reveal that all the functions satisfying Dirichlet conditions can be described by gathering uncountably infinite waveforms with varying frequencies. Therefore, the Fourier transform gives us a unique and powerful way of viewing a physical phenomenon in the time domain, in terms of the weighted integral of waveforms, where the weights are the Fourier transform function $X(j\omega)$.

WATCH Learn more about the Euler's identity @ https://384book.net/v0804

In this section, we shall investigate the properties of this marvelous tool of Fourier transform for continuous time signals and systems. We shall use the properties to switch between the continuous time and frequency domains without taking the integrals of analysis and synthesis equations. We shall study the frequency content of the aperiodic continuous time signals. We shall design and implement LTI systems in the time and frequency domains for reshaping the continuous time signals.

8.8.1 Basic Properties of Continuous Time Fourier Transform

Recall that Fourier transform is the extension of Fourier series, where an aperiodic function is represented as a periodic function of infinite period. As the period approaches infinity, the Fourier series of a periodic function converges to the Fourier transform of an aperiodic function. Consequently, most properties of Fourier transform to resemble the properties of Fourier series. A list of properties is provided in Table 8.1.

The properties of Fourier transform not only give us an insight into the frequency content of a physical phenomenon, but they also allow us to observe the similarities and distinctions between the behavior of the time domain and frequency domain functions, which represent that phenomenon. Furthermore, it helps us to compute the analysis and synthesis equations, in an efficient way.

Considering the fact that finding the Fourier transform and its inverse requires taking the integral of complex functions, computing the analysis and synthesis equations may not be easy. In order to simplify the computations, we provide the Fourier transform pairs of popular functions, as shown

Table 8.1 Basic properties of Fourier transform.

Signal	Fourier transform
$x(t) = \frac{1}{2\pi}\int_{-\infty}^{\infty} X(j\omega)e^{j\omega t}d\omega$	$X(j\omega) = \int_{-\infty}^{\infty} x(t)e^{-j\omega t}dt$
(can also be written with frequency $f = 2\pi/\omega$)	
$x(t)$	$X(j\omega)$
$y(t)$	$Y(j\omega)$
$ax(t) + by(t)$	$aX(j\omega) + bY(j\omega)$
$x(t - t_0)$	$e^{-j\omega t_0}X(j\omega)$
$e^{j\omega_0 t}x(t)$	$X(j(\omega - \omega_0))$
$x^*(t)$	$X^*(j(-\omega))$
$x(-t)$	$X(j(-\omega))$
$x(at)$	$\frac{1}{\lvert a\rvert}X(\frac{\omega}{a})$
$x(t) * y(t)$	$\frac{1}{2\pi}X(j\omega) * Y(j\omega)$
$\frac{d}{dt}x(t)$	$j\omega X(j\omega)$
$\int_{-\infty}^{t} x(t)dt$	$\frac{1}{j\omega}X(j\omega) + \pi X(0)\delta(\omega)$
$tx(t)$	$j\frac{d}{d\omega}X(j\omega)$
For real-valued $x(t)$	$\begin{cases} X(j\omega) = X^*(j(-\omega)) \\ \text{Re}\{X(j\omega)\} = \text{Re}\{X(j(-\omega))\} \\ \text{Im}\{X(j\omega)\} = \text{Im}\{X(j(-\omega))\} \\ \lvert X(j\omega)\rvert = \lvert X(j(-\omega))\rvert \\ \sphericalangle X(j\omega) = -\sphericalangle X(j(-\omega)) \end{cases}$
Even part of $x(t)$	$\text{Re}\{X(j\omega)\}$
Odd part of $x(t)$	$j\text{Im}\{X(j\omega)\}$
Duality: $f(u) = \int_{-\infty}^{\infty} g(v)e^{-juv}dv \implies$	$\begin{cases} g(t) \overset{\mathcal{F}}{\longleftrightarrow} f(j\omega) \\ f(t) \overset{\mathcal{F}}{\longleftrightarrow} 2\pi g(j(-\omega)) \end{cases}$
Parseval's relation for nonperiodic signals:	$\int_{-\infty}^{\infty} \lvert x(t)\rvert^2 dt = \frac{1}{2\pi}\int_{-\infty}^{\infty} \lvert X(j\omega)\rvert^2 d\omega$

in Table 8.2. The properties of Table 8.1 and lookups of Table 8.2 enable us to find the Fourier transforms of complicated functions and their inverse without evaluating the integrals, in most problems.

The properties can be directly proved by employing the Fourier transform analysis and synthesis equations. For this reason, we shall not provide a rigorous proof for each of the properties. Instead, we roughly show the way how they can be proved. The reader is strongly recommended to prove all the properties and solve the Fourier transform pairs, given in Tables 8.1 and 8.2.

Table 8.2 Fourier transform pairs of popular continuous time functions.

Signal $x(t)$	Fourier transform $X(j\omega)$
$\sum_{k=-\infty}^{\infty} a_k e^{jk\omega_0 t}$	$2\pi \sum_{k=-\infty}^{\infty} a_k \delta(\omega - k\omega_0)$
K	$2\pi K \delta(\omega)$
$\delta(t)$	1
$\delta(t - t_0)$	$e^{-j\omega t_0}$
$\sum_{n=-\infty}^{\infty} \delta(t - nT)$	$\frac{2\pi}{T} \sum_{k=-\infty}^{\infty} \delta\left(\omega - \frac{2\pi k}{T}\right)$
$\cos \omega_0 t$	$\pi[\delta(\omega - \omega_0) + \delta(\omega + \omega_0)]$
$\sin \omega_0 t$	$\frac{\pi}{j}[\delta(\omega - \omega_0) - \delta(\omega + \omega_0)]$
$e^{j\omega_0 t}$	$2\pi\delta(\omega - \omega_0)$
$u(t)$	$\pi\delta(\omega) + \frac{1}{j\omega}$
$u(t + T_1) - u(t - T_1)$ (or $x(t) = 1$, for $\|t\| < T_1$, 0 for $\|t\| > T_1$)	$\frac{2 \sin \omega T_1}{\omega}$
$\text{sign}(t) = \frac{t}{\|t\|}$	$\frac{1}{j\omega}$
$\frac{1}{\pi t}$	$-j\text{sign}(\omega)$
$tu(t)$	$j\pi \frac{d\delta(\omega)}{d\omega} - \frac{1}{\omega^2}$
t^n	$2\pi (j)^n \delta^{(n)}(\omega)$
$e^{-\alpha t}u(t)$, Re $\{\alpha\} > 0$	$\frac{1}{j\omega + \alpha}$
$e^{-\alpha\|t\|}$, Re $\{\alpha\} > 0$	$\frac{2\alpha}{\omega^2 + \alpha^2}$
$\frac{1}{\beta - \alpha}\left[e^{-\alpha t} - e^{-\beta t}\right] u(t)$, Re $\{\alpha\} > 0$, Re $\{\beta\} > 0$	$\frac{1}{(j\omega + \alpha)(j\omega + \beta)}$
$te^{-\alpha t}u(t)$, Re $\{\alpha\} > 0$	$\frac{1}{(j\omega + \alpha)^2}$
$\frac{t^{n-1}}{(n-1)!} e^{-\alpha t} u(t)$	$\frac{1}{(j\omega + \alpha)^n}$
$\frac{\sin Wt}{\pi t} = \frac{W}{\pi} \text{sinc} \frac{Wt}{\pi}$, Re $\{\alpha\} > 0$	$u(\omega + W) - u(\omega - W)$
$e^{-(\alpha t)^2}$, Re $\{\alpha\} > 0$	$\frac{\sqrt{\pi}}{\alpha} e^{-(\omega/2\alpha)^2}$
$e^{-\alpha t} \sin(\omega_0 t) u(t)$, Re $\{\alpha\} > 0$	$\frac{\omega_0}{(j\omega + \alpha)^2 + \omega_0^2}$
$e^{\alpha t} \sin(\omega_0 t) u(-t)$, Re $\{\alpha\} > 0$	$\frac{-\omega_0}{(\alpha - j\omega)^2 + \omega_0^2}$
$e^{-\alpha t} \cos(\omega_0 t) u(t)$, Re $\{\alpha\} > 0$	$\frac{\alpha + j\omega}{(j\omega + \alpha)^2 + \omega_0^2}$
$e^{\alpha t} \cos(\omega_0 t) u(-t)$, Re $\{\alpha\} > 0$	$\frac{\alpha - j\omega}{(\alpha - j\omega)^2 + \omega_0^2}$
$(\cos \omega_0 t)\left[u(t + T_1) - u(t - T_1)\right]$	$T_1 \left[\frac{\sin(\omega - \omega_0)T_1}{(\omega - \omega_0)T_1} + \frac{\sin(\omega + \omega_0)T_1}{(\omega + \omega_0)T_1}\right]$
Periodic square wave: $\begin{cases} 1, & \|t\| < T_1 \\ 0, & T_1 < \|t\| \leq \frac{T}{2} \end{cases}$ with period T	$\sum_{k=-\infty}^{\infty} \frac{2 \sin k\omega_0 T_1}{k\pi} \delta(\omega - k\omega_0)$

In the following, we study a selected set of properties to grasp the time and frequency domain representations and their relationship.

1) **Linearity:** Fourier transform is a linear transform. Mathematically speaking, if we have two functions and their corresponding Fourier transforms,

$$x(t) \longleftrightarrow X(j\omega), \tag{8.68}$$

and

$$y(t) \longleftrightarrow Y(j\omega), \tag{8.69}$$

then,

$$ax(t) + by(t) \longleftrightarrow aX(j\omega) + bY(j\omega). \tag{8.70}$$

Thus, the superposition is transformed from the time to the frequency domain and vice versa. In other words, if we superpose two signals in the time domain, the same superposition applies in the frequency domain. This property follows from the fact that integral is a linear operator.

2) **Time Shifting:** If the function $x(t)$ is shifted in time by a constant amount, $t_0 \in \mathbb{R}$, its Fourier transform $X(j\omega)$ is multiplied by $e^{-j\omega t_0}$, as follows:

$$x(t - t_0) \longleftrightarrow e^{-j\omega t_0}X(j\omega). \tag{8.71}$$

This property can be proved by defining $y(t) = x(t - t_0)$ and inserting the shifted signal into the analysis equation

$$Y(j\omega) = \int_{-\infty}^{\infty} x(t - t_0)e^{-j\omega t}dt. \tag{8.72}$$

Let us change the variable of integration to $\tau = t - t_0$ and get,

$$Y(j\omega) = \int_{-\infty}^{\infty} x(\tau)e^{-j\omega(\tau+t_0)t}dt = e^{-j\omega t_0}X(j\omega). \tag{8.73}$$

Note: Complex exponential, $e^{-j\omega t_0}$, has a magnitude of 1. Thus, the time delay alters the phase of the frequency domain signal, $X(j\omega)$, but not its magnitude. As a result, time delay does not cause the frequency content of $X(j\omega)$ to change at all.

In order to illustrate the use of the linearity and time-shift properties, let us solve the following example.

Exercise 8.10 Consider the following signal:

$$x(t) = \begin{cases} 1, & \text{if } 1 < t < 2 \quad \text{and} \quad 3 < t < 4 \\ 1.5, & \text{if } 2 < t < 3 \\ 0, & \text{otherwise.} \end{cases} \tag{8.74}$$

a) Plot and decompose $x(t)$ as a linear combination of two signals in the time domain using two rectangular pulse signals.
b) Use the decomposition of part a to find and plot the magnitude and phase of the Fourier transform $X(j\omega)$.

Solution

a) Figure 8.11a shows the plot of $x(t)$. We observe that $x(t)$ can be expressed as the linear combination of two signals,

$$x(t) = \frac{1}{2}x_1(t - 2.5) + x_2(t - 2.5), \tag{8.75}$$

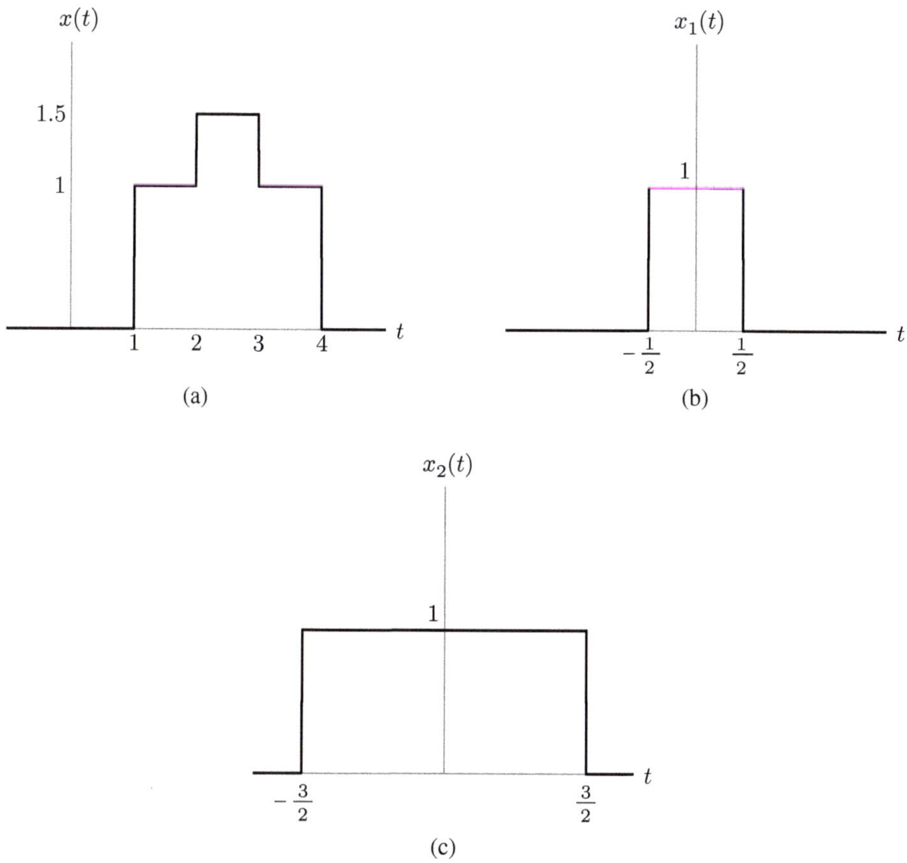

Figure 8.11 Decomposing a signal into the linear combination of two simpler signals, $x_1(t)$ and $x_2(t)$. (a) The signal $x(t) = \frac{1}{2}x_1(t-2.5) + x_2(t-2.5)$, (b) and (c) the signals, $x_1(t)$ and $x_2(t)$, which are used to represent $x(t)$.

where the signals $x_1(t)$ and $x_2(t)$ are the rectangular pulse signals shown in Figure 8.11b and 8.11c.

b) Fourier transform of rectangular pulses, $x_1(t)$ and $x_2(t)$ are

$$x_1(t) \leftrightarrow X_1(j\omega) = \frac{2\sin(\omega/2)}{\omega} \tag{8.76}$$

and

$$x_2(t) \leftrightarrow X_2(j\omega) = \frac{2\sin(3\omega/2)}{\omega}, \tag{8.77}$$

respectively.

Equation (8.75) shows that for both $x_1(t)$ and $x_2(t)$, time shift is $t_0 = 5/2$. In order to find the Fourier transform $X(j\omega)$, we multiply the Fourier transforms of $X_1(j\omega)$ and $X_2(j\omega)$ by $e^{-j\omega t_0} = e^{-j\omega\frac{5}{2}}$. Using the linearity and time-shift properties, the Fourier transform becomes

$$X(j\omega) = e^{-j5\omega/2}\left(\frac{\sin(\omega/2) + 2\sin(3\omega/2)}{\omega}\right). \tag{8.78}$$

Exercise 8.11 Consider the following signal:

$$x(t) = \cos(\omega_0 t + \phi). \tag{8.79}$$

a) Find the Fourier transform of $x(t)$, for $\phi = 0$.
b) Find the Fourier transform of $x(t)$, for $\phi \neq 0$.
c) Compare the results of parts a and b.

Solution

a) For $\phi = 0$, the function is

$$x(t) = \cos \omega_0 t. \tag{8.80}$$

Spectral coefficients of this function is $a_1 = a_{-1} = \frac{1}{2}$. Fourier transform of this function is

$$X(j\omega) = \pi(\delta(\omega + \omega_0) + \delta(\omega - \omega_0)). \tag{8.81}$$

b) For $\phi \neq 0$, let us arrange the function as follows:

$$x(t) = \cos\left(\omega_0\left(t + \frac{\phi}{\omega_0}\right)\right), \tag{8.82}$$

and apply the time shift property,

$$X(j\omega) = \pi e^{j\frac{\phi}{\omega_0}\omega}(\delta(\omega + \omega_0) + \delta(\omega - \omega_0)). \tag{8.83}$$

Note that there are only two nonzero values of $X(j\omega)$, one at $\omega = \omega_0$ and the other at $\omega = -\omega_0$. Therefore,

$$X(j\omega) = \pi(e^{-j\phi}\delta(\omega + \omega_0) + e^{j\phi}\delta(\omega - \omega_0)). \tag{8.84}$$

c) Comparing the results of parts (a) and (b), we observe that a phase shift of ϕ, in time domain, does not change the frequency content of the signal, but it multiplies the amplitude at each frequency by an amount of $e^{j\phi}$ and $e^{-j\phi}$.

3) **Time Scale:** If the time variable of the function $x(t)$ is scaled by a constant amount, $a \in \mathbb{R}$, its Fourier transform has the following form:

$$x(at) \longleftrightarrow \frac{1}{|a|}X\left(\frac{j\omega}{a}\right). \tag{8.85}$$

The Fourier analysis integral provides different solutions for positive and negative values of a for convergence. This is basically because the dummy variable of integral, $\tau = t/a$ changes the sign:
For $a > 0$, the analysis equation becomes

$$Y(j\omega) = \int_{-\infty}^{\infty} x(at)e^{-j\omega t}dt = \frac{1}{a}\int_{-\infty}^{\infty} x(\tau)e^{-j\omega\frac{\tau}{a}}d\tau = \frac{1}{a}X\left(\frac{j\omega}{a}\right). \tag{8.86}$$

For $a < 0$, the analysis equation becomes

$$Y(j\omega) = \int_{-\infty}^{\infty} x(at)e^{-j\omega t}dt = -\frac{1}{a}\int_{-\infty}^{\infty} x(\tau)e^{-j\omega\frac{\tau}{a}}d\tau = -\frac{1}{a}X\left(\frac{j\omega}{a}\right). \tag{8.87}$$

These derivations explain the reason for the absolute value, $\frac{1}{|a|}$, which multiplies the frequency-scaled Fourier transform.

Exercise 8.12 Given a signal in time and frequency domain as:

$$x(t) \leftrightarrow X(j\omega),$$

find the Fourier transform of

$$y(t) = x(3t + 7), \tag{8.88}$$

in terms of $X(j\omega)$.

Solution

The signal $y(t)$ is both time-scaled and time-shifted version of $x(t)$ and it can be written in the following form:

$$y(t) = x\left(3\left(t + \frac{7}{3}\right)\right). \tag{8.89}$$

Using the time shift and time scale properties for $a = 3$ and $t_0 = -7/3$, we find that

$$y(t) = x(3t + 7) \longleftrightarrow Y(j\omega) = \frac{e^{\frac{7}{3}j\omega}}{3} X\left(\frac{j\omega}{3}\right). \tag{8.90}$$

4) **Time Reversal:** A special case of time scale is time reversal. Applying the time scale property for $a = -1$, we observe that reversing a signal in time also reverses the frequencies in the Fourier transform:

$$x(-t) \longleftrightarrow X(-j\omega). \tag{8.91}$$

Note: Reversing the time, i.e., starting from the end of the signal does not change the magnitude of the signal, but just its phase.

Exercise 8.13 Consider the following continuous time signal:

$$x(t) = e^{-a|t|}. \tag{8.92}$$

Find the Fourier transform of this signal, without taking the integral of the analysis equation.

Solution

First, we split $x(t)$ into two parts as follows:

$$x(t) = e^{-a|t|} = e^{-at}u(t) + e^{at}u(-t). \tag{8.93}$$

Let's define,

$$x_1(t) = e^{-at}u(t) \quad \text{and} \quad x_2(t) = e^{at}u(-t). \tag{8.94}$$

Then, by using the linearity property,

$$X(j\omega) = X_1(j\omega) + X_2(j\omega). \tag{8.95}$$

By using the lookup tables for Fourier transforms, we find

$$X_1(j\omega) = \frac{1}{a + j\omega}, \tag{8.96}$$

and

$$X_2(j\omega) = X_1(-j\omega) = \frac{1}{a - j\omega}. \tag{8.97}$$

Therefore, the Fourier transform of $x(t)$ is

$$X(j\omega) = X_1(j\omega) + X_2(j\omega) = \frac{2a}{a^2 + \omega^2}. \tag{8.98}$$

5) **Conjugate Symmetry:** In most practical applications, such as biological signals, speech, music, the time domain function, $x(t)$ is real. However, when we design an electrical circuit, we face complex signals. There is an interesting relationship between the time and frequency representation of complex signals:

$$x(t) \longleftrightarrow X(j\omega) \Longleftrightarrow x^*(t) \longleftrightarrow X^*(-j\omega), \tag{8.99}$$

where $(*)$ indicates the complex conjugate operation. This result can easily be shown from the analysis equation.

Note: Conjugacy is preserved in both time and frequency domain representation of signals.

6) **Differentiation Property:** Taking the derivative of a signal in the time domain corresponds to multiplying its Fourier transform by $j\omega$ in the frequency domain:

$$\frac{dx(t)}{dt} \longleftrightarrow j\omega X(j\omega). \tag{8.100}$$

Differentiation property can be easily shown by taking the derivative of both sides of the synthesis equation, as follows:

$$\frac{dx(t)}{dt} = \frac{1}{2\pi}\frac{d}{dt}\int_{-\infty}^{\infty} X(j\omega)e^{j\omega t}d\omega = \frac{1}{2\pi}\int_{-\infty}^{\infty} (j\omega)X(j\omega)e^{j\omega t}d\omega. \tag{8.101}$$

This property can be generalized to the nth derivative as follows:

$$\frac{d^n x(t)}{dt^n} \longleftrightarrow (j\omega)^n X(j\omega). \tag{8.102}$$

Note: If we have an nth-order differential equation in the time domain, its Fourier transform gives us an nth-order algebraic equation, with the powers of $(j\omega)$. Therefore, solving a differential equation in the time domain is equivalent to solving an algebraic equation in the frequency domain.

Exercise 8.14 Solve the following differential equation using the differentiation property, when the input is $x(t) = e^{-3t}u(t)$, for the initial rest systems.

$$\frac{d^2y(t)}{dt} + \frac{dy(t)}{dt} - 2y(t) = x(t). \tag{8.103}$$

Solution

The Fourier transform of the input is

$$\mathcal{F}\{x(t)\} = \mathcal{F}[e^{-3t}u(t)] = \frac{1}{j\omega + 3}. \tag{8.104}$$

Taking the Fourier transform of both sides of the differential equation yields,

$$[(j\omega)^2 + (j\omega) - 2]Y(j\omega) = \frac{1}{j\omega + 3}. \tag{8.105}$$

Leaving the Fourier transform of the output on the left-hand side and factorizing the second-order term, we get,

$$Y(j\omega) = \frac{1}{(j\omega + 3)(j\omega - 1)(j\omega + 2)}. \tag{8.106}$$

Using the method of partial fraction expansion,

$$Y(j\omega) = \frac{1}{(j\omega + 3)(j\omega - 1)(j\omega) + 2)} = \frac{A}{(j\omega + 3)} + \frac{B}{(j\omega - 1)} + \frac{C}{(j\omega + 2)}, \tag{8.107}$$

we find $A = 1/4, \quad B = 1/12, \quad C = -1/3$.

Thus, the Fourier transform of the output is

$$Y(j\omega) = \frac{1}{(j\omega+3)(j\omega-1)(j\omega)+2)} = \frac{1/4}{(j\omega+3)} + \frac{1/12}{(j\omega-1)} - \frac{1/3}{(j\omega+2)}. \tag{8.108}$$

Taking the inverse Fourier transform we get the output as:

$$y(t) = \frac{1}{4}e^{-3t}u(t) - \frac{1}{12}e^{t}u(-t) - \frac{1}{3}e^{-2t}u(t). \tag{8.109}$$

This exercise shows that we can solve nth-order linear differential equations with constant coefficients, algebraically in the frequency domain.

7) **Integration Property:** Integrating a signal in time domain corresponds to dividing its Fourier transform by $j\omega$ in the frequency domain:

$$\int_{-\infty}^{t} x(\tau)d\tau \longleftrightarrow \frac{1}{j\omega}X(j\omega) + \pi X(0)\delta(\omega). \tag{8.110}$$

Integration property can be derived by defining the synthesis equation for $x(t)$,

$$x(t) = \frac{1}{2\pi}\int_{-\infty}^{\infty} X(j\omega)e^{j\omega t}d\omega. \tag{8.111}$$

Let us take the integral of both sides of the synthesis equation:

$$\int_{-\infty}^{t} x(t)dt = \frac{1}{2\pi}\int_{-\infty}^{t}\int_{-\infty}^{\infty} X(j\omega)e^{j\omega t}d\omega dt = \frac{1}{2\pi}\int_{-\infty}^{\infty} \frac{X(j\omega)}{j\omega}e^{j\omega t}d\omega. \tag{8.112}$$

This integral shows that for $\omega \neq 0$,

$$\int_{-\infty}^{t} x(t)dt \longleftrightarrow \frac{X(j\omega)}{j\omega}. \tag{8.113}$$

For $\omega = 0$, however, the integral in the right-hand side of Equation (8.111) approaches to ∞. Hence, the Fourier transform of the integral of function $x(t)$ in Equation (8.113) is incomplete. Using Cauchy integral theorem (Complex Analysis: A Modern First Course in Function Theory, Jerry R. Muir Jr. ISBN: 978-1-118-70522-3 April 2015, Wiley), we can evaluate the integral of Equation (8.111) for $\omega = 0$ and obtain an additive term $\pi X(0)\delta(\omega)$ to complete the Fourier transform of the integral of $x(t)$, as:

$$\int_{-\infty}^{t} x(\tau)d\tau \longleftrightarrow \frac{1}{j\omega}X(j\omega) + \pi X(0)\delta(\omega). \tag{8.114}$$

Similar to the differentiation property, the integration property converts an integral equation into an algebraic equation in the frequency domain.

Exercise 8.15 Consider an integrator system (Figure 8.12),

a) Find the impulse response of this LTI system.
b) Find the frequency response of this LTI system.

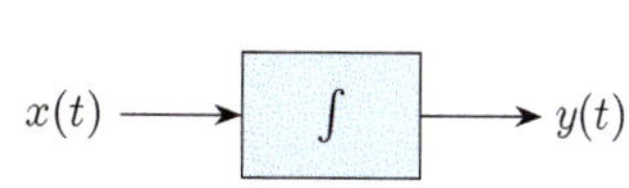

Figure 8.12 An integrator.

Solution

a) When the input of an integrator is the impulse function, the corresponding output is the impulse response,

$$h(t) = \int_{-\infty}^{t} \delta(t)d\tau = u(t), \tag{8.115}$$

which is a unit step function.

b) Integration property o reveals that the output of an integrator for a general input $x(t)$ is

$$y(t) = \int_{-\infty}^{t} x(t)d\tau \longleftrightarrow Y(j\omega) = \frac{X(j\omega)}{j\omega} + \pi X(0)\delta(\omega). \tag{8.116}$$

When we replace the input by an impulse function,

$$x(t) = \delta(t) \overset{\mathcal{F.T.}}{\longleftrightarrow} X(j\omega) = 1, \tag{8.117}$$

we obtain the frequency response as follows:

$$H(j\omega) = \frac{1}{j\omega} + \pi\delta(\omega). \tag{8.118}$$

Thus, the impulse response is

$$h(t) = u(t) \leftrightarrow \frac{1}{j\omega} + \pi\delta(\omega). \tag{8.119}$$

Exercise 8.16 Find the Fourier transform $X(j\omega)$ for the signal $x(t)$ displayed in Figure 8.13a, without evaluating the synthesis integral.

Solution

Rather than applying the Fourier integral directly to $x(t)$, we consider the signal

$$g(t) = \frac{d}{dt}x(t) \tag{8.120}$$

As illustrated in Figure 8.13b, $g(t)$ is the sum of a rectangular pulse and two impulses. The Fourier transforms of each of these component signals may be determined from Table 8.2:

$$G(j\omega) = \left(\frac{2\sin\omega}{\omega}\right) - e^{j\omega} - e^{-j\omega}. \tag{8.121}$$

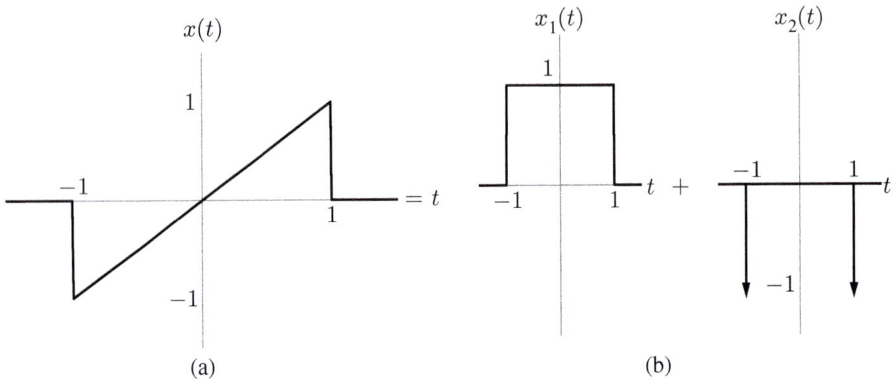

Figure 8.13 (a) A signal $x(t)$ for which the Fourier transform is to be evaluated; (b) representation of the derivative of $x(t)$ is $\frac{dx(t)}{dt} = x_1(t) + x_2(t)$.

Using the integration property, we obtain

$$X(j\omega) = \frac{G(j\omega)}{j\omega} + \pi G(0)\delta(\omega). \tag{8.122}$$

Since $G(0) = 0$, Fourier transform of $x(t)$ becomes

$$X(j\omega) = \frac{2\sin\omega}{j\omega^2} - \frac{2\cos\omega}{j\omega}. \tag{8.123}$$

Note: If we have an integral equation in the time domain, its Fourier transform gives us an algebraic equation. Therefore, taking an integral in the time domain is equivalent to dividing its Fourier transform by $j\omega$ and adding the value of $\pi X(0)\delta(\omega)$, in the frequency domain.

8) **Convolution Property:** One of the most useful properties of Fourier transform is the convolution property. This property states that convolution in the time domain corresponds to multiplication in the frequency domain. Mathematically,

$$y(t) = x(t) * h(t) \longleftrightarrow Y(j\omega) = X(j\omega)H(j\omega). \tag{8.124}$$

Convolution property follows the definition of Fourier analysis equation and convolution integral. Inserting the convolution integral,

$$x(t) * y(t) = \int_{-\infty}^{\infty} x(\tau)h(t-\tau)d\tau,$$

into the Fourier analysis equation, we obtain,

$$Y(j\omega) = \mathcal{F}\{y(t)\} = \mathcal{F}\{x(t) * h(t)\} = \int_{-\infty}^{\infty}\int_{-\infty}^{\infty} x(\tau)h(t-\tau)e^{j\omega t}d\tau dt, \tag{8.125}$$

Changing the dummy variable of integration, $t' = t - \tau$, we get

$$Y(j\omega) = \mathcal{F}[x(t) * h(t)] = \int_{-\infty}^{\infty} x(\tau)e^{j\omega\tau}d\tau \int_{-\infty}^{\infty} h(t')e^{j\omega t'}dt' = X(j\omega)H(j\omega). \tag{8.126}$$

Recall that time and frequency representations of signals and systems are one-to-one and onto. Therefore, the block diagram of an LTI system can be represented in time and frequency domains, equivalently, as shown in Figures 8.14 and 8.15.

Recall that the impulse response is the inverse Fourier transform of the frequency response,

$$h(t) = \frac{1}{2\pi}\int_{-\infty}^{\infty} H(j\omega)e^{-j\omega t}d\omega, \tag{8.127}$$

and frequency response is the Fourier transform of the impulse response,

$$H(j\omega) = \int_{-\infty}^{\infty} h(t)e^{-j\omega t}dt. \tag{8.128}$$

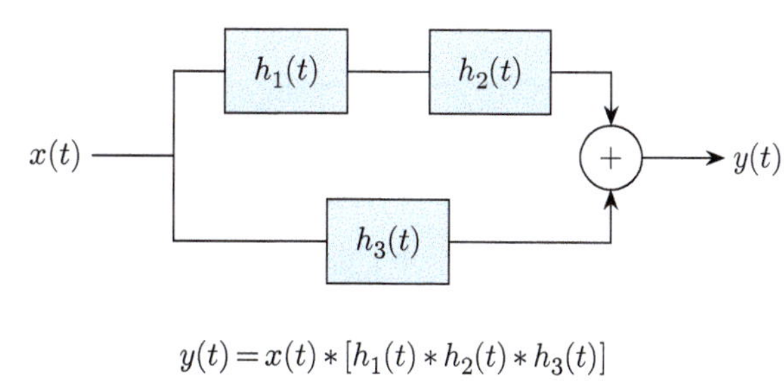

Figure 8.14 A sample block diagram representation in time domain. Note that all the operations between the input and impulse response functions are convolution.

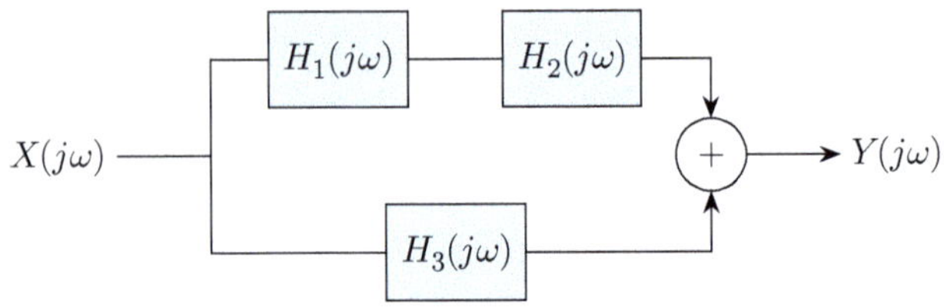

$$y(t) \longleftrightarrow Y(j\omega) = X(j\omega)[H_1(j\omega)H_2(j\omega) + H_3(j\omega)]$$

Figure 8.15 The block diagram representation of Figure 8.14 in frequency domain. Note that the convolution operations are replaced by multiplication operations.

Exercise 8.17 Consider an LTI system represented by the following shifted impulse function:

$$h(t) = \delta(t - t_0), \tag{8.129}$$

a) Find the frequency response.
b) Find the system equation, which relates the input, $X(j\omega)$, and output, $Y(j\omega)$, in the frequency domain.
c) Find the system equation, which relates the input, $x(t)$, and output, $y(t)$, in time domain.

Solution

a) Using the time shift property and Table 8.2, we obtain the frequency response as follows:

$$h(t) = \delta(t - t_0) \longleftrightarrow H(j\omega) = e^{-j\omega t_0}, \tag{8.130}$$

b) System equation in frequency domain can be directly obtained by considering the relationship among the input, output and frequency response, as follows:

$$H(j\omega) = \frac{Y(j\omega)}{X(j\omega)} = e^{-j\omega t_0}. \tag{8.131}$$

Thus, the system equation in the frequency domain is

$$Y(j\omega) = e^{-j\omega t_0} X(j\omega). \tag{8.132}$$

c) System equation in the time domain can be obtained by taking the inverse Fourier transform of the system equation in the frequency domain, found in part b,

$$y(t) = x(t - t_0). \tag{8.133}$$

9) **Multiplication Property:** Multiplication of two signals in the time domain corresponds to convolution of the Fourier transforms of the signals in the frequency domain:

$$y(t) = x(t)h(t) \longleftrightarrow Y(j\omega) = \frac{1}{2\pi} X(j\omega) * H(j\omega). \tag{8.134}$$

In order to show the multiplication property, we take the Fourier transform of the multiplication of two functions, $y(t) = x(t)h(t)$ using the analysis equation,

$$Y(j\omega) = \mathcal{F}\{(t)\} = \mathcal{F}\{(t) * h(t)\} = \int_{-\infty}^{\infty} x(t)h(t)e^{j\omega t}dt. \tag{8.135}$$

Then, we insert the inverse Fourier transform of $x(t)$,

$$x(t) = \frac{1}{2\pi} \int_{-\infty}^{\infty} X(j\omega)e^{-j\omega} td\omega, \tag{8.136}$$

into Equation (8.135),

$$Y(j\omega) = \mathcal{F}\{x(t)h(t)\} = \frac{1}{2\pi} \int_{-\infty}^{\infty} [\int_{-\infty}^{\infty} X(j\omega')e^{-j\omega' t}d\omega']h(t)e^{j\omega t}dt, \tag{8.137}$$

Finally, we arrange the integrals of Equation (8.137),

$$Y(j\omega) = \mathcal{F}\{x(t)\} = \frac{1}{2\pi}\int_{-\infty}^{\infty} X(j\omega')\int_{-\infty}^{\infty} h(t)e^{jt(\omega-\omega')}dtd\omega'. \tag{8.138}$$

Note that the integral in the right-hand side of this equation is the shifted Fourier transform of the function $h(t)$,

$$H(j(\omega-\omega')) = \int_{-\infty}^{\infty} h(t)e^{jt(\omega-\omega')}d\omega'. \tag{8.139}$$

Replacing the value of the frequency domain function $H(j(\omega-\omega_0))$ with the integral of Equation (8.139), we obtain,

$$Y(j\omega) = \mathcal{F}\{x(t)h(t)\} = \frac{1}{2\pi}\int_{-\infty}^{\infty} X(j\omega')H(j(\omega-\omega'))dt = \frac{1}{2\pi}X(j\omega) * H(j\omega). \tag{8.140}$$

This property is very useful in designing communication networks because it enables one to change the bandwidth of a signal without changing its analytical structure. The following example illustrates this fact.

Exercise 8.18 Amplitude Modulation: Suppose that we have a bandlimited signal given as follows:

$$s(t) \overset{\mathcal{F}}{\longleftrightarrow} S(j\omega), \tag{8.141}$$

where the Fourier transform has the form of Figure 8.16.

Suppose also that we have a periodic signal, $p(t) = \cos(\omega_0 t)$.

a) Find the Fourier transform, $M(j\omega)$, of the signal, $m(t) = p(t)s(t)$.
b) Compare $M(j\omega)$ to $S(j\omega)$ and comment on the bandwidth and analytical form of these two functions.

Solution

a) Let us multiply a bandlimited signal $s(t)$ with the cosine function $p(t)$ in time domain,

$$m(t) = p(t)s(t) = s(t)\cos(\omega_0 t). \tag{8.142}$$

In order to find the Fourier transform of $m(t)$, we use the multiplication property:

$$m(t) = p(t)s(t) \longleftrightarrow M(j\omega) = \frac{1}{2\pi}P(j\omega) * S(j\omega). \tag{8.143}$$

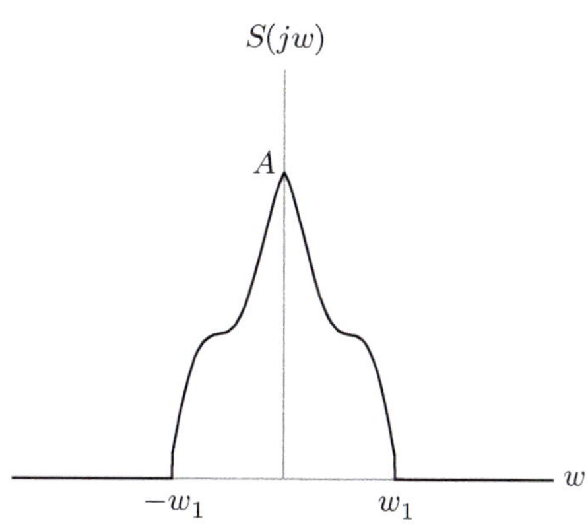

Figure 8.16 A bandlimited signal, $S(j\omega) = 0$ for $|\omega| > \omega_1$.

In general, the Fourier transform of a periodic signal, $p(t)$ is

$$P(j\omega) = \sum_{-\infty}^{\infty} 2\pi a_k \delta(\omega - k\omega_0), \tag{8.144}$$

where a_k's are the Fourier series coefficients of $p(t)$.

For this particular example, $p(t) = \cos\omega_0 t$. Then, there are two nonzero Fourier series coefficients: $a_1 = a_{-1} = 1/2$.

Thus,

$$P(j\omega) = \pi(\delta(\omega - \omega_0) + \delta(\omega + \omega_0)), \tag{8.145}$$

as shown in Figure 8.17.

Therefore, Fourier transform of $m(t)$ is

$$M(j\omega) = \frac{1}{2\pi} S(j\omega) * P(j\omega) = \frac{1}{2}\left[S(j(\omega - \omega_0)) + S(j(\omega + \omega_0))\right]. \tag{8.146}$$

b) Comparison of $M(j\omega)$ and $S(j\omega)$ shows an interesting similarity: Both signals have the same analytical form. When we multiply a low-frequency bandwidth signal, $s(t)$ with a cosine waveform of high frequency, the signal in Fourier domain preserves its analytical form, but the bandwidth is shifted toward the high frequencies. In addition, the magnitude of the signal is decreased by a factor of 0.5. This fact is depicted in Figure 8.18.

The aforementioned property of the Fourier transform has a very crucial implication. Suppose that we have a signal, $s(t)$, which has a low-frequency bandwidth. A good example is the speech signal, which ranges between 20 Hz and 20 kHz. This signal cannot be transmitted over a long distance by wireless communication technologies, in its original

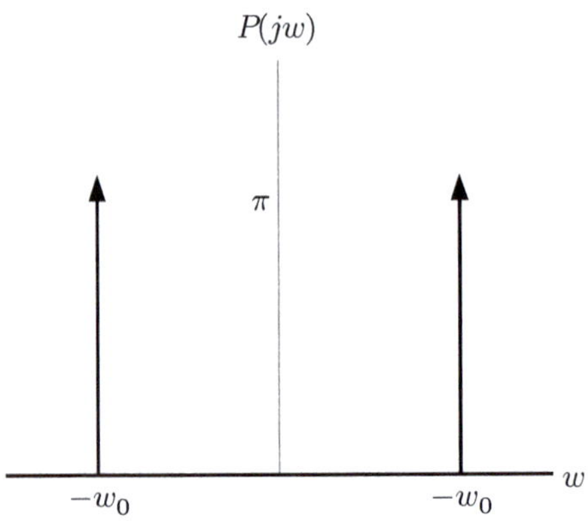

Figure 8.17 Fourier transform, $P(j\omega)$, of $p(t) = \cos(\omega_0 t)$.

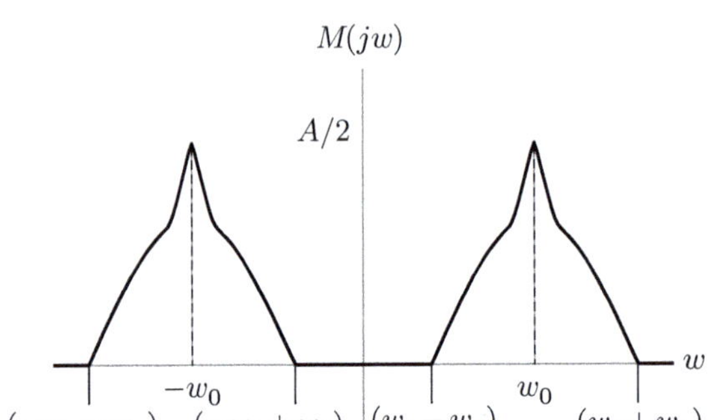

Figure 8.18 Amplitude modulation: The signal $s(t) \xrightarrow{\text{F.T.}} S(j\omega)$ is shifted to high frequencies by a carrier signal, $p(t) = \cos(\omega_0 t) \xrightarrow{\text{F.T.}} P(j\omega)$.

form. It is well known that a signal can only be transmitted over long distances if it has a very high-frequency bandwidth, such as microwaves, which require frequency bands in the order of megahertz or gigahertz.

When we need to transmit a signal over a long distance, we shift the signal to microwave bands by multiplying it with a very high-frequency periodic function, such as $p(t) = \cos(\omega_0 t)$. This operation is called **amplitude modulation**. While $m(t)$ is called the **modulated signal**, $p(t)$ is called **carrier signal**. After transmitting the modulated signal, $m(t) = p(t)s(t)$ to the final destination, a demodulation method is needed to reconstruct the original signal $s(t)$ from the modulated signal $m(t)$.

10) **Duality:** Duality is an important concept, which appears in a wide range of areas in mathematics. Fundamentally, duality refers to having a one-to-one correspondence between two mathematical objects, which represent different, but complementing characteristics of the same phenomenon.

Although the formal definition of duality changes in different areas of mathematics, the following two major properties are generally satisfied:

1) **Symmetry:** Given two mathematical objects, X and Y, if X is the dual of Y, then Y is the dual of X.
2) **Involution:** Dual of the dual of a mathematical object is the object itself. Formally speaking, let S and S' be two dual spaces of mathematical objects, and $\mathbf{F}$ be a duality mapping,

$$\mathbf{F} : S \longrightarrow S'. \tag{8.147}$$

Then,

$$\mathbf{F}(\mathbf{F}(X)) = X, \quad \forall X \in S \tag{8.148}$$

A simple everyday example of duality is a coin with two sides, which satisfies the two properties mentioned earlier:

1) **Symmetry:** The dual of head is the tail. The dual of the tail is the head.
2) **Involution:** The dual of the head is the head.

Fourier transforms exhibit the major duality properties, which link the time and frequency domain representations of the same phenomenon. Duality of Fourier transform follows from the fact that the analysis and synthesis equations are almost identical except for a factor of $\frac{1}{2\pi}$ and the difference of a minus sign in the exponential in the integral.

There are many remarkable symmetries and involution between the time and frequency domains. In the following, we overview three basic dualities of Fourier transforms.

- **Duality 1: Fourier Transform of Fourier Transform**

  The analytical form of the Fourier transform of the Fourier transform of a function is very similar to the analytical form of the function itself.

  Formally speaking, when a time domain function has an analytical function, in the form of x and its frequency domain representation has an analytical form of X, these two functions are related to each other by Fourier transform,

$$x(t) \longleftrightarrow X(j\omega). \tag{8.149}$$

  If we replace the variable $j\omega$ by t and take the Fourier transform of $\mathbf{X}(t)$, we obtain the reflected analytical form of x, scaled by 2π, in the frequency domain, as follows:

$$X(t) \longleftrightarrow 2\pi x(-\omega). \tag{8.150}$$

  In other words, taking the Fourier transform of the Fourier transform of a function nicely returns the turned-around function scaled by 2π. One consequence of this duality is that whenever we evaluate the Fourier transform of a function, the inverse can be obtained with the same algorithm, with a minor modification.

Exercise 8.19 Consider the following function:

$$x_1(t) = \begin{cases} 1, & \text{if } |t| < T \\ 0, & \text{otherwise.} \end{cases} \tag{8.151}$$

a. Find the Fourier transform of $x_1(t)$.
b. Replace the time variable t, by $j\omega$ and replace the threshold T by W in $x_1(t)$ to obtain a new frequency domain function, $X_2(j\omega)$, as follows:

$$X_2(j\omega) = \begin{cases} 1, & \text{if } |\omega| < W \\ 0, & \text{otherwise.} \end{cases} \tag{8.152}$$

and find the inverse Fourier transform of $X_2(j\omega)$.
c. Compare the analytical form of $x_2(t)$ to that of the inverse of $X_1(j\omega)$.

Solution

a. The Fourier transform of $x_1(t)$ is

$$X_1(j\omega) = \int_{-T}^{T} e^{-j\omega t} dt = \frac{1}{j\omega}(e^{j\omega T} - e^{-j\omega T}) = \frac{2\sin(\omega T)}{\omega}. \tag{8.153}$$

b. Now let us take the inverse Fourier transform of

$$X_2(j\omega) = \begin{cases} 1, & \text{if } |\omega| < W \\ 0, & \text{otherwise.} \end{cases} \tag{8.154}$$

Then, we obtain

$$x_2(t) = \frac{1}{2\pi} \int_{-W}^{W} e^{j\omega t} d\omega = \frac{\sin(\omega t)}{\pi t}. \tag{8.155}$$

c. The analytical form of $x_2(t)$ and $X_1(j\omega)$ are the same, as observed from Figure 8.19.

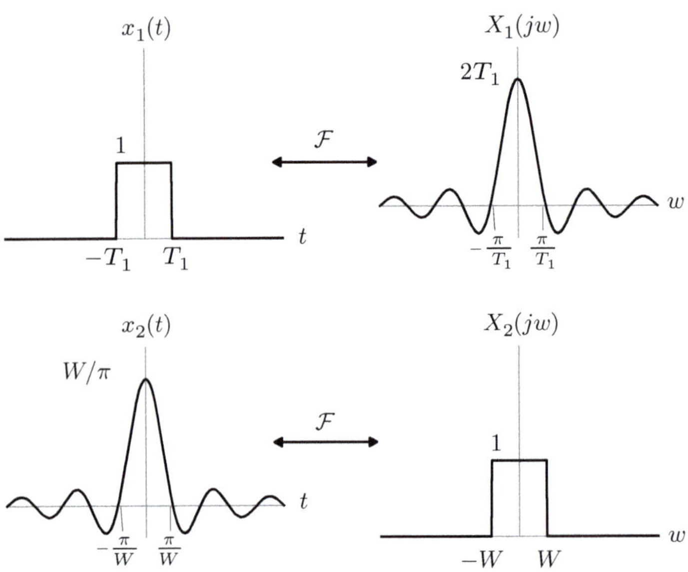

Figure 8.19 Duality property which shows the relationships between the analytical form of two functions in time and frequency domains.

- **Duality 2: Convolution vs. Multiplication Operations**
  Convolution of two functions in the time domain corresponds to multiplication in the frequency domain. Similarly, multiplication in the time domain corresponds to convolution in the frequency domains:

  Convolution in time $\longleftrightarrow$ Multiplication in frequency,
  Multiplication in time $\longleftrightarrow$ Convolution in frequency.

  Formally speaking,

$$x(t) * y(t) \longleftrightarrow X(j\omega)Y(j\omega), \tag{8.156}$$

  and

$$x(t)y(t) \longleftrightarrow \frac{1}{2\pi}X(j\omega) * Y(j\omega). \tag{8.157}$$

  This remarkable symmetry between the convolution and multiplication operations enables us to design many LTI systems, in a wide range of areas, in many disciplines of science and engineering.
- **Duality 3: Time Shift and Frequency Shift**
  A shift in the time domain corresponds to multiplication in the frequency domain. Similarly, multiplication in the time domain corresponds to a shift in the frequency domain.

  Time shift: $x(t-t_0)$ $\longleftrightarrow$ **Multiplication:** $e^{-j\omega t_0}X(j\omega)$
  Multiplication: $e^{j\omega_0}x(t)$ $\longleftrightarrow$ **Frequency shift:** $X(j(\omega-\omega_0))$.

  Note: Whenever we need a shift in one of the domains, the corresponding function in the other domain is just multiplied by a complex exponential function.

11) **Parseval's Equality:** In the aforementioned properties and examples, we observe that the representation of signals and systems in time and frequency domains have substantially different analytical forms and structures. For example, a periodic signal, which consists of sine and cosine functions, has a Fourier transform consisting of shifted impulse functions at the fundamental frequency and its harmonics. However, the energy of the signals in both domains does not change:

$$\int_{-\infty}^{\infty} |x(t)|^2 dt = \frac{1}{2\pi}\int_{-\infty}^{\infty} |X(j\omega)|^2 d\omega. \tag{8.158}$$

We can show Parseval's equality by inserting the Fourier analysis equation into the left-hand side of Equation (8.158), as follows:

$$\int_{-\infty}^{\infty} |x(t)|^2 dt = \int_{-\infty}^{\infty} x(t)x^*(t)dt \tag{8.159}$$

$$= \frac{1}{(2\pi)^2}\int_{-\infty}^{\infty}\int_{-\infty}^{\infty} X(j\omega)e^{j\omega t}d\omega \int_{-\infty}^{\infty} X^*(j\omega')e^{-j\omega' t}d\omega' dt.$$

Considering the fact that harmonically related complex exponential functions are periodic with 2π and they are orthogonal to each other,

$$\int_{0}^{2\pi} e^{j\omega t}e^{-\omega' t}dt = \begin{cases} 2\pi & \text{for } \omega = \omega' \\ 0 & \text{otherwise,} \end{cases} \tag{8.160}$$

and arranging the right-hand side of Equation (8.159), we obtain,

$$\int_{-\infty}^{\infty} |x(t)|^2 dt = \frac{1}{2\pi} \int_{-\infty}^{\infty} |X(j\omega)|^2 d\omega. \tag{8.161}$$

Parseval's equality reveals that the representation of signals in Hilbert space conserves the energy of the time domain. Note that there is a factor of $1/2\pi$, which scales the energy of the time domain.

The above properties of the Fourier transform simplify a large variety of difficult problems for designing and implementing the LTI systems, analyzing the frequency content of the signals, solving differential and integral equations, taking convolution, etc. Since time and frequency domains are one-to-one and onto, we can freely switch between them during the steps of our design and analysis processes, depending on our needs.

In Section 8.8.2, we shall study continuous time LTI systems represented by differential equations and show an algebraic method to solve the differential equations in the frequency domain. We observe that differentiation and integration properties make our lives quite easy for designing and analyzing LTI systems, in the frequency domain. Let us see how.

8.8.2 Continuous Time Linear Time-Invariant Systems in Frequency Domain

Recall that a continuous time LTI system is represented by the following ordinary constant coefficient differential equation in the time domain:

$$\sum_{k=0}^{N} a_k \frac{d^k y(t)}{dt^k} = \sum_{k=0}^{M} b_k \frac{d^k x(t)}{dt^k}. \tag{8.162}$$

In addition, recall that if the eigen function of $x(t) = e^{j\omega t}$ is the input to an LTI system represented by the impulse response, $h(t)$, then, the corresponding output is,

$$y(t) = H(j\omega)e^{j\omega t}, \tag{8.163}$$

where

$$H(j\omega) = \int_{-\infty}^{\infty} h(t)e^{-j\omega t} dt, \tag{8.164}$$

is called the frequency response. Since the right-hand side of Equation (8.164) is the Fourier transform of the impulse response, the frequency response is just the Fourier transform of the impulse response:

$$h(t) \longleftrightarrow H(j\omega). \tag{8.165}$$

If we take the Fourier transform of both sides of the nth-order differential equation given in Equation (8.162), we obtain the following equation, which represents an LTI system in the frequency domain:

$$\sum_{k=0}^{N} a_k (j\omega)^k Y(j\omega) = \sum_{k=0}^{M} b_k (j\omega)^k X(j\omega). \tag{8.166}$$

Note: All of the derivation operations in Equation (8.162) disappear in Equation (8.166) and the differential equation becomes an algebraic equation in powers of $(j\omega)$. Thus, an LTI system represented by a differential equation in the time domain is equivalently represented by an algebraic equation, in the frequency domain.

Arranging Equation (8.166), we obtain a relationship between the input and output of an LTI system in the frequency domain, as follows:

$$\frac{Y(j\omega)}{X(j\omega)} = \frac{\sum_{k=0}^{M} b_k (j\omega)^k}{\sum_{k=0}^{N} a_k (j\omega)^k}. \tag{8.167}$$

We can also obtain the frequency response of an LTI system by using the algebraic equation of Equation (8.166). When we replace the input of an LTI system with the impulse function in the time domain, the output becomes an impulse response. When the input is an impulse function in the time domain, its Fourier transform becomes

$$\mathcal{F}\{x(t)\} = \mathcal{F}\{\delta(t)\} = X(j\omega) = 1. \tag{8.168}$$

Then, the Fourier transform of the corresponding output becomes the frequency response, $H(j\omega)$. Therefore, replacing the input by $X(j\omega) = 1$, the differential equation of Equation (8.166) becomes

$$\sum_{k=0}^{N} a_k (j\omega)^k H(j\omega) = \sum_{k=0}^{M} b_k (j\omega)^k. \tag{8.169}$$

This equation provides us the **frequency response** of an LTI system, represented by an ordinary constant coefficient differential equation in the time domain and an algebraic equation in the frequency domain. Arranging Equation (8.169), we obtain the frequency response, as a rational function of $j\omega$, as follows:

$$H(j\omega) = \frac{\sum_{k=0}^{M} b_k (j\omega)^k}{\sum_{k=0}^{N} a_k (j\omega)^k}. \tag{8.170}$$

Comparing Equations (8.167) and (8.170), we observe that the right-hand sides are the same. Thus, the ratio of the Fourier transform of the output and input signals is equal to the frequency response is

$$\boxed{H(j\omega) = \frac{Y(j\omega)}{X(j\omega)} = \frac{\sum_{k=0}^{M} b_k (j\omega)^k}{\sum_{k=0}^{N} a_k (j\omega)^k}}. \tag{8.171}$$

Taking the inverse Fourier transform of the frequency response directly gives us the impulse response, without solving the differential equation, because; time and frequency domain representations are one-to-one;

$$h(t) \longleftrightarrow H(j\omega). \tag{8.172}$$

In the following example, we are going to study the relationships between the differential equations in the time domain and their algebraic representation in the frequency domain.

Exercise 8.20 Consider the following first-order differential equation:

$$\frac{dy(t)}{dt} + y(t) = x(t), \tag{8.173}$$

a) Find impulse response.
b) Find the solution $y(t)$, for the input

$$x(t) = e^{-2t}u(t), \tag{8.174}$$

when the system is initially at rest.

Solution

a) If we did not know the Fourier transforms, we would replace the input by an impulse function and solve the given differential equation for $h(t)$. However, taking the Fourier transform avoids solving the differential equation as follows: First, let us take the Fourier transform of both sides of the differential equation given above;

$$[(j\omega) + 1]Y(j\omega) = X(j\omega). \tag{8.175}$$

Then, let us replace the input by an impulse function and its Fourier transform, which is,

$$x(t) = \delta(t) \longrightarrow X(j\omega) = 1. \tag{8.176}$$

The corresponding output, which is the frequency response, can be directly obtained from the Fourier transform of the differential equation, as follows:

$$H(j\omega) = \frac{1}{1 + j\omega}. \tag{8.177}$$

The impulse response of this system can be obtained from the inverse Fourier transform of the frequency response, given in Table 8.2,

$$H(j\omega) = \frac{1}{1 + j\omega} \xrightarrow{\mathcal{F}^{-1}} h(t) = e^{-t}u(t). \tag{8.178}$$

The above method enables us to find the impulse response of an LTI system without solving the differential equation.

b) Next, we shall find the output $y(t)$ without solving the differential equation. We simply use the relationship between the input and output in the frequency domain:

$$Y(j\omega) = H(j\omega)X(j\omega). \tag{8.179}$$

The Fourier transform of the input is found in Table 8.2, as:

$$x(t) = e^{-2t}u(t) \longleftrightarrow X(j\omega) = \frac{1}{(2 + j\omega)}. \tag{8.180}$$

Then, we obtain the Fourier transform of the output,

$$\begin{aligned} Y(j\omega) &= \frac{1}{(1 + j\omega)(2 + j\omega)} \\ &= \frac{1}{1 + j\omega} - \frac{1}{2 + j\omega}. \end{aligned} \tag{8.181}$$

By using the linearity property and Table 8.2, we find the inverse Fourier transform of $Y(j\omega)$, as follows:

$$y(t) = (e^{-t} + e^{-2t})u(t). \tag{8.182}$$

8.9 Laplace Transforms as an Extension of Continuous Time Fourier Transforms

Recall that the Fourier transform of a continuous time function exists if it satisfies the Dirichlet conditions. When a time domain function is not absolutely integrable, it violates the Dirichlet condition and it is not possible to find a finite Fourier transform, in the frequency domain.

Motivating Question: Can we further generalize the Fourier transform in such a way that the transform domain representation of a time domain function exists in some predefined values of the new variable of this domain?

Laplace transform opens a door to answer the above question by defining a new domain, called, Laplace domain or s-domain. In Laplace domain, a complex variable, $s = \sigma + j\omega$, is defined as an alternative to the purely imaginary variable, $j\omega$, of the frequency domain. Therefore, Laplace transform is considered as an extension of the Fourier transform, where the purely imaginary variable of the frequency domain is generalized by defining a complex variable, s with real and imaginary parts, in the Laplace domain. Considering the two-dimensional complex plane, Fourier transform, $X(j\omega)$ can only exist on the $j\omega$ axis. In other words, while the Fourier transform maps a time domain function into the one-dimensional purely imaginary axis, $j\omega$, of the complex plane, Laplace transform maps the function into the entire two-dimensional complex plane.

Recall that Fourier transform of a time domain function is defined as the weighted integral of the complex exponential function,

$$\mathcal{F}\{x(t)\} = X(j\omega) = \int_{-\infty}^{\infty} x(t)e^{-j\omega t}dt. \tag{8.183}$$

Laplace transform can be obtained by extending the Fourier transform. This requires simply to replace the purely imaginary frequency variable $j\omega$ of the Fourier transform with a complex variable, $s = \sigma + j\omega$, which consists of a real and imaginary part.

Formally, the Laplace transform of a continuous time function $x(t)$ is defined as:

$$\mathcal{L}\{x(t)\} = X(s) = \int_{-\infty}^{\infty} x(t)e^{-st}dt, \tag{8.184}$$

where $s = \sigma + j\omega$ is a complex variable and e^{-st} is a complex exponential function. If we replace $s = \sigma + j\omega$ in the integral of Equation (8.184) , we obtain,

$$\mathcal{L}\{x(t)\} = \int_{-\infty}^{\infty} x(t)e^{-(\sigma+j\omega)t}dt = \int_{-\infty}^{\infty} x(t)e^{-\sigma t}e^{j\omega t}dt, \tag{8.185}$$

which yields a relationship between Laplace and Fourier transforms, as follows:

$$\mathcal{L}\{x(t)\} = F\{x(t)e^{-\sigma t}\}. \tag{8.186}$$

Theorem: A time domain function can be uniquely obtained from its Laplace transform by the following equation:

$$x(t) = \frac{1}{2\pi j}\int_{\sigma-j\infty}^{\sigma+j\infty} X(s)e^{st}ds, \tag{8.187}$$

provided that the function, $x(t)e^{-\sigma t}$, is absolutely integrable, i.e.,

$$\int_{-\infty}^{\infty} |x(t)|e^{-\sigma t}e^{j\omega t}dt < \infty. \tag{8.188}$$

Approximate Proof: Recall the relationship between the Laplace transform and Fourier transform is given by

$$\mathcal{L}\{x(t)\} = \mathcal{F}\{x(t)e^{-\sigma t}\} = \int_{-\infty}^{\infty} x(t)e^{-\sigma t}e^{-j\omega t}dt = X(\sigma + j\omega). \tag{8.189}$$

Let us take the inverse Fourier transform of $x(t)e^{-\sigma t}$,

$$x(t)e^{-\sigma t} = \mathcal{F}^{-1}\{X(\sigma + j\omega)\} = \frac{1}{2\pi}\int_{-\infty}^{\infty} X(\sigma + j\omega)e^{-j\omega t}d\omega. \tag{8.190}$$

Leaving $x(t)$ alone in the left-hand side of the equation, we obtain,

$$x(t) = \frac{1}{2\pi}\int_{-\infty}^{\infty} X(\sigma + j\omega)e^{-t(\sigma+j\omega)}d\omega. \tag{8.191}$$

Now, let us replace $s = \sigma + j\omega$. Then, assuming that σ is fixed, we replace $ds = jd\omega$ in the above integral equation to obtain the inverse Laplace transform for each value of σ, from the following synthesis equation:

$$\boxed{x(t) = \frac{1}{2\pi j}\int_{\sigma-j\infty}^{\sigma+j\infty} X(s)e^{-st}d\omega}, \tag{8.192}$$

where the weight $X(s)$ of the complex exponential function is called the Laplace transform of $x(t)$, obtained from the following analysis equation:

$$\boxed{X(s) = \int_{-\infty}^{\infty} x(t)e^{-st}dt}. \tag{8.193}$$

Note that finding the inverse Laplace transform, using the aforementioned equation requires contour integration, which can be done by using the Cauchy residue theorem [see Complex Analysis: A Modern First Course in Function Theory, Jerry R. Muir Jr., Wiley, ISBN: 978-1-118-70522-3 April 2015]. In the context of this book, we suffice to use lookup tables and the properties of the Laplace transform for finding the inverse Laplace transforms.

Laplace transform has several advantages compared to the Fourier transform. It is very handy to solve the differential equations. It is applicable to the functions, where the Fourier transform does not exist. It is a very powerful tool to analyze the stability of linear or nonlinear systems. It has a wide range of applications in developing Signal, image, and video Processing; computer vision; and machine learning systems.

8.9.1 One-Sided Laplace Transform

Fourier transform requires that the limit of the integral ranges $(-\infty, \infty)$. Thus, the time domain function is to be definite for both positive and negative values of time in the real axis. On the other hand, it is possible to define one-sided Laplace transform, where the time domain functions do not require negative real numbers for the time variable. For a wide range of practical problems, negative times are undefined or zero. In order to avoid negative times, we restrict the Laplace transform integral for $0 \leq t \leq \infty$.

Formally, we define one-sided Laplace transform, where the limits of the integral ranges between $(0, \infty)$, as follows:

$$X(s) = \int_{0}^{\infty} x(t)e^{-st}dt, \tag{8.194}$$

where $s = \sigma + j\omega$ is still a complex variable and e^{-st} is a complex exponential function. The time domain function can be uniquely obtained from the one-sided Laplace transform by the following equation:

$$x(t) = \frac{1}{2\pi j} \int_{\sigma - j\infty}^{\sigma + j\infty} X(s) e^{st} ds. \quad (8.195)$$

8.9.2 Region of Convergence in Laplace Transforms

The addition of the real part, σ, to the purely imaginary part, $j\omega$ of the frequency domain variable, enables us to evaluate the Laplace transform for each specific value of σ. In the Laplace domain, where the variable $s = \sigma + j\omega$ is a two-dimensional complex number, it is possible to find region(s) of the complex plane for some values of σ, such that the Laplace integral converges to a finite value.

The aforementioned capability of Laplace transform creates a great advantage of the Laplace transforms over the Fourier transform, when the function, $x(t)$ is not absolutely integrable, but, the function $x(t)e^{-\sigma t}$ is. Thus, Laplace transform, relaxes the Dirichlet condition of the Fourier transform, leaving us some regions of the complex plane, where Laplace transform exits. The regions, where the existence of the Laplace transform is assured are called region of convergence (ROC).

Definition: Region of Convergence (ROC): The ROC is defined as the regions in the complex plane, where the Laplace transform $X(s)$ of the function $x(t)$ exits for some values of $\sigma = \text{Re}\{s\}$.

ROCs of the Laplace transform is in the form of vertical stripes in the complex plane. The location(s) and width(s) of the stripes depend on the type of the time domain function, $x(t)$.

There are four major forms for the ROC of the Laplace transform:

1) If the function $x(t)$ is absolutely integrable and has a finite duration, in other words,

$$x(t) \begin{cases} \neq 0 & \text{for} \quad t_0 < t < t_1, \\ = 0 & \text{otherwise.} \end{cases} \quad (8.196)$$

for some finite values of $t_0 < t_1$, then, ROC covers the entire s-plane. Since it also covers the $j\omega$ axis, the Fourier transform of the function also exists.

2) If the function is right-sided, in other words, if there exists a finite t_0, such that

$$x(t) = 0 \quad \text{for} \quad t \leq t_0,$$

then, the ROC is the right-hand side of the complex plane with $\sigma > \sigma_0$.

3) If there exists a finite t_0, such that

$$x(t) = 0 \quad \text{for} \quad t \geq t_0,$$

then, the ROC is the left-hand side of the complex plane with $\sigma < \sigma_0$.

4) If the function is left-sided, in other words, if there exists two finite values, t_0 and t_1, such that

$$x(t) \begin{cases} \neq 0 & \text{for} \quad t < t_0 \quad \text{and} \quad t > t_1, \\ = 0 & \text{otherwise,} \end{cases} \quad (8.197)$$

then, the ROC is a stripe with $\sigma_0 < \sigma < \sigma_1$.

In order to observe the aforementioned forms of ROC and various capabilities of Laplace transform over the continuous time Fourier transform, let us solve the following exercises and investigate the existences of both Fourier and Laplace transforms.

Exercise 8.21 Consider the following continuous time right-sided signal:

$$x(t) = e^{at}u(t). \quad (8.198)$$

a) Find the Fourier transforms of this signal.
b) Find the Laplace transform of this signal and its ROC
c) Compare the range of a, which assures the existence of Fourier and Laplace transforms.

Solution

a) Fourier transform of the signal, $x(t)$ is defined as:

$$X(j\omega) = \int_{-\infty}^{\infty} x(t)e^{-j\omega t}dt = \int_{0}^{\infty} e^{at}e^{-j\omega t}dt \tag{8.199}$$

$$= \int_{0}^{\infty} e^{-t(j\omega - a)}dt = \frac{1}{j\omega - a}, \quad \text{for} \quad a < 0.$$

This integral does not converge for $a > 0$.

b) Laplace transform of the signal $x(t)$ is

$$X(s) = \int_{-\infty}^{\infty} x(t)e^{-st}dt = \int_{0}^{\infty} e^{-t(s-a)}dt = \int_{0}^{\infty} e^{-t((\sigma - a)+j\omega)}dt. \tag{8.200}$$

This integral exists for only $\sigma - a \geq 0$ or $\sigma \geq a$. Otherwise, it approaches to ∞. Taking the integral in ROC for $\sigma \geq a$, we obtain

$$\frac{1}{s-a} = \frac{1}{j\omega + (\sigma - a)}. \tag{8.201}$$

c) Comparison of the convergence properties of Fourier and Laplace transform, reveals that Fourier transform exists, for only negative values of a. On the other hand, the Laplace transform exits in the region of the complex plane, where the real part of s is greater than a. Although for positive values of a Fourier transform do not exist, Laplace transform exits in the region of the complex plain, where $\sigma > a$. The region of the complex plane, where the Laplace transform exits is the ROC.

Exercise 8.22 Consider a slightly different version of the continuous time signal of the previous example, which is left-sided,

$$x(t) = -e^{at}u(-t). \tag{8.202}$$

a) Find the Fourier transforms of this signal.
b) Find the Laplace transform of this signal with its ROC
c) Compare the range of a, which assures the existence of Fourier and Laplace transforms

Solution

a) Fourier transform of the signal, $x(t)$ is defined as:

$$X(j\omega) = -\int_{-\infty}^{0} e^{at}e^{-j\omega t}dt = \frac{1}{j\omega - a}, \quad \text{for} \quad a > 0. \tag{8.203}$$

This integral does not converge for $a < 0$.

b) Laplace transform of the signal $x(t)$ is

$$X(s) = \int_{-\infty}^{\infty} x(t)e^{-st}dt = \frac{1}{s-a} = \frac{1}{(\sigma - a) + j\omega}. \tag{8.204}$$

Therefore, this integral exists for $\sigma - a < 0$ or $\sigma < a$.

c) As in the previous example, the analytical forms of Fourier and Laplace transforms are the same. However, the conditions of convergence changes. While Fourier transform exists, for only

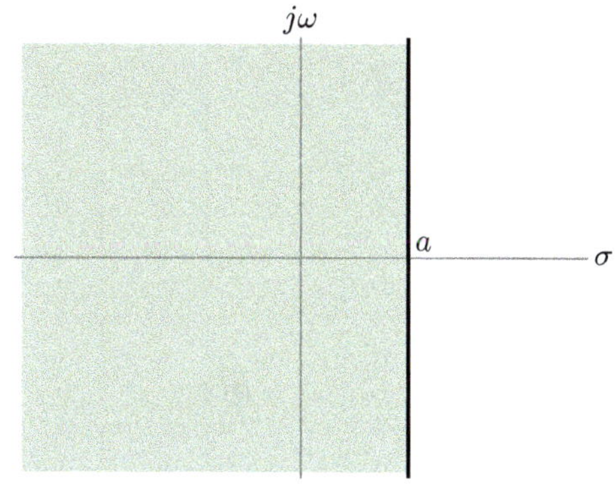

Figure 8.20 Region of convergence for the Laplace transform of $x(t) = -e^{at}u(-t)$.

positive values of a, Laplace transform exits in the region of the complex plane, where the real part of s is less than or equal to a. Since Laplace transform may exist for some restricted values of σ and for all values of $j\omega$, the ROCs of $X(s)$ consist of strips parallel to the jw-axis in the s-plane, as shown in Figure 8.20.

Exercise 8.23 Find the Laplace transform and its ROC for the following right-sided function:

$$x(t) = u(t). \tag{8.205}$$

Solution

From the definition of Laplace transform,

$$X(s) = \int_{-\infty}^{\infty} x(t)e^{-st}dt = \int_{0}^{\infty} e^{-(\sigma+j\omega)t}dt = \frac{1}{(\sigma+j\omega)} = \frac{1}{s}. \tag{8.206}$$

Note that the Laplace integral exists for $\sigma = \text{Re}\{s\} > 0$. Thus, ROC is the positive half of the complex plane, as shown in Figure: 8.21. This is the case when an absolutely integrable function $x(t)$ is right-sided.

Exercise 8.24 Find the Laplace transform and its ROC for the following limited-time duration function:

$$x(t) = u(t) - u(t - T). \tag{8.207}$$

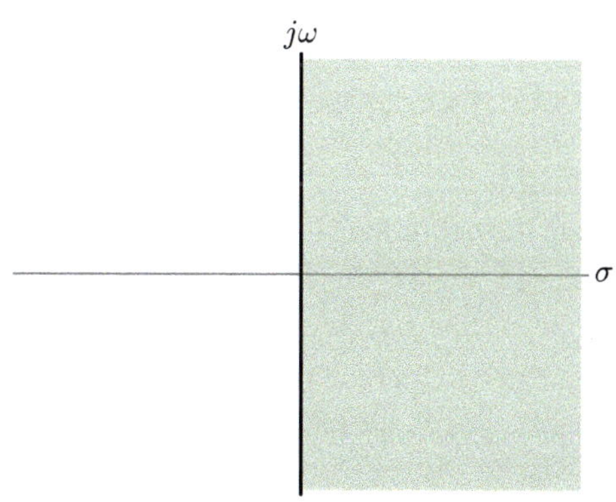

Figure 8.21 Region of convergence for the Laplace transform of the unit step function, $u(t)$.

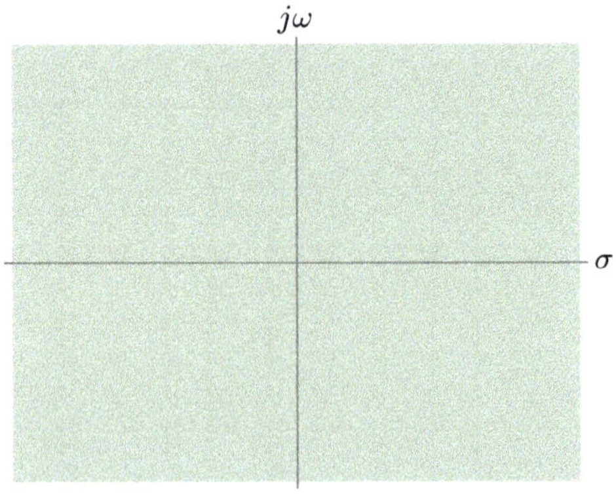

Figure 8.22 Region of convergence for the Laplace transform of $x(t) = u(t) - u(t-T)$.

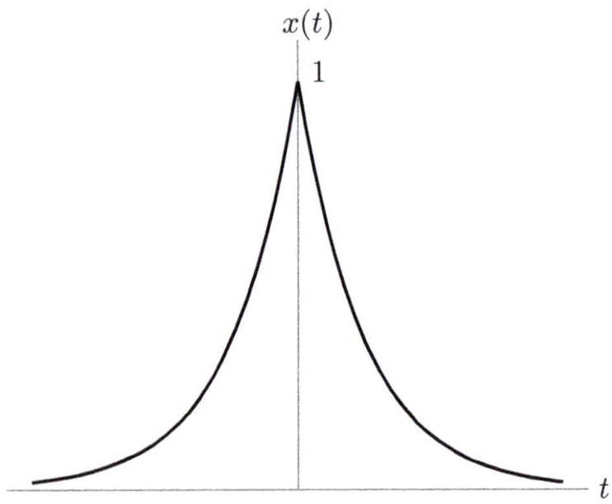

Figure 8.23 Two-sided function $x(t) = e^{-at}u(t) + e^{at}u(-t)$.

Solution

From the definition of Laplace transform,

$$X(s) = \int_{-\infty}^{\infty} x(t)e^{-st}dt = \int_{0}^{T} e^{-(\sigma+j\omega)t}dt = \frac{1}{s}[1 - e^{-sT}]. \tag{8.208}$$

Since the time duration, $t \in [0, T]$ is bounded, the Laplace integral exists for all values of σ. Thus, ROC is the entire complex plane, as shown in Figure 8.22. This is the case when an absolutely integrable function $x(t)$ has a finite duration.

Exercise 8.25 Find the Laplace transform and ROC of the following two sided function (Figure 8.23):

$$x(t) = e^{-at}u(t) + e^{at}u(-t). \tag{8.209}$$

Solution

From the definition of Laplace transform,

$$X(s) = \int_{-\infty}^{\infty} (e^{at} + e^{at})e^{-st}dt = \int_{0}^{\infty} e^{-(\sigma+j\omega)t}dt = \frac{1}{s+a} + \frac{1}{s-a}. \tag{8.210}$$

In order to find the ROC of the Laplace transform, we need to find the ROC of the first and the second term on the left-hand side of Equation (8.210):

ROC for $\frac{1}{s+a}$ is $\sigma \geq -a$ and

ROC for $\frac{1}{s-a}$ is $\sigma \leq a$.

The ROC of $X(s)$ is shown in Figure 8.24.

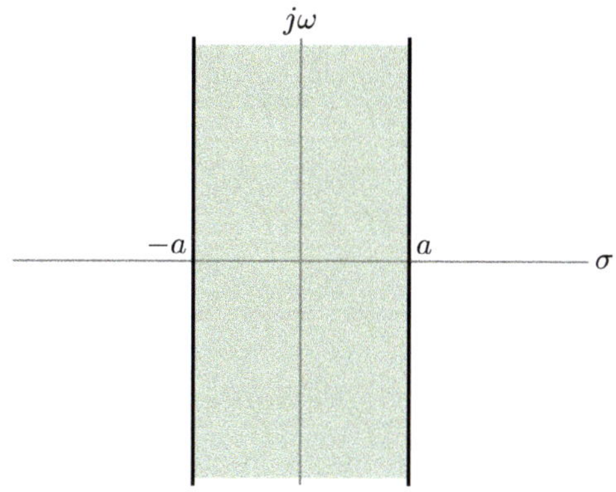

Figure 8.24 Region of convergence for the Laplace transform of $x(t) = e^{-at}u(t) + e^{at}u(-t)$.

These examples show that it is critical to determine the Region of Convergence of the Laplace transforms in the complex plane.

8.10 Inverse of Laplace Transform

As we mentioned earlier, recovering the time domain signal $x(t)$ from its Laplace transform, $X(s)$ requires the following contour integration:

$$x(t) = \frac{1}{2\pi j}\int_{\sigma-j\infty}^{\sigma+j\infty} X(s)e^{st}ds, \tag{8.211}$$

which may not be easy for a large class of functions. In order to avoid contour integration, we frequently use the lookup tables and properties of the Laplace transform. Since they are quite similar to that of the Fourier series and Fourier transformation, we suffice to provide the list of properties and look up tables for common transform pairs, $x(t) \leftrightarrow X(s)$ together with ROCs, in Tables 8.3 and 8.4. The following examples demonstrate how we utilize the tables to compute the inverse Laplace transform.

Exercise 8.26 Find the inverse Laplace transform of the following s-domain function:

$$X(s) = \frac{1}{(s+1)(s+2)}, \quad \text{ROC for} \quad \sigma < -1. \tag{8.212}$$

Solution

Let us apply partial fraction expansion to simplify the Laplace function;

$$X(s) = \frac{1}{(s+1)(s+2)} = \frac{1}{s+1} - \frac{1}{s+2}. \tag{8.213}$$

In the this equation, the inverse Laplace transformation of the first term is

$$\mathcal{L}^{-1}\left[\frac{1}{s+1}\right] = e^{-t}u(t), \tag{8.214}$$

the inverse Laplace transformation of the second term is

$$\mathcal{L}^{-1}\left[\frac{1}{s+2}\right] = e^{-2t}u(t). \tag{8.215}$$

Using the linearity property, we obtain the inverse Laplace transform of $X(s)$ as follows:

$$x(t) = [e^{-t} - e^{-2t}]u(t). \tag{8.216}$$

Table 8.3 Properties of Laplace transform.

Signal	Laplace transform	ROC
$x(t)$	$X(s)$	R
$x_1(t)$	$X_1(s)$	R_1
$x_2(t)$	$X_2(s)$	R_2
$ax_1(t) + bx_2(t)$	$aX_1(s) + bX_2(s)$	At least $R_1 \cap R_2$
$x(t - t_0)$	$e^{-st_0}X(s)$	R
$e^{-s_0 t}x(t)$	$X(s - s_0)$	Shifted version of R (i.e., s is in the ROC if $s - s_0$ is in R)
$x(at)$	$\frac{1}{\|a\|}X\left(\frac{s}{a}\right)$	Scaled ROC (i.e., s is in the ROC $\frac{s}{a}$ is in R)
$x^*(t)$	$X^*(s^*)$	R
$x_1(t) * x_2(t)$	$X_1(s)X_2(s)$	At least $R_1 \cap R_2$
$\frac{d}{dt}x(t)$	$sX(s)$	At least R
$-tx(t)$	$\frac{d}{ds}X(s)$	R
$\int_{-\infty}^{t} x(\tau)d(\tau)$	$\frac{1}{s}X(s)$	At least $R \cap \{\mathrm{Re}\{s\} > 0\}$

Initial- and final-value theorems:

If $x(t) = 0$ for $t < 0$ and $x(t)$ contains no impulses or higher-order singularities at $t = 0$, then

$$x(0^+) = \lim_{s\to\infty} sX(s)$$

$$\lim_{t\to\infty} x(t) = \lim_{s\to\infty} sX(s)$$

Exercise 8.27 Find the inverse Laplace transform of the following s-domain function:

$$X(s) = \frac{3s+2}{s^2+9}, \quad \text{ROC for} \quad \sigma \geq 0. \tag{8.217}$$

Solution

Let us separate the function into two parts:

$$X(s) = \frac{3s+2}{s^2+9} = \frac{3s}{s^2+9} + \frac{2}{s^2+9}. \tag{8.218}$$

From the Laplace transform table, we can see that the inverse Laplace transform of the first term is

$$\mathcal{L}^{-1}\left[\frac{3s}{s^2+9}\right] = [3\cos 3t]u(t), \quad \text{ROC for} \quad \sigma > 0. \tag{8.219}$$

and the inverse Laplace transform of the second term is

$$\mathcal{L}^{-1}\left[\frac{2}{s^2+9}\right] = \left[\frac{2}{3}\sin 3t\right]u(t), \quad \text{ROC for} \quad \sigma > 0. \tag{8.220}$$

Using the linearity property, we obtain the inverse Laplace transform of $X(s)$, as follows:

$$x(t) = \mathcal{L}^{-1}\left[\frac{3s}{s^2+9}\right] + \mathcal{L}^{-1}\left[\frac{2}{s^2+9}\right] = \left[3\cos 3t + \frac{2}{3}\sin 3t\right]u(t). \tag{8.221}$$

Table 8.4 Laplace transform pairs.

Signal	Laplace transform	ROC
$\delta(t)$	1	All s
$u(t)$	$\frac{1}{s}$	$\text{Re}\{s\} > 0$
$-u(-t)$	$\frac{1}{s}$	$\text{Re}\{s\} < 0$
$\frac{t^{n-1}}{(n-1)!}u(t)$	$\frac{1}{s^n}$	$\text{Re}\{s\} > 0$
$-\frac{t^{n-1}}{(n-1)!}u(-t)$	$\frac{1}{s^n}$	$\text{Re}\{s\} < 0$
$e^{-\alpha t}u(t)$	$\frac{1}{s+\alpha}$	$\text{Re}\{s\} > -\alpha$
$-e^{-\alpha t}u(-t)$	$\frac{1}{s+\alpha}$	$\text{Re}\{s\} < -\alpha$
$\frac{t^{n-1}}{(n-1)!}e^{-\alpha t}u(t)$	$\frac{1}{(s+\alpha)^n}$	$\text{Re}\{s\} > -\alpha$
$-\frac{t^{n-1}}{(n-1)!}e^{-\alpha t}u(-t)$	$\frac{1}{(s+\alpha)^n}$	$\text{Re}\{s\} < -\alpha$
$\delta(t-T)$	e^{-sT}	All s
$[\cos\omega_0 t]u(t)$	$\frac{s}{s^2+\omega^2}$	$\text{Re}\{s\} > 0$
$[\sin\omega_0 t]u(t)$	$\frac{\omega_0}{s^2+\omega^2}$	$\text{Re}\{s\} > 0$
$[e^{-\alpha t}\cos\omega_0 t]u(t)$	$\frac{s+\alpha}{(s+\alpha)^2+\omega^2}$	$\text{Re}\{s\} > -\alpha$
$[e^{-\alpha t}\sin\omega_0 t]u(t)$	$\frac{\omega_0}{(s+\alpha)^2+\omega^2}$	$\text{Re}\{s\} > -\alpha$
$u_n(t) = \frac{d^n\delta(t)}{dt^n}$	s^n	All s
$u_n(t) = \underbrace{u(t) * \cdots * u(t)}_{n \text{ times}}$	$\frac{1}{s^n}$	$\text{Re}\{s\} > 0$

Exercise 8.28 Find the inverse Laplace transform of the following s-domain function:

$$X(s) = \frac{2}{3-7s}, \quad \text{ROC for} \quad \sigma < \frac{3}{7}. \tag{8.222}$$

Solution

Let us arrange $X(s)$, as follows:

$$X(s) = \frac{2}{3-7s} = -\frac{2}{7}\frac{1}{s-\frac{3}{7}} \tag{8.223}$$

Using the linearity property and the lookup table, we get,

$$x(t) = \frac{2}{7}e^{\frac{3}{7}t}u(-t) \leftrightarrow X(s) = -\frac{2}{7}\frac{1}{s-\frac{3}{7}}, \quad \text{ROC for} \quad \sigma < \frac{3}{7}. \tag{8.224}$$

Exercise 8.29 Find the inverse Laplace transform of the following s-domain function:

$$X(s) = \frac{1}{3-4s} + \frac{3-2s}{s^2+49} \quad \text{ROC for} \quad \sigma > 3/4. \tag{8.225}$$

Solution

Let us separate the function into three parts using partial fraction expansion:

$$X(s) = \frac{1}{3-4s} + \frac{3}{s^2+49} - \frac{2s}{s^2+49}. \tag{8.226}$$

From the Laplace transform table, we can see that the inverse Laplace transform of the first term is

$$\mathcal{L}^{-1}[\frac{1}{3-4s}] = -\left(\frac{1}{4}e^{\frac{3}{4}t}\right)u(t) \quad \text{ROC for} \quad \sigma > 3/4, \tag{8.227}$$

the inverse Laplace transform of the second term is

$$\mathcal{L}^{-1}\left[\frac{3}{s^2+49}\right] = \left(\frac{3}{7}\sin 7t\right)u(t) \quad \text{ROC for} \quad \sigma > 0, \tag{8.228}$$

and the inverse Laplace transform of the third term is

$$\mathcal{L}^{-1}\left[\frac{2s}{s^2+49}\right] = (2\cos 7t)u(t) \quad \text{ROC for} \quad \sigma > 0. \tag{8.229}$$

Thus, the inverse Laplace transform of $X(s)$ is

$$x(t) = \left[-\frac{1}{4}e^{\frac{3}{4}t} + \frac{3}{7}\sin 7t - 2\cos 7t\right]u(t). \tag{8.230}$$

The aforementioned exercises show that a practical method for finding the inverse Laplace transform is to make algebraic manipulations on the s-domain function and put it into the linear combination of the known pairs of transform table. Then, use the linearity property to obtain the inverse transform.

8.11 Continuous Time Linear Time-Invariant Systems in Laplace Domain

Recall that a continuous time LTI system is represented by the following ordinary constant coefficient differential equation in the time domain:

$$\sum_{k=0}^{N} a_k \frac{d^k y(t)}{dt^k} = \sum_{k=0}^{M} b_k \frac{d^k x(t)}{dt^k}. \tag{8.231}$$

In addition, recall that if the eigen function of $x(t) = e^{j\omega t}$ is the input to an LTI system represented by the impulse response, $h(t)$, then, the corresponding output is

$$y(t) = H(j\omega)e^{j\omega t}, \tag{8.232}$$

where

$$H(j\omega) = \int_{-\infty}^{\infty} h(t)e^{-j\omega t}dt, \tag{8.233}$$

is called the frequency response. Since the right-hand side of Equation (8.233) is the Fourier transform of the impulse response, the frequency response is just the Fourier transform of the impulse response,

$$h(t) \longleftrightarrow H(j\omega). \tag{8.234}$$

Motivating Question: What if the frequency response does not exist? Can we employ Laplace transform to analyze the frequency content of an LTI system in some regions of convergences of the s-plane?

Laplace transform, indeed, provides us with a strong tool to analyze the LTI systems, which do not exist in the frequency domain.

Let us take the Laplace transform of both sides of the nth-order differential equation given in Equation (8.231):

$$\sum_{k=0}^{N} a_k s^k Y(s) = \sum_{k=0}^{M} b_k s^k X(s). \tag{8.235}$$

Note that all of the derivation operations in Equation (8.231) disappear in Equation (8.235) and the differential equation becomes an algebraic equation in powers of s Thus, an LTI system represented by a differential equation in the time domain is equivalently represented by an algebraic equation, in the Laplace domain.

Arranging Equation (8.235), we obtain a relationship between the input and output signals of an LTI system in the Laplace domain, as follows:

$$\frac{Y(s)}{X(s)} = \frac{\sum_{k=0}^{M} b_k s^k}{\sum_{k=0}^{N} a_k (s)^k}. \tag{8.236}$$

8.11.1 Eigenvalues and Transfer Functions in s-Domain

Recall that when the input of an LTI system is an exponential function the output is just the scaled version of the input. Thus, exponential functions are the eigenfunctions of the LTI systems and the scaling factor is simply the eigenvalue, computed from the convolution integral,

$$y(t) = x(t) * h(t) = \int_{-\infty}^{\infty} e^{\lambda(t-\tau)} h(\tau) d\tau = H(\lambda) e^{\lambda t}, \tag{8.237}$$

where

$$H(\lambda) = \int_{-\infty}^{\infty} h(t) e^{-\lambda t} dt. \tag{8.238}$$

$$x(t) = e^{\lambda t} \rightarrow \boxed{\text{LTI}} \rightarrow y_P(t) = H(\lambda) e^{\lambda t}. \tag{8.239}$$

In the this formulation, if we set, $\lambda = j\omega$, then, the eigenfunction at the input becomes $x(t) = e^{j\omega t}$ and the eigenvalue of the LTI system becomes the Fourier transform of the impulse response, which is called frequency response,

$$H(j\omega) = \int_{-\infty}^{\infty} h(t) e^{-j\omega t} dt. \tag{8.240}$$

If we set, $\lambda = s = \sigma + j\omega$, then, the eigenfunction at the input becomes $x(t) = e^{st}$ and the eigenvalue becomes the Laplace transform of the impulse response.

Definition: Transfer Function The Laplace transform of the impulse response is called transfer function,

$$H(s) = \int_{-\infty}^{\infty} h(t) e^{-st} dt. \tag{8.241}$$

When the frequency response of an LTI system does not converge, we cannot represent the LTI system with an eigenvalue, in the frequency domain. However, the Laplace transform enables us to find the eigenvalue of the system, which converges in some regions of the complex s-plane. In other words, Laplace transform generalizes the Fourier transform by extending the purely imaginary variable $j\omega$ to a complex variable $s = \sigma + j\omega$. This extension enabled us to find the Laplace transform of a continuous time function, even if the Fourier transform does not exist. We found the regions of the complex plane, called the ROC, where the Laplace transform exits.

For a more general representation of frequency response, instead of $j\omega$, we can define a complex number, as $s = \sigma + j\omega$, then, the frequency response becomes **transfer function**,

$$\boxed{H(s) = \frac{Y(s)}{X(s)} = \frac{\sum_{k=0}^{M} b_k s^k}{\sum_{k=0}^{N} a_k s^k}}, \tag{8.242}$$

which transfers an input signal to an output signal of an LTI system represented by a differential equation. The type of this transferring process is determined by the constant coefficients, $\{a_k\}$ and $\{b_k\}$ of the differential equation.

Similarly, taking the inverse Laplace transform of the transfer function also gives us the impulse response, without solving the differential equation because time and s-domain representations are one-to-one,

$$h(t) \longleftrightarrow H(s). \tag{8.243}$$

The following exercise demonstrates the utilization of Laplace transforms for describing various properties of the LTI systems.

Exercise 8.30 Consider an LTI system, represented by the following impulse response:

$$h(t) = [e^{2t} - e^{-3t}]u(t). \tag{8.244}$$

a) Find the frequency response of this system.
b) Find the transfer function of this system.
c) Comment on the region of convergence.

Solution

a) This system is causal, but unfortunately, the Fourier transform of the first term does not exist. Thus, $H(j\omega) \to \infty$.
b) The transfer function is the Laplace transform of the impulse response:

$$H(s) = \int_{-\infty}^{\infty} h(t)e^{-st}dt = \int_{0}^{\infty} [e^{2t} - e^{-3t}]e^{-st}dt. \tag{8.245}$$

The transfer function consists of two subsystems, which are paralleled each other,

$$H(s) = H_1(s) - H_2(s), \tag{8.246}$$

where

$$H_1(s) = \int_{0}^{\infty} e^{2t}e^{-st}dt = \frac{1}{s-2}, \quad \text{ROC for} \quad \text{Re}\{s\} > 2 \tag{8.247}$$

and

$$H_2(s) = \int_{0}^{\infty} e^{-3t}e^{-st}dt = \frac{1}{s+3}, \quad \text{for ROC} \quad \text{Re}\{s\} > -3. \tag{8.248}$$

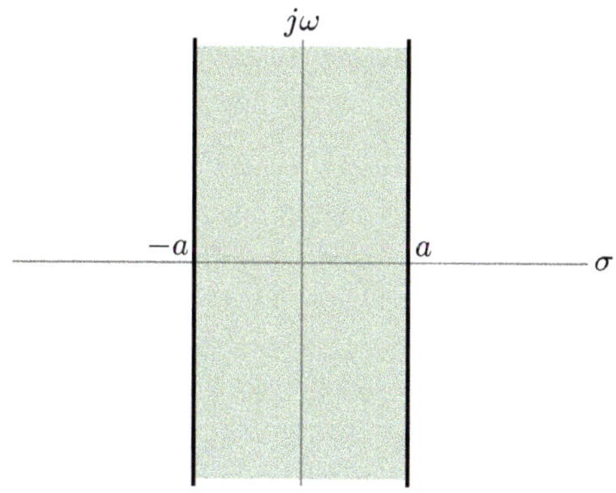

Figure 8.25 Region of convergence of the transfer function $H(s) = \frac{1}{s-2} + \frac{1}{s-3}$ for ROC $\sigma > 2$.

Thus,

$$H(s) = \frac{5}{s^2 + s - 6}, \quad \text{for ROC} \quad \mathrm{Re}\{s\} > 2. \tag{8.249}$$

c) The ROC of the overall transfer function, $H(s)$, lies in the intersection of ROCs of $H_1(s)$ and $H_2(s)$, which includes the region of the complex plane for $\sigma > 2$, as shown in Figure 8.25.

Exercise 8.31 Consider an initially at-rest LTI system given by the following differential equation:

$$\frac{d^2y(t)}{dt^2} + 4\frac{dy(t)}{dt} + 2y(t) = 5x(t) \tag{8.250}$$

a) Find the transfer function of this system.
b) Find the impulse response of this system.

Solution

a) Let us set the input to impulse function, $x(t) = \delta(t)$, then, the corresponding output of the differential equation becomes the impulse response, $h(t)$. The differential equation for impulse response is

$$\frac{d^2h(t)}{dt^2} + 4\frac{dh(t)}{dt} + 2h(t) = 5 \cdot \delta(t) \tag{8.251}$$

From the properties, we see that

$$\frac{d^n y(t)}{dt^n} \leftrightarrow s^n X(s),$$

from the Fourier transform table, we see that

$$x(t) = \delta(t) \leftrightarrow X(s) = 1.$$

Using the transform pairs, we take the Laplace transform of both sides of the differential equation (9.23), we obtain an equation for the transfer function:

$$[s^2 + 4s + 2]H(s) = 5.$$

Finally, we get the transfer function, as follows:

$$H(s) = \frac{5}{(s+2)^2} \quad \text{for ROC:} \quad \sigma > -2.$$

b) The impulse response of this system is the inverse of the transfer function. Using the Fourier transform table, we obtain the impulse response, as follows:

$$h(t) = (5te^{-2t})u(t).$$

The above example demonstrates that we can obtain the transfer function and impulse response of an LTI system, which is initially at rest, without solving the differential equation. This method is also available in Fourier domain, provided that the frequency response exits. In the case of undefined frequency responses, Laplace domain enables us to compute the transfer function and impulse response, using a simple algebraic method.

As it is observed throughout this chapter, the transform domains capture different views of the physical phenomena other than time domain representations. Furthermore, the beautiful synergy created by the representations of time and transform domains bridges the mathematics of linear algebra and differential equations.

8.12 Chapter Summary

Can we extend the Fourier series representation of continuous time periodic signals to aperiodic ones? If yes, how do we represent an aperiodic function in the frequency domain? Is it possible to represent any time domain signal in the frequency domain uniquely? Are the representations in the frequency domain and time domain one-to-one and onto? What are the necessary and sufficient conditions to transform a time domain function to the frequency domain?

In this chapter, first, we answer these questions by generalizing the Fourier series of continuous time periodic signals to aperiodic signals. We simply assume that an aperiodic signal can be considered as periodic, as the period approaches infinity, $T \to \infty$.

The generalized form of the Fourier series, which enables us to represent both continuous time-periodic and aperiodic functions in terms of its frequency content, is called the Fourier Transform. Spectral coefficients of the Fourier series and Fourier transform enable us to represent the time domain signals in the frequency domain. The existence of frequency domain representation is assured by Dirichlet conditions. Although the frequency domain of the periodic signals and that of the aperiodic signals resemble each other, they possess different characteristics. Both frequency domains enable us to represent signals in terms of their frequency content. While the frequency domain of spectral coefficients is discrete harmonics of the fundamental frequency, $k\omega_0$, the frequency domain of the Fourier transform is continuous functions of frequency, ω.

The periodic signals can be represented by both Fourier series and Fourier transform. Weighted summation of shifted impulse functions, $\delta(\omega - k\omega_0)$ for each harmonic of the fundamental frequency, $k\omega_0$, provides us the Fourier transform of periodic signals, where the weights are the scaled spectral coefficients, $2\pi a_k$. Frequency domain representations provide us with important information about the frequency content of the signals and the characteristics of LTI systems.

Next, we try to answer the following questions.

What type of an operator is Fourier transform? What are the relationships between the functions represented in the time and frequency domains? What are the properties of the signals and systems in the frequency domain? Where do we use Fourier transformations?

In order to answer these questions, we dive into the deeper meanings of Fourier analysis and synthesis equations, investigating the power of Fourier transforms in solving mathematical problems and designing LTI systems.

We studied the basic properties of Fourier transforms, such as linearity, time shifting, time scaling, derivation, and integration properties. We saw that convolution operation in the time domain is transformed into multiplication operation in the frequency domain. We noticed that the energy is preserved in time and frequency domains. We also studied an important concept, called duality. In short, we observed that the infinite-dimensional frequency domain, spanned but uncountably many complex exponential functions have many interesting properties and forms a fertile environment for understanding the frequency content of continuous time signals. We observed that differential and integral equations become algebraic equations in the frequency domain. Thus, solving them in the frequency domain is rather easier compared to solving them in the time domain. We also, show that there is one-to-one correspondence between the representation of LTI systems by impulse response, frequency response, and differential equations.

Finally, we define a new domain, called Laplace domain, where the purely imaginary frequency domain variable ($j\omega$), is extended to a complex plain variable $s = \sigma + j\omega$. This generalization enables us to find the Laplace transforms of the functions, which do not exist in the frequency domain, in which the Laplace transform exists in some regions of the complex plane called, Region of Convergence. While Fourier Transform maps a time domain function into a frequency-domain function, Laplace Transform maps a time domain function into an s-domain function, in the entire complex plane, where the transformed function exits in the ROC.

Problems

8.1 Consider the following continuous time signal:

$$x(t) = \begin{cases} e^{-2t}, & 0 \leq t \leq 1 \\ 0, & \text{otherwise} \end{cases}$$

Determine the Fourier transform of the following signals:

a) $x(-t) + x(t)$
b) $x(t) - x(-t)$
c) $x(t) + x(t+1)$
d) $tx(t)$

8.2 Find the Fourier transforms of the following continuous time signals:

a) $e^{-2(t-2)}u(t-2)$
b) $e^{-2|t-2|}$

8.3 Find the inverse Fourier transform of the following frequency domain signals:

a) $X(j\omega) = 4\pi\delta(\omega) + 10\pi\delta(\omega - 7\pi) - 10\pi\delta(\omega + 7\pi)$
b)

$$X(j\omega) = \begin{cases} -3, & 0 \leq \omega \leq 3 \\ -3, & -3 \leq \omega \leq 0 \\ 0, & |\omega| < 3 \end{cases}$$

8.4 A signal in the frequency domain is represented by polar coordinate system $X(j\omega) = |X(j\omega)|e^{j\sphericalangle X(j\omega)}$ has the following magnitude and phase:

$$|X(j\omega)| = u(\omega - 3) - u(\omega + 3)$$
$$\sphericalangle X(j\omega) = -\frac{1}{2}\omega + \pi$$

a) Find the real and imaginary part of $X(j\omega)$.
b) Find the inverse Fourier transform of this signal.
c) Find the time interval t, where $x(t) = 0$.

8.5 Use the Fourier transform properties and tables to evaluate the Fourier transform of the following signals:
a) $y(t) = te^{|t|} + e^{|t|}$
b) $y(t) = \dfrac{t}{(1+t^2)^2}$

8.6 The input of an LTI system is a continuous time signal whose Fourier transform is

$$X(j\omega) = \delta(\omega) + \delta(\omega - 2) + \delta(\omega + 2).$$

a) Find the inverse Fourier transform $x(t)$ of this signal. Is $x(t)$ periodic? If yes, find the period.
b) Given an LTI system represented by the impulse response,

$$h(t) = u(t + 1) + u(t - 1),$$

find the output $y(t)$ when the input is
i) $x(t)$
ii) $tx(t)$

8.7 Consider a causal linear time-invariant system represented by the following frequency response:

$$H(j\omega) = \frac{1}{4j\omega + 3}.$$

a) Find the impulse response of the system.
b) Find the input $x(t)$, when we observe the following output:

$$y(t) = (e^{-5t} - e^{-10t})u(t).$$

8.8 Find the Fourier transforms of the following signals:
a) $e^{-4|t|}\sin 3t$.
b) $x(t)$ given in Figure P8.8a.
c) $x(t)$ given in Figure P8.8b.

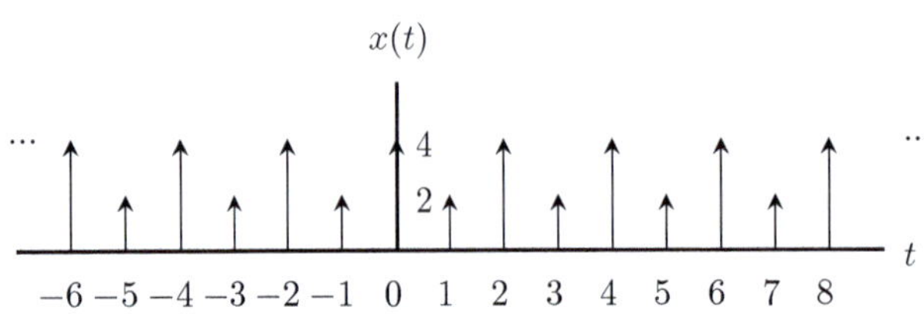

Figure P8.8a Continuous time periodic signal of Problem 8.8.b.

Figure P8.8b Continuous time signal of Problem 8.8.c.

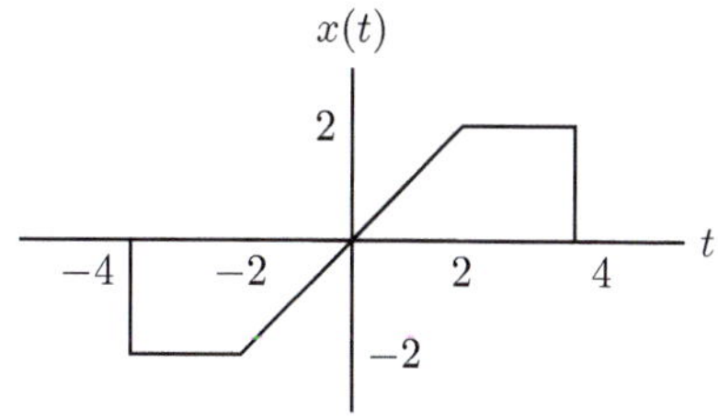

8.9 In Figure P8.9, $x(t+1)$ is given. $x(t)$ has Fourier transform $X(j\omega)$.

Figure P8.9 Continuous time signal of Problem 8.9.

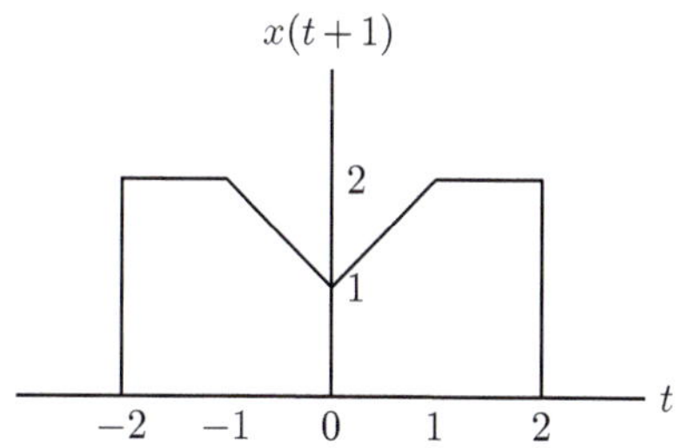

a) Find $\sphericalangle X(j\omega)$ and sketch it.
b) Calculate $X(j\omega)$ at $\omega = 0$.
c) $\displaystyle\int_{-\infty}^{\infty} X(j\omega)\, d\omega$.

8.10 Find the Fourier transform of the following signals:
a) $\sin\left(2\pi t + \frac{\pi}{4}\right)$
b) $1 + \cos\left(12\pi t + \frac{\pi}{4}\right)$.

8.11 A continuous time signal is given as follows:

$$x(t) = \sum_{k=-\infty}^{\infty} \frac{\sin(k\frac{\pi}{3})}{k\frac{\pi}{3}} p\left(t - k\frac{\pi}{3}\right).$$

where $p(t) = \delta(t)$.
a) Plot $x(t)$.
b) Find and plot $X(j\omega)$.
c) is $X(j\omega)$ periodic? If yes, what is the period?

8.12 The signal $x(t)$ has Fourier transform $X(j\omega)$. Find the Fourier transform of the Following signals in terms of $X(j\omega)$.
a) $x(2-t) + x(2+t)$
b) $x(-2t-4)$
c) $\frac{d^3}{dt^3}x(1-t)$

8.13 Find the inverse Fourier transform $x(t)$. Are these real in the time domain?
a) $X(j\omega) = u(\omega + 8) - u(\omega - 8)$

b) $X(j\omega) = \cos(\omega)\sin(\omega)$

c) $X(j\omega) = \sum_{k=-\infty}^{\infty} \left(\frac{1}{8}\right)^{|k|} \delta\left(\omega - \frac{k\pi}{4}\right)$

8.14 A continuous time signal is given as follows:

$$x(t) = (t+1)(u(t+1) - u(t-1)) + u(t-1)).$$

a) Find and plot the Fourier transform, $X(j\omega)$, of $x(t)$.
b) Find and plot the Fourier transform of the even part of $x(t)$.
c) Find and plot the Fourier transform of the odd part of $x(t)$.

8.15 A continuous time signal is given here:

$$x(t) = \frac{(t+1)}{2}(u(t+1) - u(t-1))$$

a) Find and plot the Fourier transform, $X(j\omega)$, of $x(t)$.
b) Find the energy of the signal in the time domain.
c) Find the energy of this signal in the frequency domain.

8.16 A continuous time signal is given as follows:

$$x(t) = \frac{\sin^2 t}{2\pi^2 t}.$$

a) Find and plot the Fourier transform of this signal.
b) Find the energy of this signal.

8.17 Find the inverse Fourier transform of the signals given in the frequency domain.

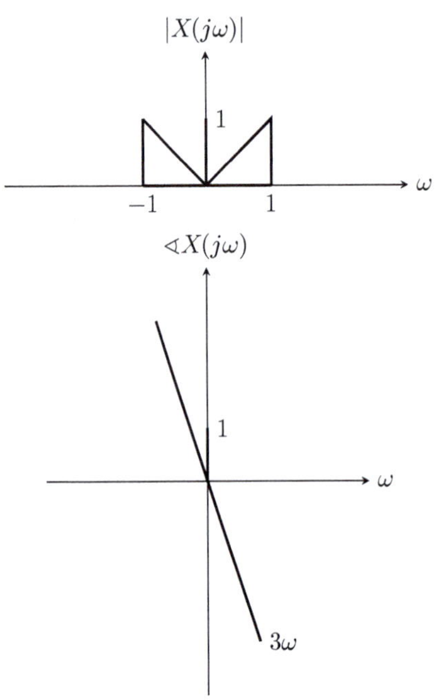

Figure P8.17a Magnitude and phase plots of $X(j\omega)$, in problem 8.17.

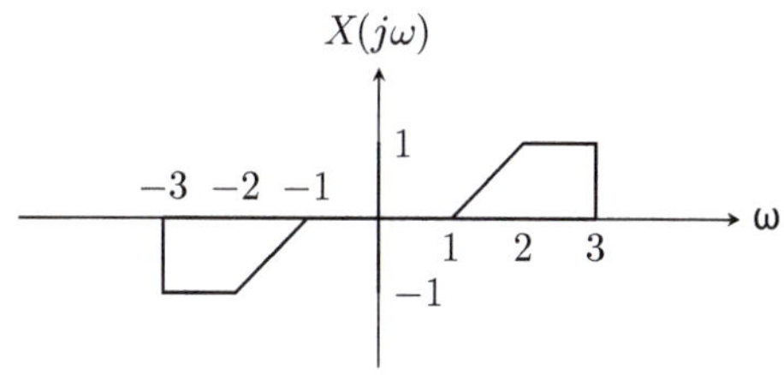

Figure P8.17b Frequency domain signal of Problem 8.17b

a) $X(j\omega)$ satisfying graphs in Figure P8.17a
b) $X(j\omega)$ satisfying graphs in Figure P8.17b
c) $X(j\omega) = \cos(8\omega + \pi/3)$
d) $X(j\omega) = \dfrac{\sin(\omega - 3\pi)}{\omega - 3\pi}$
e) $X(j\omega) = 6\{\delta(\omega + 4) - \delta(\omega - 4)\} + 4\{\delta(\omega + \pi) - \delta(\omega - \pi)\}$

8.18 It is given that

$$g(t) \longleftrightarrow G(j\omega) = \begin{cases} 1, & |\omega| \leq 2 \\ 0, & \text{otherwise} \end{cases}$$

a) Find $x(t)$ such that $x(t) = \dfrac{g(t)}{\cos t}$
b) Find $h(t)$ such that $h(t) = \dfrac{g(t)}{\cos \frac{1}{6}t}$

8.19 Calculate the response corresponding to $x(t) = \cos(t)$ of the following systems whose impulse responses are given as follows:
a) $h_1(t) = 2u(t)$
b) $h_1(t) = -4\delta(t) + 10e^{-2t}u(t)$.
c) $h_1(t) = 2te^{-t}u(t)$.

8.20 The signal $x(t)$ has the Fourier transform $X(j\omega)$ and let $g(t)$ be a periodic signal whose fundamental frequency is ω_0. Its Fourier series representation is as follows:

$$g(t) = \sum_{n=-\infty}^{\infty} a_n e^{jn\omega_0 t}.$$

a) Find the Fourier transform of $y(t) = x(t)g(t)$.
b) For each of the following $g(t)$, sketch the spectrum of $y(t)$.
   i) $g(t) = \cos(4t)$
   ii) $g(t) = \sin(2t)\sin(4t)$
   iii) $g(t) = \sum_{n=-\infty}^{\infty} \delta(t - 8\pi n)$
   iv) $g(t)$ as drawn in Figure P8.20
   v) $g(t) = \frac{1}{2}\sum_{n=-\infty}^{\infty} \delta(t - \pi n) - \sum_{n=-\infty}^{\infty} \delta(t - 2\pi n)$

Figure P8.20 Frequency domain signal of Problem 8.20.

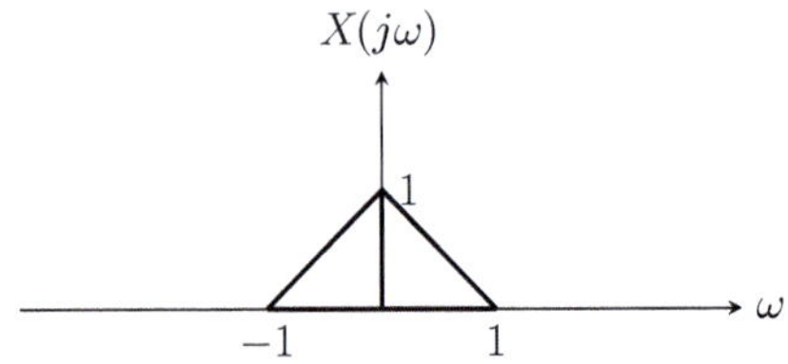

8.21 The following signal is fed to a continuous time LTI system:

$$x(t) = u(t-3) - 2u(t-4) + u(t-5).$$

a) Find and plot the Fourier transform, $X(j\omega)$, of $x(t)$.
b) Find and plot the Fourier transform of the corresponding output

$$y(t) = \sum_{k=-\infty}^{\infty} x(t-kT).$$

c) Find the frequency response of this system.

8.22 a) Calculate the convolution of the following signals. (*Hint*: Use Fourier Transform.)
i) $x_1(t) = te^{-3t}u(t)$, $x_2(t) = e^{-6t}u(t)$
ii) $x_1(t) = te^{-3t}u(t)$, $x_2(t) = te^{-6t}u(t)$
iii) $x_1(t) = e^{-2t}u(t)$, $x_2(t) = e^{2t}u(-t)$
b) Consider $x(t) = e^{2-t}u(t-2)$ and $g(t)$ as drawn in Figure P8.22. Find the Fourier transform of $x(t) * g(t)$ and then find $X(j\omega)G(j\omega)$.

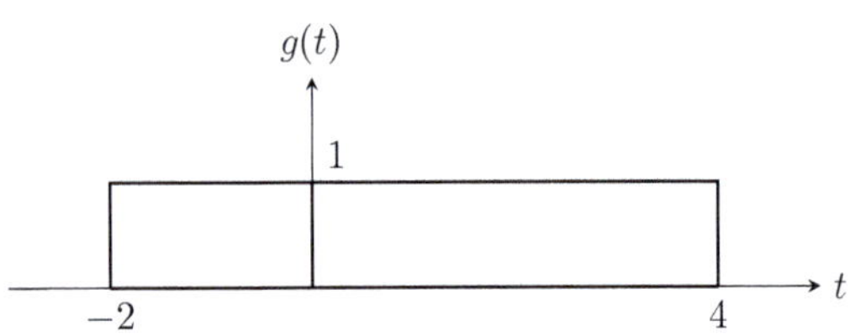

Figure P8.22 Continuous time signal of Problem 8.22b.

8.23 Find the Laplace transform and the ROC for the following signals in time domain.
a) $x(t) = (e^{-5t} + e^{-6t})u(t)$
b) $x(t) = (e^{-7t} + e^{-8t}\sin(8t))u(t)$
c)

$$x(t) = \begin{cases} t & 0 \le t \le 1 \\ 2-t & 1 \le t \le 2 \end{cases}$$

d) $x(t) = \delta(2t) + u(2t)$

8.24 Find the functions $x(t)$ whose Laplace transforms and ROCs are given.
a) $\dfrac{s}{s^2+25}$, $\mathrm{Re}\{s\} > 0$

b) $\dfrac{s+1}{s^2+5s+6}, -3 < \text{Re}\{s\} < -2$

c) $\dfrac{s^2+2s+1}{s^2-s+1}, \ \text{Re}\{s\} > \dfrac{1}{2}$

8.25 Consider a continuous time LTI system which is represented by the following second-degree differential equation:

$$\frac{d^2y(t)}{d^2t} - \frac{dy(t)}{dt} - 2y(t) = x(t).$$

a) Find the transfer function, $H(s)$ of the impulse response of the system.
b) Find the frequency response of this system.
c) Find the impulse response of this system.
d) Find a block diagram representation of this system.

8.26 The transfer function of the causal LTI system is given as:

$$H(s) = \frac{s+2}{s^2+4s+5}.$$

a) Find the impulse response of the system.
b) Find the frequency response of this system.
c) Find the differential equation, which represents this system.
d) Find the response, $y(t)$, when the input is $x(t) = e^{-2|t|}$ for $-\infty < t < \infty$.
e) Find a block diagram representation of this system.

8.27 The transfer function of the causal LTI system, S_1, is given as:

$$H_1(s) = \frac{s^2+2s-3}{s^2+3s+2}.$$

Another system, S_2, has the system function

$$H_2(s) = \frac{2}{s^2+3s+2}.$$

Assume that both systems have the same input, $x(t)$. Let corresponding output of S_1 be $y_1(t)$ and that of S_2 be $y_2(t)$. Find $y_1(t)$ in terms of the followings:

a) $y_2(t)$

b) $\dfrac{dy_2(t)}{dt}$

c) $\dfrac{d^2y_2(t)}{d^2t}$

8.28 Find the Laplace transform of the following signals with their region of convergences:

a) $2\delta(t+2) - \delta(t-3)$

b) $\dfrac{d}{dt}\{u(-1-t) + u(t-1)\}$

8.29 Consider the continuous time LTI system, represented by the following block diagram in the s-domain (Figure P8.29):

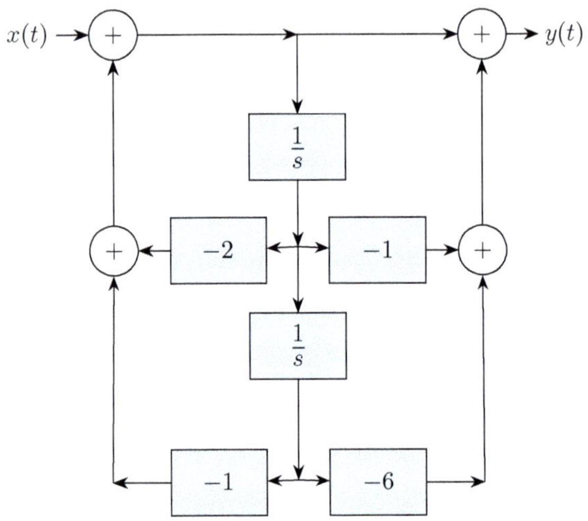

Figure P8.29 Block diagram representation of an LTI system in Problem 8.29.

a) Find the differential equation representing the system.
b) Is the system stable? Explain briefly.

9

Discrete Time Fourier Transform and Its Extension to z-Transforms

WATCH

Listen to the sounds of periodic functions @ https://384book.net/v0901

Historically, the continuous world converged into a digital world gradually, starting with F. Gauss, who invented the fast Fourier transform (FFT), in the 1800s. After the pioneering work of C. Shannon in 1949, which bridges continuous and discrete time functions through sampling theorem, J. Cooley and J. Tukey published an efficient method for digital implementation of FFT, in 1966. Since then, the entire information and telecommunication technology has been smoothly converted from analog to digital systems. The related theoretical background is developed to extend the continuous time Fourier series and transform to their discrete counterparts.

In this chapter, we carry the methodology and intuition lying behind the continuous time Fourier transform to the discrete domain.

Recall that we extended the continuous time Fourier series representation of periodic signals to aperiodic signals by using Fourier transforms.

Motivating Question: How did we make such a generalization?

The idea was simple: We assumed that an aperiodic signal could be considered as a signal with an infinite period.

In this chapter, we use the same idea to extend the discrete time Fourier series representation to discrete time Fourier transforms. We study the basic properties of discrete time Fourier transform. Furthermore, we generalize the discrete time Fourier transform to z-transform by extending the complex exponential basis functions with unit magnitude to complex variable $z = |r|e^{j\omega}$ with arbitrary magnitude, where $r \in R$.

9.1 Fourier Series Extension to Discrete Time Aperiodic Functions

Recall that a discrete time function is periodic if there exists an **integer value**, N, such that,

$$x[n] = x[n + N]. \tag{9.1}$$

We can extend the integer value, N as large as we need, in Equation (9.1). When a discrete time function has an infinite period, we assume that it repeats itself at every integer value $N \to \infty$.

Signals and Systems: Theory and Practical Explorations with Python,
First Edition. Fatoş Tunay Yarman Vural and Emre Akbaş.

Companion website: www.wiley.com/go/signalsandsystems

The aforementioned approach enables us to define a discrete time aperiodic function, $x[n]$, as a periodic function for $N \to \infty$, in the limit. Interestingly, as $N \to \infty$, the sum operation of the Fourier series synthesis equation converges to an integral operation, allowing us to represent aperiodic functions in terms of their frequency content.

Let us see how.

9.1.1 Discrete Time Fourier Transform

Theorem 9.1 A discrete time function, $x[n]$, can be uniquely represented as a weighted integral of complex exponential function by the following **synthesis equation:**

$$x[n] = \frac{1}{2\pi} \int_{2\pi} X(e^{j\omega}) e^{j\omega n} d\omega, \tag{9.2}$$

where the weight, called the Fourier transform, is a continuous function of frequency, which can be uniquely obtained from the time domain function by the following **analysis equation:**

$$X(e^{j\omega}) = \sum_{n=-\infty}^{\infty} x[n] e^{-j\omega n}. \tag{9.3}$$

The **synthesis equation** states that a discrete time function, $x[n]$, can be uniquely represented by the weighted integral of waves, i.e., complex exponentials. The weight function, $X(e^{j\omega})$, called the **discrete time Fourier transform of** $x[n]$, is a continuous function of frequency variable, ω, which measures the amount of wave with a particular frequency band in the signal.

The **analysis equation** shows us how to obtain the Fourier transform, $X(e^{j\omega})$ of $x[n]$, which represents the signal as a function of frequencies, in the frequency domain. The Fourier transform representation of a signal enables us to decompose an aperiodic discrete time signal into its frequency components, which are embedded in the signal.

The aforementioned representation of a physical phenomenon by a function in discrete time domain and continuous frequency domain is one-to-one and onto:

$$x[n] \longleftrightarrow X(e^{j\omega}). \tag{9.4}$$

Approximate Proof: Consider the Fourier series representation of a periodic signal, $\tilde{x}[n]$ and its spectral coefficients, as follows:

$$\tilde{x}[n] = \sum_{k=<N>} a_k e^{jk\omega_0 n}. \tag{9.5}$$

and

$$a_k = \frac{1}{N} \sum_{n=<N>} \tilde{x}[n] e^{-jk\omega_0 n}, \tag{9.6}$$

where $<N>$ indicates the coverage of one full period in the limits of the summations. Consider, also, a finite duration discrete time aperiodic function, $x[n]$, which corresponds to the center part of the periodic function, $\tilde{x}[n]$, for one full period,

$$x[n] = \begin{cases} \tilde{x}[n] & -N_1 < n < N_2 \\ 0 & \text{otherwise,} \end{cases} \tag{9.7}$$

In this formulation, the periodic function $\tilde{x}[n]$ is generated by repeating an aperiodic function, $x[n]$, with the fundamental period, N. For the time being, let us assume that the nonzero range in the interval, $N_1 + N_2 < N$ is finite, in Equation (9.7), as shown in Figure 9.1.

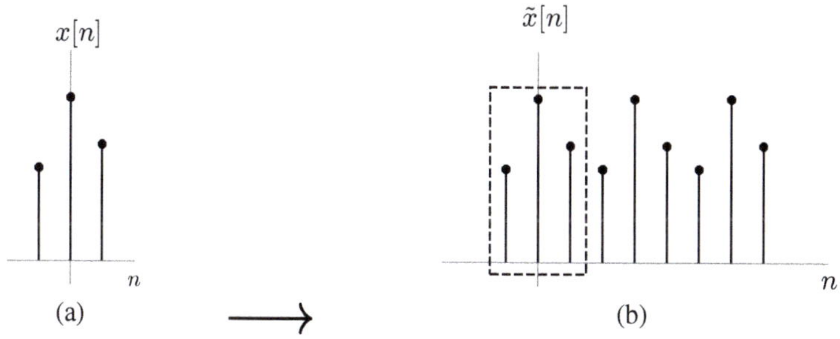

Figure 9.1 (a) A finite duration signal $x[n]$. (b) It is repeated at every fundamental period N, to generate a periodic signal, $\tilde{x}[n]$.

Now, let us define a new function, in the frequency domain:

$$X(e^{jk\omega_0}) = Na_k, \tag{9.8}$$

and replace it by Na_k in the analysis equation, to obtain the following equation:

$$X(e^{jk\omega_0}) = \sum_{n=<N>} \tilde{x}[n]e^{-jk\omega_0 n}. \tag{9.9}$$

Since a_k is periodic, with the fundamental period N, $X(e^{jk\omega_0})$ is also periodic with the same fundamental period.

Replacing a_k by $X(e^{jk\omega_0})/N$ in the synthesis equation, Fourier series representation of $\tilde{x}[n]$ becomes

$$\tilde{x}[n] = \frac{1}{N}\sum_{k=<N>} X(e^{jk\omega_0})e^{jk\omega_0 n}. \tag{9.10}$$

Now, let us take the limit to stretch the period N to infinity. Then, the angular frequency converges to an infinitesimal interval,

$$\omega_0 = \lim_{N\to\infty}\frac{2\pi}{N} \to d\omega, \tag{9.11}$$

and the discrete variable $k\omega_0$ converges to a continuous variable,

$$\lim_{N\to\infty} k\omega_0 \to \omega. \tag{9.12}$$

In the limit, the periodic function, $\tilde{x}[n]$ converges to the aperiodic function, $x[n]$, which repeats itself at every $N \to \infty$. The summation operation of the synthesis equation converges to integral operation, yielding a continuous frequency domain function as follows:

$$\begin{aligned} \lim_{N\to\infty}\tilde{x}[n] = x[n] &= \lim_{N\to\infty}\frac{1}{2\pi}\sum_{k=<N>}\omega_0 X(e^{jk\omega_0})e^{jk\omega_0 n}, \\ x[n] &= \frac{1}{2\pi}\int_0^{2\pi} X(e^{j\omega})e^{j\omega n}\,d\omega, \end{aligned} \tag{9.13}$$

where

$$X(e^{j\omega}) = \sum_{n=-\infty}^{\infty} x[n]e^{-j\omega n}. \tag{9.14}$$

Interestingly, due to the limit of $N \to \infty$, the Fourier transform of a discrete time function converges to a continuous frequency function. Moreover, the transform domain function, $X(e^{j\omega})$ becomes periodic, as shown by the following Lemma.

Lemma 9.1 Discrete time Fourier transform, $X(e^{j\omega})$ of a function $x[n]$ is always periodic, with 2π.

Proof: Recall that the discrete time complex exponential,

$$e^{-j\omega n} = e^{j\omega(n+2\pi)} = e^{-j\omega n}(\cos 2\pi + j \sin 2\pi).$$

is periodic with $N = 2\pi$. Since the linear combination of periodic functions is also periodic, the Fourier transform, $X(e^{j\omega})$ is periodic with $N = 2\pi$.

Motivating Question: Why does the Fourier transform, $X(e^{j\omega})$, of a discrete time function have argument $e^{j\omega}$ instead of $j\omega$ of the continuous time Fourier transform, $X(j\omega)$?

This is basically because of the analytical form of the synthesis equation of the discrete time Fourier transform, which has a summation operation instead of the integral operation of the continuous time Fourier transform. The integral operation of the continuous time synthesis equation changes the analytical form of the complex exponential basis functions. On the other hand, the sum operation of the discrete time synthesis equation keeps them as is. Thus, the Fourier transform of a discrete time functions are always functions of the complex exponentials, $e^{j\omega}$. Keep in mind that we can always use the Euler formula,

$$e^{j\omega} = \cos\omega + j \sin\omega,$$

to convert the Fourier transform into trigonometric form.

Note: The Fourier transform $X(e^{j\omega})$ of a discrete time aperiodic function $x[n]$ is continuous and periodic with 2π. Thus, the integral of the synthesis equation covers only one full period of 2π.

9.2 Dirichlet Conditions Are Relaxed for the Existence of Discrete Time Fourier Transform

Since the discrete time functions carry a finite number of samples in a finite interval, we do not bother with bounded variations and the finiteness of the discontinuities in a finite interval. However, the existence of the discrete time Fourier transform still requires some constraints to the class of functions to be represented in the frequency domain. We must consider the convergence of the infinite sum in the analysis equation. Mathematically speaking, the Fourier transform, $X(e^{j\omega})$, exists if and only if

$$\lim_{K\to\infty} |X(e^{j\omega}) - X_K(e^{j\omega})| \to 0, \tag{9.15}$$

where

$$X_K(e^{j\omega}) = \sum_{n=-K}^{K} x[n]e^{-j\omega n}, \tag{9.16}$$

is a truncated Fourier transform for a finite K.

A sufficient condition (but not necessary) for the existence of the discrete time Fourier transform is that the function should be absolutely summable. Mathematically, if a function is absolutely summable,

$$\sum_{n=-\infty}^{\infty} |x[n]| < \infty. \tag{9.17}$$

then, its Fourier transform exists, i.e.,

$$X(e^{j\omega}) = \sum_{n=-\infty}^{\infty} x[n]e^{-j\omega n} < \infty. \tag{9.18}$$

However, even if the function is not absolutely summable, the Fourier transform of this function may exist. In this case, it may be possible to represent the Fourier transform in terms of continuous time impulse functions.

In the following exercises, let us practice finding the Fourier transform of popular discrete time functions.

Exercise 9.1 Consider the following discrete time signal:

$$x[n] = a^n u[n], \text{ where } |a| < 1. \tag{9.19}$$

a) Is this function absolutely summable?
b) If your answer is yes, find the Fourier transform of $x[n]$ and plot its magnitude and phase spectra.

Solution

a) This function is absolutely summable, since,

$$\sum_{n=-\infty}^{\infty} |x[n]| = \sum_{n=0}^{\infty} |a^n| = \frac{1}{1-|a|} < \infty. \tag{9.20}$$

b) Let us use the synthesis equation to find the one full period of Fourier transform of this signal:

$$X(e^{j\omega}) = \sum_{n=0}^{\infty} a^n e^{-j\omega n} = \frac{1}{1-ae^{-j\omega}},$$
$$X(e^{j\omega}) = \frac{1-a(\cos\omega + j\sin\omega)}{(1-a\cos\omega)^2 + a^2\sin^2\omega}. \tag{9.21}$$

Let us express this complex function in the rectangular coordinate system:

$$X(e^{j\omega}) = \frac{1-a\cos\omega}{1-2a\cos\omega + a^2} + j\frac{a\sin\omega}{1-2a\cos\omega + a^2}. \tag{9.22}$$

Then, the magnitude and phase spectra of the complex signal, $X(e^{j\omega})$ are computed as follows:

$$|X(e^{j\omega})| = \frac{1}{\sqrt{1-2a\cos\omega + a^2}}, \quad \measuredangle X(e^{j\omega}) = \tan^{-1} - \frac{a\sin\omega}{1-a\cos\omega}. \tag{9.23}$$

The behavior of the magnitude and phase plot depends on the value of the parameter, a. Figures 9.2 and 9.3 show two plots of the magnitude and phase spectra for positive and negative values of a.

Exercise 9.2 Consider the following absolutely summable discrete time signal:

$$x[n] = a^{|n|}, \qquad |a| < 1. \tag{9.24}$$

a) Plot the signal, $x[n]$ for $0 < a < 1$.
b) Find and plot the discrete time Fourier transform, $X(e^{j\omega})$.

Solution

a) This signal is sketched for $0 < a < 1$ in Figure 9.4a.

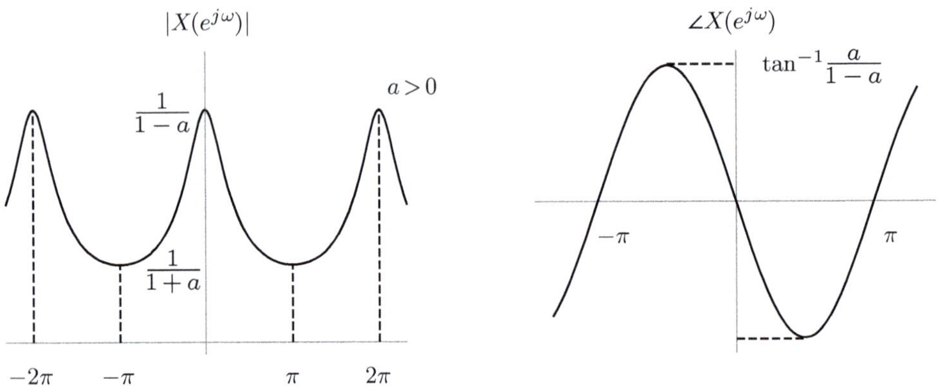

Figure 9.2 Magnitude and phase plots of $X(e^{j\omega})$, for $a > 0$.

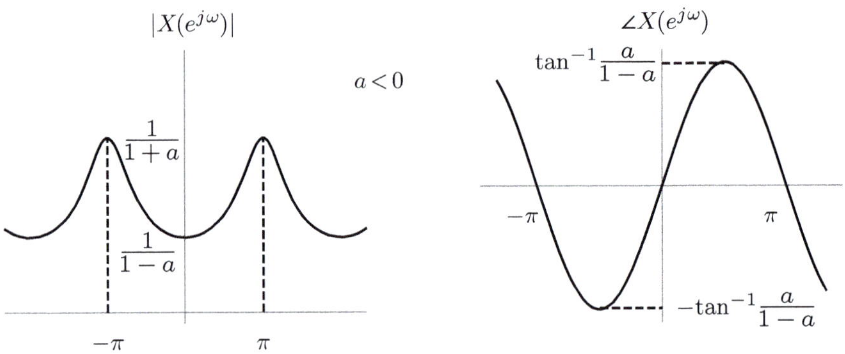

Figure 9.3 Magnitude and phase plots of $X(e^{j\omega})$, for $a < 0$.

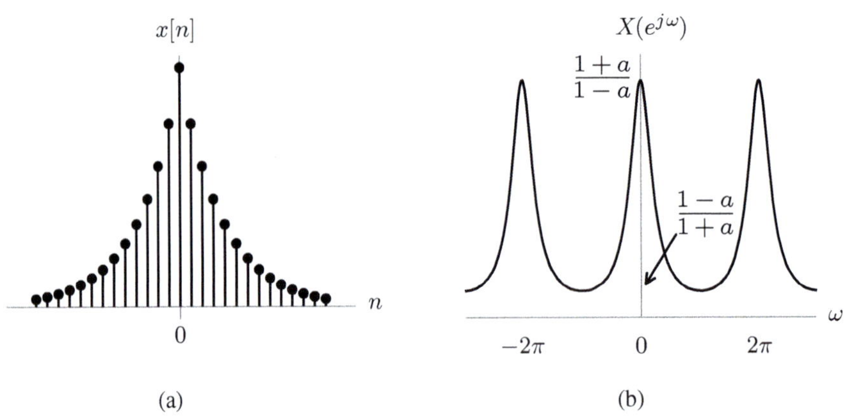

Figure 9.4 (a) Signal $x[n] = a^{|n|}$ and (b) its Fourier transform $(0 < a < 1)$.

b) Noting that $-1 < a < 1$, we can split the analysis equation into two sums, to obtain the Fourier transform of this signal as follows:

$$\begin{aligned} X(e^{j\omega}) &= \sum_{n=-\infty}^{+\infty} a^{|n|} e^{-j\omega n} \\ &= \sum_{n=0}^{+\infty} a^n e^{-j\omega n} + \sum_{n=-\infty}^{-1} a^{-n} e^{-j\omega n}. \end{aligned} \tag{9.25}$$

Making the substitution of variables $m = -n$ in the second summation, we obtain:

$$X(e^{j\omega}) = \sum_{n=0}^{+\infty} (ae^{-j\omega})^n + \sum_{m=1}^{\infty} (ae^{j\omega})^m. \tag{9.26}$$

Both of these summations are infinite geometric series that we can evaluate in closed form, yielding

$$\begin{aligned} X(e^{j\omega}) &= \frac{1}{1 - ae^{-j\omega}} + \frac{ae^{j\omega}}{1 - ae^{j\omega}} \\ &= \frac{1 - a^2}{1 - 2a\cos\omega + a^2}. \end{aligned} \tag{9.27}$$

The Fourier transform, $X(e^{j\omega})$ is a real, periodic and continuous frequency function, where the period is 2π, as illustrated in Figure 9.4b, for $0 < a < 1$.

Exercise 9.3 Consider the general form of a symmetric rectangular pulse, given as follows:

$$x[n] = \begin{cases} 1, & |n| \leq N_1, \\ 0, & |n| > N_1 \end{cases} \tag{9.28}$$

a) Plot the signal for $N_1 = 2$.
b) Find the discrete time Fourier transform of this signal, and plot the magnitude and phase spectrum for $N_1 = 2$.

Solution

a) The plot of $x[n]$ is given in Figure 9.5a, for $N_1 = 2$.

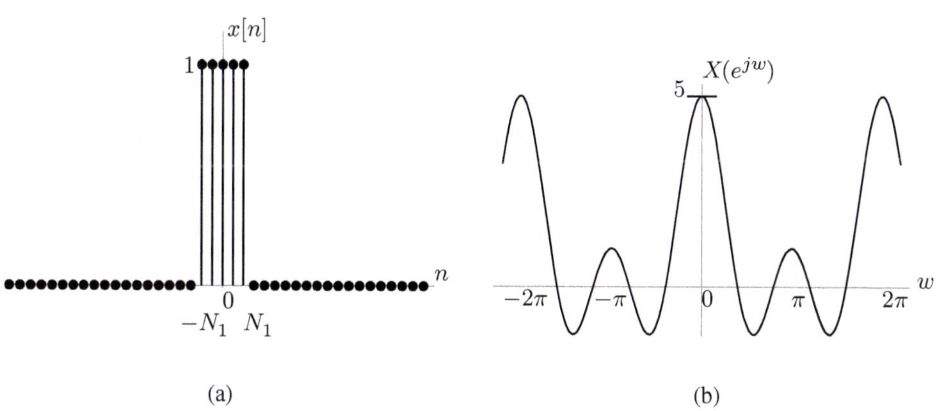

Figure 9.5 (a) Rectangular pulse for $N_1 = 2$ and (b) its Fourier transform.

b) Discrete time Fourier transform of this signal is

$$X(e^{j\omega}) = \sum_{n=-N_1}^{N_1} e^{-j\omega n}. \tag{9.29}$$

Using the finite sum formula,

$$\sum_{n=0}^{N-1} \alpha^n = \begin{cases} N, & \text{for} \quad \alpha = 1, \\ \frac{1-\alpha^N}{1-\alpha}, & \text{otherwise.} \end{cases} \tag{9.30}$$

and changing the limits of the summation, the finite sum of Fourier transform can be written in a compact form as follows:

$$X(e^{j\omega}) = \frac{\sin \omega \left(N_1 + \frac{1}{2}\right)}{\sin(\omega/2)}. \tag{9.31}$$

For $N_1 = 2$, the Fourier transform can be written as:

$$X(e^{j\omega}) = \frac{\sin \frac{5}{2}\omega}{\sin(\omega/2)}. \tag{9.32}$$

This Fourier transform, $X(e^{j\omega})$, is a real, symmetric and periodic function with period 2π, and it is sketched in Figure 9.5b.

Exercise 9.4 Consider a physical phenomenon represented by the discrete time impulse function,

$$x[n] = \delta[n]. \tag{9.33}$$

a) Find and plot the discrete time Fourier transform of $x[n]$.
b) Comment on the frequency content of $X(e^{j\omega})$.

Solution

a) We employ the analysis equation:

$$x[n] = \delta[n] \leftrightarrow X(e^{j\omega}) = \sum_{n=0}^{\infty} \delta[n] e^{-j\omega n} = 1. \tag{9.34}$$

As it can be seen from Figure 9.6, the Fourier transform of discrete impulse function is a continuous time constant function. This function can be considered periodic, with the fundamental period of 2π.

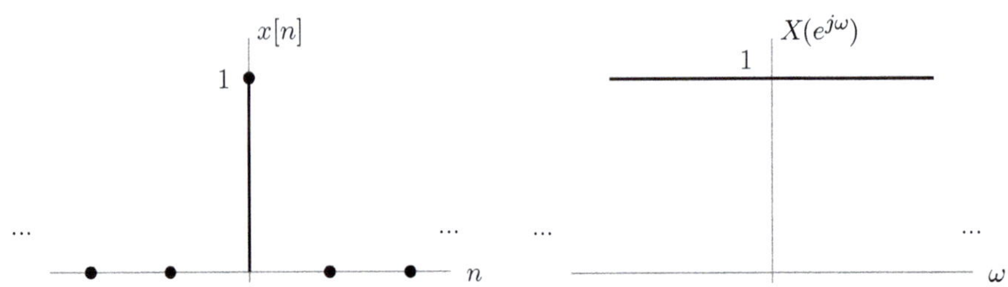

Figure 9.6 Fourier transform of a discrete impulse function, $x[n] = \delta[n]$ is constant. The transform function, $X(e^{j\omega})$ can be considered as periodic with 2π, where it is constant at every interval of $2k\pi$, for all k.

b) Discrete time Fourier transform of the impulse function consists of all frequencies with the same amount.

Exercise 9.5 Consider a physical phenomenon represented by the discrete time-shifted impulse function,

$$x[n] = \delta[n - n_0]. \tag{9.35}$$

a) Find the discrete time Fourier transform of $x[n]$.
b) Compare your results with the previous example.

Solution

a) We employ the analysis equation,

$$x[n] = \delta[n - n_0] \leftrightarrow X(e^{j\omega}) = \sum_{n=-\infty}^{\infty} \delta[n - n_0]e^{-j\omega n} = \sum_{n=-\infty}^{\infty} \delta[n]e^{-j\omega(n+n_0)} = e^{-j\omega n_0}. \tag{9.36}$$

b) Let us consider the magnitude and the phase of this complex function and compare it to the Fourier transform of the impulse function, which was a real function, $X(e^{j\omega}) = 1$. The magnitude of the shifted impulse is the same as that of the unshifted impulse,

$$|X(e^{j\omega})| = 1. \tag{9.37}$$

However, the phase is

$$\measuredangle X(e^{j\omega})| = -\omega n_0. \tag{9.38}$$

Exercise 9.6 Consider a physical phenomenon represented by the discrete time Fourier transform, given as a continuous impulse train, in the frequency domain,

$$X(e^{j\omega}) = 2\pi \sum_{l=-\infty}^{\infty} \delta(\omega - 2\pi l). \tag{9.39}$$

a) Find and plot the inverse discrete time Fourier transform of $X(e^{j\omega})$.
b) Compare time and frequency domain representation of the underlying physical phenomenon.
c) Give a real-life example, which is represented by an impulse train.

Solution

a) Let us use the synthesis equation:

$$x[n] = \frac{1}{2\pi} \int_0^{2\pi} X(e^{j\omega})e^{j\omega n} d\omega = \sum_{l=-\infty}^{\infty} \int_0^{2\pi} \delta(\omega - 2\pi l)e^{j\omega n} d\omega = 1, \forall n. \tag{9.40}$$

Equivalently,

$$x[n] = \sum_{l=-\infty}^{\infty} \delta[n - l]. \tag{9.41}$$

b) Interestingly, the discrete time impulse train in the time domain, is a continuous frequency impulse train with amplitude 2π and period 2π, in the frequency domain (see Figure 9.7).
c) Neurons in the brain produce action potentials, which travel along the axons to govern the communication all over the brain. The signals generated by these electrochemical activities are in the form of impulse trains. Thus, our brain can be considered as a massively parallel impulse train generator and processor.

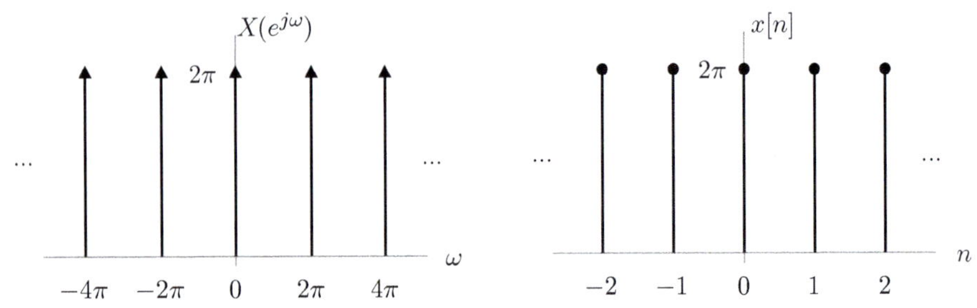

Figure 9.7 Impulse train preserves its analytic form in both time and frequency domains. However, while it is a discrete time impulse train, in the time domain, it becomes a continuous frequency impulse train in the frequency domain. Note that it is scaled by 2π in amplitude and has the fundamental period 2π.

Exercise 9.7 Consider a physical phenomenon represented by the discrete time Fourier transform, given as a **shifted** continuous impulse train, in the frequency domain (see Figure 9.8),

$$X(e^{j\omega}) = 2\pi \sum_{l=-\infty}^{\infty} \delta(\omega - \omega_0 - 2\pi l). \tag{9.42}$$

a) Find and plot the inverse discrete time Fourier transform of $X(e^{j\omega})$.
b) Study the effect of shift by comparing your results with the previous example of an unshifted impulse train.

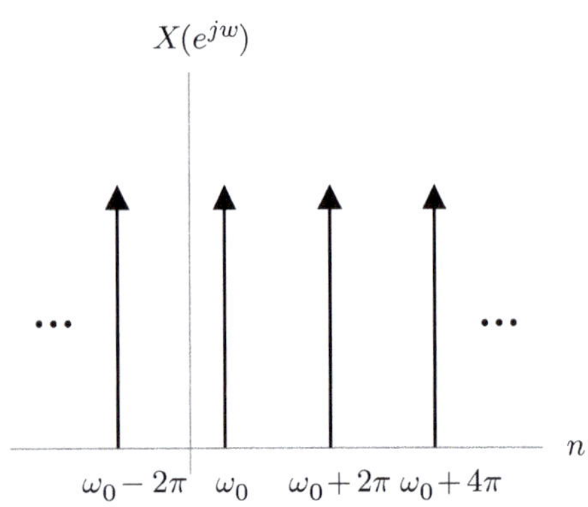

Figure 9.8 Impulse train, shifted by ω_0.

Solution

a) Let us directly use the synthesis equation:

$$\begin{aligned} x[n] &= \int_{\omega_0-\pi}^{\omega_0+\pi} \delta(\omega - \omega_0) e^{j\omega n} d\omega = e^{j\omega_0 n} \\ x[n] &= e^{j\omega_0 n} \leftrightarrow 2\pi \sum_{l=-\infty}^{\infty} \delta(\omega - \omega_0 - 2\pi l). \end{aligned} \tag{9.43}$$

b) The inverse Fourier transform of the shifted impulse train is a **discrete time complex function.** Therefore, the function $x[n] = e^{j\omega_0 n}$, involves two plots, namely, the magnitude and the phase spectra are

$$|x[n]| = 1, \forall n \quad \text{and} \quad \measuredangle x[n] = \omega_0 n, \quad \forall n, \tag{9.44}$$

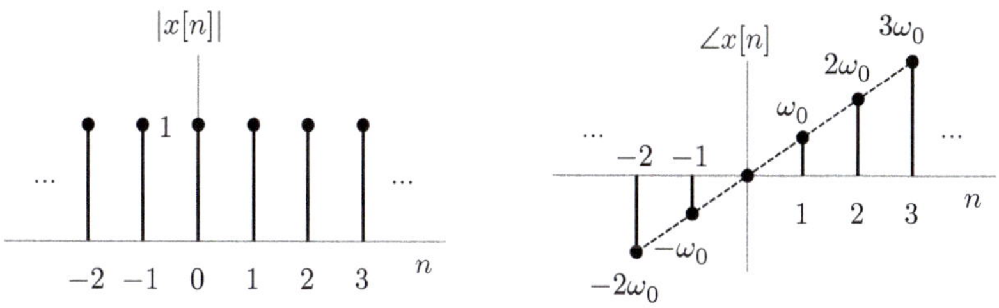

Figure 9.9 Magnitude and phase spectrum of the complex discrete time domain function, $x[n] = e^{j\omega_0 n}$.

respectively. Analysis of Figure 9.9 reveals that a shift in frequency domain does not change the magnitude spectrum of the unshifted version of the signal, but it creates a phase spectrum that is linear with respect to time.

Exercise 9.8 Consider the following discrete time signal:

$$x[n] = u[n] - u[n - N]. \tag{9.45}$$

a) Find the Fourier transform of the $x[n]$.
b) Plot the magnitude and phase spectra of the Fourier transform for $N = 4$.

Solution

a) Let us find the discrete time Fourier transform of $x[n]$ by using the synthesis equation for one full period in the interval, $-\pi \leq \omega \leq \pi$:

$$X(e^{j\omega}) = \sum_{n=-\infty}^{\infty} x[n]e^{-j\omega n} = \sum_{n=0}^{N-1} e^{-j\omega n} = \frac{1 - e^{-j\omega N}}{1 - e^{-j\omega}}. \tag{9.46}$$

We have to keep in mind that $X(e^{j\omega})$ is periodic with 2π.

b) There is an easy way of computing the magnitude and phase spectrum, in the interval, $-\pi \leq \omega \leq \pi$, by arranging Equation (9.46) and putting it in a polar form, as follows:

$$X(e^{j\omega}) = \frac{1 - e^{-j\omega N}}{1 - e^{-j\omega}} = \frac{e^{-j\omega N/2}}{e^{-j\omega/2}} \cdot \frac{e^{j\omega N/2} - e^{-j\omega N/2}}{e^{j\omega/2} - e^{-j\omega/2}} = e^{-j\omega(N-1)/2} \frac{\sin(\omega N/2)}{\sin(\omega/2)}. \tag{9.47}$$

Recall that a complex function in polar form is represented in terms of its magnitude and phase as:

$$X(e^{j\omega}) = |X(e^{j\omega})|e^{j\measuredangle X(e^{j\omega})}. \tag{9.48}$$

Therefore, the magnitude spectrum of $X(e^{j\omega})$ is

$$|X(e^{j\omega})| = \left|\frac{\sin(\omega N/2)}{\sin[\omega/2]}\right|. \tag{9.49}$$

The analytical form of the phase of $X(e^{j\omega})$ is a little tricky: For the values of ω, where the multiplicative factor is,

$$\frac{\sin(\omega N/2)}{\sin(\omega/2)} \geq 0, \tag{9.50}$$

the phase equation has the following analytical form;

$$\measuredangle X(e^{j\omega}) = -\frac{\omega(N-1)}{2}. \tag{9.51}$$

However, for some values of ω, the multiplicative factor becomes negative,

$$\frac{\sin(\omega N/2)}{\sin(\omega/2)} < 0. \tag{9.52}$$

In this case, the minus sign, which can be represented as $e^{j\pi}$, changes the analytical form of $X(e^{j\omega})$, as follows:

$$X(e^{j\omega}) = e^{-j\omega(N-1)/2} e^{j\pi} \left| \frac{\sin(\omega N/2)}{\sin(\omega/2)} \right|. \tag{9.53}$$

Hence, for ω values, where the inequality of Equation (9.52) is satisfied, the phase equation becomes,

$$\measuredangle X(e^{j\omega}) = \pi - \frac{\omega(N-1)}{2}, \tag{9.54}$$

which is repeated at every 2π period. The plot of the magnitude and phase for $N = 4$, is shown in Figure 9.10.

For $N = 4$, the magnitude of $X(e^{j\omega})$ is

$$|X(e^{j\omega})| = \left| \frac{\sin(2\omega)}{\sin[\omega/2]} \right|, \tag{9.55}$$

and the phase of $X(e^{j\omega})$ is

$$\measuredangle X(e^{j\omega}) = \frac{3\omega}{2}, \tag{9.56}$$

as shown in Figure 9.10.

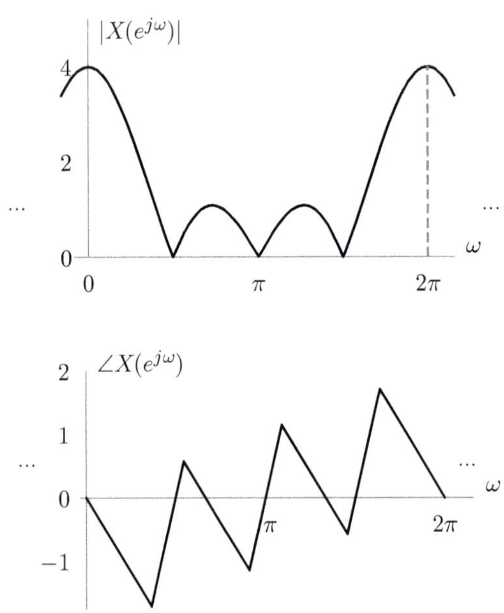

Figure 9.10 Magnitude and phase spectra of $X(e^{j\omega})$, in Exercise 9.8.

Exercise 9.9 Consider the discrete time unit step function, $u[n]$.

a) Is this function absolutely summable?
b) Can you find the discrete time Fourier transform of this function?

Solution

a) This function is **not** absolutely summable

$$\sum_{n=0}^{\infty} |u[n]| \to \infty. \tag{9.57}$$

b) Finding the Fourier transform of the unit step function is a little tricky. First let us represent the unit step function in terms of the summation of two functions,

$$u[n] = f[n] + g[n], \tag{9.58}$$

where

$$f[n] = \frac{1}{2}, \quad \forall n, \tag{9.59}$$

and

$$g[n] = \begin{cases} \frac{1}{2}, & \text{if } n \geq 0 \\ -\frac{1}{2}, & \text{if } n < 0. \end{cases} \tag{9.60}$$

In order to find the Fourier transform of $u[n]$, we find the Fourier transform of $f[n]$ and $g[n]$ and then add them. Mathematically,

$$u[n] = f[n] + g[n] \longleftrightarrow U(e^{j\omega}) = F(e^{j\omega}) + G(e^{j\omega}). \tag{9.61}$$

Considering the fact that the inverse Fourier transform of the shifted impulse function, $\delta(\omega - k)$, is

$$x[n] = \frac{1}{2\pi} \int_0^{2\pi} \delta(\omega - k) e^{j\omega n} d\omega = e^{-jk\omega}, \tag{9.62}$$

the Fourier transform of $f[n]$ can be obtained as the sum of shifted impulses as follows:

$$F(e^{j\omega}) = \sum_{n=-\infty}^{\infty} f[n] e^{-j\omega n} = \frac{1}{2} \sum_{n=-\infty}^{\infty} e^{-j\omega n} = \pi \sum_{n=-\infty}^{\infty} \delta(\omega - 2\pi n), \tag{9.63}$$

and the Fourier transform of $g[n]$ is

$$G(e^{j\omega}) = \sum_{n=-\infty}^{\infty} g[n] e^{-j\omega n} = -\frac{1}{2} \sum_{n=-\infty}^{1} e^{-j\omega n} + \frac{1}{2} \sum_{n=0}^{\infty} e^{-j\omega n} = \frac{1}{1 - e^{-j\omega}}. \tag{9.64}$$

Therefore, the discrete time Fourier transform of the unit step function is

$$U(e^{j\omega}) = F(e^{j\omega}) + G(e^{j\omega}) = \frac{1}{1 - e^{-j\omega}} + \pi \sum_{n=-\infty}^{\infty} \delta[\omega - 2\pi n]. \tag{9.65}$$

Note: The aforementioned exercises demonstrate that although the unit step function $u[n] = 0$ for $n < 0$, we need to evaluate the analysis equation in the interval of $-\infty < n < \infty$. The continuous frequency Fourier transform of discrete time unit step function covers one period in $-\pi < \omega < \pi$ and it repeats itself at every 2π period.

9.3 Fourier Transform of Discrete Time Periodic Functions

Let us now investigate the relationship between the Fourier series and Fourier transform representation of discrete time periodic functions.

Recall that the Fourier series representation of a discrete time periodic function is given by the following synthesis and analysis equation pair:

$$x[n] = \sum_{k=<N>} a_k e^{jk\omega_0 n} \overset{\text{F.S.}}{\longleftrightarrow} a_k = \frac{1}{N} \sum_{n=<N>} x[n]e^{-jk\omega_0 n}, \tag{9.66}$$

where $x[n]$ and a_k are both periodic with $N = \frac{2\pi}{\omega_0}$.

Recall also that the Fourier transform of a discrete time signal is

$$x[n] = \frac{1}{2\pi} \int_0^{2\pi} X(e^{j\omega})e^{j\omega n} \overset{\text{F.T.}}{\longleftrightarrow} X(e^{j\omega}) = \sum_{n=-\infty}^{\infty} x[n]e^{-j\omega n}. \tag{9.67}$$

Motivating Question: What is the relationship between the spectral coefficients a_k and Fourier transform of $X(e^{j\omega})$, when the discrete time signal $x[n]$ is periodic?

As we show in the previous example, the discrete time complex exponential for the kth harmonic has the Fourier transform as the shifted impulse train, as follows:

$$x[n] = e^{jk\omega_0 n} \longleftrightarrow X(e^{j\omega}) = \sum_{l=-\infty}^{\infty} 2\pi\delta(\omega - k\omega_0 - 2\pi l). \tag{9.68}$$

Let us replace $x[n]$ in the Fourier transform equation by its Fourier series representation. Each term in the Fourier series equation of $x[n]$ and its Fourier transform will be as follows:

$$x[n] = \sum_{k=<N>} a_k e^{jk\omega_0 n}. \tag{9.69}$$

Each term in the right-hand side of the summation in Equation (11.64) has the following Fourier transform:

$$\begin{aligned}
a_0 &\longleftrightarrow a_0 \sum 2\pi\delta(\omega - 2\pi l), \\
a_1 e^{j\omega_0 n} &\longleftrightarrow a_1 \sum 2\pi\delta(\omega - \omega_0 - 2\pi l), \\
a_2 e^{j2\omega_0 n} &\longleftrightarrow a_2 \sum 2\pi\delta(\omega - 2\omega_0 - 2\pi l), \\
&\vdots \\
a_{N-1} e^{j(N-1)\omega_0 n} &\longleftrightarrow a_{N-1} \sum 2\pi\delta(\omega - (N-1)\omega_0 - 2\pi l).
\end{aligned} \tag{9.70}$$

If we add all the terms in the left-hand side of Equation (9.70), we obtain the Fourier series representation of the discrete time signal $x[n]$. If we add the transforms in the right-hand side of Equation (9.70), we obtain the superposition of the shifted impulse functions, where the superposition parameters are $2\pi a_k$. In order to get an idea about the behavior of this superposition, let us plot the Fourier transform of the first term a_0 and that of the second term $a_1 e^{j\omega n}$, as shown in Figures 9.11 and 9.12.

Figure 9.13 indicates that the two superposed terms in time domain generate an impulse train, with two different amplitudes, in the frequency domain,

$$a_0 + a_1 e^{j\omega n} \longleftrightarrow 2\pi \left[a_0\delta(\omega) + a_1\delta\left(\omega - \frac{2\pi}{N}(N+1)\right)\right],$$

which repeats at every 2π period.

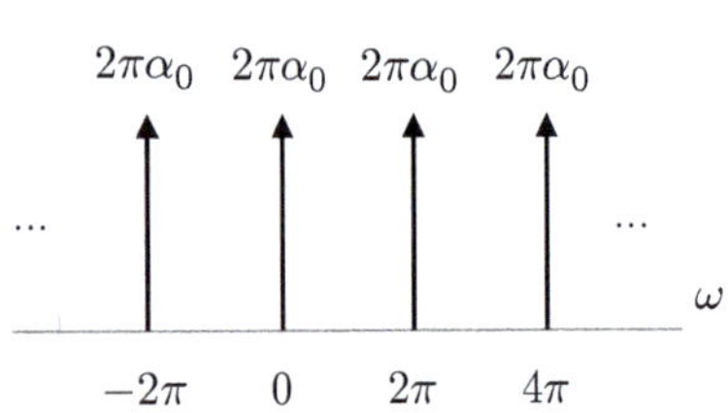

Figure 9.11 Fourier transform of a_0.

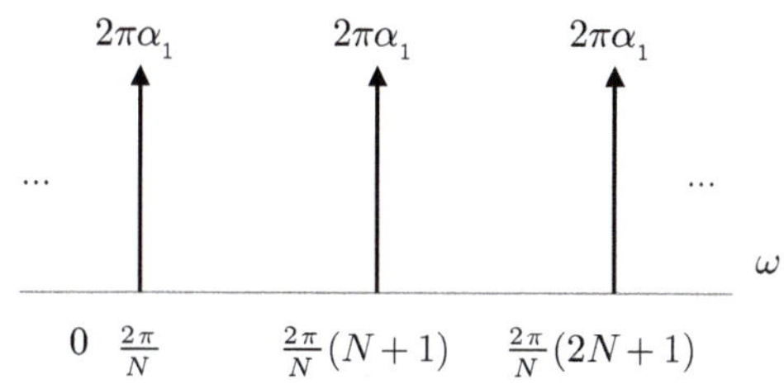

Figure 9.12 Fourier transform of a_1.

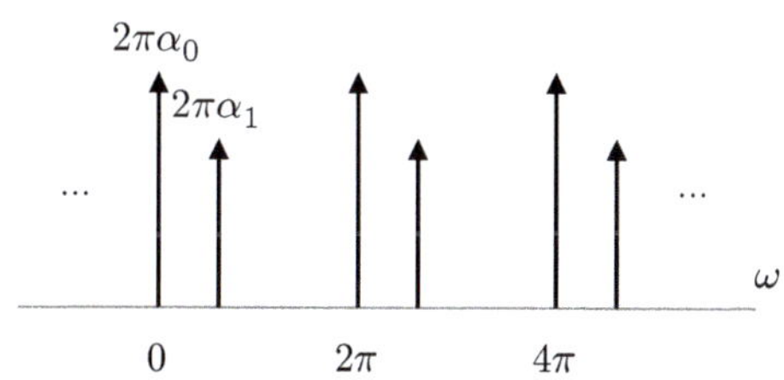

Figure 9.13 Fourier transform of $a_0 + a_1 e^{j\omega}$.

If we add all the terms on the left-hand side and right-hand side of the Fourier transoms of Equation (9.70), we obtain the Fourier transform of a discrete time periodic signal $x[n]$ in terms of the spectral coefficients as follows:

$$X(e^{j\omega}) = \sum_{k=0}^{N} \sum_{l=-\infty}^{\infty} 2\pi a_k \delta\left(\omega - \frac{2\pi}{N}(k + lN)\right). \tag{9.71}$$

We can further simplify Equation (9.71) to obtain the relationship between the Fourier transform and Fourier series representation of a periodic signal as follows:

$$X(e^{j\omega}) = \sum_{k=-\infty}^{\infty} 2\pi a_k \delta\left(\omega - \frac{2\pi k}{N}\right), \tag{9.72}$$

where $\omega_0 = \frac{2\pi}{N}$ and

$$\begin{aligned} x[n] &= \sum_{k=<N>} a_k e^{jk\omega_0 n}, \\ a_k &= \frac{1}{N} \sum_{n=<N>} x[n] e^{-jk\omega_0 n}. \end{aligned} \tag{9.73}$$

This equation reveals that the Fourier transform of a periodic signal $x[n]$ converts the discrete time spectral coefficients of weighted and shifted impulses into continuous time weighted and shifted impulses.

Note: While the period of the spectral coefficients is $N = 2\pi/\omega_0$, the period of the Fourier transform is 2π. Therefore, the Fourier transform axis is scaled by ω_0 in the frequency domain.

Exercise 9.10 Consider the following discrete time periodic signal:

$$x[n] = \cos \omega_0 n = \frac{1}{2} e^{j\omega_0 n} + \frac{1}{2} e^{-j\omega_0 n}. \tag{9.74}$$

a) Find the spectral coefficients of $x[n]$.
b) Find the Fourier transform of $x[n]$.
c) Compare the Fourier series and Fourier transform representations of $x[n]$.

Solution

a) The spectral coefficients of $x[n]$ are

$$a_1 = a_{-1} = \frac{1}{2}, \quad N = \frac{2\pi}{\omega_0}. \tag{9.75}$$

b) Fourier transform of $x[n]$ is

$$X(e^{j\omega}) = \sum_{k=-\infty}^{\infty} 2\pi a_k \delta\left(\omega - \frac{2\pi k}{N}\right) \tag{9.76}$$
$$\pi[\delta(\omega - \omega_0) + \delta(\omega + \omega_0)], \quad \text{for } -\pi \leq \omega \leq \pi.$$

and it repeats at every 2π.

c) While the spectral coefficients are discrete time impulse train with period N, the Fourier transform is a continuous impulse train with period 2π, as shown in Figure 9.14. In other words, the Fourier transform, $X(e^{j\omega})$ repeats itself with period 2π. The spectral coefficients a_k repeat with period N:

$$X(e^{j\omega}) = X(e^{j\omega \pm 2k\pi}), \quad \text{and} \quad a_k = a_{k \pm N}, \quad \forall k. \tag{9.77}$$

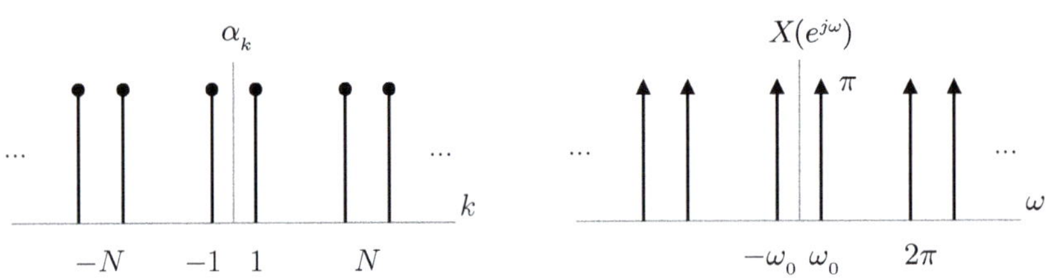

Figure 9.14 Fourier series coefficients and Fourier transform of the periodic signal, $x[n] = \cos\omega_0 n$.

Exercise 9.11 Consider an arbitrary pulse signal, defined in a finite interval,

$$x[n] = \sum_{k=-N_1}^{N_2} c_k \delta[n-k]. \tag{9.78}$$

a) Is this function absolutely summable?
b) If your answer is yes, find the discrete time Fourier transform of this signal.

Solution

a) Since the summation is bounded, this function is absolutely summable, provided that all the values of c_k are bounded:

$$x[n] = \sum_{k=-N_1}^{N_2} c_k \delta[n-k] < \infty. \tag{9.79}$$

b) Discrete time Fourier transform of this function can be obtained from the analysis equation,

$$X(e^{j\omega}) = \sum_{n=-\infty}^{\infty} \sum_{k=-N_1}^{N_2} c_k \delta[n-k] e^{-jn\omega} = \sum_{k=-N_1}^{N_2} c_k e^{-jk\omega}. \tag{9.80}$$

INTERACTIVE

An example application: removing unwanted noise from audio @ https://384book.net/i0901

9.4 Properties of Fourier Transforms for Discrete Time Signals and Systems

So far, we has seen that discrete time Fourier transforms map a discrete time domain function into a continuous and periodic frequency domain function. In the frequency domain, the time variable disappears and the function are represented in terms of a continuous variable of frequencies by the following analysis equation:

$$X(e^{j\omega}) = \sum_{n=-\infty}^{\infty} x[n]e^{-j\omega n}, \tag{9.81}$$

where the time domain function can be uniquely obtained from its frequency domain representation by the synthesis equation,

$$x[n] = \frac{1}{2\pi}\int_{2\pi} X(e^{j\omega})e^{j\omega n}d\omega. \tag{9.82}$$

Discrete time Fourier analysis and synthesis equations reveal that the class of absolutely summable discrete time functions can be represented by uncountably infinite waveforms, namely complex exponentials, with continuously varying frequencies. Therefore, the Fourier transform, $X(e^{j\omega})$, gives us a unique and powerful way of viewing a physical phenomenon in terms of weighted summation of waveforms, where the weights are the time domain function $x[n]$. Furthermore, the time domain function can be uniquely recovered from its Fourier transform. In other words, time and frequency domain representation of a physical phenomenon is one-to-one and onto,

$$x[n] \longleftrightarrow X(e^{j\omega}). \tag{9.83}$$

In Section 9.4.1, we shall investigate the properties of discrete time Fourier transform. We shall use the properties to go back and forth between the discrete time and continuous frequency domains. We shall study the frequency content of the aperiodic discrete time signals. We shall design and implement LTI systems in the time and frequency domains for filtering the discrete time signals.

9.4.1 Basic Properties of Discrete Time Fourier Transform

Recall that discrete time Fourier transform is the extension of discrete time Fourier series, where an aperiodic function is represented as a periodic function of infinite period. As we stretch the period N to infinity, the Fourier series of a discrete time periodic function converges to a continuous time aperiodic function. Furthermore, since the complex exponential basis functions $\{e^{-j\omega n}\}_{n=-\infty}^{\infty}$ are periodic with 2π, the superposition of the complex exponential functions, which makes the Fourier transform $X(e^{j\omega n})$ is also periodic with 2π. Mathematically,

$$X(e^{j\omega}) = X(e^{j(\omega+2\pi)}). \tag{9.84}$$

Although some properties of discrete time Fourier series resemble the properties of discrete time and continuous time Fourier transform, there are substantial differences imposed by taking the limit of the integer period, $N \to \infty$.

As we did for the continuous time Fourier transforms, we provide the basic properties and transform pairs in Tables 9.1 and 9.2, for the discrete time counterparts. These tables simplify the computations for finding the Fourier transforms and/or their inverse.

Table 9.1 Properties of the discrete time Fourier transform.

Nonperiodic signal	Fourier transform
$x[n] = \frac{1}{2\pi}\int_{2\pi} X(e^{j\omega})e^{j\omega n}d\omega$	$X(e^{j\omega}) = \sum_{n=-\infty}^{\infty} x[n]e^{-j\omega n}$
$x[n]$	$X(e^{j\omega})$, periodic with period 2π
$y[n]$	$Y(e^{j\omega})$, periodic with period 2π
$ax[n] + by[n]$	$aX(e^{j\omega}) + bY(e^{j\omega})$
$x[n - n_0]$	$e^{-j\omega n_0}X(e^{j\omega})$
$e^{j\omega_0 n}x[n]$	$X(e^{j(\omega-\omega_0)})$
$x^*[n]$	$X^*(e^{-j\omega})$
$x[-n]$	$X(e^{-j\omega})$
$x_{(m)}[n] = \begin{cases} x[n/m], & n \text{ is multiple of } m \\ 0, & \text{otherwise} \end{cases}$	$X(e^{jm\omega})$
$x[n] * y[n]$	$X(e^{j\omega})Y(e^{j\omega})$
$x[n]y[n]$	$\frac{1}{2\pi}\int_{2\pi} X(e^{j\theta})Y(e^{j(\omega-\theta)})d\theta$
$x[n] - x[n-1]$	$(1 - e^{j\omega})X(e^{j\omega})$
$\sum_{k=-\infty}^{n} x[k]$	$\frac{1}{1-e^{j\omega}}X(e^{j\omega}) + \pi X(0)\sum_{k=-\infty}^{\infty}\delta(\omega - 2\pi k)$
$nx[n]$	$j\frac{d}{d\omega}X(e^{j\omega})$
For real-valued $x[n]$	$\begin{cases} X(e^{j\omega}) = X^*(e^{-j\omega}) \\ \text{Re}\left\{X(e^{j\omega})\right\} = \text{Re}\left\{X(e^{-j\omega})\right\} \\ \text{Im}\left\{X(e^{j\omega})\right\} = -\text{Im}\left\{X(e^{-j\omega})\right\} \\ \lvert X(e^{j\omega})\rvert = \lvert X(e^{-j\omega})\rvert \\ \sphericalangle X(e^{j\omega}) = -\sphericalangle X(e^{-j\omega}) \end{cases}$
Even part of $x[n]$	$\text{Re}\left\{X(e^{j\omega})\right\}$
Odd part of $x[n]$	$\text{Im}\left\{X(e^{j\omega})\right\}$
Parseval's relation for nonperiodic signals: $\sum_{n=-\infty}^{\infty} \lvert x[n]\rvert^2 = \frac{1}{2\pi}\int_{2\pi} \lvert X(e^{j\omega})\rvert^2 d\omega$	

The properties can be proven by directly employing the analysis and synthesis equations. The reader is strongly recommended to prove all the properties and solve the Fourier transform pairs, as given in Tables 9.1 and 9.2.

In the following, we study some of the selective properties to grasp the discrete time and continuous frequency domain representations of signals and systems and their relationship.

1) **Linearity:** Like the continuous time Fourier transform, the discrete time Fourier transform is a linear operator. Mathematically speaking, if we have two functions and their corresponding Fourier transforms,

$$x[n] \longleftrightarrow X(e^{j\omega}), \tag{9.85}$$

Table 9.2 Fourier transform pairs of popular discrete time functions.

$x[n]$	$X(e^{j\omega})$
$\delta[n]$	1
$\delta[n-n_0]$	$e^{-j\omega n_0}$
$\sum_{k=-\infty}^{\infty}\delta(n-kN)$	$\frac{2\pi}{N}\sum_{k=-\infty}^{\infty}\delta\left(\omega-\frac{2\pi k}{N}\right)$
1	$2\pi\sum_{k=-\infty}^{\infty}\delta(\omega-2\pi k)$
$e^{j\omega_0 n}$	$2\pi\sum_{k=-\infty}^{\infty}\delta(\omega-\omega_0-2\pi k)$
$\cos(\omega_0 n)$	$\pi\sum_{k=-\infty}^{\infty}\left[\delta(\omega-\omega_0-2\pi k)+\delta(\omega+\omega_0-2\pi k)\right]$
$\sin(\omega_0 n)$	$\frac{\pi}{j}\sum_{k=-\infty}^{\infty}\left[\delta(\omega-\omega_0-2\pi k)-\delta(\omega+\omega_0-2\pi k)\right]$
$u[n]$	$\frac{1}{1-e^{-j\omega}}+\pi\sum_{k=-\infty}^{\infty}\delta(\omega-2\pi k)$
$a^n u(n),\quad \lvert a\rvert<1$	$\frac{1}{1-ae^{-j\omega}}$
$(n+1)a^n u[n],\quad \lvert a\rvert<1$	$\frac{1}{(1-ae^{-j\omega})^2}$
$\frac{(n+m-1)!}{n!(m-1)!}a^n u[n],\quad \lvert a\rvert<1$	$\frac{1}{(1-ae^{-j\omega})^m}$
$\frac{1}{1-a^2}a^{\lvert n\rvert},\quad \lvert a\rvert<1$	$\frac{1}{1+a^2-2a\cos\omega}$
$\begin{cases}1, & \lvert n\rvert\le N_1\\ 0, & N_1<\lvert n\rvert\le\frac{N}{2}\end{cases}$, period N	$2\pi\sum_{k=-\infty}^{\infty}a_k\delta\left(\omega\frac{2\pi k}{N}\right)$
$\begin{cases}1, & \lvert n\rvert\le N_1\\ 0, & \lvert n\rvert>N_1\end{cases}$	$\frac{\sin\omega\left(N_1+\frac{1}{2}\right)}{\sin\frac{\omega}{2}}$
$\begin{cases}\frac{\sin Wn}{\pi n}=\frac{W}{\pi}\mathrm{sinc}\frac{Wn}{\pi}\\ 0<W<\pi\end{cases}$	$\begin{cases}1, & \lvert\omega\rvert\le W\\ 0, & W<\lvert\omega\rvert\le\pi\end{cases}$, period 2π

and

$$y[n] \longleftrightarrow Y(e^{j\omega}), \tag{9.86}$$

then,

$$ax[n]+by[n] \longleftrightarrow aX(e^{j\omega})+bY(e^{j\omega}). \tag{9.87}$$

This property follows from the fact that both summation and integral operators, required to take the Fourier transform and its inverse, are linear operators.

2) **Time Shifting:** If the function $x[n]$ is shifted in time by a constant amount, n_0, its Fourier transform $X(e^{j\omega})$ is multiplied by the complex exponential function, $e^{-j\omega n_0}$, in the frequency domain:

$$x[n - n_0] \longleftrightarrow e^{-j\omega n_0} X(e^{j\omega}). \tag{9.88}$$

This property can be shown by defining $y[n] = x[n - n_0]$ and inserting the shifted signal into the analysis equation,

$$Y(e^{j\omega)} = \sum_{n=-\infty}^{\infty} x[n - n_0]e^{-j\omega n}. \tag{9.89}$$

Let us change the variable of summation to $n' = n - n_0$ and get the time shifting property:

$$Y(j\omega) = \sum_{-\infty}^{\infty} x(n')e^{-j\omega(n'+n_0)t} = e^{-j\omega n_0} X(j\omega). \tag{9.90}$$

Note that, since the multiplicative complex exponential function, $e^{-j\omega n_0}$, has a magnitude of 1, the time delay alters the phase of $X(e^{j\omega})$, but not its magnitude. As a result, time delay does not cause the frequency content of $X(e^{j\omega})$ to change at all.

Linearity and time-shifting properties enable us to determine the response of LTI systems without solving the representative linear constant-coefficient difference equations, as illustrated by the following example.

Exercise 9.12 Consider the difference equation of a discrete time LTI system, which is initially at rest:

$$y[n] + \frac{1}{4}y[n-1] - \frac{1}{8}y[n-2] = x[n] - x[n-1]. \tag{9.91}$$

a) Find the frequency response of this system.
b) Find the impulse response of this system.

Solution

a) Let us take the Fourier transform of both sides using the time shifting property,

$$Y(e^{j\omega})\left[1 + \frac{1}{4}e^{-j\omega} - \frac{1}{8}e^{-2j\omega}\right] = X(e^{j\omega})[1 - e^{-j\omega}]. \tag{9.92}$$

Let us replace the input by the impulse function,

$$x[n] = \delta[n] \longleftrightarrow X(e^{j\omega}) = 1.$$

Then, the corresponding output is to be replaced by the frequency response. In this case, the system equation in the frequency domain becomes

$$H(e^{j\omega})[1 + \frac{1}{4}e^{-j\omega} - \frac{1}{8}e^{-2j\omega}] = 1 - e^{-j\omega}. \tag{9.93}$$

The frequency response of this system can be directly obtained from Equation (9.93), as follows:

$$H(e^{j\omega}) = \frac{1 - e^{-j\omega}}{1 + \frac{1}{4}e^{-j\omega} - \frac{1}{8}e^{-2j\omega}}. \tag{9.94}$$

b) Impulse response of this system is the inverse Fourier transform of the frequency response. By using partial fraction expansion method, we can simplify the frequency response,

$$H(e^{j\omega}) = \frac{2}{1 + \frac{1}{2}e^{-j\omega}} - \frac{1}{1 - \frac{1}{4}e^{-j\omega}}. \tag{9.95}$$

Using the Fourier transform pairs of Table 9.2, we obtain the impulse response as follows:

$$h[n] = 2\left(-\frac{1}{2}\right)^n u[n] - \left(\frac{1}{4}\right)^n u[n]. \tag{9.96}$$

The aforementioned exercise shows how the time difference property converts a difference equation of time domain into an algebraic equation in the frequency domain. This handy property allows us to avoid the cumbersome recursions for finding the output $y[n]$ and the impulse response $h[n]$ of an LTI system.

Exercise 9.13 Find the Fourier transform of the following signal:

$$x[n] = \delta[n-1] + \delta[n+1]. \tag{9.97}$$

Solution

Using the Fourier transform of the impulse function together with the time shift property, we obtain,

$$\begin{aligned} \delta[n] &\leftrightarrow 1, \\ \delta[n-n_0] &\leftrightarrow e^{-j\omega n_0}, \\ X(e^{j\omega}) &= e^{-j\omega} + e^{j\omega} = 2\cos\omega. \end{aligned} \tag{9.98}$$

Two discrete time impulses located at $n = 1$ and $n = -1$, are represented by a continuous time periodic cosine function in the frequency domain, where the period is 2π.

Exercise 9.14 Find the inverse Fourier transform of the following signal:

$$Y(e^{j\omega}) = e^{-j\omega}\cos\omega. \tag{9.99}$$

Solution

Recall, from the previous example,

$$x[n] = \delta[n-1] + \delta[n+1] \overset{\text{F.T.}}{\longleftrightarrow} X(e^{j\omega}) = 2\cos\omega. \tag{9.100}$$

In order to get a multiplicative factor $e^{-j\omega}$ in the frequency domain, we need to get a time shift with $n_0 = 1$, in the time domain,

$$x[n-1] \longleftrightarrow e^{-j\omega}X(e^{j\omega}). \tag{9.101}$$

Let us shift $x[n]$ by $n_0 = 1$,

$$x[n-1] = \delta[n-2] + \delta[n] \longleftrightarrow 2e^{-j\omega}\cos\omega. \tag{9.102}$$

Hence,

$$y[n] = \frac{1}{2}x[n-1] = \frac{1}{2}(\delta[n-2] + \delta[n]) \longleftrightarrow e^{-j\omega}\cos\omega. \tag{9.103}$$

Note: We avoided taking the integral to find the inverse Fourier transform. Instead we used the properties. Why? Because taking a complex integral is not an easy task in most cases. It may require sophisticated methods, which is beyond the scope of this book.

3) **Frequency Shift:** A shift in frequency domain corresponds to scaling the time domain function by the complex exponential:

$$e^{j\omega_0 n}x[n] \leftrightarrow X(e^{j(\omega-\omega_0)}). \tag{9.104}$$

This property can be shown directly by defining $y[n] = e^{-j\omega_0 n}x[n]$ and finding its Fourier transform using the analysis equation,

$$Y(e^{j\omega)} = \sum_{n=-\infty}^{\infty} x[n]e^{j\omega_0 n}e^{-j\omega n} = \sum_{n=-\infty}^{\infty} x[n]e^{-j(\omega-\omega_0)n} = X(e^{j(\omega-\omega_0)}). \tag{9.105}$$

Comparison of time and frequency shift properties uncovers an elegant symmetry between the shifts in time and frequency domains. A shift in one domain corresponds to a multiplication in the other domain. This is one of the duality properties of Fourier transforms.

4) **Time Scale:** Time scale in discrete time functions requires a special care, since the domain should remain integer valued after the scaling. When we scale time by a factor of m, we need to define a new function,

$$y[n] = \begin{cases} x[n/m], & \text{for } n \text{ is integer multiple of } m \\ 0 & \text{otherwise.} \end{cases} \tag{9.106}$$

In other words, when $m > 1$, we stretch the signal $x[n]$ by inserting zero values into the discrete time function $y[n]$ for the noninteger values of n/m. When $m < 1$, we squish the signal by skipping some of the values of the function $x[n]$.
Taking the discrete time Fourier transform of $y[n]$, we obtain,

$$Y(e^{j\omega}) = \begin{cases} \sum_{n=-\infty}^{\infty} x[n/m]e^{-j\omega n}, & \text{for } n \text{ is integer multiple of } m \\ 0, & \text{otherwise.} \end{cases} \tag{9.107}$$

Changing the dummy variable of summation to $n' = n/m$, we obtain,

$$Y(e^{j\omega}) = \begin{cases} \sum_{n=-\infty}^{\infty} x[n']e^{-j\omega mn'}, & \text{for } n \text{ is integer multiple of } m \\ 0, & \text{otherwise.} \end{cases} \tag{9.108}$$

Therefore,

$$Y(e^{j\omega}) = \begin{cases} X(e^{jm\omega}), & \text{for } n \text{ is integer multiple of } m \\ 0, & \text{otherwise.} \end{cases} \tag{9.109}$$

Note: When we stretch the function, $x[n]$ in time domain, for $m > 1$, the frequency of the Fourier transform is increased by mn. On the other hand, when we squish the time domain signal, for $m < 1$, the frequency of the Fourier transform is decreased by mn.

5) **Time Reversal:** A special case of time scale property is the time reverse property, where $m = -1$. Replacing the value of m in Equation (9.106), we obtain,

$$x[-n] \longleftrightarrow X(e^{-j\omega}). \tag{9.110}$$

This equation states that reversing the time in time domain corresponds to reversing the frequency in the frequency domain.

6) **Discrete Time Fourier Transform of Even and Odd Functions:** An even function in the time domain has a real Fourier transform, in the frequency domain. Similarly, an odd function in the time domain has a purely imaginary Fourier transform in the frequency domain. Mathematically,

$$\begin{aligned} \text{Ev}\{x[n]\} &\leftrightarrow \text{Re}\{X(e^{j\omega})\}, \\ \text{Odd}\{x[n]\} &\leftrightarrow \text{Im}\{X(e^{j\omega})\}. \end{aligned} \tag{9.111}$$

This property directly follows the definition of even and odd parts of the functions. Suppose that the Fourier transform pair of a function $x[n]$ is given by,

$$x[n] \leftrightarrow X(e^{j\omega}), \tag{9.112}$$

where the Fourier transform can be represented in Cartesian coordinate system, as follows:

$$X(e^{j\omega}) = \text{Re}\{X(e^{j\omega})\} + j\text{Im}\{X(e^{j\omega})\}. \tag{9.113}$$

Recall that even part of a function $x[n]$ is defined as:

$$\text{Ev}\{x[n]\} = \frac{1}{2}[x[n] + x[-n]]. \tag{9.114}$$

From the time reverse property, we get,

$$x[-n] \longleftrightarrow X(e^{-j\omega}) = \text{Re}\{X(e^{j\omega})\} - j\text{Im}\{X(e^{j\omega})\}. \tag{9.115}$$

Taking the Fourier transform of both sides of Equation (9.114), we obtain,

$$\text{Ev}\{x[n]\} \leftrightarrow \tfrac{1}{2}[\text{Re}\{X(e^{j\omega})\} + j\text{Im}\{X(e^{j\omega})\} + \text{Re}\{X(e^{j\omega})\} - j\text{Im}\{X(e^{j\omega})\}]. \tag{9.116}$$

Hence,

$$\text{Ev}\{x[n]\} \leftrightarrow \text{Re}\{X(e^{j\omega})\} \tag{9.117}$$

Similarly, the odd part of a function $x[n]$ is defined as:

$$\text{Odd}\{x[n]\} = \tfrac{1}{2}[x[n] - x[-n]]. \tag{9.118}$$

Taking the Fourier transform of both sides of Equation (9.118), we obtain,

$$\text{Odd}\{x[n]\} \leftrightarrow \tfrac{1}{2}[\text{Re}\{X(e^{j\omega})\} + j\text{Im}\{X(e^{j\omega})\} - \text{Re}\{X(e^{j\omega})\} + j\text{Im}\{X(e^{j\omega})\}]. \tag{9.119}$$

Hence,

$$\text{Odd}\{x[n]\} \leftrightarrow \text{Im}\{X(e^{j\omega})\}. \tag{9.120}$$

Since the Fourier transform of an even function is real, there is no imaginary part. Thus, the phase is 0 for all the frequencies. Mathematically, the magnitude of an even function is

$$|X(e^{j\omega}| = \text{Re}\{X(e^{j\omega})\}$$

and the phase of an even function is

$$\angle X(e^{j\omega}) = 0.$$

On the other hand, the Fourier transform of an odd function is a purely imaginary function. Thus, the phase spectra is a constant value at $\pi/2$. The magnitude is the imaginary part itself. Mathematically, the magnitude of an even function is

$$|X(e^{j\omega}| = \text{Im}\{X(e^{j\omega})\},$$

and the phase of an even function is

$$\angle X(e^{j\omega}) = \pi/2.$$

7) **Convolution Property:** Convolution of two functions in the time domain corresponds to multiplication of their Fourier transform in the frequency domain.

$$y[n] = x[n] * h[n] \longleftrightarrow Y(e^{j\omega}) = X(e^{j\omega})H(e^{j\omega}), \tag{9.121}$$

Convolution property follows the definition of Fourier analysis equation and convolution summation. Inserting the convolution summation,

$$x[n] * y[n] = \sum_{k=-\infty}^{\infty} x[k]h[n-k],$$

into the Fourier analysis equation, we obtain,

$$Y(e^{j\omega}) = \mathcal{F}[y(t)] = \mathcal{F}[x[n] * h[n]] = \sum_{k=-\infty}^{\infty} \sum_{n=-\infty}^{\infty} x[k]h[n-k]e^{-j\omega n}. \tag{9.122}$$

Changing the dummy variable of summation, $n' = n - k$, we get,

$$Y(e^{j\omega}) = \mathcal{F}[x[n] * h[n]] = \sum_{k=-\infty}^{\infty} x[k]e^{-j\omega k} \sum_{n'=-\infty}^{\infty} h[n']e^{-j\omega n'} = X(j\omega)H(j\omega). \tag{9.123}$$

Convolution property is a direct consequence of the fact that the Fourier transform decomposes a signal into a linear combination of complex exponential functions, $\{e^{j\omega n}\}_{n=-\infty}^{\infty}$ each of which is an eigenfunction of a linear, time-invariant system,

$$y[n] = H(e^{j\omega})e^{j\omega n}.$$

The frequency response $H(e^{j\omega})$ corresponds to the eigenvalue of the LTI system. Filtering a discrete time signal is a direct consequence of the convolution property.

8) **Modulation Property:** This property is similar to the continuous time modulation. The only difference is to deal with periodic convolution.

Modulation property states that the multiplication of two signals in the time domain corresponds to **circular convolution** of their Fourier transform in the frequency domain, which is evaluated for one full period. Formally speaking,

$$y[n] = x[n] \cdot h[n] \longleftrightarrow Y(e^{j\omega}) = \frac{1}{2\pi} X(e^{j\omega}) \otimes H(e^{j\omega}) = \frac{1}{2\pi} \int_{2\pi} X(e^{j\theta})H(e^{j(\omega-\theta)})d\theta. \tag{9.124}$$

Since the Fourier transform of a discrete time function is periodic with 2π, in the frequency domain, we apply circular convolution operation over 2π, which is indicated by the $\otimes$ symbol. Circular convolution brings, also, a scaling factor of $\frac{1}{2\pi}$.

In order to show the multiplication property, we take the Fourier transform of the multiplication of two functions, $y[n] = x[n]h[n]$ using the analysis equation:

$$Y(e^{j\omega}) = \mathcal{F}[y[n]] = \mathcal{F}[x[n] \cdot h[n]] = \sum_{n=-\infty}^{\infty} x[n]h[n]e^{-j\omega n}. \tag{9.125}$$

Then, we insert the inverse Fourier transform of $x[n]$:

$$x[n] = \frac{1}{2\pi} \int_0^{2\pi} X(e^{j\omega})e^{j\omega n} d\omega, \tag{9.126}$$

into Equation (9.125),

$$Y(e^{j\omega}) = \mathcal{F}[x[n] \cdot h[n]] = \frac{1}{2\pi} \sum_{n=-\infty}^{\infty} \int_0^{2\pi} X(e^{j\omega'})e^{j\omega' n}h[n]e^{-j\omega n} d\omega', \tag{9.127}$$

Finally, we arrange the Equation (9.127),

$$Y(e^{j\omega}) = \mathcal{F}[x[n] \cdot h[n]] = \frac{1}{2\pi} \sum_{n=-\infty}^{\infty} \int_{-\infty}^{\infty} X(e^{j\omega'})h[n]e^{-jn(\omega-\omega')} d\omega'. \tag{9.128}$$

Note that the summation in the right-hand side of Equation (9.128) is the shifted Fourier transform of the function $h(t)$,

$$H(e^{j(\omega-\omega')}) = \sum_{n=-\infty}^{\infty} h[n]e^{-jn(\omega-\omega')}. \tag{9.129}$$

Inserting the shifted frequency response $H(j(\omega - \omega_0))$ into Equation (9.128), we obtain,

$$Y(j\omega) = \mathcal{F}[x(t)h(t)] = \frac{1}{2\pi}\int_{-\infty}^{\infty} X(j\omega')H(j(\omega - \omega'))dt = \frac{1}{2\pi}X(j\omega) * H(j\omega). \tag{9.130}$$

Modern communication systems rely on discrete time modulation techniques, rather than their continuous counterparts. Modulation property is extensively used to increase or decrease the frequency bandwidth of signals.

9) **Parseval's Equality:** In all of the properties and examples, we observe that the representation of signals and systems in time and frequency domains, have substantially different analytical forms and structures. One striking difference is that a discrete time aperiodic function is represented by a continuous periodic function in the frequency domain. For example, a discrete time complex exponential signal has a Fourier transform consisting of continuous time impulse train function.

An important invariant between the two domains is the energy of a signal. Formally speaking, the energy of the signals in both domains does not change:

$$\sum_{n=-\infty}^{\infty} |x[n]|^2 = \frac{1}{2\pi}\int_{2\pi} |X(e^{j\omega})|^2 d\omega. \tag{9.131}$$

This relation, called Parseval's equality, shows that the energy of a signal in time and frequency domains are preserved.

As we did in the continuous time functions, we can show Parseval's equality by inserting the analysis equation into the left-hand side of Equation (9.131),

$$\begin{aligned}\sum_{n=-\infty}^{\infty} |x(t)|^2 dt &= \sum_{-\infty}^{\infty} x(t)x^*(t)dt \\ &= \frac{1}{(2\pi)^2}\sum_{n=-\infty}^{\infty}\int_{2\pi} X(e^{j\omega})e^{-j\omega n}d\omega \int_{2\pi} X^*(e^{j\omega'})e^{-j\omega' n}d\omega' \\ &= \frac{1}{(2\pi)^2}\sum_{n=-\infty}^{\infty}\int_{2\pi}\int_{2\pi} X(e^{j\omega})X^*(e^{j\omega})e^{-jn(\omega-\omega')}d\omega d\omega'.\end{aligned} \tag{9.132}$$

We can show that

$$\int_0^{2\pi} \sum_{n=-\infty}^{\infty} e^{-jn(\omega-\omega')}d\omega = \begin{cases} 2\pi & \text{for } \omega = \omega' \\ 0 & \text{otherwise.}\end{cases} \tag{9.133}$$

Inserting the right-hand side of this equality into Equation (9.131), we obtain,

$$\int_T |x(t)|^2 dt = \frac{1}{2\pi}\int_{-\infty}^{\infty} |X(j\omega)|^2 d\omega. \tag{9.134}$$

Parseval's equality reveals that representation of signals in Hilbert space conserves the energy of time domain. Note that there is a factor of $1/2\pi$, which scales the energy of time domain.

Exercise 9.15 Find the energy of the following discrete time impulse response:

$$h[n] = \frac{\sin \omega_c n}{\pi n}. \tag{9.135}$$

Solution

From the definition of the energy,

$$E = \sum_{n=-\infty}^{\infty} |h[n]|^2 = \sum_{n=-\infty}^{\infty} \left|\frac{\sin \omega_c n}{\pi n}\right|^2. \tag{9.136}$$

It is very difficult to evaluate this infinite summation. However, considering the Fourier transform of $h[n]$, as:

$$h[n] = \frac{\sin \omega_c n}{\pi n} \longleftrightarrow H(e^{j\omega}) = \begin{cases} 1 & \text{for } |\omega| < \omega_c. \\ 0 & \text{otherwise.} \end{cases} \tag{9.137}$$

and using the Parseval's equality, we can write,

$$E = \sum_{n=-\infty}^{\infty} |\frac{\sin \omega_c n}{\pi n}|^2 = \frac{1}{2\pi} \int_{-\omega_c}^{\omega_c} |1|^2 d\omega = \omega_c / \pi. \tag{9.138}$$

Note that this is the energy of an ideal low-pass filter. Thus, the energy of the ideal low pass filter is proportional to its cutoff frequency.

10) **Duality:** Although the analytical forms of the time and frequency domain functions are mostly different, there are elegant dualities between the time and frequency domain representations of physical phenomenon.

Duality properties between the time and frequency domain representations of discrete time signals and systems are very similar to that of the continuous time counterpart. In the following, we study three striking dualities of discrete time Fourier transform:

- **Duality Between Time and Frequency Shifts:** A shift in the time domain corresponds to multiplication in the frequency domain. Similarly, multiplication in the time domain corresponds to a shift in frequency domain.
  Time shift: $x[n - n_0] \longleftrightarrow$ **Multiplication:** $e^{-j\omega n_0} X(e^{j\omega})$
  Multiplication: $e^{j\omega_0 n} x[n] \longleftrightarrow$ **Frequency shift:** $X(e^{j(\omega - \omega_0)})$.
  In summary, whenever we need a shift in one of the domains, the corresponding function in the other domain is just multiplied by a complex exponential function.
- **Duality Between the Convolution and Multiplication Operations:** As in the continuous case, convolution in the time domain corresponds to multiplication in the frequency domain and vice versa. However, the convolution operation is replaced by the circular convolution for the discrete time Fourier transform:

$$\begin{aligned} x[n] * h[n] &\longleftrightarrow X(e^{j\omega})H(e^{j\omega}), \\ x[n]h[n] &\longleftrightarrow \frac{1}{2\pi} X(e^{j\omega}) \otimes H(e^{j\omega}). \end{aligned} \tag{9.139}$$

- **Duality Between Discrete Time Fourier Transform and Continuous Time Fourier Series:** There is an interesting duality between the **discrete time Fourier transform** and **continuous time Fourier series**, as explained in the following text.

  Suppose that we are given a continuous time periodic signal, $x(t)$ with its Fourier series representation and a discrete time aperiodic signal $x[n]$, with its Fourier transform representation;

$$x(t) \longleftrightarrow a_k \quad \text{and} \quad x[n] \longleftrightarrow X(e^{j\omega}). \tag{9.140}$$

  A comparison of the continuous time Fourier series synthesis equation and the discrete time Fourier transform analysis equation shows that they have the same analytical forms.

The continuous time periodic signal $x(t)$ and discrete time Fourier transform, $X(e^{j\omega})$ of the aperiodic signal $x[-n]$ are the same, for $k\omega_0 = \omega$. Both functions are periodic and continuous. However, $x(t)$ is in the time domain, whereas $X(e^{j\omega})$ is in the frequency domain.

Similarly, the spectral coefficients, a_k, of the continuous time function $x(t)$ correspond to the discrete time aperiodic signal $x[n]$. This duality states that if we are given a continuous time periodic signal, we can find a dual discrete time signal $x[-n]$, which corresponds to the spectral coefficients of the continuous time function $x(t)$. Surprisingly, the Fourier transform, $X(e^{j\omega})$, of the aperiodic signal $x[n]$ corresponds to the continuous time periodic signal $x(t)$. This fact is depicted in Figure 9.15.

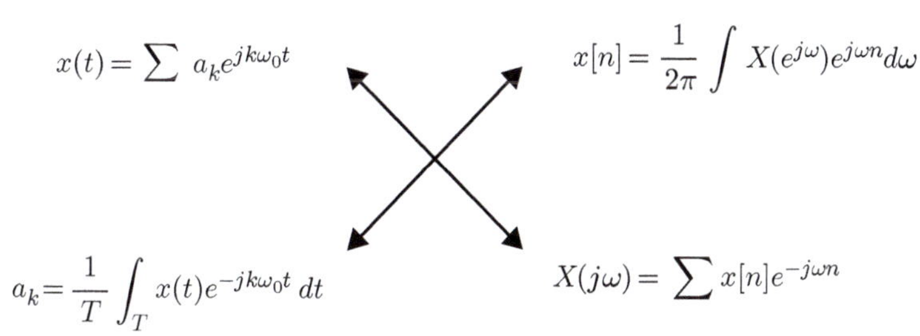

Figure 9.15 If we replace ω by $k\omega_0$ in the discrete time Fourier transform of $x[n]$, the reversed signal, $x[-n]$ becomes the spectral coefficients of the continuous time signal $x(t)$.

Since the Fourier transform $X(e^{j\omega})$ of a discrete time signal, $x[n]$, is a periodic and continuous function, we can expand it into Fourier series as follows:

$$X(e^{j\omega}) = \sum_{n=-\infty}^{\infty} c_n e^{j\omega n}. \tag{9.141}$$

The corresponding Fourier transform is

$$X(e^{j\omega}) = \sum_{n=-\infty}^{\infty} x[n] e^{-j\omega n}. \tag{9.142}$$

Equating the right-hand sides of the Fourier series and Fourier transform of Equations (9.141) and (9.142), we get a relationship between the Fourier series coefficients, $\{c_n\}$ of the discrete time Fourier transform, $X(e^{j\omega})$ and its inverse time domain signal, $x[n]$,

$$\boxed{c_n = x[-n]}. \tag{9.143}$$

Therefore, the Fourier series coefficients of a discrete time Fourier transform is the time domain signal itself, with a reversed time direction.

9.5 Discrete Time Linear Time-Invariant Systems in Frequency Domain

Recall that a discrete time LTI system can be represented by the following constant coefficient difference equation in the time domain:

$$\sum_{k=0}^{N} a_k y[n-k] = \sum_{k=0}^{M} b_k x[n-k]. \tag{9.144}$$

Also, recall that if the eigenfunction of $x[n] = e^{j\omega_0 n}$ is fed as an input to an LTI, then, the corresponding output is

$$y[n] = h[n] * x[n] = H(e^{j\omega_0})e^{j\omega_0 n}, \tag{9.145}$$

where

$$H(e^{j\omega_0}) = \sum_{n=-\infty}^{\infty} h[n]e^{-j\omega_0 n} \tag{9.146}$$

is called the eigenvalue of the system corresponding to the eigenfunction $x[n] = e^{j\omega_0 n}$.

Note that, for $\omega_0 \to \omega$, the eigenvalue, $H(e^{j\omega_0})$, converges to the discrete time Fourier transform of the impulse response. Fourier transform of the impulse response. As in the continuous time case, discrete time Fourier transform of the impulse response uniquely represents the LTI system in the frequency domain, as defined in the following text.

Definition: Frequency response of a discrete time LTI system is defined as the Fourier transform of the impulse response. Mathematically, the frequency response of a discrete time LTI system is

$$H(e^{j\omega}) = \sum_{n=-\infty}^{\infty} h[n]e^{-j\omega n}. \tag{9.147}$$

Therefore, impulse response and frequency response,

$$h(t) \leftrightarrow H(e^{j\omega}), \tag{9.148}$$

are one-to-one and onto representation of the same LTI system in two different domains, namely, in time and frequency domains. While the time-dependent properties of the LTI system are investigated by its impulse response or by the corresponding difference equation, the frequency-shaping properties of the LTI system are investigated by analyzing the frequency response.

Note that the eigenvalues $H(e^{jk\omega_0})$ of a discrete time LTI system for each harmonic frequency $k\omega_0$ for all integer values of k are specific values of the frequency response $H(e^{j\omega})$. In other words, the eigenvalues are the values of the frequency response at $\omega = k\omega_0, \forall k$.

The frequency response of a discrete time LTI system is represented by the following polar coordinate form:

$$H(e^{j\omega}) = |H(e^{j\omega})|e^{j\measuredangle H(e^{j\omega})}, \tag{9.149}$$

where the real-valued functions $|H(e^{j\omega})|$ and $\measuredangle H(e^{j\omega})$ are called the magnitude and phase spectrum respectively. Analysis of Fourier transform of a function requires the analysis of magnitude and phase spectrum.

Generally speaking, the magnitude and phase spectrum of the frequency response $H(e^{j\omega})$ indicate the frequency content of the impulse response function, $h[n]$.

Let us now take the discrete time Fourier transform of both sides of the nth order difference equation given in Equation (9.144). We obtain the following equation, which represents a discrete time LTI system in the frequency domain:

$$\sum_{k=0}^{N} a_k e^{-j\omega k} Y(e^{j\omega}) = \sum_{k=0}^{M} b_k e^{-j\omega k} X(e^{j\omega}). \tag{9.150}$$

Thus, an LTI system represented by a difference equation in the time domain is equivalently represented by an algebraic equation, in the frequency domain.

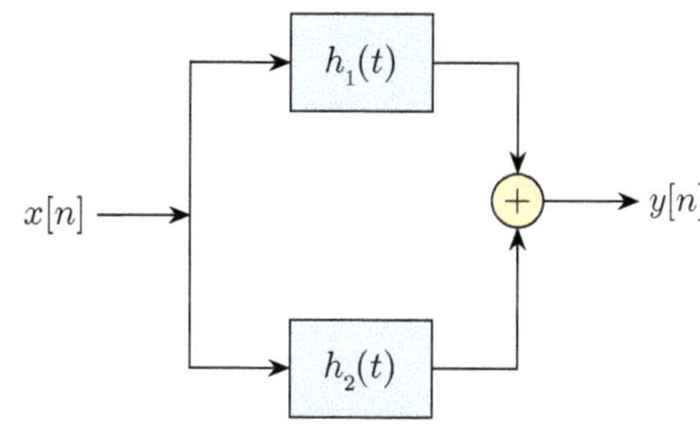

Figure 9.16 A discrete time linear time-invariant system with two parallel impulse responses, $h_1[n]$ and $h_2[n]$, joined by an adder.

Let us find the frequency response of an LTI system by using the algebraic equation given in Equation (9.150). Recall, that the Fourier transform of the impulse function is

$$x[n] = \delta[n] \longleftrightarrow X(e^{j\omega}) = 1. \quad (9.151)$$

When the input is a discrete time impulse function in the time domain, the output becomes impulse response and the Fourier transform of the output becomes frequency response. Therefore, replacing the input by $X(e^{j\omega}) = 1$, the output becomes the frequency response:

$$\sum_{k=0}^{N} a_k e^{-j\omega k} H(e^{j\omega}) = \sum_{k=0}^{M} b_k e^{-j\omega k}. \quad (9.152)$$

This equation provides us the **frequency response** of an LTI system, represented by an ordinary constant coefficient difference equation in time domain and an algebraic equation in the frequency domain. Arranging this equation, we obtain the frequency response, as follows:

$$H(e^{j\omega}) = \frac{\sum_{k=0}^{M} b_k e^{-j\omega k}}{\sum_{k=0}^{N} a_k e^{-j\omega k}}. \quad (9.153)$$

The right-hand side of Equation (9.149) is equal to the ratio between the Fourier transforms of the input and output:

$$\boxed{H(e^{j\omega}) = \frac{Y(e^{j\omega})}{X(e^{j\omega})} = \frac{\sum_{k=0}^{M} b_k e^{-j\omega k}}{\sum_{k=0}^{N} a_k e^{-j\omega k}}}. \quad (9.154)$$

Taking the inverse Fourier transform of the frequency response directly gives us the impulse response, without solving the differential equation. Because,

$$h[n] \longleftrightarrow H(e^{j\omega}). \quad (9.155)$$

In the following examples, let us study the representation of a discrete time LTI system in time and frequency domains.

Exercise 9.16 Consider a discrete time LTI system represented by the block diagram of Figure 9.16.

Given that $h_1[n] = \left(\frac{1}{3}\right)^n u[n]$, and the frequency response of the overall system is $H(e^{j\omega}) = \frac{5e^{-j\omega} - 12}{e^{-2j\omega} - 7e^{-j\omega} + 12}$,

a) Find $H_2(e^{j\omega})$ and $h_2[n]$.
b) Find the difference equation, which represents this system.

Solution

a) The overall impulse response of this system is

$$h[n] = h_1[n] + h_2[n].$$

Taking the Fourier transform of both sides of this equation, we obtain:

$$H(e^{j\omega}) = \frac{5e^{-j\omega} - 12}{e^{-2j\omega} - 7e^{-j\omega} + 12} = H_1(e^{j\omega}) + H_2(e^{j\omega}). \tag{9.156}$$

Fourier transform of $h_1[n]$ is

$$\begin{aligned} H_1(e^{j\omega}) &= \sum_{n=0}^{\infty}\left(\frac{1}{3}\right)^n e^{-j\omega n} = \sum_{n=0}^{\infty}\left(\frac{1}{3}e^{-j\omega}\right)^n \\ &= \frac{1}{1-\frac{1}{3}e^{-j\omega}} = \frac{3}{3-e^{-j\omega}}. \end{aligned} \tag{9.157}$$

Insert $H_1(e^{j\omega})$ into the overall frequency response equation to find the Fourier transform of $h_2[n]$ as:

$$\begin{aligned} H_2(e^{j\omega}) = H(e^{j\omega}) - H_1(e^{j\omega}) &= \frac{5e^{-j\omega} - 12}{e^{-2j\omega} - 7e^{-j\omega} + 12} - \frac{3}{3-e^{-j\omega}} \\ &= \frac{2e^{-2j\omega} + 24e^{-j\omega} - 72}{e^{-3j\omega} - 4e^{-2j\omega} - 9e^{-j\omega} + 36} \\ &= -2\frac{1}{1-\frac{1}{4}e^{-j\omega}}. \end{aligned} \tag{9.158}$$

Take the inverse Fourier transform of $H_2(e^{j\omega})$ to obtain,

$$h_2[n] = -2\left(\frac{1}{4}\right)^n u[n]. \tag{9.159}$$

b) The difference equation of this system can be obtained by taking the inverse Fourier transform of the frequency response,

$$H(e^{j\omega}) = \frac{5e^{-j\omega} - 12}{e^{-2j\omega} - 7e^{-j\omega} + 12} = \frac{Y(e^{j\omega})}{X(e^{j\omega})}, \tag{9.160}$$

which yields,

$$y[n-2] - 7y[n-1] + 12y[n] = 5x[n-1] - 12x[n]. \tag{9.161}$$

9.6 Representation of Discrete Time LTI Systems

Until now we used the word "**representation**" frequently to formally describe a physical phenomenon.

Q: What does representation mean?

Representation is a general concept in mathematics. In system theory, representation means expressing or describing a system by some mathematical objects, such as **equations, relations, functions, graphs, trees, matrices, vectors, groups, sets,** and **manifolds**. The representation of a system is not unique and depends on the design goal(s) of the system.

So far, we have seen a variety of representations for LTI systems, as summarized as follows:

A Discrete time LTI system can be represented by:

1) Impulse response, $h[n]$

2) Unit step response, $s[n]$
3) Frequency response, $H(e^{j\omega})$
4) Difference equation
5) Block diagram

These representations are all related and one-to-one, except for the block diagram representation. Since the realization of an LTI system in a physical environment requires a set of hardware components together with some driving software, it is possible to implement it in a variety of design forms. In the following, we summarize the relationships among the representations of an LTI system.

9.6.1 Impulse Response

Recall the definition of impulse response, which is the response of an LTI system, when the input signal is a unit impulse function:

$$x[n] = \delta[n] \rightarrow y[n] = \delta[n] * h[n] = h[n]. \tag{9.162}$$

Exercise 9.17 Given the following impulse response of a discrete time LTI system,

$$h[n] = K_0\delta[n] + K_1\delta[n-1], \tag{9.163}$$

a) Find the frequency response.
b) Find the difference equation, which represents this system.
c) Find the unit step response.

Solution

a) Frequency response is just the Fourier transform of the impulse response,

$$H(e^{j\omega}) = K_0 + K_1 e^{-j\omega}. \tag{9.164}$$

b) Recall that

$$H(e^{j\omega}) = \frac{Y(e^{j\omega})}{X(e^{j\omega})}. \tag{9.165}$$

Hence,

$$Y(e^{j\omega}) = X(e^{j\omega})[K_0 + K_1 e^{-j\omega}]. \tag{9.166}$$

Taking the inverse Fourier transform of both sides of Equation (9.166), we obtain,

$$y[n] = K_0 x[n] + K_1 x[n-1]. \tag{9.167}$$

c) Unit step response of this LTI system is

$$s[n] = u[n] * h[n] = u[n] * [K_0\delta[n] + K_1\delta[n-1]]. \tag{9.168}$$

Hence,

$$s[n] = K_0 u[n] + K_1 u[n-1]. \tag{9.169}$$

9.6.2 Unit Step Response

Recall the definition of unit step response, which is the response of an LTI system when the input signal is a unit step function:

$$x[n] = u[n] \rightarrow y[n] = s[n] = u[n] * h[n]. \tag{9.170}$$

The relationship between the impulse response and unit step response is

$$h[n] = s[n] - s[n-1]. \tag{9.171}$$

Exercise 9.18 Consider a discrete time LTI system represented by the following unit step response:

$$s[n] = \sum_{k=0}^{n} \left(\frac{1}{2}\right)^k. \tag{9.172}$$

a) Find the impulse response.
b) Find the frequency response.
c) Find the difference equation, which represents this system.

Solution

a)

$$h[n] = s[n] - s[n-1] = \sum_{k=0}^{n} \left(\frac{1}{2}\right)^k - \sum_{k=0}^{n-1} \left(\frac{1}{2}\right)^k = \left(\frac{1}{2}\right)^n u[n]. \tag{9.173}$$

b) Frequency response is

$$H(e^{j\omega}) = \mathcal{F}\{h[n]\} = \frac{1}{1 - \frac{1}{2}e^{-j\omega}}. \tag{9.174}$$

c) Recall that

$$H(e^{j\omega}) = \frac{Y(e^{j\omega})}{X(e^{j\omega})}. \tag{9.175}$$

Hence,

$$Y(e^{j\omega})[1 - \frac{1}{2}e^{-j\omega}] = X(e^{j\omega}). \tag{9.176}$$

Taking the inverse Fourier transform of both sides of Equation (9.176), we find the following difference equation:

$$y[n] - \frac{1}{2}y[n-1] = x[n]. \tag{9.177}$$

9.6.3 Frequency Response

Recall the definition of the frequency response, which is the Fourier transform of the impulse response:

$$H(e^{j\omega}) = \mathcal{F}\{h[n]\}$$
$$y[n] = h[n] * x[n] \leftrightarrow Y(e^{j\omega}) = X(e^{j\omega})H(e^{j\omega}) \tag{9.178}$$
$$H(e^{j\omega}) = \frac{Y(e^{j\omega})}{X(e^{j\omega})}.$$

Exercise 9.19 Consider the following frequency response of a discrete time LTI system:

$$H(e^{j\omega}) = e^{-j\omega n_0}. \tag{9.179}$$

a) Find the impulse response of this system.

b) Find the system equation, in time domain.
c) Find the unit step response.

Solution

a) Frequency response of this system can be written as,

$$H(e^{j\omega}) = e^{-j\omega n_0} = \sum_{n=-\infty}^{\infty} \delta[n - n_0]e^{-j\omega n}. \tag{9.180}$$

The frequency response of Equation (9.180) is the Fourier transform of the shifted impulse function, $\delta[n - n_0]$. Thus, the impulse response is

$$h[n] = \delta[n - n_0]. \tag{9.181}$$

b) Using the convolution equation, we get

$$y[n] = x[n] * h[n] = x[n] * \delta[n - n_0]. \tag{9.182}$$

Thus, the system equation is

$$y[n] = x[n] * h[n] = x[n - n_0]. \tag{9.183}$$

c) The unit step response of this LTI system is

$$s[n] = u[n] * h[n] = u[n - n_0]. \tag{9.184}$$

9.6.4 Difference Equation

A discrete time LTI system can be uniquely represented by a difference equation:

$$\sum_{k=0}^{N} a_k y[n-k] = \sum_{k=0}^{N} b_k x[n-k]. \tag{9.185}$$

The relationship between the constant coefficients of this difference equation and impulse response can be obtained by simply replacing the input signal with the unit impulse function $x[n] = \delta[n]$, and the output signal by the impulse response, $y[n] = h[n]$, as follows:

$$\sum a_k h[n-k] = \sum b_k \delta[n-k]. \tag{9.186}$$

If we take the Fourier transform of both sides of this difference equation, we obtain the relationship between the frequency response and the constant coefficients of the difference equation, as follows:

$$H(e^{j\omega}) = \frac{\sum a_k e^{-j\omega k}}{\sum b_k e^{-j\omega k}} = \frac{Y(e^{j\omega})}{X(e^{j\omega})}. \tag{9.187}$$

Note: The coefficients a_k and b_k of the difference equation determine the structure of the filter represented by the frequency response, $H(e^{j\omega})$.

Exercise 9.20 Consider the following second-order difference equation, which represents an LTI system:

$$y[n] - (2b\cos\beta)y[n-1] + b^2 y[n-2] = x[n], \tag{9.188}$$

a) Find the frequency response of this system.
b) Find the impulse response for $b = 0.5$ and $\beta = \pi/4$.

Solution

a) Taking the Fourier transform of both sides, we get the frequency response as follows:

$$H(e^{j\omega}) = \frac{Y(e^{j\omega})}{X(e^{j\omega})} = \frac{1}{1-(2b\cos\beta)e^{-j\omega}+b^2e^{-2j\omega}}. \tag{9.189}$$

b) In order to find the impulse response, we need to take the inverse Fourier transform of the frequency response.

Inserting the Euler formula,

$$\cos\beta = \frac{e^{j\beta}+e^{-j\beta}}{2}$$

into Equation (9.189) and arranging it, we can factorize the denominator as follows:

$$H(e^{j\omega}) = \frac{1}{1-b(e^{j\beta}+e^{-j\beta})e^{-j\omega}+b^2e^{-2j\omega}} = \frac{1}{(1-be^{-j(\omega-\beta)})(1-be^{-j(\omega+\beta)})}. \tag{9.190}$$

Inserting the values for $b = 0.5$ and $\beta = \pi/4$ and using partial fraction expansion, we obtain:

$$H(e^{j\omega}) = \frac{A}{1-(0.5e^{j\pi/4})e^{-j\omega}} + \frac{B}{1-(0.5e^{-j\pi/4})e^{-j\omega}}, \tag{9.191}$$

where $A = -j\sqrt{2}e^{j\pi/4}$, and $B = j\sqrt{2}e^{-j\pi/4}$.

Taking the inverse Fourier transform of Equation (9.191), we obtain the impulse response as follows:

$$h[n] = \frac{1}{2}[A(0.5e^{j\pi/4})^n + B(0.5e^{-j\pi/4})^n]u[n]. \tag{9.192}$$

or equivalently,

$$h[n] = \frac{1}{j\sqrt{2}}[(e^{j\pi/4})(0.5e^{j\pi/4})^n - (e^{-j\pi/4})(0.5e^{-j\pi/4})^n]u[n]. \tag{9.193}$$

This impulse response consists of complex exponential terms. However, reorganizing Equation (9.193) and using the Euler formula, we obtain:

$$h[n] = \frac{(0.5)^n}{j\sqrt{2}}\left[e^{\frac{\pi}{4}j(n+1)} - e^{\frac{\pi}{4}j(n+1)}\right]u[n] = \sqrt{2}(0.5)^n \sin\frac{\pi}{4}(n+1)u[n]. \tag{9.194}$$

Note that, in the aforementioned exercise, the coefficients A and B are complex. However, all the complex derivations yield a real discrete time impulse response.

9.6.5 Block Diagram Representation

Block diagram representation of an LTI system enables us to realize the system in a real-life application environment. Depending on the quality and the cost of the individual components in the diagram, we can design a variety of versions of the same LTI system.

A popular way of realization of an nth-order difference equation is the direct form, as shown in Figure 9.17. In this representation $N + M$ delay operators together with M adders are used.

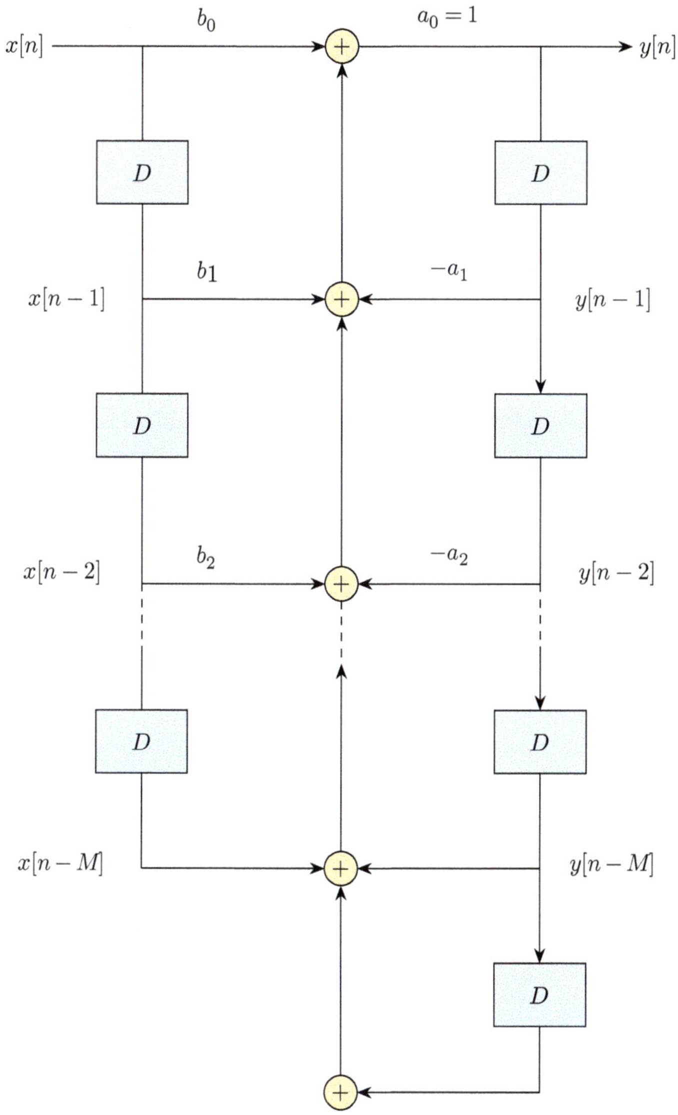

Figure 9.17 A block diagram representation of discrete time LTI systems, with unit delay operators, *D*, and adders.

Exercise 9.21 Let us study different representations of a general first-order LTI system of the following difference equation,

$$y[n] - ay[n-1] = x[n], \quad 0 < a < 1,$$

assuming that the system is initially at rest.

a) Find and plot the frequency response.
b) Find and plot the impulse response.

c) Find and plot the unit step response.
d) Find a block diagram representation.

Solution

a) Frequency response can be obtained directly by taking the Fourier transform of both sides of the difference equation and arranging it as follows:

$$(1 - ae^{-j\omega})Y(e^{j\omega}) = X(e^{j\omega})$$

$$H(e^{j\omega}) = \frac{Y(e^{j\omega})}{X(e^{j\omega})} = \frac{1}{1 - ae^{-j\omega}} = \frac{1 - a\cos\omega}{1 - 2a\cos\omega + \omega^2} + j\frac{a\sin\omega}{1 - 2a\cos\omega + \omega^2} \quad (9.195)$$

The magnitude and the phase spectra of the frequency response is as follows:

$$|H(e^{j\omega})| = \frac{1}{1 - 2a\cos\omega + a^2}, \quad \measuredangle H(e^{j\omega}) = \tan^{-1}\frac{a\sin\omega}{1 - a\cos\omega}. \quad (9.196)$$

Note that the magnitude and phase plots of frequency response, shown in Figure 9.18, is a continuous and periodic function, with period $\omega = 2\pi$.

$$\begin{aligned} \omega = 0 \quad &: H(e^{j\omega}) = \frac{1}{1 - 2a + a^2}, \quad \measuredangle\tan^{-1}0 \\ \omega = \frac{\pi}{2} \quad &: H(e^{j\omega}) = \frac{1}{1 + a^2}, \quad \measuredangle\tan^{-1}a \\ \omega = \pm\pi \quad &: H(e^{j\omega}) = \frac{1}{1 + 2a + a^2}, \quad \measuredangle\tan^{-1}0 \end{aligned} \quad (9.197)$$

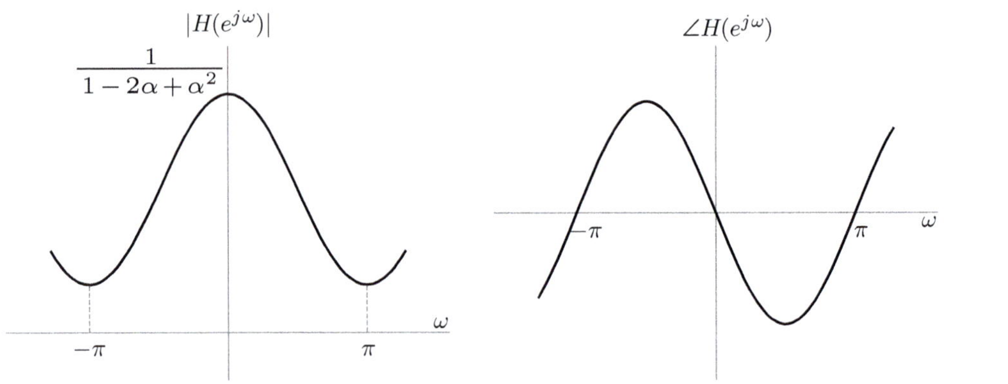

Figure 9.18 Magnitude and phase spectrum of the frequency response of a first order difference equation.

b) Since there is a one-to-one correspondence between the impulse response and frequency response, $h[n] \longleftrightarrow H(e^{j\omega})$, the impulse response of this system can be obtained by taking the inverse Fourier transform of the frequency response, using Table 9.2, $h[n] = a^n u[n]$, as shown in Figure 9.19.

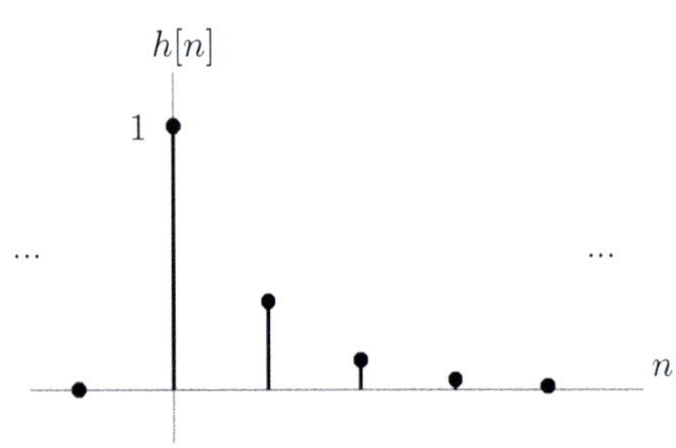

Figure 9.19 Impulse response of a first-order difference equation for $0 < a < 1$.

c) Unit step response of this system is

$$s[n] = h[n] * u[n] = \sum_{k=0}^{\infty} a^{n-k} = a^n \sum_{k=0}^{\infty} a^{-k}$$
$$= \frac{a^n}{1-a^{-1}} u[n] = \frac{a^{n+1}}{a-1} u[n]. \tag{9.198}$$

The plot of the unit step function is given in Figure 9.20.

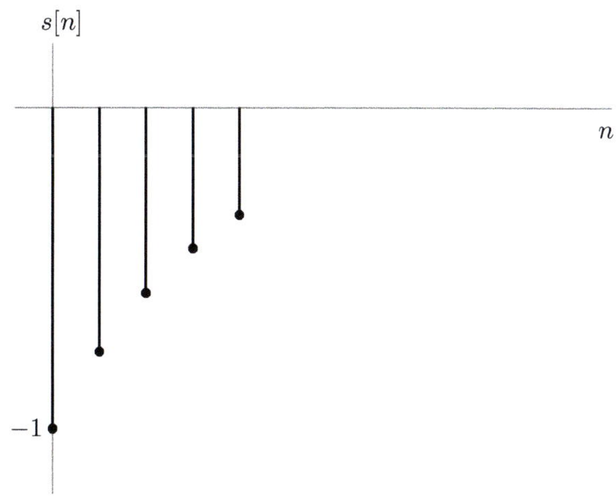

Figure 9.20 Plot of the unit step response for $a = 0.5$.

d) A block diagram representation of this system is given in Figure 9.21.

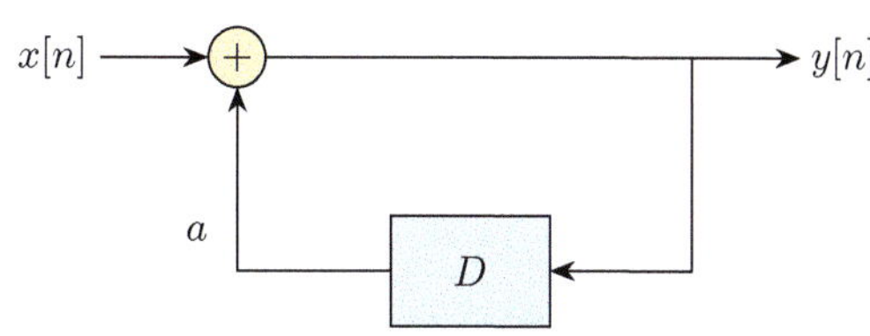

Figure 9.21 Block diagram representation of a feedback control system, represented by a first-order difference equation, using a unit delay operator, *D*, and an adder.

INTERACTIVE

How to reconstruct a 2D image using only sine functions @ https://384book.net/i0902

9.7 z-Transforms as an Extension of Discrete Time Fourier Transforms

Although the Dirichlet conditions are relaxed in the discrete time signals, we still need a sufficient condition for the existence of the discrete time Fourier transform. Recall that the existence of the discrete time Fourier transform is assured, when a time domain function is absolutely summable. If this condition is violated, the discrete time Fourier transform may or may not exist. In this case, it may not be possible to find a finite Fourier transform, in the frequency domain.

Motivating Question: Can we generalize the discrete time Fourier transform in such a way that the transform domain representation of a time domain function exists in some predefined values of the new variable of this domain?

As we did in the continuous time Fourier transform, we define a new domain, called z-domain, where a complex variable,

$$z = re^{j\omega} = \text{Re}\{\{\}z\} + \text{Im}\{\{\}z\},$$

is defined as an alternative to the complex exponential function with magnitude one.

Recall that the discrete time Fourier transform of a function $x[n]$ is defined as the weighted summation of the complex exponential function:

$$\mathcal{F}\{x(t)\} = X(e^{j\omega}) = \sum_{n=-\infty}^{\infty} x[n]e^{-j\omega n}. \tag{9.199}$$

z-transform can be obtained by extending the discrete time Fourier transform. This requires simply replacing the frequency variable $e^{j\omega}$ of discrete time Fourier transform with a complex variable, $z = re^{j\omega}$.

Formally, the z-transform of a function $x[n]$ is defined as:

$$\mathcal{Z}\{x[n]\} = X(z) = \sum_{n=-\infty}^{\infty} x[n](re^{j\omega})^{-n}. \tag{9.200}$$

If we replace $z = re^{j\omega}$ in Equation (9.200), we obtain,

$$\mathcal{Z}\{x[n]\} = \sum_{n=-\infty}^{\infty} x[n](re^{j\omega})^{-n} = \sum_{n=-\infty}^{\infty} x[n]z^{-n}, \tag{9.201}$$

which yields a relationship between z-transform and discrete time Fourier transforms as follows:

$$\mathcal{Z}\{x[n]\} = \mathcal{F}\{x[n]r^{-n}\}. \tag{9.202}$$

Note that z-transform reduces the discrete time Fourier transform for $r = 1$. Therefore, z-transform is considered as an extension of the discrete time Fourier transform, where the complex exponential variable z has a magnitude $r \in R$. In the two-dimensional complex plane, z represents a variable, with a trajectory of a circle with radius r. Comparing the definitions of the discrete time Fourier transform and that of z-transform, we observe that discrete time Fourier transform, $X(e^{j\omega})$ is only defined over the unit circle of radius $r = 1$. In other words, while the discrete time Fourier transform maps a time domain function into the unit circle with radius $r = 1$, in the complex plane, z-transform maps the discrete time function into uncountably many circles with radius $r \in R$, in two-dimensional complex plane.

Theorem: A discrete time function can be uniquely obtained from its z-transform by the following contour integration in the complex plane:

$$x[n] = \frac{1}{2\pi j}\oint_C X(z)z^{n-1}dz, \tag{9.203}$$

provided that the function, $x[n]$, is absolutely summable,

$$\sum_{n=-\infty}^{\infty} |x[n]| < \infty, \tag{9.204}$$

for which the function $X(z)$ exists in the region C.

Approximate Proof: Recall that the relationship between the z-transform and discrete time Fourier transform is given by

$$X(z) = X(re^{j\omega}) = \mathcal{F}\{x[n]r^{-n}\}. \tag{9.205}$$

Taking the inverse discrete time Fourier transform of both sides of Equation (9.205), we obtain,

$$x[n]r^{-n} = \frac{1}{2\pi}\int_{2\pi} X(re^{j\omega})e^{-j\omega n}d\omega. \tag{9.206}$$

Leaving $x[n]$ alone in the left-hand side of the equation, we obtain,

$$x[n] = \frac{1}{2\pi}\int_{2\pi} X(re^{j\omega})r^{n}e^{-j\omega n}d\omega. \tag{9.207}$$

Now, let us change the dummy variable of integral by defining $z = re^{j\omega}$ and assuming that r is fixed, we obtain,

$$dz = rje^{j\omega}d\omega.$$

The change of dummy variable defines a region of convergence (ROC) for the existence of Equation (9.207) in the shape of a ring around the origin as r varies. Then, the integral becomes the contour integral in the ROC of the complex plain,

$$x[n] = \frac{1}{2\pi j}\oint_{C} X(z)z^{-1}z^{n}dz, \tag{9.208}$$

where C is a counter-clock-wise closed path, lying entirely in the ROC. Hence, the inverse z-transform is

$$x[n] = \frac{1}{2\pi j}\oint_{C} X(z)z^{n-1}dz. \tag{9.209}$$

Note that finding the inverse z-transform, using Equation (9.209) requires sophisticated methods for contour integration [see Complex Analysis: A Modern First Course in Function Theory, Jerry R. Muir Jr., Wiley, ISBN: 978-1-118-70522-3, April 2015]. In the context of this book, we suffice to use look-up tables and the properties of z-transforms for finding the inverse of z-transform.

z-transform has several advantages over the discrete time Fourier transform. It is very handy to solve the difference equations. It is applicable to the functions, where the discrete time Fourier transform does not exist. It is a very powerful tool to analyze the stability of linear or nonlinear discrete time systems. It has a wide range of applications in developing digital systems and storing, transmitting, and processing digital signals.

9.7.1 One-Sided z-Transform

Discrete time Fourier transform requires the time domain function to be defined in the interval of $n \in (-\infty, \infty)$. However, in most real time systems, there are no negative values of time. As in the Laplace transform of continuous time functions, it is possible to define one-sided z-transform, where the discrete time functions do not require negative integer numbers for the time variables. Hence, in order to avoid negative times, we restrict the z-transform summation for $0 \le n \le \infty$, as follows:

$$X(z) = \sum_{n=0}^{\infty} x[n]z^{-n}, \tag{9.210}$$

where $z = re^{j\omega}$ is a complex variable.

The time domain function can be uniquely obtained from the one sided Laplace transform by the following equation:

$$x[n] = \frac{1}{2\pi j} \oint_C X(z) z^{n-1} dz, \tag{9.211}$$

where C is the counterclockwise closed path which lie in the region of convergence, specific to the one-sided signal $x[n]$.

9.7.2 Region of Convergence in z-Transforms

The magnitude $r \in R$ of the complex variable $z = re^{j\omega}$ enables us to evaluate the z-transform for each specific value of r. In the z-domain, where the variable z represents a circular trajectory with radius r, it is possible to find a ring of the complex plane around the origin for some values of $r_0 < r < r_1$, such that the z-transform summation converges to a finite value. In some cases, the lowest value of $r_0 = 0$, then, the ROC is inside of the circle with radius $r = r_1$. Or for some functions $r_1 \to \infty$, then, the ring is the region, which covers outside of the circle with radius $r = r_0$.

The aforementioned capability of z-transform creates a great advantage over the discrete time Fourier transform when the function, $x[n]$ is not absolutely summable, but it is absolutely summable for some values of r. Thus, z-transform, relaxes the absolute summability condition of the discrete time Fourier transform, leaving us a ring-like region of the complex plane, where the z-transform exits. The region where the existence of the z-transform is assured is called the ROC.

Definition: Region of Convergence (ROC): The ROC is defined as the set of points in the complex plane, where the z-transform $X(z)$ of the function $x[n]$ exits for some values of $r = |z|$.

ROCs of the z-transform are in the form of a ring centered about the origin, in the complex plane. The radius and width of the ring depend on the type of the time domain function, $x[n]$.

There are four major forms of the ring for the ROC of z-transform:

1) If the function $x[n]$ has finite duration, in other words,

$$x[n] \begin{cases} \neq 0 & \text{for} \quad n_0 < n < n_1, \\ = 0 & \text{otherwise.} \end{cases} \tag{9.212}$$

for some finite values of $n_0 < n_1$, then, ROC covers the entire z-plane, except $z = 0$, as shown in Figure 9.22. Since it also covers the circle with radius $r = 1$, the discrete time Fourier transform of the function also exists.

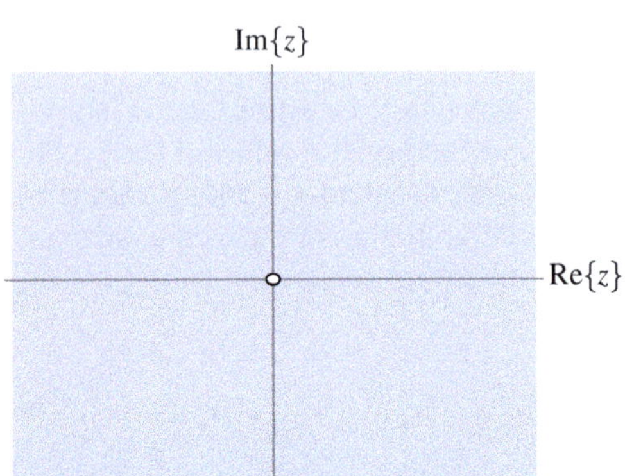

Figure 9.22 ROC for the z-transform of finite duration signals.

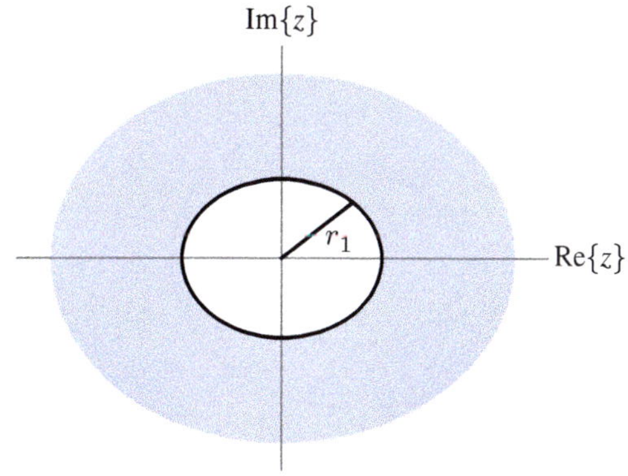

Figure 9.23 ROC for the z-transform of right-sided signals.

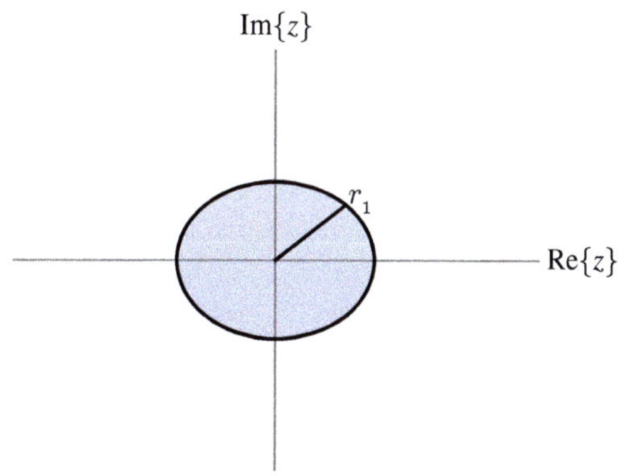

Figure 9.24 ROC for the z-transform of left-sided signals.

2) If the function $x[n]$ is right-sided, in other words, there exists a finite n_0, such that

$$x[n] = 0 \quad \text{for} \quad n \leq n_0,$$

then, there exits an $r = r_0$, such that ROC is outside of the circle, $r > r_0$, as shown in Figure 9.23.

3) If the function $x[n]$ is left-sided, in other words, if there exists a finite n_0, such that

$$x[n] = 0 \quad \text{for} \quad n \geq n_0,$$

then, the ROC is inside of the circle, $0 < r < r_1$, as shown in Figure 9.24.

4) If the function $x[n]$ is two-sided, in other words, there exists two finite values, n_0 and n_1, such that

$$x[n] \begin{cases} \neq 0 & \text{for} \quad n < n_0 \quad \text{and} \quad n > n_1, \\ = 0 & \text{otherwise.} \end{cases} \tag{9.213}$$

then, the ROC is in the shape of the ring about the origin, $r_0 < r < r_1$ as shown in Figure 9.25.

In order to observe the capabilities of z-transform over the discrete time Fourier transform, let us solve the following exercises and investigate the existences of both discrete time Fourier transform and z-transforms.

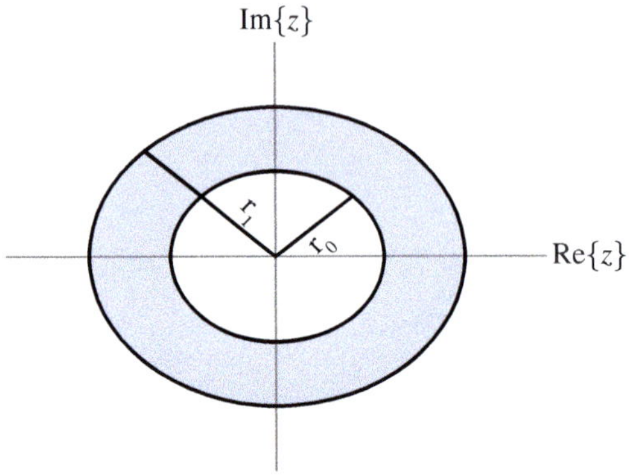

Figure 9.25 ROC for the z-transform of two-sided signals.

Exercise 9.22 Consider the following discrete time right-sided signal:

$$x[n] = a^n u[n]. \tag{9.214}$$

a) Find the discrete time Fourier transforms of this signal.
b) Find the values of a, which assures the existence of the discrete time Fourier transform.
c) Find the z-transform of this signal and its ROC, which assures the existence of the Laplace transform.
d) Compare the range of a, which assures the existence of discrete time Fourier and z-transforms.

Solution

a) Discrete time Fourier transform of the signal, $x[n]$ is defined as:

$$X(e^{j\omega}) = \sum_{n=-\infty}^{\infty} x[n]e^{-j\omega n} = \sum_{n=0}^{\infty} a^n e^{-j\omega n}. \tag{9.215}$$

b) The aforementioned summation diverges for $a \geq 1$. Thus, it only exists for $a < 1$ with the following equation:

$$X(e^{j\omega}) = \frac{1}{1 - ae^{-j\omega}}, \quad \text{for} \quad |a| < 1. \tag{9.216}$$

Hence, the discrete time Fourier transform does not exist for $|a| \geq 1$.

c) The z-transform of the signal $x[n]$ is

$$X(z) = \sum_{n=-\infty}^{\infty} x[n]z^{-n} = \sum_{n=0}^{\infty} a^n z^{-n} = \frac{1}{1 - az^{-1}}. \tag{9.217}$$

The aforementioned summation converges to a finite value if it is absolutely summable,

$$X(z) = \sum_{n=0}^{\infty} |az^{-1}|^n < \infty. \tag{9.218}$$

This is only possible if $|az^{-1}| < 1$, which implies that $|z| > a$. Hence the z-transform exits for the ROC is $r > a$, as shown in Figure 9.26.

d) Comparison of the convergence properties of discrete time Fourier and z-transform reveals that discrete time Fourier transform exists, for only $a < 1$. On the other hand, z-transform exits for the ROC is $r > a$. Thus, the ROC depends on the value of a. For example, given a discrete time function,

$$x[n] = 2^n u[n],$$

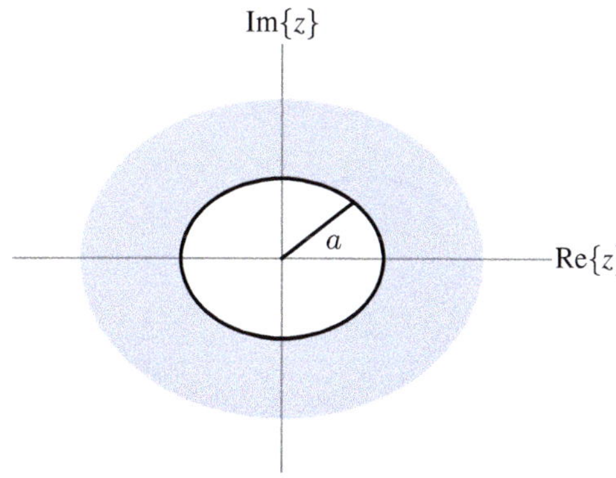

Figure 9.26 ROC for the z-transform of $x[n] = a^n u[n]$.

the base $a = 2$. Then, the discrete time Fourier transform does not exist. However, z-transform exits for $r > 2$, as shown in Figure 9.26.

Exercise 9.23 Consider a slightly different version of the discrete time signal of the previous example, which is a left-sided function,

$$x[n] = a^n u[-n]. \tag{9.219}$$

a) Find the discrete time Fourier transforms of this signal.
b) Find the values of a, which assures the existence of the discrete time Fourier transform.
c) Find the z-transform and the ROC of this signal, which assures the existence of the z-transform.
d) Compare the discrete time Fourier and z-transforms of this signal.

Solution

a) Discrete time Fourier transform of the signal, $x[n]$ is defined as,

$$X(e^{j\omega}) = \sum_{n=-\infty}^{\infty} x[n]e^{-j\omega n} = \sum_{n=-\infty}^{0} a^n e^{-j\omega n} = \sum_{n=0}^{\infty} a^{-n} e^{j\omega n}. \tag{9.220}$$

b) The aforementioned summation diverges for $a \leq 1$. Thus, it only exists for $a > 1$ with the following equation:

$$X(e^{j\omega}) = \frac{1}{1 - a^{-1}e^{j\omega}}, \quad \text{for} \quad |a| > 1. \tag{9.221}$$

Hence, the discrete time Fourier transform does not exit for $|a| \leq 1$.

c) The z-transform of the signal $x[n]$ is

$$X(z) = \sum_{n=-\infty}^{\infty} x[n]z^{-n} = \sum_{n=-\infty}^{0} a^n z^{-n} = \frac{a}{a - z}. \tag{9.222}$$

The aforementioned summation converges to a finite value if it is absolutely summable,

$$X(z) = \sum_{n=-\infty}^{0} |az^{-1}|^n = \sum_{n=0}^{\infty} |a^{-1}z|^n < \infty. \tag{9.223}$$

This is only possible if $|a^{-1}z| > 1$, which implies that $|z| < |a|$. Hence the z-transform exits for the ROC, $r < |a|$ as shown in Figure 9.27.

d) Comparison of the convergence properties of discrete time Fourier and z-transform reveals that discrete time Fourier transform exists, for only $a > 1$. On the other hand, z-transform exits for the ROC is $r < |a|$. The ROC depends on the value of a.

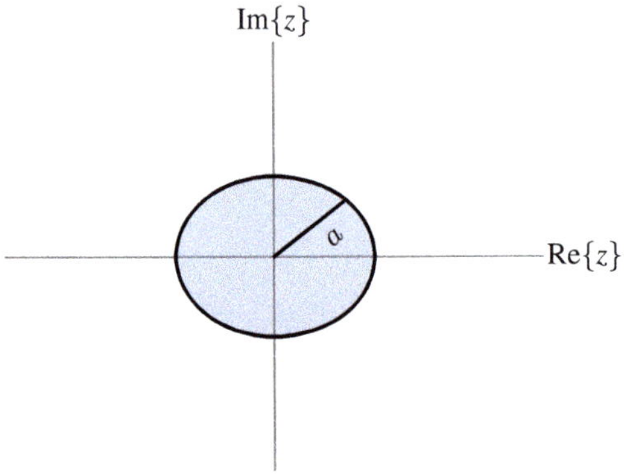

Figure 9.27 ROC for the z-transform of $x[n] = a^n u[-n]$.

Exercise 9.24 Find the z-transform and its ROC for the following right-sided function:

$$x[n] = u[n]. \tag{9.224}$$

Solution

From the definition of z-transform,

$$X(z) = \sum_{n=-\infty}^{\infty} x[n]z^{-n} = \sum_{n=0}^{\infty} z^{-n} = \frac{1}{1-z^{-1}}. \tag{9.225}$$

The ROC is $|z| > 1$.

Exercise 9.25 Find the z-transform and its ROC for the following limited-time duration function:

$$x[n] = a^n(u[n] - u[n - n_0]). \tag{9.226}$$

Solution

From the definition of z-transform;

$$X(z) = \sum_{n=-\infty}^{\infty} x[n]z^{-n} = \sum_{n=0}^{n_0-1} a^n z^{-n} = \frac{1-(az^{-1})^{n_0}}{1-az^{-1}}. \tag{9.227}$$

Since the time duration, $n \in [0, n_0]$ is bounded, the summation of the z-transform is finite for all values of $n_0 < \infty$. Thus, ROC is the entire complex plane. This is the case, when an absolutely summable function $x[n]$ has finite duration.

Exercise 9.26 Find the z-transform and ROC of the following two-sided function:

$$x[n] = a^{-n}u[n] + a^n u[-n], \quad \text{for} \quad a > 0. \tag{9.228}$$

Solution

From the definition of z-transform,

$$X(z) = \sum_{n=-\infty}^{\infty} x[n]z^{-n} = \sum_{n=0}^{\infty} a^{-n}z^{-n} + \sum_{n=-\infty}^{0} a^n z^{-n} = \frac{1}{1-a^{-1}z^{-1}} + \frac{1}{1-az^{-1}}. \tag{9.229}$$

In order to find the ROC of $X(z)$, we need to find the ROC of the first and the second term in the left-hand side of the Equation (9.229)

For the first term,

$$\frac{1}{1-a^{-1}z^{-1}} \quad \text{ROC is} \quad |z| > 1/a.$$

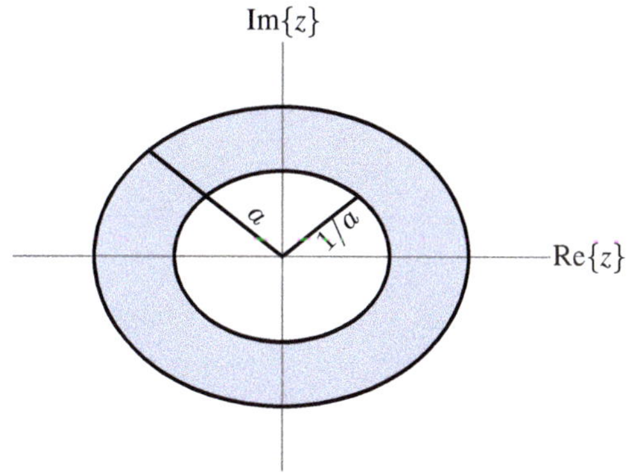

Figure 9.28 ROC for the z-transform of $x[n] = a^{-n}u[n] + a^{n}u[-n]$.

For the second term,

$$\frac{1}{1 - az^{-1}} \quad \text{ROC is} \quad |z| < a.$$

Hence, the ROC is a ring in

$$1/a < r < a,$$

as shown in Figure 9.28.

These examples show that it is critical to determine the ROC of the z-transforms in the complex plane.

9.8 Inverse of z-Transform

As we mentioned earlier, recovering the time domain signal $x[n]$ from its z-transform, $X(z)$ requires the following contour integration:

$$x[n] = \frac{1}{2\pi j} \oint_C X(z) z^{n-1} dz, \tag{9.230}$$

which may not be easy for a large class of functions. In order to avoid contour integration, we frequently use the look-up tables and properties of the z-transform. Since they are quite similar to that of the discrete time Fourier transformation, we suffice to provide the list of properties and look-up tables for common transform pairs, $x[n] \leftrightarrow X(z)$ together with ROCs, in Tables 9.3 and 9.4. The following examples demonstrate how we utilize the tables to compute the inverse z-transform.

Exercise 9.27 Find the inverse z-transform of the following function in the z-domain:

$$X(z) = \frac{0.2z}{(z - 0.5)(z - 0.3)}, \tag{9.231}$$

for three different ROCs given as follows:

a) ROC for $|z| > 0.5$, as in Figure 9.29.
b) ROC for $|z| < 0.3$, as in Figure: 9.30.
c) ROC for $0.3 < |z| < 0.5$, as in Figure 9.31.

Table 9.3 Properties of z-transform.

Signal	z-transform	ROC
$x[n]$	$X(z) = \sum_{n=-\infty}^{\infty} x[n]z^{-n}$	R_x
$y[n]$	$Y(z)$	R_y
$ax[n] + by[n]$	$aX(z) + bY(z)$	Contains $R_x \cap R_y$
$x[n - n_0]$	$z^{-n_0}X(z)$	R_x, except possible addition or deletion of the origin or ∞
$e^{j\omega_0 n}x[n]$	$X\left(e^{-j\omega_0}z\right)$	R_x
$z_0^n x[n]$	$X(z/z_0)$	$\lvert z_0\rvert R_x$
$x[-n]$	$X\left(z^{-1}\right)$	Inverted R (i.e., R^{-1} = the set of points z^{-1}, where z is in R)
$x^*[n]$	$X^*(z^*)$	R_x
$x^*[-n]$	$X^*(1/z^*)$	$1/R_x$
$x[n] * y[n]$	$X(z)Y(z)$	Contains $R_x \cap R_y$
$x[n] - x[n-1]$	$(1 - z^{-1})X(z)$	At least the intersection of R and $\lvert z\rvert > 0$
$\sum_{k=-\infty}^{n} x[k]$	$\frac{1}{1 - z^{-1}}X(z)$	At least the intersection of R and $\lvert z\rvert > 1$
$nx[n]$	$-z\frac{d}{dz}X(z)$	R_x, except possible addition or deletion of the origin or ∞
$\text{Re}\,\{x[n]\}$	$\frac{1}{2}[X(z) + X^*(z^*)]$	Contains R_x
$\text{Im}\,\{x[n]\}$	$\frac{1}{2j}[X(z) - X^*(z^*)]$	Contains R_x

Solution

Firstly, let us apply partial fraction expansion to simplify the z-transform function,

$$X(z) = \frac{0.2z}{(z-0.5)(z-0.3)} = \frac{z}{z-0.5} - \frac{z}{z-0.3} \tag{9.232}$$

$$X(z) = X_1(z) - X_2(z) = \frac{1}{1-0.5z^{-1}} - \frac{1}{1-0.3z^{-1}}.$$

The inverse of the z-transform of $X(e^{j\omega})$ depends on the ROCs defined in parts a), b), and c).

a) In order to obtain the inverse z-transform of the given function $X(z)$ for ROC $|z| > 0.5$, we need to get the ROC of $X_1(z)$ as $|z| > 0.5$ and the ROC of $X_2(z)$ as $|z| > 0.3$, so that the intersection of both ROCs becomes $|z| > 0.5$. Hence, we obtain the inverse z-transform of the first term as,

$$X_1(z) = \frac{1}{1-0.5z^{-1}} \longleftrightarrow x_1[n] = 0.5^n u[n] \quad \text{ROC for} \quad |z| > 0.5. \tag{9.233}$$

Similarly, the inverse z-transformation of the second term is

$$X_2(z) = \frac{1}{1-0.3z^{-1}} \longleftrightarrow x_2[n] = 0.3^n u[n] \quad \text{ROC for} \quad |z| > 0.3. \tag{9.234}$$

Using the linearity property of z-transform, we obtain the inverse z-transform of $X(z)$ as follows:

$$x[n] = x_1[n] - x_2[n] = [0.5^n - 0.3^n]u[n], \tag{9.235}$$

where the ROC is the intersection of $|z| > 0.5$. and $|z| > 0.3$, which is $|z| > 0.5$.

Note: This is a right-sided function.

Table 9.4 z-transform pairs for popular functions.

$x[n]$	$X(z)$	ROC
$\delta[n]$	1	All z
$\delta[n-n_0]$	z^{-n_0}	All z, except $0\ (n_0>0)$ or $\infty (n_0<0)$
$u[n]$	$\dfrac{1}{1-z^{-1}}$	$\lvert z\rvert > 1$
$-u[-n-1]$	$\dfrac{1}{1-z^{-1}}$	$\lvert z\rvert < 1$
$a^n u[n]$	$\dfrac{1}{1-az^{-1}}$	$\lvert z\rvert > a$
$-a^n u[-n-1]$	$\dfrac{1}{1-az^{-1}}$	$\lvert z\rvert < a$
$na^n u[n]$	$\dfrac{az^{-1}}{(1-az^{-1})^2}$	$\lvert z\rvert > a$
$-na^n u[-n-1]$	$\dfrac{az^{-1}}{(1-az^{-1})^2}$	$\lvert z\rvert < a$
$[\cos\omega_0 n]u[n]$	$\dfrac{1-[\cos\omega_0]z^{-1}}{1-[2\cos\omega_0]z^{-1}+z^{-2}}$	$\lvert z\rvert > 1$
$[\sin\omega_0 n]u[n]$	$\dfrac{1-[\sin\omega_0]z^{-1}}{1-[2\cos\omega_0]z^{-1}+z^{-2}}$	$\lvert z\rvert > 1$
$[r^n\cos\omega_0 n]u[n]$	$\dfrac{1-[r\cos\omega_0]z^{-1}}{1-[2r\cos\omega_0]z^{-1}+r^2z^{-2}}$	$\lvert z\rvert > r$
$[r^n\sin\omega_0 n]u[n]$	$\dfrac{1-[r\sin\omega_0]z^{-1}}{1-[2r\cos\omega_0]z^{-1}+r^2z^{-2}}$	$\lvert z\rvert > r$
$\begin{cases} a^n, & 0\le n\le N-1 \\ 0, & \text{otherwise}\end{cases}$	$\dfrac{1-a^Nz^{-N}}{1-az^{-1}}$	$\lvert z\rvert > 0$

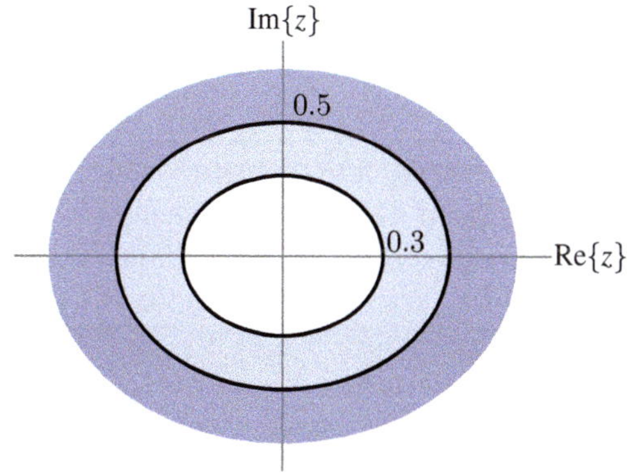

Figure 9.29 ROC for $|z| > 0.5$.

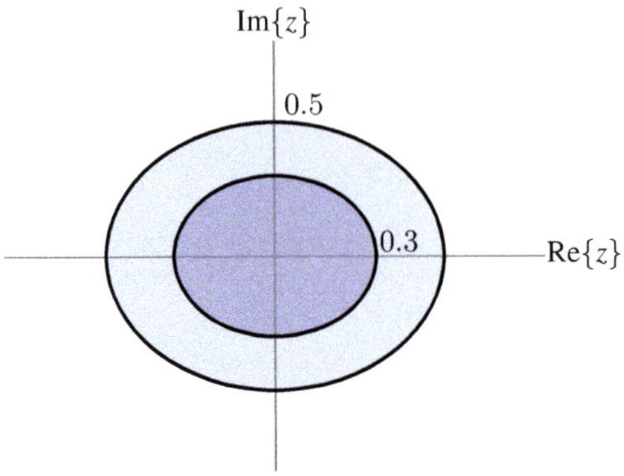

Figure 9.30 ROC for $|z| < 0.3$.

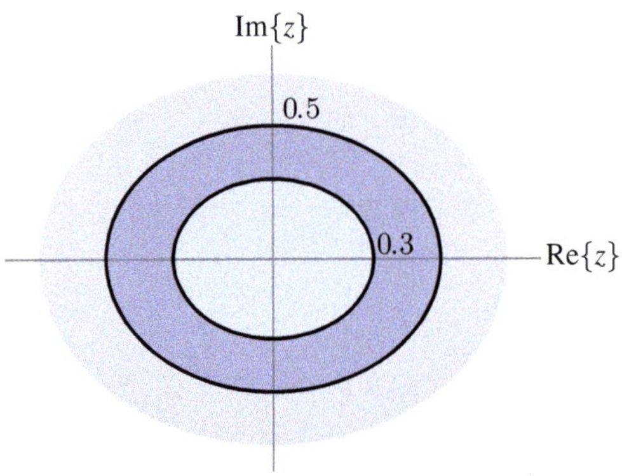

Figure 9.31 ROC for $0.3 < |z| < 0.5$.

b) In order to obtain the inverse z-transform of the given function $X(z)$ for ROC $|z| < 0.3$, we need to get the ROC of $X_1(z)$ as $|z| < 0.5$ and the ROC of $X_2(z)$ as $|z| < 0.3$, so that the intersection of both ROCs becomes $|z| < 0.3$. Hence, the inverse z-transform of the first term is

$$X_1(z) = \frac{1}{1 - 0.5z^{-1}} \longleftrightarrow x_1[n] = 0.5^n u[-n-1] \quad \text{ROC for} \quad |z| > 0.5. \tag{9.236}$$

The inverse z-transformation of the second term is

$$X_2(z) = \frac{1}{1 - 0.3z^{-1}} \longleftrightarrow x_2[n] = 0.3^n u[-n-1] \quad \text{ROC for} \quad |z| > 0.3. \tag{9.237}$$

Finally, the inverse z-transform of $X(z)$ as follows:

$$x[n] = x_1[n] - x_2[n] = [0.5^n - 0.3^n]u[-n-1], \tag{9.238}$$

where the ROC is the intersection of $|z| < 0.5.$ and $|z| < 0.3$, which is $|z| < 0.3$.

Note: This is a left-sided function.

c) In order to obtain the inverse z-transform of $X(z)$ for ROC $0.3 < |z| < 0.5$, we need to get the ROC of $X_1(z)$ as $|z| < 0.5$ and the ROC of $X_2(z)$ as $|z| > 0.3$, so that we obtain a ring-shaped region.

Hence, the inverse z-transform of the first term is

$$X_1(z) = \frac{1}{1 - 0.5z^{-1}} \longleftrightarrow x_1[n] = 0.5^n u[-n-1] \quad \text{ROC for} \quad |z| > 0.5, \tag{9.239}$$

the inverse z-transformation of the second term is

$$X_2(z) = \frac{1}{1 - 0.3z^{-1}} \longleftrightarrow x_2[n] = 0.3^n u[n] \quad \text{ROC for} \quad |z| > 0.3. \tag{9.240}$$

Finally, the inverse z-transform of $X(z)$ is as follows:

$$x[n] = x_1[n] - x_2[n] = [0.5^{-n-1} - 0.3^n]u[n], \tag{9.241}$$

where the ROC is the intersection of $|z| < 0.5.$ and $|z| < 0.3$, which is $|z| < 0.3$.

Note: This is a two-sided function.

Exercise 9.28 Find the inverse z-transform of the following z-domain function:

$$X(z) = \frac{z+2}{z}, \quad \text{ROC for all} \quad z \neq 0. \tag{9.242}$$

Solution

Let us arrange the function as follows:

$$X(z) = 1 + \frac{1}{z} = 1 + z^{-1}. \tag{9.243}$$

From the z-transform table, we can see that the inverse z-transform of the first term is

$$\mathcal{Z}^{-1}[1] = \delta[n], \quad \text{ROC for all} \quad z \tag{9.244}$$

and the inverse z-transform of the second term is

$$\mathcal{Z}^{-1}[\delta[n-1]] \quad \text{ROC for all} \quad z \neq 0 \tag{9.245}$$

Using the linearity property, we obtain the inverse Laplace transform of $X(s)$, as follows:

$$x[n] = \delta[n] - \delta[n-1]. \tag{9.246}$$

The aforementioned exercises show that a practical method for finding the inverse z-transform is to make algebraic manipulations on the z-domain function and put it into the linear combination of the known pairs of transform table. Then, use the linearity property to obtain the inverse transform.

9.9 Discrete Time Linear Time-Invariant Systems in z-Domain

Recall that a discrete time LTI system is represented by difference equation and impulse response, in time domain and it is represented by the frequency response in the frequency domain.

Motivating Question: What if the frequency response does not exit? Can we employ z-transform to analyze the frequency content of an LTI system in some ROCs of the z-plane?

z-Transform, indeed, provides us with a strong tool to analyze the LTI systems, which do not exist in the frequency domain.

9.9.1 Eigenvalues and Transfer Functions in z-Domain

Recall that, when the input of a discrete time LTI system is an exponential function, the output is just the scaled version of the input. Thus, exponential functions are the eigenfunctions of the LTI systems and the scaling factor is simply the eigenvalue, computed from the convolution summation,

$$y[n] = x[n] * h[n] = \sum_{k=-\infty}^{\infty} e^{\lambda(n-k)}h[k] = H(\lambda)e^{\lambda n}, \tag{9.247}$$

where

$$H(\lambda) = \sum_{k=-\infty}^{\infty} h[k]e^{-\lambda k}, \tag{9.248}$$

$$x(t) = e^{\lambda t} \rightarrow \boxed{\text{LTI}} \rightarrow y_P(t) = H(\lambda)e^{\lambda t}. \tag{9.249}$$

In the aforementioned formulation, if we set, $\lambda = j\omega$, then, the eigenfunction at the input becomes $x[n] = e^{j\omega n}$ and the eigenvalue of the LTI system becomes the discrete time Fourier transform of the impulse response, which is called frequency response,

$$H(e^{j\omega}) = \sum_{n=-\infty}^{\infty} h[n]e^{-j\omega n}. \tag{9.250}$$

Now let us extend the discrete time Fourier transform, $H(e^{j\omega})$, to the z-transform of the impulse response, by defining $z = re^{j\omega}$. In this case, the eigenfunction at the input becomes $x[n] = re^{j\omega n}$ and the eigenvalue becomes the z-transform of the impulse response.

Definition: Transfer Function of Discrete time Systems The z-transform of the impulse response is called transfer function,

$$H(z) = \sum_{n=-\infty}^{\infty} h[n]z^{-n}. \tag{9.251}$$

When the frequency response of a discrete time LTI system does not converge, we cannot represent the LTI system with an eigenvalue, in the frequency domain. However, z-transform enables us to find the eigenvalue of the system, which converges in some regions of the complex z-plane.

Let us now establish the relationship between the transfer function and the difference equation of an LTI system,

$$\sum_{k=0}^{N} a_k y[n-k] = \sum_{k=0}^{M} b_k x[n-k]. \tag{9.252}$$

Taking the z-transform of both sides of Equation (9.252), we obtain,

$$\sum_{k=0}^{N} a_k z^{-k} Y(z) = \sum_{k=0}^{M} b_k z^{k} X(z). \tag{9.253}$$

Let us find the transfer function of an LTI system by using the algebraic equation of Equation (9.253). Recall the z-transform of the impulse function is,

$$x[n] = \delta[n] \longleftrightarrow X(z) = 1. \tag{9.254}$$

When the input is a discrete time impulse function in the time domain, the output becomes impulse response and the z-transform of the output becomes transfer function. Therefore, replacing the input by $X(z) = 1$, the output becomes the transfer function,

$$\sum_{k=0}^{N} a_k z^{-k} H(z) = \sum_{k=0}^{M} b_k z^{-k}. \tag{9.255}$$

Equation (9.255) provides us the **transfer function** of an LTI system, represented by an ordinary constant coefficient difference equation in time domain and an algebraic equation in z-domain. Arranging this equation, we obtain the transfer function as follows:

$$\boxed{H(z) = \frac{Y(z)}{X(z)} = \frac{\sum_{k=0}^{M} b_k z^{-k}}{\sum_{k=0}^{N} a_k z^{-k}}}, \tag{9.256}$$

which transfers an input signal to an output signal of an LTI system represented by a difference equation. The type of this transferring process is determined by the constant coefficients, $\{a_k\}$ and $\{b_k\}$ of the differential equation.

Taking the inverse z-transform of the transfer function directly gives us the impulse response, without solving the differential equation. Because

$$h[n] \longleftrightarrow H(z). \tag{9.257}$$

Therefore, impulse response and transfer function are one-to-one and onto representation of the same LTI system in two different domains, namely in time and z-domains.

Note that the eigenvalues $H(e^{jk\omega_0})$ of a discrete time LTI system for each harmonic frequency $k\omega_0$ for all integer values of k are specific instances of the transfer function at $z = e^{jk\omega_0}$. Furthermore, frequency response $H(e^{j\omega})$ is a specific form of the transfer function for $z = e^{j\omega}$.

Transfer function of a discrete time LTI system is represented by the following polar coordinate form,

$$H(z) = |H(z)|e^{j\measuredangle H(z)}, \tag{9.258}$$

where the real-valued functions $|H(z)|$ and $\measuredangle H(z)$ are called the magnitude and phase of the transfer function, respectively. Analysis of z-transform of a discrete time function requires the analysis of magnitude and phase spectrums.

The following exercises demonstrate the utilization of discrete time transfer function for describing various properties of the LTI systems.

Exercise 9.29 Consider a discrete time LTI system represented by the following impulse response:

$$h[n] = [(0.5)^n u[n] + (0.5)^{n-1} u(n-1]. \tag{9.259}$$

a) Find the transfer function of this system.
b) Comment on the ROC.
c) Find the difference equation, which represents this system.

Solution

a) The transfer function is the z-transform of the impulse response. Using the look-up table and linearity property, we obtain

$$H(z) = \frac{1}{1-0.5z^{-1}} + \frac{z^{-1}}{1-0.5z^{-1}}. \tag{9.260}$$

b) Transfer function consists of two subsystems, which are paralleled to each other,

$$H(z) = H_1(z) + H_2(z), \tag{9.261}$$

where

$$\text{the ROC for } H_1(z) : \ |z| < 0.5, \tag{9.262}$$

and

$$\text{the ROC for } H_2(z) : \ |z| < 1. \tag{9.263}$$

The ROC for the transfer function $H(z)$ lies in the intersection of both ROCs, hence it is $|z| < 1$.

c) Recall that the transfer function provides a relationship between the input and output of an LTI system, in the z-domain, as follows:

$$H(z) = \frac{Y(z)}{X(z)} = \frac{1}{1 - 0.5z^{-1}} + \frac{z^{-1}}{1 - 0.5z^{-1}}. \tag{9.264}$$

Arranging Equation (9.264), we get,

$$[1 - 0.5z^{-1}]Y(z) = [1 + z^{-1}]X(z). \tag{9.265}$$

Taking the inverse z-transform of both sides of this equation, we obtain,

$$y[n] - 0.5y[n-1] = x[n] + x[n-1]. \tag{9.266}$$

Exercise 9.30 Consider an LTI system at initial rest, given by the following difference equation:

$$y[n-2] - 4y[n-1] + 4y[n] = 2x[n-1] \tag{9.267}$$

a) Find the transfer function of this system.
b) Find the impulse response of this system.

Solution

a) Let us set the input to impulse function, $x[n] = \delta[n]$, then, the corresponding output of the difference equation becomes the impulse response, $h[n]$. The difference equation of Equation (9.267) for impulse response is

$$h[n-2] - 4h[n-1] + 4y[n] = 2\delta[n-1]. \tag{9.268}$$

From the z-transform properties, we see that

$$x[n - n_0] \leftrightarrow z^{-n_0}X(z).$$

From the z-transform table, we see that

$$x[n] = \delta[n] \leftrightarrow X(z) = 1.$$

Using the transform pairs, we take the z-transform of both sides of the difference equation,

$$[z^{-2} - 4z^{-1} + 4]H(z) = 2z^{-1}.$$

Arranging the aforementioned equation, we get the transfer function, as follows:

$$H(z) = \frac{2z^{-1}}{(z^{-1}+2)^2} = \frac{0.5z^{-1}}{(1 - 0.5z^{-1})^2}.$$

Note that taking the z-transform of the difference equation, given in Equation (9.267) does not provide the ROC for the transfer function. In fact, there are two alternatives for the ROC of this transfer function; The first alternative is $|z| < 0.5$ and the second alternative is $|z| > 0.5$.

b) From the z-transform table, we observe that depending on the selection of ROC, there are two impulse responses, which correspond to the same difference equation: for ROC, $|z| < 0.5$,

$$h[n] = ((0.5)^n)u[-n-1].$$

Since the system is causal, the impulse response, $h[n]$, for $|z| > 0.5$, is eliminated. Hence, the impulse response can be obtained for the second alternative of ROC, $|z| > 0.5$,

$$h[n] = (0.5)^n)u[n],$$

The aforementioned example demonstrates that we can obtain the transfer function and impulse response of an LTI system, which is initially at rest, without solving the differential equation. This method is also available in Fourier domain, provided that the frequency response exits. In case of undefined frequency responses, Laplace domain enables us to compute the transfer function and impulse response, using a simple algebraic method.

As it is observed throughout this chapter, the transform domains capture different view of the physical phenomena other than time domain representations. Furthermore, the beautiful synergy created by the representations of time and transform domains bridges the mathematics of linear algebra and recursive equations.

9.10 Chapter Summary

Can we extend the Fourier series representation of discrete time periodic functions to the aperiodic ones? If yes, how can we do that? What are the conditions for the existence of such transformations? What is the relationship between the Fourier transform and Fourier series representations of discrete time periodic signals? What type of an operator is a discrete time Fourier transform? What are the similarities and distinctions between the continuous time and discrete time Fourier transforms and Fourier series? What are the relationships between the functions represented in the discrete time and continuous frequency domain? What are the properties of the discrete time signals and systems in the frequency domain?

In this chapter, first, we studied the discrete time Fourier transforms by extending the Fourier series representation of periodic functions to aperiodic functions assuming that an aperiodic function can be considered as a periodic function of infinite period. Then, we derived the Fourier analysis and synthesis equations for discrete time functions by stretching the period of a discrete time function to infinity. Interestingly, when we take the limit as the period approaches infinity, the Fourier transform of a discrete time function became a continuous frequency function.

While taking the discrete time Fourier transform, we replace the integral operation of the continuous time Fourier transform with the summation operation. In this case, we do not have to bother with the existence of the complex integrals of the continuous time Fourier transform. This replacement relaxed the Dirichlet conditions for existence of the discrete time Fourier transform, where the only constraint we need is finite summability of the discrete time functions. Furthermore, the fact that the superposition of periodic complex functions of the analysis equation is periodic, makes the discrete time Fourier transformations also periodic with period 2π. We also established the relationship between the discrete time Fourier series and Fourier transforms for periodic functions.

Second, we investigate the power of discrete time Fourier transforms in manipulating the frequency content of discrete time signals. We studied basic properties of discrete time Fourier transforms, such as linearity, time shifting, time scaling, and difference properties. We observe many similarities between the continuous and discrete time Fourier transforms. As in the continuous time Fourier transforms, the energy of a discrete time signals is also preserved in continuous frequency domain. We also studied several duality properties between the time and frequency domain.

We observe that difference equations become algebraic equations in the frequency domain. Thus, solving them in the frequency domain is rather easier compared to solving them in time domain. We also, show that there is one-to one correspondence between the representation of LTI systems by impulse response, frequency response, and difference equations.

WATCH

Noise reduction using the Fourier transform @ https://384book.net/v0902

Problems

9.1 Find and plot the discrete time Fourier transforms of the following signals in polar coordinate system:

a) $x_1[n] = \left(\frac{1}{2}\right)^{n+1} u[n+1]$,

b) $x_2[n] = \left(\frac{1}{2}\right)^{|n+1|}$.

9.2 Find and plot the discrete time Fourier transforms of the following signals in Cartesian coordinate system:

a) $x_1[n] = (0.5)^{-n}u[-n+2]$

b) $x_2[n] = x[n-5]$, and $x[n] = u[n] - u[n-3]$

c) $x_3[n] = \left(\frac{2}{5}\right)^{|n|} u(5n-2)$.

9.3 Find and plot the even parts and odd parts of the discrete time Fourier transforms of the following signals:

a) $x_1[n] = \dfrac{\delta[n-2]}{2} + \dfrac{\delta[n+2]}{2}$

b) $x_2[n] = \dfrac{\delta[n+1]}{2} - \dfrac{\delta[n-1]}{2}$

c) $x_3[n] = \cos\omega_0 n + \cos 2\omega_0 n$.

9.4 Find and plot the discrete time Fourier transforms of the following signals and comment about the frequency content of these signals:

a) $x_1[n] = \sin\left(\frac{\pi}{2}n\right)$

b) $x_2[n] = \cos\left(\frac{\pi}{4}n + \frac{\pi}{2}\right)$

c) $x_3[n] = 2\sin\left(\frac{\pi}{6}n\right) + \pi\cos\left(\frac{\pi}{3}n + \frac{\pi}{4}\right)$

9.5 Consider the following discrete time Fourier transform of a signal $x[n]$, in one full period:

$$X(e^{j\omega}) = \begin{cases} -j & 0 < \omega \le \pi \\ j & -\pi < \omega \le 0. \end{cases}$$

a) Find and plot the inverse Fourier transform $x[n]$.
b) Find and plot the even and odd parts of $x[n]$.
c) Find and plot the even and odd parts of $X(e^{j\omega})$.

9.6 Consider the following discrete time Fourier transform of a signal $x[n]$:

$$X(e^{j\omega}) = \sum_{k=-\infty}^{\infty} \pi\delta(\omega + 2\pi k) - 4\pi\delta\left(\omega + \frac{\pi}{3} + 2\pi k\right) - 4\pi\delta\left(\omega - \frac{\pi}{3} + 2\pi k\right).$$

a) Plot this signal in the frequency domain.
b) Find and plot the inverse Fourier transform $x[n]$.
c) Find and plot the even and odd parts of $x[n]$.

9.7 Consider the following discrete time Fourier transform of a signal $x[n]$:

$$X(e^{j\omega}) = \begin{cases} e^{-0.5j\omega} & 0 \le |\omega| < \frac{\pi}{3} \\ 0 & \frac{\pi}{3} \le |\omega| < \pi. \end{cases}$$

a) Find and plot the magnitude and the phase of this function.
b) Find and plot the real and imaginary part of this function.
c) Find and plot the inverse Fourier transform $x[n]$.

9.8 Consider an LTI system represented by the following impulse response and frequency response pair:

$$h[n] \longleftrightarrow H(e^{j\omega}),$$

where the frequency response $H(e^{j\omega}) \neq 0$ in $0 \le \omega \le \pi$ and it is zero otherwise.
a) Given that $H(e^{j(\omega/3)}) = \pi$, find the frequency response $H(e^{j\omega})$.
b) Find the impulse response of this system.
c) When the input to this system is $x[n] = \left(\frac{1}{\pi}\right)^n u[n]$, find the output $y[n]$.

9.9 Consider an LTI system represented by the following impulse response and frequency response pair:

$$h[n] \longleftrightarrow H(e^{j\omega}),$$

with the following input–output pair;

$$x[n] = \left(\frac{2}{3}\right)^n u[n],$$

$$y[n] = n\left(\frac{2}{3}\right)^{n+1} u[n].$$

a) Find and plot the frequency response $H(e^{j\omega})$.
b) Find the difference equation relating the input $x[n]$ and output $y[n]$.

9.10 Consider a discrete time LTI system with impulse response $h[n] = \left(\frac{1}{3}\right)^n u[n]$. Find the output signal $y[n]$ for all the following inputs to this system:

a) $x[n] = \left(\frac{1}{4}\right)^n u[n]$
b) $x[n] = (n-2)\left(\frac{2}{5}^n u[n]\right)$
c) $x[n] = \cos(\pi n)$.

9.11 Consider an initially at rest discrete time LTI system with impulse response,

$$h[n] = (0.5)^{(n+2)} u[n].$$

a) Find the frequency response of this system.
b) Find the difference equation, which represents this system.
c) Find the discrete time Fourier transform of the output, when the input is

$$x[n] = \sin\left(\frac{\pi}{2} n\right).$$

9.12 Consider an initially at rest discrete time LTI system with impulse response,

$$h[n] = \left(\frac{1}{3}\right)^n \cos\left(\frac{\pi n}{2}\right) u[n].$$

a) Find the frequency response.
b) Find the Fourier transform of the output signal $y[n]$, when the input signal is $x[n] = \cos\left(\frac{\pi n}{2}\right)$.
c) Find the output $y[n]$.

9.13 Consider a causal LTI system, whose input and output are related by the difference equation

$$y[n] - 0.2y[n-1] = x[n]$$

Find the outputs $y_1[n]$ and $y_2[n]$ for each of the following inputs defined in one period:

a) $X_1(e^{j\omega}) = \dfrac{1 - 0.2e^{j\omega}}{1 + \frac{1}{2}e^{-j\omega}}$

b) $X_2(e^{-j\omega}) = \dfrac{1}{(1 + 0.3e^{-j\omega})(1 - 0.2e^{-j\omega})}$.

9.14 Find the inverse discrete time Fourier transform of the signals given in one period, as follows:

a) $X(e^{j\omega}) = \sum_{k=1}^{15} e^{j\omega k} \cos(k\omega)$

b) $X(e^{j\omega}) = j \sin\left(\frac{\pi}{2}\omega\right)$.

9.15 Find and plot the inverse discrete time Fourier transform of the following signals:

a) $X_1(e^{j\omega}) = e^{j\omega} \sin(2\omega + \pi)$, for $-\pi < \omega \leq \pi$

b) $X_2(e^{j\omega}) = j \tan\left(\frac{\pi}{3}\omega\right)$, for $-\pi < \omega \leq \pi$.

9.16 Find the inverse $x[n]$ of the following discrete time Fourier transform:

$$X(e^{j\omega}) = \frac{\sin(\frac{\omega}{3})}{(1 - 0.5e^{-2j\omega})}, \quad \text{for } -\pi < \omega \leq \pi.$$

9.17 Find the discrete time Fourier transform of the following signal:

$$x[n] = \left(\frac{\sin(\frac{\pi}{3}n)}{2\pi n}\right) * \left(\frac{\sin(\frac{\pi}{6}n)}{2\pi n}\right),$$

where (*) indicates the convolution operation.

9.18 An LTI system is defined by its impulse response, which is $h[n] = h'[n] + \left(\frac{2}{5}\right)^n u[n]$. The frequency response of this system is given as follows:

$$H(e^{j\omega}) = \frac{60 - 20e^{-j\omega}}{2 - 9e^{-j\omega} + 10e^{-2j\omega}}.$$

a) Find and plot $h[n]$.

b) Find and plor $h'[n]$.
c) Find and plot $H'(e^{j\omega})$.

9.19 Let $x[n]$ be a discrete time signal defined as follows:

$$x[n] = \frac{\sin(\omega_0 n)}{\pi n}.$$

a) Find the total energy of $x[n]$.
b) If the discrete time Fourier transform of $x[n]$ is $X(e^{j\omega}) = 2/3$, for $-\pi < \omega \leq \pi$, find the value of ω_0.

9.20 Does the following discrete time function satisfy the Dirichlet conditions? Verify your answer.

$$x[n] = 2^n u[n]. \tag{9.269}$$

9.21 Consider the z-transform of a discrete time signal $x[n]$ given as follows:

$$X(z) = \sum_{k=0}^{3} \frac{(0.5)^k}{4 - e^{-\frac{\pi}{2}k} z^{-1}}.$$

a) Find $x[n]$.
b) Find the z-transform of $g[n] = a^n x[n]$.

9.22 Consider a discrete time initially at rest LTI system, represented by the following difference equation:

$$y[n] - 0.2y[n-1] + 0.1y[n-2] = x[n].$$

a) Find and plot the frequency response $H(e^{j\omega})$.
b) Find and plot the impulse response $h[n]$ of this system.
c) Find a block diagram representation of this system.

9.23 Find the inverse discrete time Fourier transforms of the following transforms:

a) $X_1(e^{j\omega}) = \sum_{k=-\infty}^{\infty} \left(\frac{1}{2}\right)^k \delta\left(\omega - \frac{2\pi}{3}k\right)$

b) $X_2(e^{j\omega}) = \dfrac{1 + \frac{1}{8}e^{j\omega}}{1 - \frac{1}{6}e^{j\omega} - \frac{1}{6}e^{2j\omega}}.$

9.24 The discrete time Fourier transform of the signal $x[n]$ is given in the following figure:

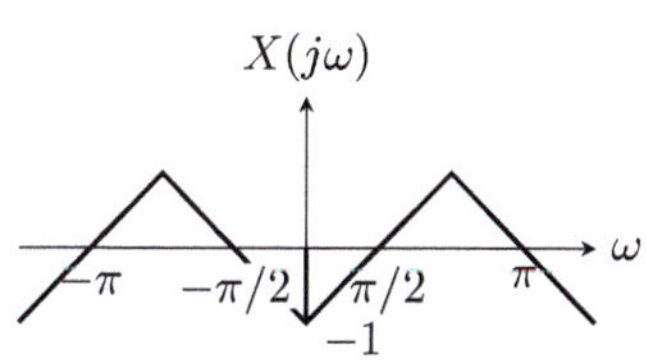

Find and plot the Fourier transforms of the following functions:

a) $x_1[n] = x[n]\left(\cos\left(\frac{\pi}{2}n\right)\right)$

b) $x_2[n] = \pi x[n]\left(\sin\left(\frac{2\pi}{3}n\right)\right)$

c) $x_3[n] = x[n]\sum_{k=-\infty}^{\infty}\delta(n-3k).$

9.25 Consider a discrete time LTI system with the following Fourier transforms of the input signal $x[n]$ and the impulse response $h[n]$, respectively:

$$X(e^{j\omega}) = 6e^{-j\omega} + e^{-2j\omega} - 2e^{-j\omega}$$
$$H(e^{j\omega}) = 2 - e^{-j\omega} + 4e^{-3j\omega}.$$

a) Find and plot the discrete time Fourier transform $Y(e^{j\omega})$ of the output.
b) Find and plot the output $y[n]$, in the time domain.

9.26 Consider a system consisting of parallel connection of two subsystems with the following impulse responses:

$$h_1[n] = \left(\frac{j}{2}\right)^n u[n+5]$$
$$h_2[n] = -\left(\frac{j}{2}\right)^n u[n-5].$$

a) Find and plot the frequency response of $h_1[n]$.
b) Find and plot the frequency response of $h_2[n]$.
c) Find and plot the frequency response of the overall system $h[n] = h_1[n] + h_2[n]$.

9.27 Consider an initially at rest LTI system represented by the following difference equation:

$$y[n] + y[n-1] + 0.5y[n-2] = x[n] - 0.25x[n-1].$$

a) Find and plot the frequency response of this system.
b) Find and plot the impulse response of this system.
c) Find and plot the inverse $h^{-1}[n]$ of this system.

9.28 Consider a discrete time LTI system, represented by the following impulse response:

$$h[n] = (0.5)^n u[n] + (0.25)^n u[n].$$

a) Find and plot the frequency response of this system.
b) Find the difference equation, which represents this system.
c) Find and plot the output $y[n]$, when the input is

$$x[n] = (0.5)^n u[n] - 0.5(0.5)^{(n-1)}u[n-1].$$

9.29 Find and plot the z-transform of the following signal, and specify the corresponding ROC of the following signal:

$$x[n] = \left(\frac{1}{3}\right)^n u[n-4].$$

9.30 Let $x[n]$ to be defined as follows:

$$x[n] = \left(\frac{-1}{4}\right)^n u[n] + \alpha u[-n-2].$$

Given that the ROC of $X(z)$ is $1 < |z| < 2$, find the possible values of the magnitude of the complex value α.

9.31 Find the z-transforms of the following signals and their region of convergences:

a)

$$x[n] = \left(\frac{2}{5}\right)^n \sin\left(\frac{\pi}{3}n\right) u[-n]$$

b)

$$x[n] = \left(\frac{2}{5}\right)^n \sin\left(\frac{\pi}{3}n\right) u[n].$$

9.32 Find the inverse z-transform of the following signal:

$$X(z) = \frac{1 + \frac{1}{4}z^{-1}}{(1 + z^{-1})(1 - \frac{1}{2}z^{-1})}, \quad |z| > 1.$$

9.33 Consider a discrete time LTI system represented by the following equation:

$$y[n] = x[2-n] - x[-n-1].$$

a) Find the transfer function of this system.
b) Find the impulse response of this system.
c) Find the z-transform of the output and its ROC, when the input is $x[n] = \delta(n)$.

9.34 Consider a discrete time system, represented by the following equation:

$$y[n] = (n+2)x[n].$$

a) Find the output $y[n]$, when the discrete time Fourier transform of the signal $x[n]$ is

$$X(e^{j\omega}) = \frac{2}{2 - e^{-j\omega}}, \quad \text{for } -\pi \le \omega \le \pi.$$

b) Find and plot the discrete time Fourier transform $Y(e^{j\omega})$.
c) Find the z-transform of the output $Y(z)$.

9.35 Programming – Frequency Domain Encoding

- **Introduction**
  We are not using text messages anymore. They are very boring. Now, almost all mainstream messaging apps support voice messages. Therefore, as of this moment, you and I will communicate with voice messages, but I have a problem. I am paranoid about privacy. I do not trust any Big Tech company, so I encoded my voice message with a special encoding that only you and I know. Your task is to decode and write my message. Don't worry. I will give you the decoding recipe.
- **Decoding Recipe** (Encoding is also same, so you can send me encoded messages using the same recipe.)

a) Transform the given voice signal to the Fourier domain using FFT.
b) Split the Fourier domain representation into two parts, positive and negative frequencies, reverse both parts and concatenate them again. For example, if FFT of x is $X[j\omega] = [a, b, c, d, e, f, g, h]$, then the encoded list must be $X'[j\omega] = [d, c, b, a, h, g, f, e]$
c) Return to the time domain using Inverse FFT and listen the message.

- **Fast Fourier Transform (FFT)**

DTFT and CTFT are great tools for theoretical purposes and filter designs, but they are not so practical for digital signals because they are defined in infinite domain. On the other hand, there is a better tool to analyze the finite sampled signals in a more elegant way, called discrete Fourier transform (DFT). The method is the bridge between continuous time signals and signals. In addition, it is not possible to catch some frequencies in a sampled signals (see Nyquist–Shannon Sampling Theorem).

DFT is one of the fundamental tools for analyzing signals in frequency domain. DFT, $X[k]$, of a signal $x[n]$ is defined as follows:

$$\textit{Discrete Fourier transform: } X[k] = \sum_{n=0}^{N-1} x[n] e^{-j2\pi \frac{k}{N} n}, \qquad k = [0, \dots, N-1].$$

$$\textit{Inverse discrete Fourier transform: } x[n] = \frac{1}{N} \sum_{k=0}^{N-1} X[k] e^{j2\pi \frac{k}{N} n}, \qquad n = [0, \dots, N-1].$$

The complexity of naive DFT algorithm is $O(n^2)$. Therefore, a lot effort was spent to improve the efficiency of DFT algorithm family. The result is elegant divide-and-conquer FFT Algorithm, which is chosen as one of the most important 10 algorithms in the 20th century by Science. Although the algorithm was invented by Carl Friedrich Gauss in 1805 when he needed it to interpolate the orbit of asteroids Pallas and Juno from sample observations, it was reinvented and popularized during 60s. The complexity of the algorithm is $O(N \log N)$. After that point, lots of FFT variants were proposed.

You will implement the best-known FFT algorithm. The main idea is to divide DFT algorithm into odd and even parts. It first computes the DFTs of the even-indexed inputs ($x_{2m} = x_0, x_2, \dots, x_{N-2}$) and of the odd-indexed inputs ($x_{2m+1} = x_1, x_3, \dots, x_{N-1}$), and then combines those two results to produce the DFT of the whole sequence. This idea can then be performed recursively to reduce the overall runtime to $O(N \log N)$.

$$X[k] = \sum_{n=0}^{N/2-1} x[2n] e^{\frac{-j2\pi}{N}(2n)k} + \sum_{n=0}^{N/2-1} x[2n+1] e^{\frac{-j2\pi}{N}(2n+1)k}.$$

We can simplify the procedure of the formula as follows:

$$X[k] = O[k] + E[k] e^{\frac{-j2\pi}{N}(k-1)},$$

where $O[k]$ and $E[k]$ are the discrete Fourier Transforms of elements with odd and even indices, respectively. Moreover, since we know that the discrete Fourier transform of a signal is periodic, we do not have to calculate two periods in the summations; we can calculate only the first period and then concatenate the result with itself. The only concern is that we are multiplying $E[k]$ with $e^{\frac{-j2\pi}{N}}$. However, it has a nice property that:

$$e^{\frac{-j2\pi}{N}(k-1+N/2)} = e^{\frac{-j2\pi}{N}(k-1)}.$$

Therefore, we can write this equation as:

$$X[k] = \begin{cases} O[k] + E[k]e^{\frac{-j2\pi}{N}(k-1)}, & \text{if } k \leq N/2 \\ O[k - N/2] - E[k - N/2]e^{\frac{-j2\pi}{N}(k-1-N/2)}, & \text{if } k > N/2. \end{cases}$$

You can implement ifft() function by using fft() function. Think about that.

- **Hints**
  a) You can check your fft() function by comparing numpy.fft.fft(). If the individual differences are below 10^{-7} for our input, your function works correctly.
  b) For simplicity, you can assume that the length of the input file is 2^n, where $n \in \mathbb{N}$
  c) Complexity of FFT algorithm is **O(N logN)**. Please be careful about the complexity.
  d) To read the sound data, you can use **scipy.io**. It also returns the sample rate of the audio file. It is very useful to determine the frequency bins.
  e) Please be careful about the frequency bins of your implementation even if they are not required to complete the problem. They may be out of order.
- **Regulations**
  a) You should add the plot of frequency domain magnitude of encoded and decoded signal and the time domain plots to your solutions to see the difference between two signals. That means your solution must contain **4** different plots.
  b) You should write the secret message to your solution. You can find the encoded message in **encoded.wav**
  c) You should use **Python3** during the problem.
  d) You are not allowed to use any library other than **numpy**, **matplotlib.pyplot** and **scipy.io**
  e) You are not allowed to use **numpy.fft** in the problem, you should implement your own **fft()** and **ifft()** function.

10

Linear Time-Invariant Systems as Filters

Until now, we have studied various representations of linear time-invariant (LTI) systems. We defined impulse responses in the time domain and the corresponding frequency responses in the frequency domain,

$$h(t) \longleftrightarrow H(j\omega) \quad \text{and} \quad h[n] \longleftrightarrow H(e^{j\omega}),$$

for continuous time and discrete time LTI systems, respectively. Then, we extend the Fourier transforms to Laplace transforms for continuous time systems and z-transforms for discrete time systems to generalize the frequency response to transfer functions,

$$h(t) \longleftrightarrow H(s) \quad \text{and} \quad h[n] \longleftrightarrow H(z),$$

for continuous time and discrete time LTI systems, respectively.

We represented the dynamic behavior of continuous time and discrete time LTI systems by a differential equation,

$$\sum_{k=0}^{N} a_k \frac{d^k y(t)}{dt^k} = \sum_{k=0}^{M} b_k \frac{d^k x(t)}{dt^k}, \tag{10.1}$$

and difference equations,

$$\sum_{k=0}^{N} a_k y[n-k] = \sum_{k=0}^{M} b_k x[n-k], \tag{10.2}$$

respectively.

We studied the basic properties of LTI systems, such as memory, causality, stability and convertibility. We studied how an LTI system relates the input and output signals. But where and when do we use LTI systems? What do they do to the input signals? How do they change the input signal to produce an output signal?

The answers to these questions are manyfold and depend on the application areas. One very important area, where the LTI systems are widely used is the **filtering**. LTI systems act as a filter to change the structure of the input signals.

Motivating Question: What is a filter?

In general, **filters** are devices that separate the *unwanted* stuff from a *pool* of objects to obtain the *wanted* stuff. The *pool* may contain anything, such as water, air, chemicals, and soils.

In the context of signals and systems, the "pool" contains signals. LTI systems are considered as filters to "clean-up" the input signals.

Signals and Systems: Theory and Practical Explorations with Python,
First Edition. Fatoş Tunay Yarman Vural and Emre Akbaş.

Companion website: www.wiley.com/go/signalsandsystems

Recall that Fourier series representation enables us to decompose a periodic input signal into harmonically related complex exponentials. The amount of each harmonic frequency is measured by the spectral coefficients of the Fourier series, $\{a_k\}$. We may want to eliminate some of the *unwanted* harmonics of complex exponential functions. These components may correspond to a type of noise in a speech recording or some unwanted objects in an image.

Later, we extended the Fourier series representation to Fourier transforms, where we could represent a time domain signal by a continuous frequency spectrum. In order to manipulate and/or reshape the frequency content of an input signal to generate a desired signal at the output, we can design an LTI system that suppresses some of the frequencies or emphasizes some others. For example, we may want to change the frequency content of a signal to isolate some musical instruments in an orchestra or accentuate the voice of the singer.

Motivating Question: How can we design an LTI System to filter the input signal, which generates an output signal in a desired form? In other words, how can we design an LTI system that outputs a signal with a prescribed frequency content?

At the core of the answers to these questions resides the frequency response or transfer function of the LTI systems.

10.1 Filtering the Periodic Signals by Frequency Response

Recall that a periodic input signal of a **discrete time LTI system** can be represented by the spectral coefficients, a_k, using Fourier series. The corresponding output is also periodic and, hence, can be represented by the Fourier series with the spectral coefficients of b_k.

The spectral coefficients of the output signal, b_k, can be uniquely obtained from the spectral coefficients of the input signal a_k by scaling them with the kth eigenvalue of the **discrete time LTI system**,

$$b_k = a_k H(e^{jk\omega_0}) \tag{10.3}$$

as shown in Figure 10.1. In Equation (10.3), the kth eigenvalue is the frequency response of an LTI system at $\omega = k\omega_0$ and can be calculated from the impulse response as follows:

$$H(e^{jk\omega_0}) = \sum_{n=-\infty}^{\infty} h[n]e^{-j\omega_0 nk}. \tag{10.4}$$

Similarly, a **continuous time LTI system** scales the complex exponential eigenfunction $e^{jk\omega_0 t}$ by its eigenvalue $H(jk\omega_0)$, which is the frequency response of the LTI system at $\omega = k\omega_0$.

As in the discrete time LTI systems, we observe that the spectral parameters of the output are the scaled version of the spectral parameters of the input,

$$b_k = a_k H(jk\omega_0), \tag{10.5}$$

$$x[n] = \sum_{k=<N>} a_k e^{jk\omega_0 n} \rightarrow \boxed{h[n]} \rightarrow y[n] = \sum_{k=<N>} a_k H(e^{jk\omega_0}) e^{jk\omega_0 n}.$$

Figure 10.1 When the input of a discrete time LTI system is the superposition of the eigenfunctions, the output is the superposition of the input, scaled by the kth eigenvalue, which is the frequency response $H(e^{j\omega})$, for $\omega = k\omega_0$.

$$x(t) = \sum_{k=-\infty}^{\infty} a_k e^{jk\omega_0 t} \rightarrow \boxed{h(t)} \rightarrow y(t) = \sum_{k=-\infty}^{\infty} a_k e^{jk\omega_0 t} H(jk\omega_0)$$

Figure 10.2 When the input of a continuous time LTI system is the superposition of the eigenfunctions, $e^{jk\omega_0 t}$, the output is the superposition of the same eigenfunctions. However, each weight, a_k at the input is scaled by the kth eigenvalue of the LTI system at the output, where the scaling factor is the frequency response at $\omega = k\omega_0$.

where the scaling factor is the kth eigenvalue of the continuous time LTI system, which is defined as the frequency response at $\omega = k\omega_0$,

$$H(jk\omega_0) = \int_{-\infty}^{\infty} h(t)e^{-jk\omega_0 t}dt, \tag{10.6}$$

as shown in Figure 10.2.

As in the discrete time case, kth eigenvalue of a continuous time LTI system scales each spectral coefficient at the output, i.e., $H(jk\omega_0)$ multiplies a_k to generate b_k.

Therefore, frequency response of an LTI system reshapes the frequency content of periodic signals by scaling the spectral coefficients of the input signal. For example, the complex exponential functions $e^{jk\omega_0}$ for the frequencies $k\omega_0$ may correspond to the noise embedded in a discrete time input signal. In this case, we design an LTI system, where the frequency response $H(e^{jk\omega_0}) = 0$ for the corresponding k-values. At the output of the filter, we obtain the spectral coefficients,

$$b_k = H(e^{jk\omega_0})a_k = 0,$$

which corresponds to the noise, vanishes, yielding a "clean" signal. This process of reshaping the spectral coefficients of the input signals is called **filtering.**

10.2 Filtering the Aperiodic Signals by Frequency Response

Equations (10.3), (10.4), (10.5), and (10.6), reveals that each spectral coefficient of the input, a_k, is scaled by an eigenvalue of an LTI system, to generate the spectral coefficient of the output, b_k. We can relax this assumption and take the limiting case to define the eigenfunction of a continuous time LTI system as:

$$x(t) = e^{j\omega t},$$

and eigenfunction of a discrete time LTI system as:

$$x[n] = e^{j\omega n},$$

where $k\omega_0 \rightarrow \omega$. In this case, the output of the LTI system for the continuous time LTI system becomes

$$y(t) = H(j\omega)e^{j\omega t}. \tag{10.7}$$

where the eigenfunction $e^{j\omega t}$ is scaled by the eigenvalue of an LTI system, which is the frequency response,

$$H(j\omega) = \int_{-\infty}^{\infty} h(t)e^{-j\omega t}dt. \tag{10.8}$$

Similarly, the output of a discrete time LTI system becomes

$$y[n] = H(e^{j\omega})e^{j\omega n},$$

where the eigenfunction $e^{j\omega n}$ is scaled by the eigenvalue of an LTI system, which is the frequency response,

$$H(e^{j\omega}) = \sum_{n=-\infty}^{\infty} h[n]e^{-j\omega n}. \tag{10.9}$$

In the aforementioned approach, rather than defining the eigenvalue of an LTI system for each integer multiple of the fundamental frequency, $k\omega_0$, we define a function with the continuous frequency variable ω. This generalization allows us to represent the eigenvalues of LTI systems by the frequency response in the frequency domain. Hence, in the limit,

$$\lim_{k\omega_0 \to \omega} H(e^{jk\omega_0}) \to H(e^{j\omega}), \tag{10.10}$$

the countably infinite eigenvalues of an LTI system, at the harmonics, $k\omega_0$, for all k converges to the frequency response with continuous frequencies, ω. Note that the eigenvalue of an LTI system, corresponding to the kth harmonic of a periodic signal, is a special value of the frequency response at $\omega = k\omega_0$.

Recall that the impulse response uniquely defines an LTI system. Thus, the frequency response also uniquely defines a discrete time and continuous time LTI systems. The impulse response and frequency response of an LTI system is one-to-one and onto for discrete time LTI systems,

$$h[n] \leftrightarrow H(e^{j\omega}), \tag{10.11}$$

and for a continuous time LTI systems,

$$h(t) \leftrightarrow H(j\omega). \tag{10.12}$$

Recall also that the relationship between an aperiodic input and output signal pair for the continuous time LTI system is given by

$$Y(j\omega) = H(j\omega)X(j\omega),$$

and the relationship between the input and output signal for the discrete time LTI system is given by

$$Y(e^{j\omega}) = H(e^{j\omega})X(e^{j\omega}).$$

Hence, the frequency content of an aperiodic input signal can be easily scaled by the frequency response to increase or decrease the amount of predefined frequency range in the signal and change the frequency content of the signal to generate a desired signal at the output. All we need to do is to design an appropriate frequency response, $H(j\omega)$ for a continuous time system and $H(e^{j\omega})$ for a discrete time system, which attenuates the undesired frequency ranges and amplifies the desired ones to shape up the frequency content of the input signal.

We can further represent the eigenvalues of an LTI system by the transfer function $H(s)$ in Laplace domain for continuous time systems and by the transfer functions $H(z)$ in z-domain for discrete time systems. However, in the rest of the chapter, we use frequency response to represent the eigenvalues of the LTI systems, assuming that the Fourier transform of the impulse response exists.

10.3 Frequency Ranges of Frequency Response

When an LTI system is designed as a filter, it passes some frequency ranges and blocks the rest of the frequencies. These frequency ranges are characterized by two major quantities, namely cutoff frequency and bandwidth. These quantities shape the analytical form of the frequency response, as defined in the following text.

Definition 10.1 Cutoff Frequency of the Frequency Response: The cutoff frequency ω_c, is the boundary of the frequency response, at which the energy flow of the LTI system is stopped or significantly reduced.

In the aforementioned definition, the cutoff frequency is the angular frequency $\omega_0 = \frac{2\pi}{T} = 2\pi f$, which is measured by radians. The fundamental period T is measured by seconds, and the fundamental frequency f is measured by Hertz (cycle/second).

Definition 10.2 Bandwidth of the Frequency Response: The interval of the frequency response between the lowest and highest cutoff frequencies, $\omega_{bw} = \omega_{c1} - \omega_{c2}$, is called the **bandwidth**.

For example, the human ear can hear the sounds between 20 Hz and 20 kHz. The lowest cutoff frequency is $\omega_{c1} = 20$ Hz and the highest cutoff frequency is $\omega_{c2} = 20,000$ Hz. Thus, the bandwidth of the human auditory system is $\omega_{bw} = 20,000 - 20 = 19,980$ Hz. The frequency response of the human auditory system is nonzero only within the bandwidth. Outside this bandwidth, both the magnitude and the phase of the frequency response are zero.

Selecting the cutoff frequencies and the bandwidths of a frequency response is an important design issue, and it depends on the application domains. For example, if we need to chop an additive noise corresponding to the high-frequency components of the input signal, we first identify the bandwidth of the noise. Then, we set the cutoff frequency to the lowest limit of the bandwidth. The frequency response of the LTI system with this cutoff frequency removes the noise in the input signal by eliminating the spectral coefficients corresponding to the noise at the output.

The cutoff frequencies of the frequency response of the continuous time systems can be selected as high as the dynamic range of the equipment. However, since the frequency response of discrete time systems is periodic with $= 2\pi$, the cutoff frequency of the discrete time systems should lie within the limits of the fundamental period, $|\omega_c| \leq \pi$.

10.4 Filtering with LTI Systems

In order to **filter an input signal**, we design the frequency response of an LTI system with a predefined cutoff frequencies and bandwidths, such that the output signal consists of the spectral coefficients of "desired form." During the design of a filter, different representations show different properties of the filter, which complement each other:

1) Differential equations for continuous time filters and difference equations for discrete time filters give an implicit relationship between the change of input and output relative to each other, in the time domain.

2) Impulse response gives the output of a system, when the input is a unit impulse function. It also provides the nonzero time duration of the filter, which is an important design issue.
3) Frequency response shows the response of an LTI system to the frequency components of the input signals. It is the most informative representation of the filters about how they shape the frequency content of the input signals. Therefore, in most practical applications, filter design starts in the frequency domain.

There is a large variety of filters for both discrete time and continuous time systems in signal processing literature. Depending on the physical realizability of the frequency response and/or impulse response in real-life applications, the filters can be categorized as follows:

1) **Ideal Filters,** where the frequency response takes only binary values, either 0 or 1. The abrupt switch between 0 and 1 creates discontinuities at cutoff frequencies. These filters pose challenges in implementation, which is why they are called ideal.
2) **Real Filters**, which are designed with a smooth frequency response to avoid discontinuities. These filters gradually attenuate the undesired frequencies and reach the cutoff frequencies smoothly. Real filters avoid most of the problems of ideal filters, such as the Gibbs problem, which creates undesired fluctuations at discontinuities.

Depending on the frequency content, LTI filters can be classified under four headings:

1) **Low-pass filters**, which suppress the high-frequency ranges of the input signal and pass relatively lower frequencies. In other words, a low-pass filter has a frequency response, which has high magnitudes in low frequencies and low magnitudes in high frequencies to suppress or eliminate the high-frequency ranges of the input signal at the output.
2) **High-pass filters**, which is rather the complement of the low-pass filters. They suppress the low-frequency ranges and pass relatively higher frequency ranges of the input signal at the output.
3) **Band-pass filters**, which pass the frequency ranges of the input signal corresponding to the desired interval of frequencies.
4) **Band-stop filters**, which suppress the frequency ranges of the input signal in a desired interval of frequencies.

WATCH

Learn more about the effect of different types of filters on music signals @ `https://384book.net/v1001`

Filters can be designed for both discrete time systems or continuous time systems.

Sections 10.5 and 10.6 overview the ideal and real filters for low-pass, high-pass, band-pass, and band-reject filters. We illustrate how an LTI system behaves as a filter and shapes the frequency content of an input signal for both continuous time and discrete time cases.

10.5 Ideal Filters for Discrete Time and Continuous Time LTI Systems

The magnitude of the ideal filters takes either 1 or 0 values to either pass or vanish the frequency band of an input signal at the output of the filter. Hence, the ideal filters either retain a frequency band of an input signal as is, or remove it, depending on the value of the magnitude spectrum of the frequency response.

Mathematically, an ideal filter outputs the following spectral coefficients for periodic inputs, for continuous time systems,

$$b_k = \begin{cases} a_k & \text{for} \quad H(jk\omega_0) = 1 \\ 0 & \text{for} \quad H(jk\omega_0) = 0. \end{cases} \tag{10.13}$$

Similarly, an ideal low-pass filter outputs the following spectral coefficients for periodic inputs, for the discrete time systems,

$$b_k = \begin{cases} a_k & \text{for} \quad H(e^{jk\omega_0}) = 1 \\ 0 & \text{for} \quad H(e^{jk\omega_0}) = 0. \end{cases} \tag{10.14}$$

When the input signal is aperiodic, an ideal filter yields the following output, for continuous time systems,

$$Y(j\omega) = \begin{cases} X(j\omega) & \text{for} \quad H(jk\omega) = 1 \\ 0 & \text{for} \quad H(jk\omega) = 0. \end{cases} \tag{10.15}$$

Similarly, when the input signal is aperiodic, an ideal filter yields the following output for the discrete time systems:

$$Y(e^{j\omega}) = \begin{cases} X(e^{j\omega}) & \text{for} \quad H(e^{jk\omega}) = 1 \\ 0 & \text{for} \quad H(e^{jk\omega}) = 0. \end{cases} \tag{10.16}$$

In Sections 10.5.1, 10.5.2, and 10.5.3, we overview the ideal filters for discrete time and continuous time systems.

10.5.1 Ideal Low-Pass Filters

An ideal low-pass filter passes all the frequencies below the cutoff frequency for $|\omega| \leq \omega_c$ and suppresses the rest of the frequencies. Mathematically, the frequency response of an ideal low-pass filter for a continuous time system is

$$H(j\omega) = \begin{cases} 1 & \text{for} \quad |\omega| \leq \omega_c \\ 0 & \text{otherwise}. \end{cases} \tag{10.17}$$

and for discrete time system is

$$H(e^{j\omega}) = \begin{cases} 1 & \text{for} \quad |\omega| \leq \omega_c \\ 0 & \text{otherwise}, \end{cases} \tag{10.18}$$

for one period, in $-\pi \leq \omega \leq \pi$ and repeats at every period of 2π, in other words, $H(e^{j\omega}) = H(e^{j(\omega+2k\pi)})$ for all k. Therefore, the cutoff frequency should also lie in $-\pi < \omega_c < \pi$ (see Figures 10.3 and 10.4).

The cutoff frequency ω_c determines the number of nonzero spectral coefficients of the periodic output signal. Lower cutoff frequencies allow fewer spectral coefficients to pass, chopping the spectral components associated with higher-frequency harmonics of the periodic input signal.

10.5.2 Ideal High-Pass Filters

The bandwidth of a high-pass filter is the complement of that of the low-pass filter. A high-pass filter passes all the frequencies, when $|\omega| \geq \omega_c$.

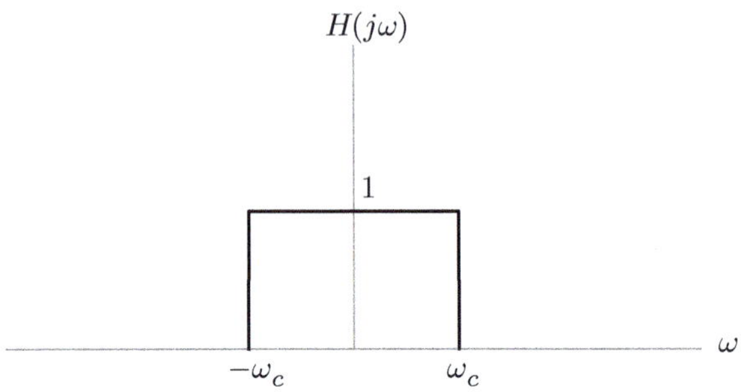

Figure 10.3 Frequency response of continuous time ideal low-pass filter with cutoff frequency, $|\omega_c|$.

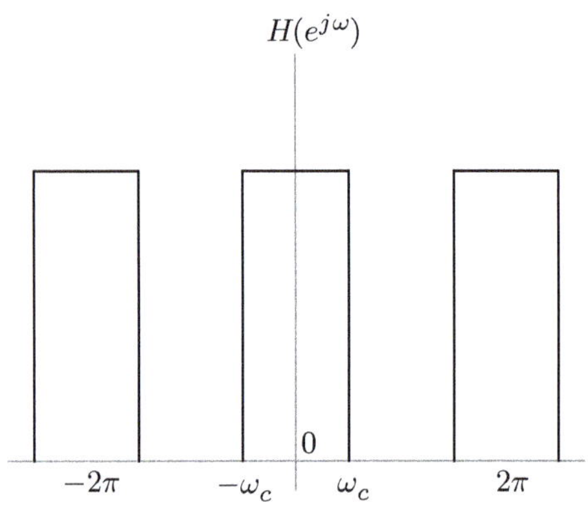

Figure 10.4 Frequency response of discrete time ideal low-pass filter with cutoff frequency, $|\omega_c|$, which is periodic with $\omega = 2\pi$.

Mathematically, the frequency response of an ideal high-pass filter for continuous time system is

$$H(j\omega) = \begin{cases} 1 & \text{for} \quad |\omega| \geq \omega_c \\ 0 & \text{otherwise.} \end{cases} \tag{10.19}$$

and for discrete time system is

$$H(e^{j\omega}) = \begin{cases} 1 & \text{for} \quad |\omega| \geq \omega_c \\ 0 & \text{otherwise.} \end{cases} \tag{10.20}$$

for one period, in $-\pi \leq \omega \leq \pi$ and repeats at every period of 2π. Note that, $H(e^{j\omega}) = H(e^{j(\omega+2k\pi)})$ for all k, as shown in Figure: 10.6. (see Figures 10.5 and 10.8).

10.5.3 Ideal Band-Pass and Band-Reject Filters

For the band-pass and band-reject filters, the **cutoff frequencies**, $(\omega_{c1}, \omega_{c2})$ are defined for pass-band and reject-band intervals for passing or suppressing the frequencies.

A band-pass filter passes the frequencies, for $\omega_{c1} \leq |\omega| \leq \omega_{c2}$. Mathematically, the frequency response of an ideal band-pass filter for a continuous time system (see Figure 10.7) is

$$H(j\omega) = \begin{cases} 1 & \text{for} \quad \omega_{c1} \leq |\omega| \leq \omega_{c2}, \\ 0 & \text{otherwise.} \end{cases} \tag{10.21}$$

and for discrete time system (see Figure 10.8) is

$$H(e^{j\omega}) = \begin{cases} 1 & \text{for} \quad \omega_{c1} \leq |\omega| \leq \omega_{c2}, \\ 0 & \text{otherwise.} \end{cases} \tag{10.22}$$

for one period, in $-\pi \leq \omega \leq \pi$ and repeats at every period of 2π.

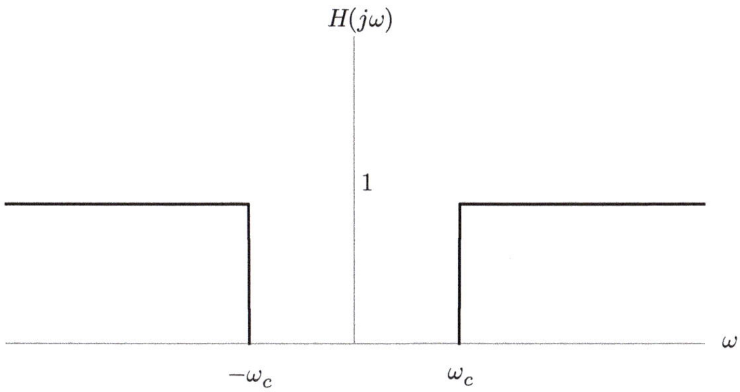

Figure 10.5 Frequency response of continuous time ideal high-pass filter with cutoff frequency, ω_c.

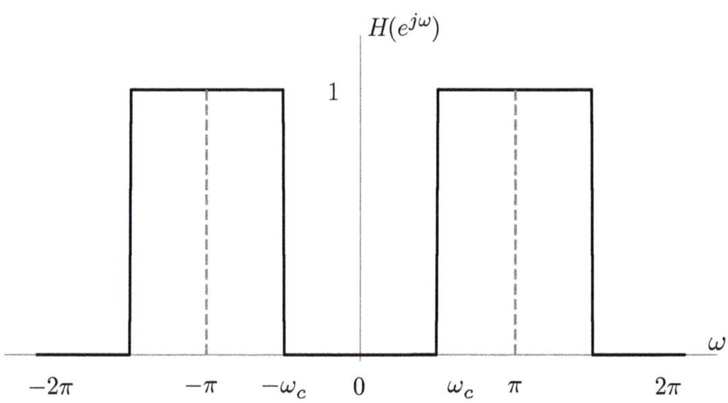

Figure 10.6 Frequency response of discrete time ideal high-pass filter with cutoff frequency, ω_c, which repeats at every 2π period.

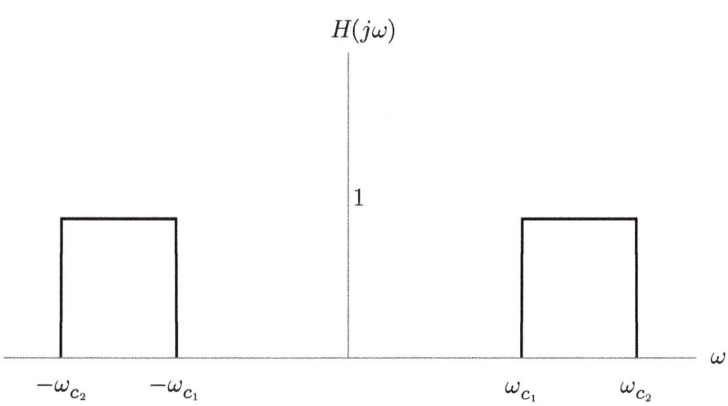

Figure 10.7 Frequency response of continuous time ideal band-pass filter between the cutoff frequencies $|\omega_{c_1}|$ and $|\omega_{c_2}|$.

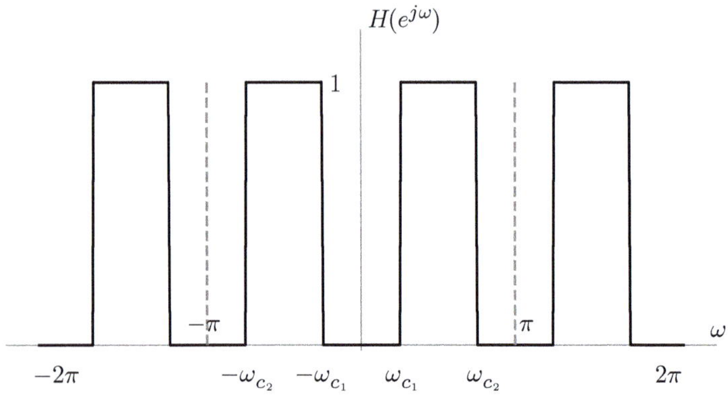

Figure 10.8 Frequency response of discrete time ideal band-pass filter between the cutoff frequencies $|\omega_{c_1}|$ and $|\omega_{c_2}|$, which repeats at every 2π period.

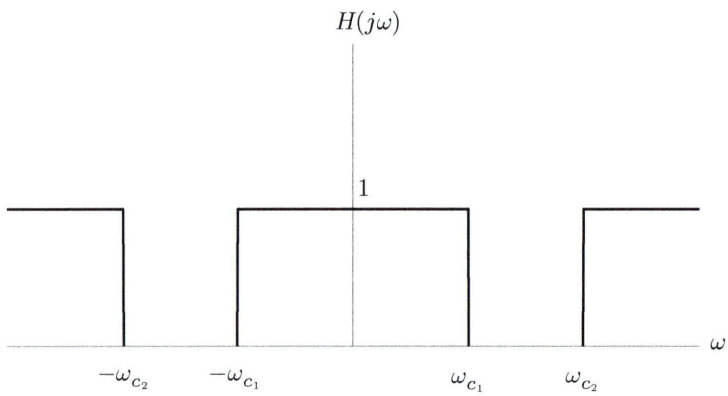

Figure 10.9 Frequency response of continuous time ideal band-reject filter, which eliminates the frequencies between the cutoff frequencies $|\omega_{c_1}|$ and $|\omega_{c_2}|$.

A band reject filter is the complement of the band-pass filter. It passes the frequencies, when $|\omega| \leq \omega_{c1}$ and $|\omega| \geq \omega_{c2}$. Mathematically, the frequency response of an ideal band-reject filter for a continuous time system (see Figure 10.9) is

$$H(j\omega) = \begin{cases} 1, & \text{for} \quad |\omega| \leq \omega_{c1} \quad \text{and} \quad |\omega| \geq \omega_{c2}, \\ 0 & \text{otherwise.} \end{cases} \tag{10.23}$$

and for discrete time system (see Figure 10.10) is

$$H(e^{j\omega}) = \begin{cases} 1, & \text{for} \quad |\omega| \leq \omega_{c1} \quad \text{and} \quad |\omega| \geq \omega_{c2}, \\ 0, & \text{otherwise.} \end{cases} \tag{10.24}$$

for one period, in $-\pi \leq \omega \leq \pi$ and repeats at every period of 2π.

Unfortunately, ideal filters have discontinuities at the cutoff frequencies, in the frequency response, while they switch between 0 and 1. Due to the discontinuities in the frequency domain, the time domain representations of ideal filters by impulse response, range in infinite duration. This property of ideal filters is named as IIR (infinite impulse response) filters. Although the ideal

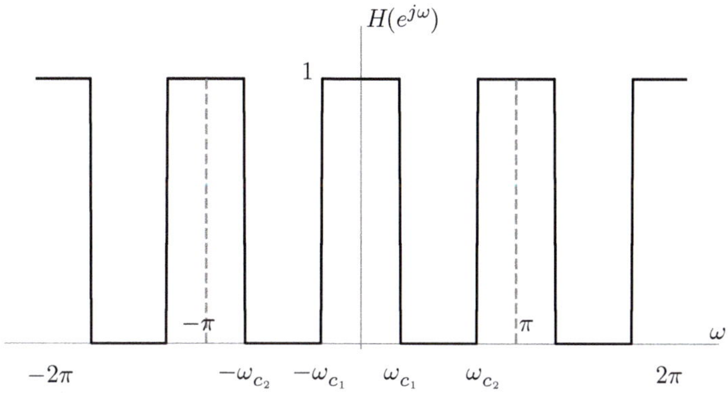

Figure 10.10 Frequency response of discrete time ideal band-reject filter cutoff frequencies $|\omega_{c_1}|$ and $|\omega_{c_2}|$, which repeats at every 2π period.

filters have simple analytical forms, they cannot be realized perfectly in real life applications by using physical system components.

In the following exercises, we study the effect of the ideal filters on the input signals.

Exercise 10.1 Consider a discrete time **ideal low-pass filter**, given by the following frequency response:

$$H_{lp}(e^{j\omega}) = \begin{cases} 1 & \text{for } |\omega| < \omega_c \\ 0 & \text{otherwise} \end{cases} \qquad (\omega_c \text{ is the cutoff frequency}). \tag{10.25}$$

Note: $H_{lp}(e^{j\omega})$ is periodic with 2π. Hence, $|\omega_c| \leq \pi$

a) Find the impulse response of this filter.
b) Plot the Fourier transform of the output for the input of Figure 10.11, where the bandwidth of the signal is $2\omega_m > 2\omega_c$.
c) Compare the input and output signals in frequency domain.

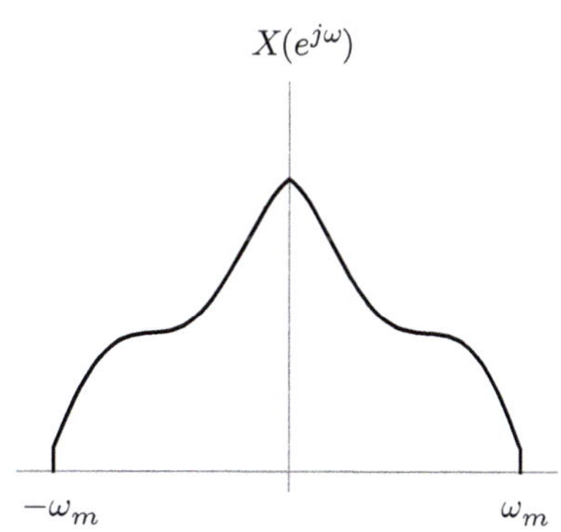

Figure 10.11 One full period of the Fourier transform of an input, for a discrete time band-limited signal, with the bandwidth, $2\omega_m$. Keep in mind that this Fourier transform is periodic and repeats itself at every 2π.

Solution

This filter is depicted in Figure 10.4.

In order to find the impulse response, we can easily take the inverse Fourier transform of the frequency response:

$$h[n] = \frac{1}{2\pi}\int_{-\omega_c}^{\omega_c} e^{j\omega n}d\omega = \frac{1}{2\pi jn}(e^{j\omega_c n} - e^{-j\omega_c n}) = \frac{\sin\omega_c n}{\pi n}. \tag{10.26}$$

Note: Ideal low-pass filters are NOT causal: $h[n] \neq 0$ for $n < 0$ (see Figure 10.12).

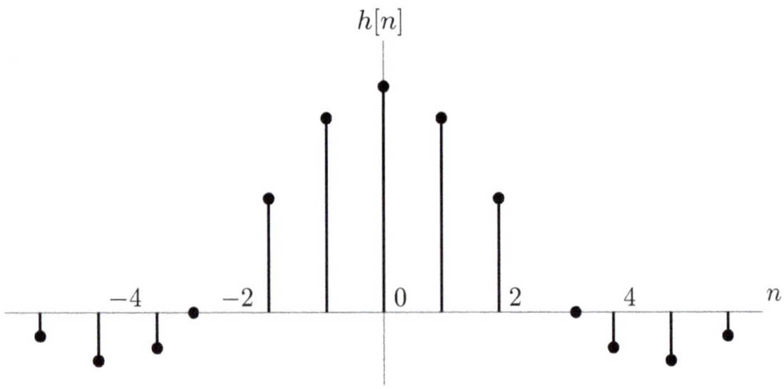

Figure 10.12 Impulse response of a discrete time ideal low-pass filter.

b) Recall the convolution property,

$$y[n] = x[n] * h[n] \longleftrightarrow Y(e^{j\omega}) = X(e^{j\omega})H_{lp}(e^{j\omega}). \tag{10.27}$$

In order to find the output of the ideal low-pass filter to the input signal given in Figure 10.11, we multiply the frequency response with the Fourier transform of the input:

$$Y(e^{j\omega}) = X(e^{j\omega})H_{lp}(e^{j\omega}). \tag{10.28}$$

Therefore, at the output, the input signal is chopped for the frequencies higher than the cutoff frequency, ω_c as indicated in Figure 10.13.

c) Comparison of the input and output signals in the frequency domain reveals that the ideal low-pass filters chop the frequencies higher than the cutoff frequency $|\omega_c|$.

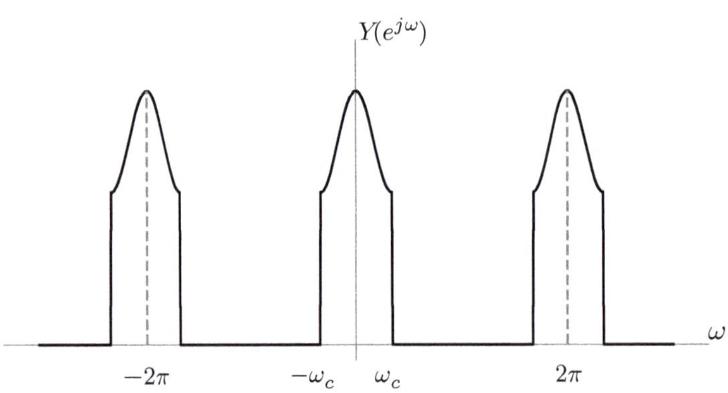

Figure 10.13 The frequency domain representation of the output signal, when an input signal is filtered by a low-pass filter, with cutoff frequency, ω_c.

Exercise 10.2 Find the impulse response of the **discrete time high-pass filter**, represented by the following frequency response:

$$H_{hp}(e^{j\omega}) = \begin{cases} 1, & \text{for} \quad |\omega| \geq \omega_c \\ 0, & \text{otherwise.} \end{cases} \tag{10.29}$$

Note: $H(e^{j\omega})$ is periodic with 2π. Hence, $|\omega_c| < \pi$

Solution

The frequency response of the high-pass filter can be represented as the complement of the low-pass filter, as follows:

$$H_{hp}(e^{j\omega}) = \begin{cases} 1, \text{ for } |\omega| \geq \omega_c \\ 0, \text{ otherwise.} \end{cases} = \begin{cases} 1 - H_{lp} & \text{for} \quad |\omega| < \omega_c \\ 0, & \text{otherwise.} \end{cases} \tag{10.30}$$

We already obtained the impulse response of the low-pass filter in the previous example. Thus, taking the inverse Fourier transform of the frequency response of Equation (10.30) gives

$$h[n] = \delta[n] - \frac{\sin \omega_c n}{\pi n}. \tag{10.31}$$

Note: Ideal high-pass filters are also non causal. In fact, all the ideal filters are non causal. Therefore, they are not realizable in the real-life applications. However, they are very useful to design real filters as they form a quality metric, when the designed filter is compared to its ideal counterpart.

Exercise 10.3 Consider the frequency response of the following **continuous time low-pass filter:**

$$H(j\omega) = \begin{cases} 1, & |\omega| < \omega_c \\ 0, & |\omega| > \omega_c \end{cases}. \tag{10.32}$$

a) Find the impulse response $h(t)$ of this system.
b) What is the bandwidth of this filter?
c) Find the output of this filter, when the input is

$$X(j\omega) = \begin{cases} 1, & |\omega| < \omega_m \\ 0, & |\omega| > \omega_m \end{cases}, \tag{10.33}$$

where $\omega_c < \omega_m$.

d) Compare the input and output of this filter in time and frequency domains.

Solution

a) The frequency response of this filter is illustrated in Figure 10.14a. Using the synthesis equation, we can determine the impulse response, $h(t)$, as follows:

$$h(t) = \frac{1}{2\pi} \int_{--\omega_c}^{\omega_c} e^{j\omega t} d\omega = \frac{\sin \omega_c t}{\pi t}, \tag{10.34}$$

which is a sinc function, depicted in Figure 10.14b.

b) The bandwidth of this continuous time low-pass filter is $2\omega_c$.

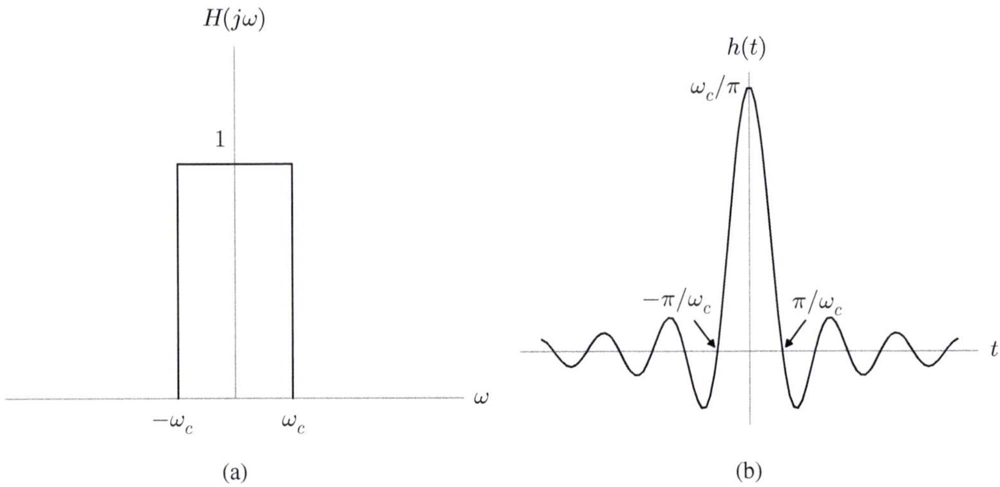

Figure 10.14 (a) Frequency response $H(j\omega)$ of a continuous time low-pass filter with the cutoff frequency $|\omega_c|$, (b) the corresponding impulse response, $h(t)$.

c) The output of the low-pass filter is

$$Y(j\omega) = X(j\omega)H(j\omega) = \begin{cases} X(j\omega), & |\omega| < \omega_c \\ 0, & |\omega| > \omega_c \end{cases}. \tag{10.35}$$

Since the cutoff frequency of the frequency response is smaller than that of the input signal, $\omega_c < \omega_m$, the low-pass filter keeps the input signal for $\omega < \omega_c$ and chops the frequencies for $\omega > \omega_c$.

The analysis of Figure 10.14 reveals that the frequency response consists of low frequencies for $|\omega| < \omega_c < \omega_m$. The impulse response, $h(t)$ alternates and attenuates, as $t \to \pm\infty$. Notice that as ω_c approaches to 0, the impulse response $h(t)$ gets flatter.

The impulse response is maximum at $t = 0$ in time domain. It keeps attenuating as $t \to \pm\infty$. Hence, the continuous time low-pass filter with the cutoff frequency ω_c is represented in time and frequency domain as follows:

$$h(t) = \frac{\sin \pi\omega_c t}{\pi t} \leftrightarrow H(j\omega) = \begin{cases} 1, & |\omega| < \omega_c \\ 0, & |\omega| > \omega_c \end{cases}. \tag{10.36}$$

d) The only difference between the input and output signals in the frequency domain is the bandwidth. While the bandwidth of the input signal is $2\omega_m$, the bandwidth of the output signal is the same as the bandwidth of the filter, which is $2\omega_c < 2\omega_m$.

In the frequency domain, the bandwidth of the input signal is decreased at the output. Hence, the sinc function of the input signal in time domain, becomes flatter around the origin at the output signal.

Exercise 10.4 Consider the following **continuous time band-pass filter:**

$$H_{bp}(j\omega) = \begin{cases} 1 & \text{for } \omega_{c1} \le |\omega| \le \omega_{c2}, \\ 0 & \text{otherwise.} \end{cases} \tag{10.37}$$

a) Find the impulse response of this filter.
b) Find the output $y(t)$ of this filter, when the input is $X(j\omega) = 1$.
c) Compare the input and output pair of this filter in the time and frequency domains.

Solution

a) Band-pass filter can be represented by a shifted low-pass filter,

$$H_{bp} = H_{lp}\left(j\left(\omega - \frac{\omega_{c1} + \omega_{c2}}{2}\right)\right) + H_{lp}\left(j\left(\omega + \frac{\omega_{c1} + \omega_{c2}}{2}\right)\right).$$

From the Fourier transform properties, we know that

$$e^{j\omega_0 t}h(t) \longleftrightarrow H(j(\omega - \omega_0)).$$

The shift for the band-pass filter is

$$\omega_0 = \frac{\omega_{c1} + \omega_{c2}}{2}.$$

Hence,

$$h_{bp} = \left(e^{j\frac{\omega_{c1}+\omega_{c2}}{2}} + e^{-j\frac{\omega_{c1}+\omega_{c2}}{2}}\right) h_{lp}(t).$$

The bandwidth of the low-pass filter H_{lp} is

$$BW_{bp} = \omega_{c2} - \omega_{c1}.$$

The cutoff frequency of the corresponding low-pass filter is

$$\omega_c = \frac{\omega_{c2} - \omega_{c1}}{2}.$$

Therefore, the impulse response of the band-pass filter is

$$h_{bp} = 2\cos\left(\frac{\omega_{c1} + \omega_{c2}}{2}\right) \cdot \frac{\sin \pi \left(\frac{\omega_{c1}-\omega_{c2}}{2}\right)}{\pi t}.$$

b) From the Fourier transform table,

$$x(t) = \delta(t) \longleftrightarrow X(j\omega) = 1.$$

Hence,

$$y(t) = h_{bp}(t) * x(t) = h_{bp}(t) = 2\cos\left(\frac{\omega_{c1} + \omega_{c2}}{2}\right) \cdot \frac{\sin \pi \left(\frac{\omega_{c1}-\omega_{c2}}{2}\right)}{\pi t}.$$

c) In the frequency domain, we note that the input signal, $X(j\omega) = 1$, is NOT band limited. On the other hand, the output signal is

$$Y(j\omega) = X(j\omega).H_{bp}(j\omega) = H_{bp}(j\omega) = \begin{cases} 1 & \text{for} \quad \omega_{c1} \le |\omega| \le \omega_{c2}, \\ 0 & \text{otherwise}, \end{cases} \tag{10.38}$$

which is a band limited signal.

In the time domain the output becomes the impulse response of the filter.

As can be observed from the aforementioned examples, ideal filters consist of discontinuities, at the cutoff frequencies, ω_c, ω_{c1}, and ω_{c2}, which gives noncausal impulse responses in the time domain, ranging between $(-\infty, \infty)$. These types of filters are called IIR filters. The discontinuities, also, result in undesired noise effect, when the systems and signals are reconstructed in time domain. Furthermore, the discontinuities are not easy to realize in real life applications. For this reason, the real-life filters are mostly designed with smooth corners, as explained in the following sections.

INTERACTIVE

Filtering with low, band and high-pass @ `https://384book.net/i1001`

10.6 Discrete Time Real Filters

Discrete time real filters approximate the ideal filters by smoothing the discontinuities of the frequency response. The smoothing effect in frequency domain truncates the IIR to make a finite impulse response (FIR) function.

Let us study a few discrete time real filters and their frequency and impulse responses.

10.6.1 Discrete Time Low-Pass and High-Pass Real Filters

Design methodologies of low-pass and high-pass filters are very similar to each other. Both of them can be represented by difference equations. The cutoff frequencies and the bandwidths can be adjusted by the constant parameters of the difference equations. In the following exercises, we provide simple examples of discrete time real low-pass and high-pass filters.

Exercise 10.5 Discrete Time Low-Pass FIR Filter

Suppose that we have the following discrete time LTI system, which averages two consecutive input signals, as follows:

$$y[n] = \frac{1}{2}(x[n] + x[n-1]). \tag{10.39}$$

a) Find the impulse response of this filter.
b) Find the frequency response of this filter.
c) Find the spectral coefficients of the output signal in terms of the spectral coefficients of the input signal, with period $T = 2$.
d) Comment on the effect of the filter on the output signal. What type of a filter is this?
e) Find the output, when the input is

$$x[n] = \sin 0.005\pi n + 0.1 \cos \pi n. \tag{10.40}$$

Solution

a) Impulse response, $h[n]$ can be easily obtained by replacing the input with the impulse function, as follows:

$$h[n] = \frac{1}{2}(\delta[n] + \delta[n-1]). \tag{10.41}$$

b) Frequency response $H(e^{j\omega})$ can be obtained from its definition,

$$H(e^{j\omega}) = \sum_{k=-\infty}^{\infty} h[k]e^{-j\omega k} = \frac{1}{2}(1 + e^{-j\omega}) = \frac{1}{2}e^{-j\frac{\omega}{2}}(e^{j\frac{\omega}{2}} + e^{-j\frac{\omega}{2}}) = e^{-j\frac{\omega}{2}} \cos \frac{\omega}{2}. \tag{10.42}$$

Equivalently, we could obtain the frequency response from the difference equation by replacing the input with the eigenfunction, $x[n] = e^{j\omega n}$ and the corresponding output, $y[n] = H(e^{j\omega})e^{j\omega n}$.

Then, the difference equation of Equation (10.41) becomes

$$y[n] = H(e^{j\omega})e^{j\omega n} = \frac{1}{2}(e^{j\omega n} + e^{j\omega[n-1]}). \tag{10.43}$$

Finally, factorizing the right-hand side of this equation by $e^{j\omega n}$ and using the Euler formula, we obtain the frequency response of this LTI system, as follows:

$$H(e^{j\omega}) = e^{-j\frac{\omega}{2}} \cos \frac{\omega}{2}. \tag{10.44}$$

The impulse response and frequency response of this LTI System is one-to-one and onto, e.g.,

$$h[n] = \frac{1}{2}(\delta[n] + \delta[n-1]) \leftrightarrow H(e^{j\omega}) = e^{-j\frac{\omega}{2}} \cos\frac{\omega}{2}. \tag{10.45}$$

Note: Since the impulse response of this filter has finite time duration between $0 \le n \le 1$, this is a FIR filter.

c) The spectral coefficients, b_k of the output signal, $y(t)$, is obtained from,

$$b_k = a_k H(e^{jk\omega_0}) = \frac{a_k}{2}(1 + e^{-jk\pi}), \tag{10.46}$$

where $\{a_k\}$ are the spectral coefficients and $\omega_0 = \pi$ is the angular frequency of the input signal, $x(t)$, corresponding to the fundamental period, $T = 2$.

d) Now, let us investigate the effect of $H(e^{j\omega})$ on changing the spectral coefficients of the input, $\{a_k\}$ to generate the spectral coefficients of the output, $\{b_k\}$.

In order to observe how this filter shapes the input signal, we can plot the magnitude and phase of the frequency response, given as follows:

Magnitude of the frequency response: $|H(e^{j\omega})| = \cos\frac{\omega}{2}$.
Phase of the frequency response: $\angle H(e^{j\omega}) = \frac{-\omega}{2}$.

The plot of the magnitude and phase spectrum of this filter is given in Figure 10.15.

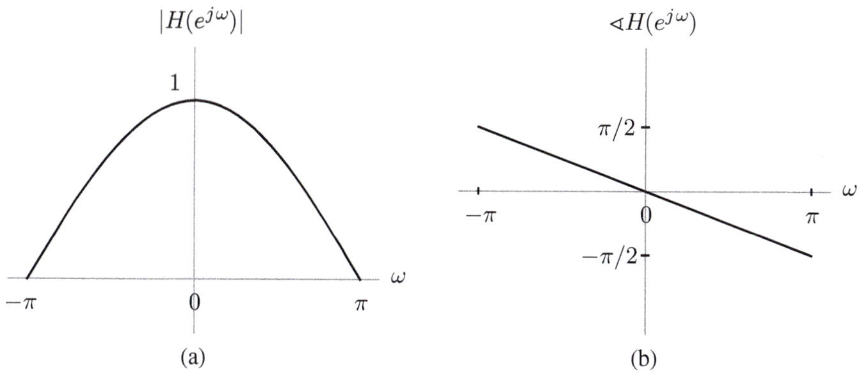

Figure 10.15 (a) Magnitude and (b) phase spectrum of a low-pass filter, represented by the frequency response, $H(e^{j\omega}) = e^{-j\frac{\omega}{2}} \cos\frac{\omega}{2}$. Note that both magnitude and phase plots are periodic, with periods of the angular frequency, $\omega = 2\pi$. The plots show only one full period of the magnitude and phase.

Recall that frequency response scales the spectral parameters of the input to generate the spectral parameters of the output by

$$b_k = a_k H(e^{jk\omega_0}). \tag{10.47}$$

The type of the filter is, basically, specified by the magnitude of the frequency response. When the magnitude of the frequency response is high, at a particular harmonic, $k\omega_0$, the corresponding spectral parameter gets relatively larger, increasing the contribution of that harmonic frequency to the signal. Conversely, when the magnitude of the frequency response gets low, the corresponding harmonics get attenuated.

Phase of the frequency response delays the harmonics, depending on its value at a certain frequency. Large phase shifts result in more delays in the corresponding harmonics.

An analysis of the magnitude plot reveals that the filter of this example attenuates the high-frequency components of the input signal, as the frequency approaches to $|\pi|$. Therefore, this is a **low-pass filter**. Unlike an ideal low-pass filter with a discontinuity a cutoff frequency, ω_c, this filter gradually suppresses the high-frequency components, as $\omega \rightarrow \pi$. For example, if the input is a voice recording, the filter decreases the treble voices, making the sound more bass.

The phase plot of the frequency response shows simply the phase angle we get between the output and input, as a function of frequency, ω. In Figure 10.15, we observe that the phase shift between the input and output signals increases as the frequency increases.

e) The output of this LTI system can be easily obtained by the convolution of the input and impulse response, as follows:

$$y[n] = h[n] * x[n],$$

where the input and impulse response are given as follows:

$$x[n] = \sin 0.005\pi n + 0.1 \cos \pi n, \tag{10.48}$$

$$h[n] = \frac{1}{2}(\delta[n] + \delta[n-1]), \tag{10.49}$$

respectively.

The input signal consists of superposition of two signals,

$$x[n] = x_1[n] + 0.1x_2[n], \tag{10.50}$$

where $x_1[n] = \sin 0.005\pi n$ and $x_2[n] = \cos \pi n$ (see Figure 10.16).

Note that the fundamental period of $x_1[n]$ is $T_1 = 40$ and it is relatively larger than the fundamental period of $x_2[n]$, which is $T_2 = 2$. As a result, $x_2[n]$ adds ripples to $x_1[n]$ to generate $x[n]$.

Inserting the input and the impulse response into the convolution operation, we obtain,

$$\begin{aligned} y[n] &= \left(\frac{1}{2}(\delta[n] + \delta[n-1])\right) * (\sin 0.005\pi n + \cos \pi n) \\ &= \frac{1}{2}(\sin 0.005\pi n + \sin 0.005\pi(n-1)). \end{aligned} \tag{10.51}$$

Comparison of Figures 10.16 and 10.17 shows that the ripples of the input are nicely smoothed by the low-pass filter, which takes the average of the consecutive signals, in time domain.

Loosely speaking, a low-pass filter smooths the input signal, depending on the cutoff frequency, ω_c. The lower the cutoff frequency results in smoother signal, obtained at the output.

Exercise 10.6 Example: Discrete Time Low-Pass IIR Filter

In this example, we shall investigate the filtering properties of discrete time LTI systems, represented by a first difference equation. We assume that the system is initially at rest:

$$y[n] - ay[n-1] = x[n]. \tag{10.52}$$

a) Find and plot the frequency response of this filter and comment on the type of the filter.
b) Find the impulse response and unit step response of this filter.

Solution

As we did before, in the continuous time case, we need to investigate the properties of frequency response.

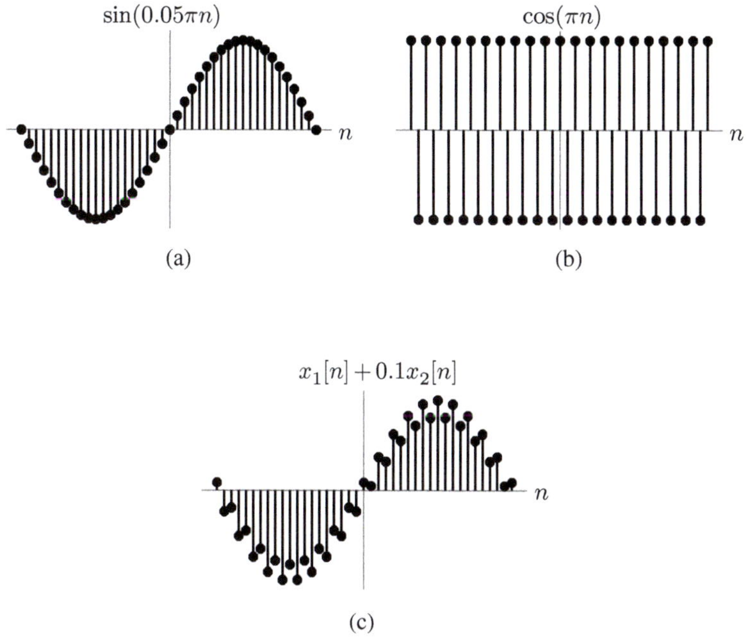

Figure 10.16 The input signal is defined as the addition of two signals: (a) $x_1[n]$ and (b) $x_2[n]$. (c) Addition of $x_1[n]$ and $x_2[n]$ yields $x_1 n] = \sin 0.005\pi n + \cos \pi n$. The signal, $x_1[n]$ has a relatively low fundamental frequency, which is $\omega_0 = 0.05\pi$, compared to the signal $x_2[n]$, which has the fundamental frequency of $\omega_0 = \pi$.

Figure 10.17 The plot of the output signal $y[n] = h[h] * x[n]$ when the input and impulse responses are $x[n] = \sin 0.005\pi n + \cos \pi n$, and $h[n] = \frac{1}{2}(\delta[n] + \delta[n-1])$, respectively.

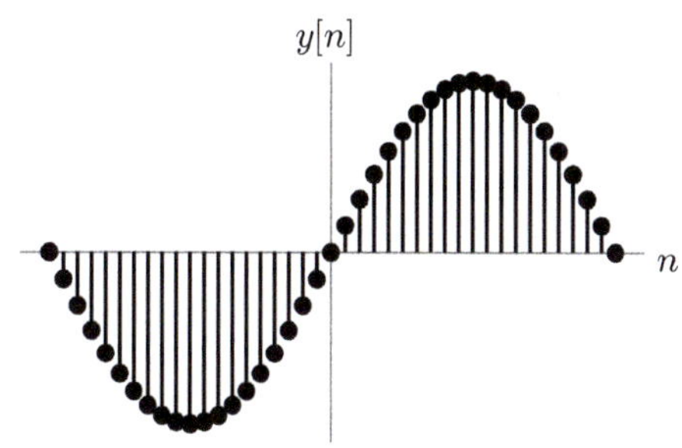

a) Let us feed the eigenfunction of the system at the input, $x[n] = e^{j\omega n}$. The corresponding output is

$$y[n] = e^{j\omega n} H(e^{j\omega}), \tag{10.53}$$

Replacing the input and output in the difference equation, $y[n] - ay[n-1] = x[n]$, we get,

$$\left[1 - ae^{-j\omega}\right] H(e^{j\omega}) e^{j\omega n} = e^{j\omega n}. \tag{10.54}$$

Arranging this equation, we obtain the frequency response,

$$H(e^{j\omega}) = \frac{1}{1 - ae^{-j\omega}}. \tag{10.55}$$

We need two plots for the frequency response:

1) Magnitude of the frequency response: $|H(e^{j\omega})| = \frac{1}{\sqrt{a^2+1-2a\cos(\omega)}}$,
2) Phase of the frequency response: $\angle H(e^{j\omega}) = -\tan^{-1}\left(\frac{a \sin \omega}{1 - a \cos \omega}\right)$.

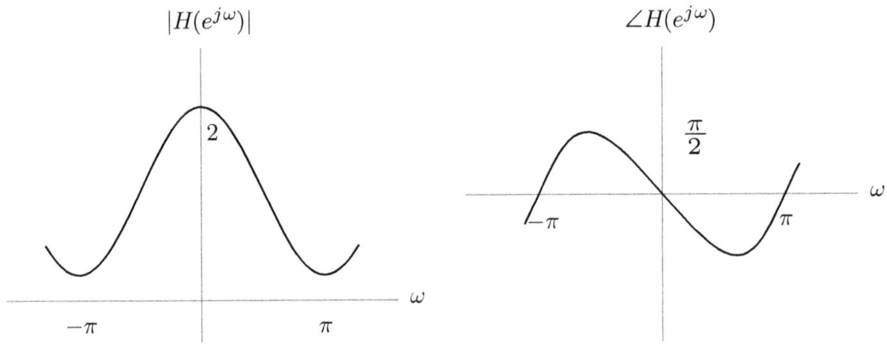

Figure 10.18 One full period of the magnitude and phase plots of the frequency response, $H(e^{j\omega}) = \frac{1}{1-ae^{-j\omega}}$, for the difference equation, $y[n] - ay[n-1] = x[n]$, in the interval, $-\pi \leq \omega \leq \pi$.

Analysis of the plot of magnitude spectrum of this filter, in Figure 10.18, reveals that the high-frequency components an input signal is gradually attenuated as $|\omega| \rightarrow \pi$. Thus, this is a low-pass filter.

b) Impulse response and unit step response can be easily obtained as follows:

$$h[n] = a^n u[n], \tag{10.56}$$

$$s[n] = u[n] * h[n] = \sum_{k=0}^{n} a^k = \frac{1-a^{n+1}}{1-a} u[n], \tag{10.57}$$

respectively.

Note: The impulse response of this filter has an infinite time duration in $0 \leq n \leq \infty$. Hence, it is an IIR filter. However, considering the exponential decay of the function, it approaches very close to zero for large values of n. Furthermore, it does not have any discontinuities in one full period. Therefore, it can be realized with practically good approximation.

Exercise 10.7 Discrete Time High-Pass FIR Filter

Consider the following difference equation:

$$y[n] = \frac{1}{2}(x[n] - x[n-1]). \tag{10.58}$$

a) Find the impulse response.
b) Find and plot the frequency response.
c) Comment on this LTI system. What type of a filter is this?

Solution

a) Impulse response is

$$h[n] = \frac{1}{2}(\delta[n] - \delta[n-1]). \tag{10.59}$$

This a FIR filter, with only two nonzero values at $n = 0$ and $n = 1$.

b) Frequency response is

$$H(e^{j\omega}) = \frac{1}{2} - \frac{1}{2}e^{-j\omega} = j\sin(\omega/2)e^{-j\omega/2}. \tag{10.60}$$

Magnitude and phase spectrum of this frequency response are as follows:

1) The magnitude spectrum: $|H(e^{j\omega})| = |\sin(\omega/2)|$
2) Phase spectrum: $\angle H(e^{j\omega}) = \tan^{-1}(\cot(\omega/2))$.

The analysis of Figure 10.19 reveals that this filter suppresses low-frequency component and gradually passes the high-frequency components of the spectral coefficients of the input signal. Therefore, it is a **high-pass filter**.

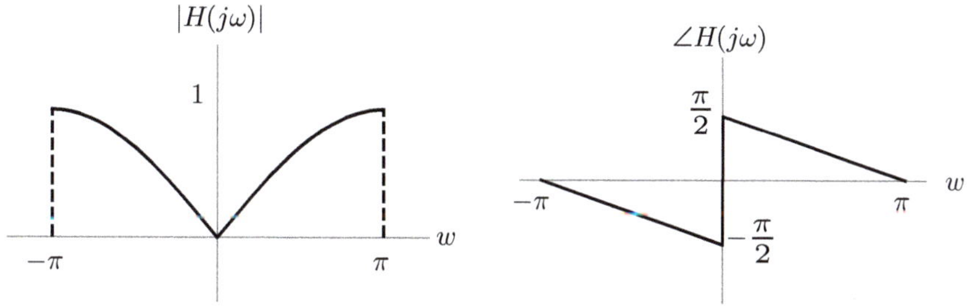

Figure 10.19 Magnitude and phase plot of the frequency response, $H(e^{j\omega}) = j\sin(\omega/2)e^{-j\omega/2}$.

10.6.2 Band-Stop Filters for Filtering Well-Defined Frequency Bandwidths

In the aforementioned examples, we have seen that a **low-pass filter** essentially passes signals with a frequency lower than a selected **cutoff frequency** and attenuates those with a frequency bandwidth higher than the cutoff frequency. Similarly, a high-pass filter passes high-frequency bandwidths, while attenuating the low frequencies.

What if we want to filter a well-defined frequency band? In this case, we design a band-stop filter, which suppresses a predefined frequency band as shown in Figure 10.20.

The following example explains a specific type of band-stop filter, called notched filter.

Exercise 10.8 Example: Autoregressive Models as Notched Filters

The following LTI system is called first-order autoregressive model:

$$y[n] - ay[n-1] = x[n] - bx[n-1], \tag{10.61}$$

where the present value of the output depends on the present and previous value of the input, and previous value of the output. The parameters a and b indicate the degree of dependency of the current value of the output to the past values of output and input, respectively.

Figure 10.20 The frequency response of a band-stop filter.

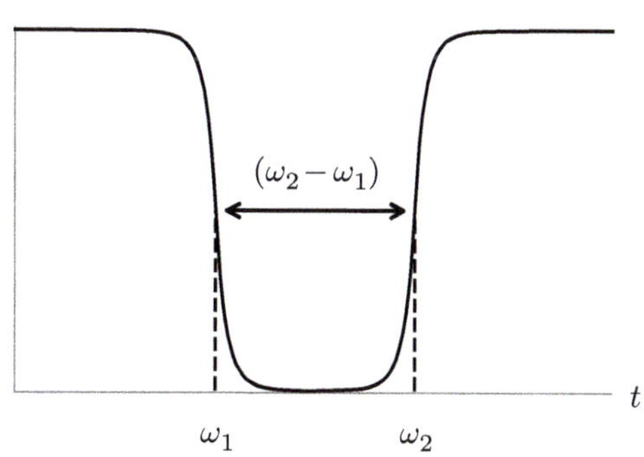

Suppose that the system is initially at rest and we select particular values of the parameters as follows:

$$b = e^{j\omega_c} \quad \text{and} \quad a = 0.99e^{j\omega_c}. \tag{10.62}$$

a) Find the impulse response of this model.
b) Find the frequency response of this model.
c) What type of a filter is this?

Solution

a) Impulse response of this model can be obtained by setting $y[n] = h[n]$ and $x[n] = \delta[n]$, and leaving $h[n]$ in the left-hand side of the equation alone,

$$h[n] = ah[n-1] + \delta[n] - b\delta[n-1]. \tag{10.63}$$

Considering the fact that the system is initially at rest, $h[n] = 0$ for $n < 0$ and using the recursive method, we obtain the impulse response, as follows:

$$h[n] = a^n u[n] - ba^{n-1}u[n-1]. \tag{10.64}$$

This is an IIR filter in $0 \le n \le \infty$.

b) From the definition of the frequency response of discrete time LTI systems, we obtain,

$$H(e^{j\omega}) = \sum_{n=0}^{\infty} h[n]e^{-j\omega} = \frac{1 - be^{-j\omega}}{1 - ae^{-j\omega}}. \tag{10.65}$$

c) Let us replace,

$$b = e^{j\omega_c} \quad \text{and} \quad a = 0.99e^{j\omega_c}, \tag{10.66}$$

in the frequency response,

$$H(e^{j\omega}) = \frac{1 - e^{j\omega_c}e^{-j\omega}}{1 - 0.99e^{j\omega_c}e^{-j\omega}} = \frac{1 - e^{-j(\omega-\omega_c)}}{1 - 0.99e^{-j(\omega-\omega_c)}}. \tag{10.67}$$

This filter has a very interesting property, which eliminates the frequency at $\omega = \omega_c$. In this case, the numerator becomes

$$1 - e^{j(\omega c - \omega c)} = 1 - 1 = 0. \tag{10.68}$$

When $\omega \neq \omega c$, the numerator and denominator get very close to each other,

$$H(e^{j\omega}) = \frac{1 - e^{-j(\omega-\omega_c)}}{1 - 0.99e^{-j(\omega-\omega_c)}} \approx 1. \tag{10.69}$$

This filter is called **notched filter**. It eliminates a very narrow band around the frequency, $\omega = \omega_c$. Notch filters are special types of **band-stop filters**, which attenuates a signal in a predefined frequency interval around, ω_c to very low levels and pass the rest unaltered.

A notch filter is basically a band-stop filter with a narrow **stopband**. When we take a look at its frequency response, we can see a very narrow "V" shape (see Figure 10.21). Hence the name "Notch."

Suppose we have a noisy audio recording. Suppose also that the actual signal lies in the interval of $f = 60$–160 Hz and the noise occurs just around 90 Hz. Low-pass or high-pass filters do not offer a solution to remove the noise. That is where the **notch filters** come in handy. If we set, $\omega_c = 2\pi \times 90$, the notch filter of Equation (10.69) eliminates the noise at 90 Hz.

WATCH

Learn more about designing a notch filter @ https://384book.net/v1002

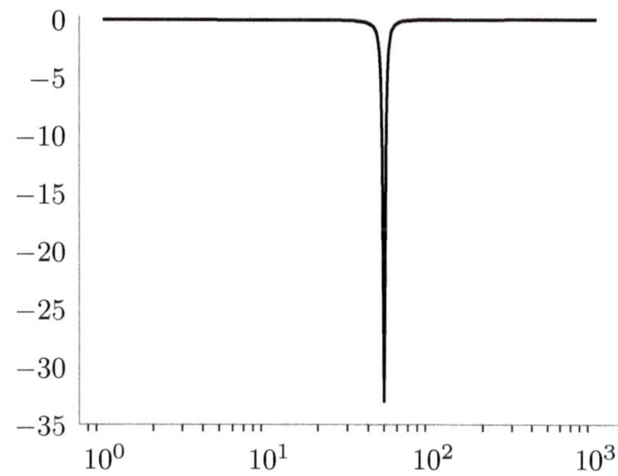

Figure 10.21 The frequency response of a notch filter.

10.7 Continuous Time Real Filters

When the input–output pairs of an LTI system are represented by continuous time signals, we need to design a continuous time filter. Recall that the frequency response of a continuous time system is not periodic with respect to the frequency variable, ω.

Exercise 10.9 First Derivative Filter

Consider the continuous time LTI represented by a simple derivative operator.

$$x(t) \rightarrow \boxed{\frac{d}{dt}} \rightarrow y(t) = \frac{dx(t)}{dt}. \tag{10.70}$$

a) Find the frequency response of this filter.
b) Find the spectral coefficients of the output signal in terms of the spectral coefficients of the input signal.
c) Comment on the effect of the filter on the output signal. What type of a filter is this?

Solution

a) A simple way of finding the frequency response is to find the generalized eigenvalue of the LTI system for a continuous domain of frequency, ω. This task is achieved by finding the output, $y(t)$, of an LTI system, when the input is $x(t) = e^{j\omega t}$,

$$y(t) = (j\omega)e^{j\omega t}, \tag{10.71}$$

which directly gives us the frequency response of the system as the scaling factor of the complex exponential as follows:

$$H(j\omega) = j\omega. \tag{10.72}$$

In order to analyze the effect of this filter on an input signal, we need to find and plot the magnitude and phase of it.

1) Magnitude of the frequency response: $|H(e^{j\omega})| = |\omega|$
2) Phase of the frequency response: $\angle H(e^{j\omega}) = |\pi/2|$,

The analysis of the magnitude plot of the frequency response of the first derivative filter, in Figure 10.22 reveals that this filter linearly attenuates the low frequency components of the signal. For example, if the input signal is a voice recording, the filter will trim the low-frequency components yielding a more treble voice at the output.

The phase plot shows that there is a constant phase shift of $|\pi|$, at all frequencies, which results in a constant delay between the input and output signals.

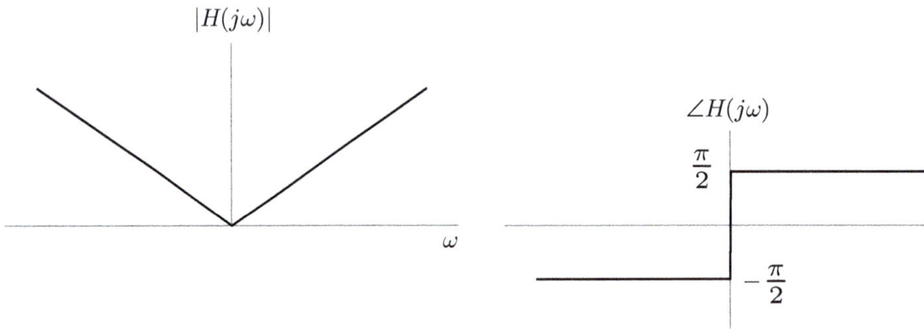

Figure 10.22 Magnitude and phase plots of the frequency response, $H(j\omega) = j\omega$. of the first derivative filter.

Exercise 10.10 Continuous Time Low-Pass Filter

In this example, we shall explore the properties of a first-order differential equation as a filter. We shall see that a continuous time LTI system represented by the following first-order differential equation, is **a low-pass filter,** provided that it is initially at rest.

$$a\frac{dy(t)}{dt} + y(t) = x(t). \tag{10.73}$$

This filter attenuates the high-frequency components of an input signal at the output of the system. The degree of attenuation is determined by the constant coefficient, a, of the differential equation.

Let us answer the following questions to investigate the behavior of the first-order constant coefficient differential equation given in Equation (10.73).

a) Find the block diagram representation of this filter.
b) Find the frequency response of this filter.
c) Find the real and imaginary part of the frequency response. Can you comment on the type of the filter by analyzing real and imaginary part of the frequency response?
d) Find and plot the magnitude and phase of the frequency response and comment on the type of filter.
e) Find the impulse response and unit step response of this filter.

Solution

a) In order to build such a filter, we need an adder and an integrator, as depicted in Figure 10.23.
b) Let us first find the frequency response, as a scaling factor, $H(j\omega)$ of the eigenfunction $x(t) = e^{j\omega t}$, of this system, as follows:

$$y(t) = H(j\omega)e^{j\omega t}. \tag{10.74}$$

Let us take the derivative of the both sides of the aforementioned equation,

$$\frac{dy(t)}{dt} = (j\omega)H(j\omega)e^{j\omega t}, \tag{10.75}$$

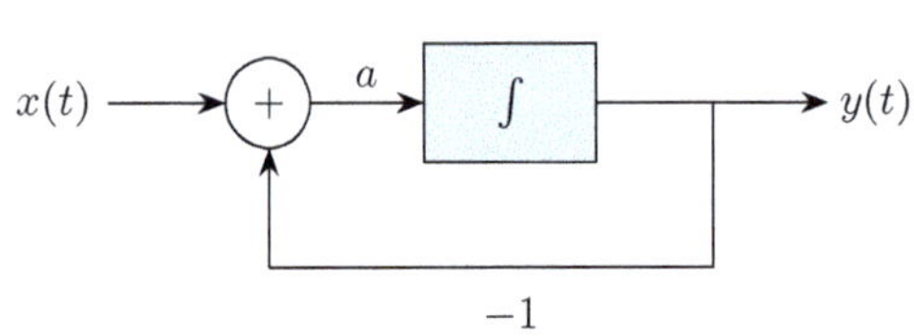

Figure 10.23 Block diagram representation of a first-order differential equation.

and insert it in the differential equation,

$$(aj\omega + 1)H(j\omega)e^{j\omega t} = e^{j\omega t}, \tag{10.76}$$

to find the frequency response,

$$H(j\omega) = \frac{1}{1 + aj\omega} = \frac{1 - aj\omega}{1 + a^2\omega^2}. \tag{10.77}$$

c) Real and imaginary part of this complex frequency response can be obtained, as follows:

$$Re\{H(j\omega)\} = \frac{1}{1 + a^2\omega^2}, \tag{10.78}$$

and

$$Im\{H(j\omega)\} = \frac{-aj\omega}{1 + a^2\omega^2}. \tag{10.79}$$

By analyzing the real and imaginary part, it is not easy to observe the type of the filter.

d) Using the definitions, we compute the magnitude, given as follows:

$$|H(j\omega)| = (\text{Re }\{H(j\omega)\}^2 + \text{Im }\{H(j\omega)\}^2)^{\frac{1}{2}}, \tag{10.80}$$

and phase,

$$\angle H(j\omega) = \tan^{-1}\frac{\text{Im}\{H(j\omega)\}}{\text{Re}\{H(j\omega)\}}, \tag{10.81}$$

we compute the magnitude of the frequency response as follows:

$$|H(j\omega)| = \frac{1}{(1 + a^2\omega^2)^2} + \frac{(a\omega)^2}{(1 + a^2\omega^2)^2} = \sqrt{\frac{1 + a^2\omega^2}{(1 + a^2\omega^2)^2}}. \tag{10.82}$$

Simplifying the aforementioned equation, we obtain the magnitude and the phase of the frequency response, as follows:

$$|H(j\omega)| = \frac{1}{\sqrt{1 + a^2\omega^2}}, \tag{10.83}$$

and

$$\angle H(j\omega) = \tan^{-1} - a\omega, \tag{10.84}$$

respectively.

The magnitude plot of the frequency response, in Figure 10.24 shows that a first-order differential equation is a low-pass filter, which smoothly attenuates the high-frequency components of the input signal. The phase plot shows that the signal is smoothly delayed as the frequency increases.

e) The impulse response can be obtained by solving the differential equation for unit impulse input.

$$a\frac{dh(t)}{dt} + h(t) = \delta(t), \tag{10.85}$$

Homogeneous solution:

$$h_H(t) = Ke^{-t/a}u(t), \tag{10.86}$$

$$a\int_{0^-}^{0^+} dh(t) + \int_{0-}^{0+} h(t)dt = \int \delta(t)dt = 1, \tag{10.87}$$

$$ah(0^+) - ah(0^-) = 1, \tag{10.88}$$

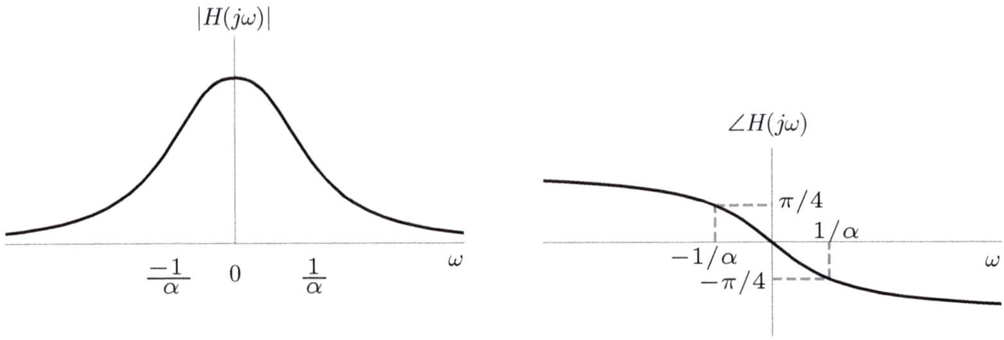

Figure 10.24 Magnitude and the phase plot of the frequency response of the first-order differential equation.

we know $h(0^-) = 0$ (initially at rest), therefore:

$$h(0^+) = \frac{1}{a}, \tag{10.89}$$

$$h(t) = \frac{1}{a}e^{-t/a}u(t), \tag{10.90}$$

Note: This is an IIR filter with an exponential decay and it has no discontinuities. Thus, it can be realized with satisfactory approximations for large t.

The unit step response can be easily obtained by taking the integral of the impulse response, as follows:

$$\begin{aligned} s(t) &= \int_{-\infty}^{t} h(\tau)d\tau = \frac{1}{a}\int_{0}^{t} e^{-\frac{\tau}{a}}d\tau \\ &= (1 - e^{-\frac{t}{a}})u(t). \end{aligned} \tag{10.91}$$

Figure 10.25 comperes the impulse response and unit step response of a low pass filter.

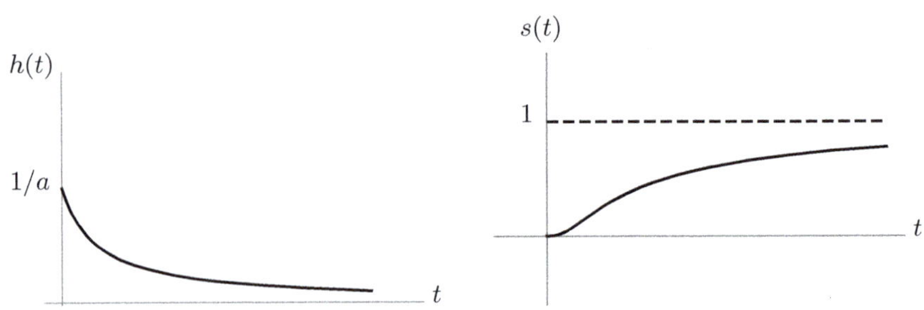

Figure 10.25 The impulse response and unit step response of a low-pass filter represented by a first-order differential equation.

Exercise 10.11 Continuous Time High-Pass Filter

This time, we shall investigate the continuous time LTI systems, represented by a first-order differential equation, where in the right-hand side, we have the derivative of the input signal, as given as follows:

$$\frac{dy(t)}{dt} + ay(t) = \frac{dx(t)}{dt}. \tag{10.92}$$

a) Find the block diagram representation of this filter.
b) Find and plot the frequency response of this filter and comment on the type of the filter.
c) Find the impulse response and unit step response of this filter.

Solution

a) The block diagram representation of this filter consists of a differentiator, an integrator, and an adder, as shown in Figure 10.26.

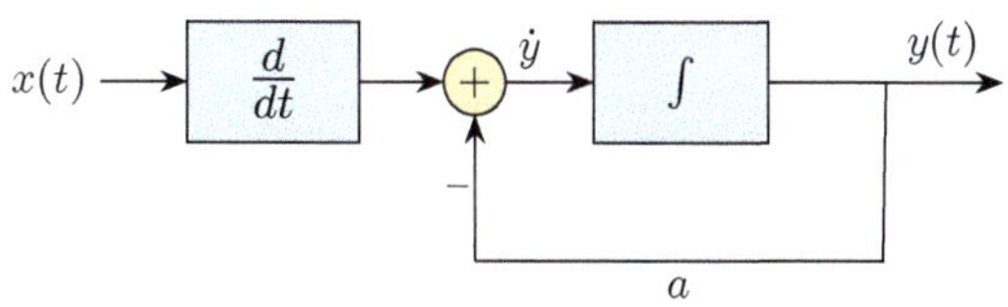

Figure 10.26 We take the derivative of the input signal by a differentiator. Then, we subtract the scaled output, $ay(t)$ from the derivative of the input to obtain the derivative of the output. The integrator returns the output, $y(t)$ from its derivative, $\dot{y}(t)$.

b) The frequency response of this system can be directly obtained by replacing the input with the complex exponential eigenfunction,

$$x(t) = e^{j\omega t}, \tag{10.93}$$

and obtain the corresponding output, as the scaled version of the eigenfunction, as follows:

$$y(t) = H(j\omega)e^{j\omega t}. \tag{10.94}$$

Taking the derivative of the input, $\dot{x}(t) = (j\omega)e^{j\omega t}$ and the derivative of the output, $\dot{y}(t) = (j\omega)H(j\omega)e^{j\omega t}$, and replacing them in the differential equation, we get,

$$(j\omega + a)H(j\omega) = j\omega,$$
$$H(j\omega) = \frac{j\omega}{a + j\omega}. \tag{10.95}$$

Next, we need to find and plot the magnitude of the frequency response.

Magnitude of the frequency response, $|H(j\omega)| = \frac{\omega}{\omega^2+a^2}$

Phase of the frequency response is $\angle H(j\omega) = \tan^{-1}\frac{a}{\omega}$.

As it is observed from Figure 10.27, the magnitude and phase plots of the frequency response of this filter attenuates the low-frequency components of the input signal. The degree of attenuation is determined by the constant coefficient, a, of the differential equation.

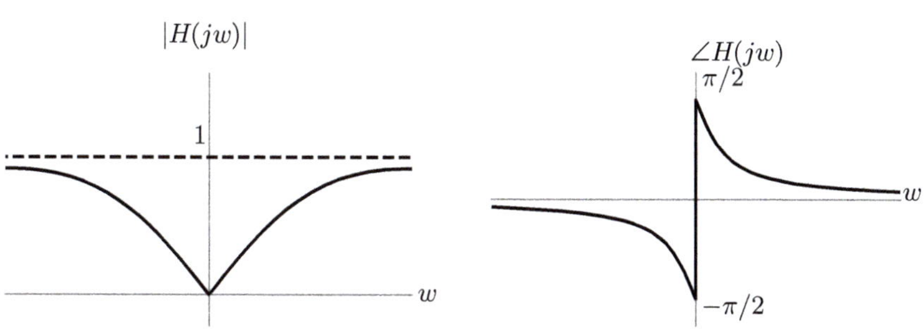

Figure 10.27 Magnitude and the phase plots of the frequency response, $H(j\omega) = \frac{j\omega}{a+j\omega}$.

10.8 Chapter Summary

Fourier series representation enables us to decompose a periodic signal into its harmonically related frequency components. In this representation, spectral coefficients can be considered as the measures, which show the amount of a particular frequency in the signal. Small spectral coefficients indicate relatively less proportion of the corresponding frequency harmonics, contained in the signal compared to large spectral coefficients. Fourier transforms take the limit of harmonically related frequency components and define a continuous spectrum of frequencies. Therefore, Fourier transform of a signal manifests the frequency content of a time domain signal.

In the LTI systems, the spectral coefficients of the output signal are just the scaled version of the spectral coefficients of the periodic input signal. When the input signal is aperiodic, the output signal is just the scaled version of the input signal, in the Fourier domain. The scaling factor is an eigenvalue of the system and it is called the frequency response. Therefore, it is possible to design the frequency response of an LTI system to reshape the input signals. This process is called filtering in the frequency domain. Depending on a pre-defined goal, it is possible to suppress the undesired frequency components or accentuate some others. Design of the filters is an important area in digital signal processing. In the context of this book, we just study special forms of high-pass, low-pass, band-pass, and band-stop filters.

Please keep in mind that designing filters is not an easy task and it requires elaborate methods, which are beyond the scope of this book.

Problems

10.1 A discrete time filter, which averages K consecutive input signals, is given as follows:

$$y[n] = \frac{1}{K}\sum_{k=0}^{K-1} x[n-k].$$

a) Find and plot the magnitude and phase of the frequency response of this filter.
b) What type of filter is this?
c) Find the output, when the input is $x[n] = \sin(0.2n)$ for $K = 2$ and $K = 3$.
d) What happens to the output $y[n]$, as we increase K?

10.2 A continuous time ideal low-pass filter is represented by the following frequency response:

$$H(j\omega) = \begin{cases} 1 & |\omega| < \omega_c \\ 0 & |\omega| \geq \omega_c. \end{cases}$$

a) Find and plot the impulse response of this filter, for $|\omega_c| = 50\pi$ radians/second. Indicate the cutoff frequency on the plot.
b) Find impulse response as $|\omega_c| \to \infty$. What type of a filter is this?
c) Given that the input signal $x(t)$ is periodic with coefficients a_k and fundamental period of the input is $T = 1$ seconds, what are the spectral coefficients of the output in terms of a_k, for $\omega_c = 50\pi$ radians/second?
d) Suppose that $y(t) = x(t)$, find the Fourier series coefficients of the input $x(t)$ for $\omega_c = 20\pi$ radians/second?

10.3 A continuous time ideal high-pass filter is represented by the following frequency response

$$H(j\omega) = \begin{cases} 1 & |\omega| \geq \omega_c \\ 0 & |\omega| < \omega_c. \end{cases}$$

a) Find and plot the impulse response of this filter, for $|\omega_c| = 50\pi$ radians/second. Indicate the cutoff frequency on the plot.
b) What happens to the impulse response of the high-pass filter, as we increase the cutoff frequency to $|\omega_c| = 100\pi$ radians/second?
c) Given that the input signal $x(t)$ is periodic with coefficients a_k and fundamental period of the input is $T = 0.1$, what are the spectral coefficients of the output in terms of a_k, for $\omega_c = 60\pi$ radians/second?
d) Suppose that $y(t) = x(t)$, find the Fourier series coefficients of the input $x(t)$ for $\omega_c = 40\pi$ radians/second?

10.4 A continuous time ideal band-pass filter is represented by the following frequency response:

$$H_{bp}(j\omega) = \begin{cases} 1 & \text{for} \quad \omega_{c1} \leq |\omega| \leq \omega_{c2}, \\ 0 & \text{otherwise.} \end{cases} \tag{10.96}$$

a) Find and plot the impulse response of this filter, for $|\omega_{c1}| = 50\pi$ radians/second and $|\omega_{c2}| = 100\pi$ radians/second. Indicate the cutoff frequencies on the plot.
b) What happens to the analytical form of the impulse response of the filter, as we increase the cutoff frequency to $|\omega_{c1}| = 150\pi$ radians/second and $|\omega_{c2}| = 200\pi$ radians/second?
c) Given that the input signal $x(t)$ is periodic with coefficients a_k and fundamental period of the input is $T = 0.01$, what are the spectral coefficients of the output in terms of a_k, for $\omega_c = 50\pi$ radians/second?

10.5 A continuous time real filter is represented by the following frequency response:

$$H(j\omega) = \frac{200 + 2\omega^2}{-\omega^2 - 110j\omega + 1000}.$$

a) Find the differential equation that represents this filter.
b) Find and plot the magnitude and phase of the frequency response of this filter.
c) What type of a filter is this?
d) Find and plot the output $Y(j\omega)$, when the input is $x(t) = e^{-t}u(t)$.

10.6 A continuous time filter is represented by the following frequency response:

$$H(j\omega) = \frac{8 + 8j\omega}{(4 + j\omega)(2 + j\omega)}.$$

a) Find the differential equation that represents this filter.
b) Find and plot the magnitude and phase of the frequency response.
c) What type of a filter is this?
d) Find and plot the output $y(t)$, when the input is $X(j\omega) = e^{-2j\omega}$.

10.7 A continuous time LTI system is represented by the following equation:

$$y(t) = 6x(t - 2\pi).$$

a) Find the Fourier transform of the output, when the input is $x(t) = \sin(\omega_0 t + \phi_0)$.
b) Find and plot the magnitude and phase of the frequency response of this system.
c) Comment on the behavior of the system.
d) Find and plot the output $y(t)$, when the input is $X(j\omega) = e^{-2j\pi\omega}$.

10.8 A discrete time LTI system is represented by the following equation:

$$y[n] = 2x[n-2].$$

a) Find and plot the magnitude and phase of the frequency response of this system.
b) Find the spectral coefficients of the output $y(t)$, when the input is $x[n] = \cos(\omega_0 n + \phi_0)$.
c) Comment on the behavior of the system.

10.9 An initially at rest discrete time filter is represented by the following difference equation:

$$y[n] + y[n-1] = x[n] - x[n-1].$$

a) Find and plot the magnitude and phase of the frequency response of this system.
b) Find the impulse response of this filter.
c) Suggest an ideal high-pass filter that approximates this system.

10.10 An initially at rest discrete time LTI system is represented by the following difference equation:

$$y[n] = 3y[n-1] - 2y[n-2] + x[n] - 4x[n-1].$$

a) Find and plot the magnitude and phase of the frequency response of this system.
b) What type of a filter is this system?
c) Suggest an ideal filter that approximates this system. Be specific about the bandwidth and cutoff frequencies.

10.11 Consider the impulse response of the low-pass Gaussian filter given as follows:

$$h(t) = \frac{1}{\sqrt{2\pi\sigma}} e^{-\frac{t^2}{2\sigma^2}},$$

where σ is the standard deviation of the filter.
a) Find the transfer function of this filter.
b) Find the equation that represents this filter.
c) What type of a filter is this? Discuss about the effect of the parameter σ on the structure of the filter.
d) Find the frequency response and impulse response of an ideal filter that approximates this system.

10.12 Consider the transfer function of a second-order Butterworth filter, given as follows:

$$H(s) = \frac{1}{(s + e^{j\pi/4})(s + e^{-j\pi/4})}.$$

a) Find the differential equation that represents this filter.
b) Find the impulse response of this filter.
c) What type of a filter is this?

10.13 The following differential equation represents an initially at rest continuous time LTI system:

$$\frac{d^2y(t)}{dt^2} + 5\frac{dy(t)}{dt} + 4y(t) = x(t).$$

a) Find and plot the magnitude and phase of the frequency response of this system.
b) Find and plot the impulse response.
c) What type of a filter is this system?
d) Suggest the frequency response and impulse response of an ideal filter that approximates this system.

10.14 An initially at rest continuous time LTI system is represented by the following differential equation:

$$5\frac{d^2y(t)}{dt^2} + 2\frac{dy(t)}{dt} + 5y(t) = 3x(t).$$

a) Find and plot the magnitude and phase of the frequency response of this system.
b) Plot a block diagram representation of this filter.
c) What type of a filter is this system?
d) Suggest an ideal filter, which approximates the aforementioned filter. Be specific about the bandwidth and cutoff frequencies.

10.15 An initially at rest LTI system is represented by the following differential equation:

$$\frac{d^2y(t)}{dt^2} + 5\frac{dy(t)}{dt} + 4y(t) = \frac{d^2x(t)}{dt^2} + 9\frac{dx(t)}{dt} + 6x(t).$$

a) Find and plot the magnitude and phase of the frequency response of this system.
b) Find the unit step response of this system.
c) What type of a filter is this system?
d) Suggest the frequency response and impulse response of an ideal filter that approximates this filter. Be specific about the bandwidth and cutoff frequencies.

10.16 The following frequency response represents an initially at rest discrete time LTI system:

$$H(e^{j\omega}) = 1 + 0.5e^{-j\omega}.$$

a) Plot the magnitude and phase of the frequency response of this system.
b) Find and plot the step response of this system.
c) Find the difference equation, which represents this system.
d) What type of a filter is this system?

10.17 The following frequency response represents an initially at rest discrete time LTI system:

$$H(e^{j\omega}) = \frac{1 + 2e^{-2j\omega}}{1 + 0.5e^{-j\omega}}.$$

a) Plot the magnitude and phase of the frequency response of this system.
b) Find and plot the impulse response of this system.
c) What type of a filter is this system?
d) Suggest the frequency response and impulse response of an ideal filter that approximates this filter. Be specific about the bandwidth and cutoff frequencies.

10.18 An initially at rest discrete time LTI system is represented by the following frequency response:

$$H(e^{j\omega}) = \frac{1}{(1-0.25e^{-j\omega})(1+0.75e^{-j\omega})}.$$

a) Plot the magnitude and phase of the frequency response of this system.
b) Find and plot the impulse response of this system.
c) What type of a filter is this system?
d) Plot a block diagram representation of this system.

11

Continuous Time Sampling

"I visualize a time, when we will be to robots what dogs are to humans, and I'm rooting for the machines."

Claude Shannon

Until now, we have defined signals and systems in two types:

- Continuous time signals and systems,
- Discrete time signals and systems.

We represented and analyzed these signals and systems in time and frequency domains separately. A system with its input and output could be either discrete or continuous, but it could not be both. There was a great wall between the discrete time and continuous time signals and systems.

Motivating Question: Is it possible to break down this wall? Given a continuous time signal, can we find its discrete version without the loss of any information? Or can we convert a discrete time signal into a continuous time counterpart?

In order to answer these questions, let us investigate the continuous time function in Figure 11.1. No matter how close we select a set of discrete points on this function, there are infinitely many continuous time functions that pass from a finite set of discrete samples. Therefore, the intuitive answer to the aforementioned questions would be "no!"

However, C. Shannon showed that it is possible to represent a continuous time, band-limited function with finitely many discrete samples under certain conditions. This pioneering discovery, called the sampling theorem, opened the door to the age of digital revolution. This week, we shall see that fitting a unique continuous time function to a set of discrete samples is possible through the sampling theorem.

The sampling theorem bridges the continuous time and discrete time functions. It enables us to convert a continuous time function into a discrete time function by using **sampling techniques**. Interestingly, we can convert the discrete time function to a continuous time function, without losing information by using some **reconstruction techniques**.

Loosely speaking, given a continuous time signal, we can uniquely find its discrete counterpart, provided that the continuous time signal is band-limited. Recall that a continuous time signal is band-limited when the bandwidth of its Fourier transform is finite. Similarly, given a discrete time signal, we can uniquely find its continuous counterpart, provided that the discrete time signal is also band-limited. In both cases sampling theorem provides the necessary and sufficient conditions of sampling and reconstruction, without losing any information about the signals and the systems.

Signals and Systems: Theory and Practical Explorations with Python,
First Edition. Fatoş Tunay Yarman Vural and Emre Akbaş.

Companion website: www.wiley.com/go/signalsandsystems

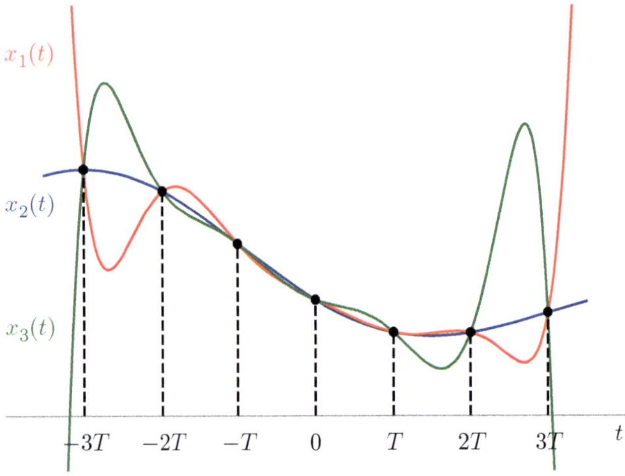

Figure 11.1 Given finitely many discrete points, we can define infinitely many functions, such as $x_1(t), x_2(t), x_3(t), \ldots$, etc., passing from these discrete points. Intuitively, it looks impossible to define finitely many samples to represent a continuous time signal without losing information.

Mathematically speaking, a continuous time signal and its sampled counterpart is one-to-one and onto,

$$x(t) \longleftrightarrow x[n], \tag{11.1}$$

under the conditions of the Sampling theorem, which will be stated later.

In the following, first, we provide the formal definition of sampling and reconstruction. Then, we state the sampling theorem, similar to the original paper of C. Shannon, published in 1948.

11.1 Sampling

A continuous time signal $x(t)$ is sampled by multiplying it with an impulse train of period T,

$$p(t) = \sum_{n=-\infty}^{\infty} \delta(t - nT),$$

as shown in Figure 11.2.

Sampled signal is defined as:

$$x_p(t) = x(t)p(t) = \sum_{n=-\infty}^{\infty} x(nT)\delta(t - nT),$$

where the **sampling period** is T.

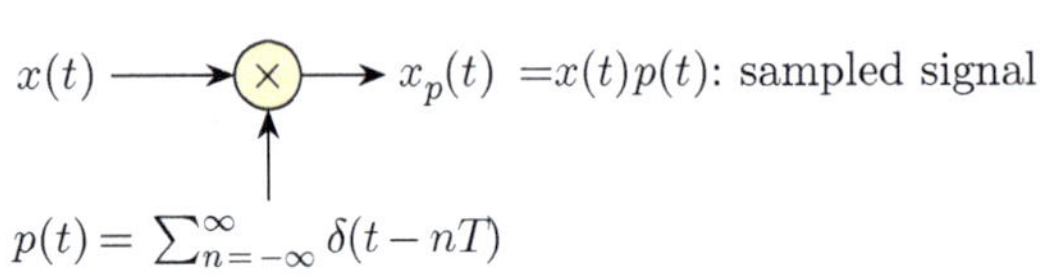

Figure 11.2 Block diagram representation of sampling process. A continuous time signal $x(t)$ is multiplied by an impulse train. The output signal is called sampled signal, $x_p(t)$, with sampling period T.

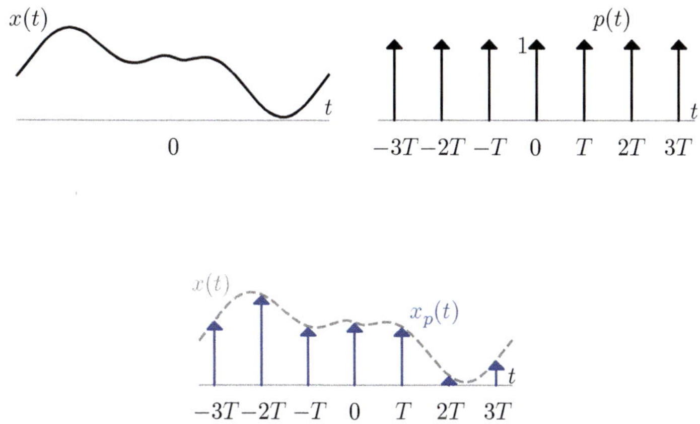

Figure 11.3 Continuous time signal $x(t)$ is multiplied by an impulse train $p(t)$ to obtain the sampled signal $x_p(t)$. That is, $x_p(t) = x(t)p(t)$. Note that the sampled signal $x_p(t)$ skips all the points of the original continuous time signal $x(t)$ between two impulses, which repeats at every period nT for all $-\infty < n < \infty$.

Since it uses an impulse train, this type of sampling is sometimes called **impulse train sampling**. Figure 11.3 shows the continuous time signal $x(t)$ and its sampled version $x_p(t)$, when sampled by an impulse train signal, $p(t)$.

11.2 Properties of the Sampled Signal in Time and Frequency Domains

Let us investigate the properties of the signal,

$$x(t) \rightarrow X(j\omega)$$

and its sampled version

$$x_p(t) \rightarrow X_p(j\omega),$$

in time and frequency domain, where the signal is band limited with bandwidth $2\omega_M$, as depicted in Figure 11.4.

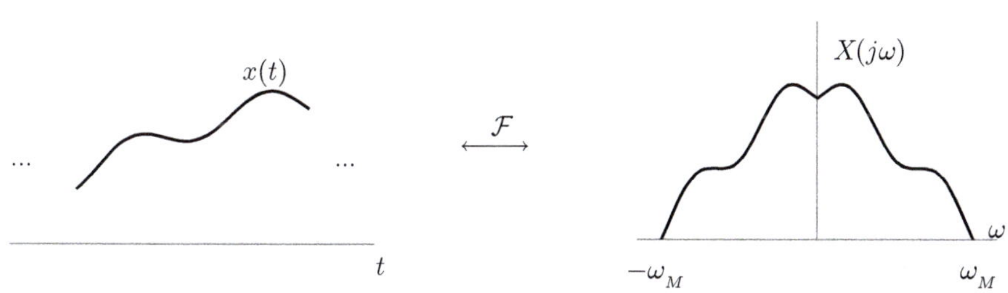

Figure 11.4 A band-limited signal $x(t)$ ranges $-\infty < t < \infty$, in time domain. However, since it is band-limited in the frequency domain, $X(j\omega) = 0$ outside the interval $-\omega_M < \omega < \omega_M$.

In the time domain, sampling operation involves the multiplication of the signal $x(t)$ with impulse train $p(t)$, which gives the sampled signal $x_p(t)$;

$$x_p(t) = x(t)p(t) = \sum_{k=\infty}^{\infty} x(kT)\delta(t - kT). \tag{11.2}$$

In the frequency domain, the multiplication operations correspond to the convolution of the Fourier transform of the signal and that of the impulse train,

$$X_p(j\omega) = \frac{1}{2\pi}X(j\omega) * P(j\omega) = \frac{1}{2\pi}\int_{-\infty}^{\infty} X(j\theta)P(j(\omega - \theta))d\theta. \tag{11.3}$$

Thus, impulse train sampling involves the convolution of the Fourier transform of the signal with that of the impulse train, in the frequency domain.

Recall that impulse train preserves its analytical form in both time and the frequency domain (see Exercise 8.9). Hence, the Fourier transform of the impulse train is also an impulse train, given as follows:

$$p(t) = \sum_{k=-\infty}^{\infty} \delta(t - kT) \longleftrightarrow P(j\omega) = \frac{2\pi}{T}\sum_{k=-\infty}^{\infty} \delta(\omega - k\omega_s), \tag{11.4}$$

where $T = \frac{2\pi}{\omega_s}$ is the sampling period and ω_s is the sampling frequency. Note that time domain representation of impulse train repeats itself at every period T, whereas the frequency domain impulse train repeats itself at every period $2\pi/T$. The amplitude of the impulse train is also scaled by $2\pi/T$ in the frequency domain, as shown in Figure 11.5.

Inserting the Fourier transform of impulse train $P(j\omega)$ into the convolution integral of Equation (11.3), we obtain the Fourier transform of the sampled signal, as follows:

$$X_p(j\omega) = \frac{1}{T}\sum_{k=-\infty}^{\infty} X(j(\omega - k\omega_s)). \tag{11.5}$$

Interestingly, the sampled signal $X(j\omega)$, in the frequency domain is a periodic function, which is generated by shifting the original function, $X(j\omega)$ with $k\omega_s$ for all integer values of $-\infty < k < \infty$.

Figure 11.6 shows the impulse train sampling in time and frequency domain. As it can be observed from this figure, the sampled signal in time domain consists of the superposition of the impulse train, weighted with the amplitude of the signal $x(t)$.

On the other hand, the sampled signal in the frequency domain is just the repetition of the Fourier transform of the continuous time signal at every sampling frequency, $\omega_s = 2\pi/T$. Note that the original function, $X(j\omega)$ is scaled by $1/T$, during the sampling in the frequency domain.

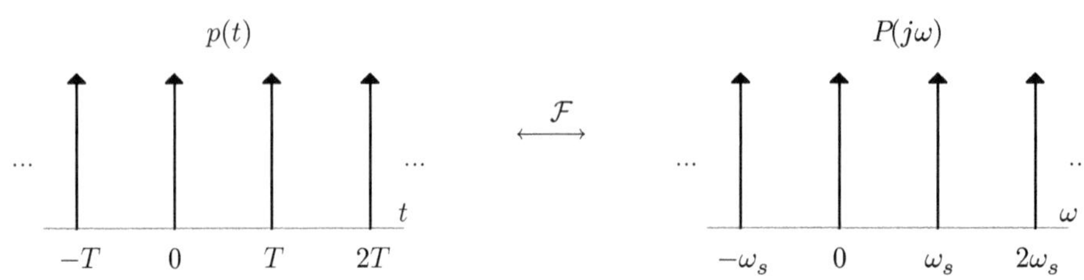

Figure 11.5 Impulse train in time domain (left) and its Fourier transform: $p(t) \leftrightarrow P(j\omega)$ (right). While the fundamental period of $p(t)$ is T and its amplitude is 1, in time domain; the fundamental frequency of $P(j\omega)$ is $\omega_s = 2\pi/T$ and the amplitude is $2\pi/T$, in frequency domain.

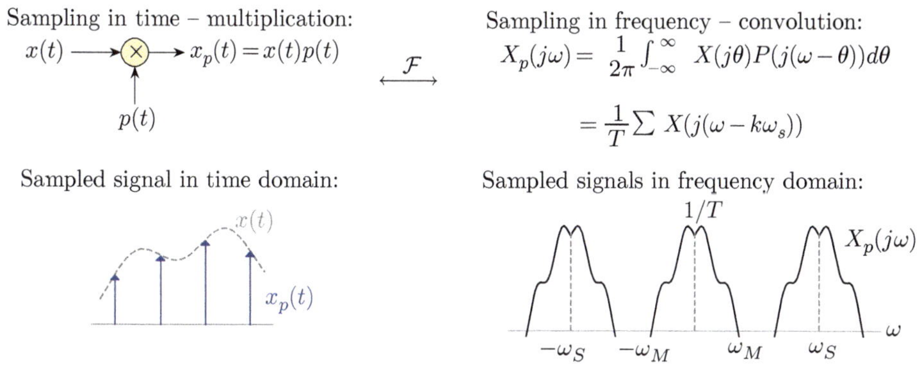

Figure 11.6 Sampled signal in time and frequency domains.

In the time domain, it looks as if we lose information about the signal by skipping infinitely many values of the continuous time signal $x(t)$ in between each sampling period, T. However, we observe that the sampling process carries all the information about the signal $X(j\omega)$, in the frequency domain. Moreover, the sampled signal creates a periodic signal in the frequency domain, where at each period, it carries the original signal, $X(j\omega)$, scaled by $1/T$.

In the following exercise, we study the behavior of the sampled signal in time and frequency domains.

Exercise 11.1 Consider an impulse train sampling,

$$x_p(t) = x(t)p(t),$$

where the input signal is

$$x(t) = \sin \omega_0 t$$

and the time domain impulse train is

$$p(t) = \sum_{k=-\infty}^{\infty} \delta\left(t - \frac{k}{3}\right).$$

a) What is the sampling period T and sampling frequency ω_s?
b) Find $x_p(t)$, for $\omega_0 = \pi/2$.
c) Find $X_p(j\omega)$, for $\omega_0 = \pi/2$.

Solution

a) Sampling period of the impulse train function is $T = \frac{1}{3}$ second, whereas the sampling frequency is $\omega_s = \frac{2\pi}{T} = 6\pi$ rad/s.
b) The sampled signal in time domain is

$$x_p(t) = x(t)p(t) = \sum_{k=-\infty}^{\infty} x(kT)\delta(t - kT). \tag{11.6}$$

Inserting the function $x(t) = \sin \omega_0 t$ for $\omega_0 = \pi/2$ and $T = 1/3$, we obtain

$$x_p(t) = x(t)p(t) = \sum_{k=-\infty}^{\infty} \sin\left(\frac{\pi}{6}k\right) \delta\left(t - \frac{k}{3}\right). \tag{11.7}$$

$x(t)$, $p(t)$ and $x_p(t)$ are illustrated in Figure 11.7.

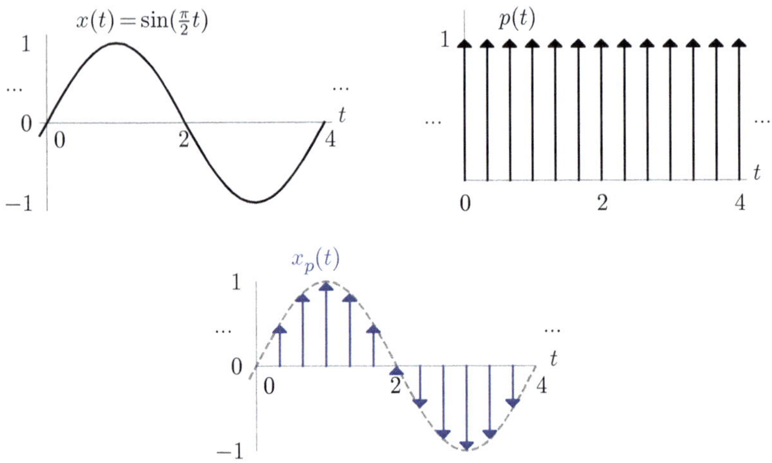

Figure 11.7 Plots of $x(t)$, $p(t)$, and $x_p(t)$ in Exercise 11.1.

c) The sampled signal in transform domain is

$$X_p(j\omega) = \frac{1}{T}\sum_{k=-\infty}^{\infty} X(j(\omega - k\omega_s)). \tag{11.8}$$

Recall that Fourier transform pair for the function $\sin \omega_0 t$ is

$$x(t) = \sin(\omega_0 t) \longleftrightarrow X(j\omega) = j\pi(\delta(\omega + \omega_0) - \delta(\omega - \omega_0)). \tag{11.9}$$

Inserting the Fourier transform of $x(t) = \sin \omega_0 t$ in Equation (11.8), for $\omega_0 = \pi/2$ we obtain

$$X_p(j\omega) = 3\pi j \sum_{k=-\infty}^{\infty} \left(\delta\left(\omega + \frac{\pi}{2} - 6k\pi\right) - \delta\left(\omega - \frac{\pi}{2} - 6k\pi\right)\right), \tag{11.10}$$

which is plotted in Figure 11.8.

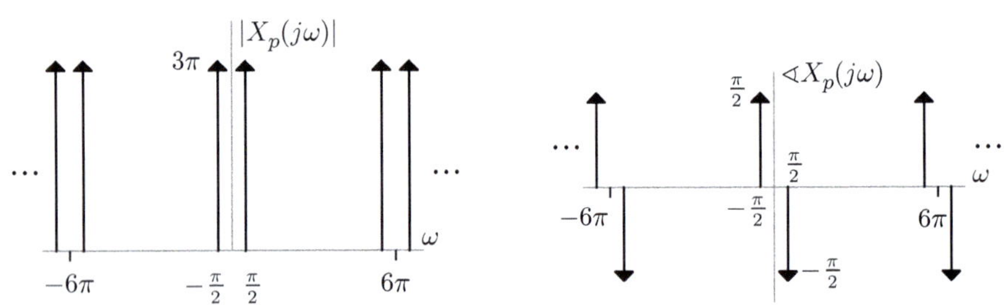

Figure 11.8 Plot of the magnitude and phase of $X_p(j\omega)$ in Exercise 11.1.

The aforementioned exercise shows that the sampled signal in time domain and frequency domain both consist of impulse trains. The sampled signal in time domain is weighted by the amplitude of the sine function at every sampling instance $k/3$. Hence, the envelope of the sampled signal is a sinusoidal function, $x(t) = \sin \frac{\pi}{2}t$. In the frequency domain, the impulse train of the sampled signal is weighted with the same scalar, which is $3\pi j$ at every sampling frequency, $k\omega_s = 6k\pi$ for all $k \in (-\infty, \infty)$.

Exercise 11.2 Suppose that an impulse train sampling,

$$x_p(t) = x(t)p(t),$$

generates the following sampled signal:

$$x_p(t) = \sum_{k=-\infty}^{\infty} (-1)^k \delta(t - 10^{-3}k), \tag{11.11}$$

where the input signal is

$$x(t) = \cos \omega_0 t, \tag{11.12}$$

a) What is the sampling period and sampling frequency of the signal $x(t)$?
b) Find the smallest value of the angular frequency, ω_0 of $x(t)$ to get the sampled signal $x_p(t)$.
c) Find the sampled signal $X_p(j\omega)$, in the frequency domain.

Solution

a) The sampled signal in time domain is

$$x_p(t) = x(t)p(t) = \sum_{k=-\infty}^{\infty} x(kT)\delta(t - kT) = \sum_{k=-\infty}^{\infty} (-1)^k \delta(t - 10^{-3}k). \tag{11.13}$$

Thus, the sampling period is $T = 10^{-3}$ seconds and the sampling frequency is $\omega_s = \frac{2\pi}{T} = 2 \times 10^3\pi$ rad/s.

b) Inserting the function $x(t) = \cos \omega_0 t$ and noting that the sampling period is $T = 10^{-3}$ seconds, we obtain,

$$x_p(t) = x(t)p(t) = \sum_{k=-\infty}^{\infty} \cos(10^{-3}\omega_0 k)\delta(t - 10^{-3}k). \tag{11.14}$$

Hence, we need

$$x(kT) = \cos kT\omega_0 = \cos 10^{-3}\omega_0 k = (-1)^k. \tag{11.15}$$

The smallest value should satisfy $10^{-3}\omega_0 = \pi$. Hence, the angular frequency of $x(t)$ should be at least $\omega_0 = 10^3\pi$ rad/s.

c) The sampled signal in transform domain is

$$X_p(j\omega) = \frac{1}{T} \sum_{k=-\infty}^{\infty} X(j(\omega - k\omega_s)). \tag{11.16}$$

Recall that Fourier transform pair for the function $\cos \omega_0 t$ is

$$x(t) = \cos(\omega_0 t) \longleftrightarrow X(j\omega) = \pi(\delta(\omega + \omega_0) + \delta(\omega - \omega_0)). \tag{11.17}$$

Inserting the Fourier transform of $x(t)$ in Equation (13.16), for $\omega_0 = 10^3\pi$ rad/s, we obtain,

$$X_p(j\omega) = 10^3\pi \sum_{k=-\infty}^{\infty} (\delta(\omega + 10^3\pi - 2 \times 10^3 k\pi) + \delta(\omega - 10^3\pi - 2 \times 10^3 k\pi)). \tag{11.18}$$

Motivating Question: How can we reconstruct the original signal,

$$x(t) \longleftrightarrow X(j\omega)$$

from its sampled version,

$$x_p(t) \longleftrightarrow X_p(j\omega)$$

without losing any information about the original signal $x(t)$?

Figure 11.6 gives us a clue about the answer to this question. Notice that the central part of the sampled signal $X_p(j\omega)$ in the interval of the bandwidth $(-\omega_M, \omega_M)$ is nothing but the scaled version of the original signal, $X(j\omega)$, where the scale is $1/T$. Therefore, all we need to do is to design a filter, which re-scales and passes the central part of the periodic signal $X_p(j\omega)$, in the frequency domain by multiplying it with T and suppresses the rest of the periodic signal $X_p(j\omega)$, which is somewhat redundant and can be omitted.

11.3 Reconstruction

Reconstruction of the original signal,

$$x(t) \longleftrightarrow X(j\omega)$$

from its sampled version,

$$x_p(t) \longleftrightarrow X_p(j\omega)$$

can be accomplished by designing a low-pass filter in the frequency domain, so that when we filter $X_p(j\omega)$ we obtain $X(j\omega)$.

Comparing the continuous time signal $X(j\omega)$ and its sampled version $X_p(j\omega)$, in the frequency domain, we can easily design the following ideal low-pass filter for reconstruction:

$$H(j\omega) = \begin{cases} T, & \text{for } |\omega| < \omega_c, \\ 0, & \text{otherwise,} \end{cases} \tag{11.19}$$

where ω_c is the cutoff frequency of the filter, as shown in Figure 11.9. The sampling period, T of the impulse train function, $p(t) = \sum_{k=-\infty}^{\infty} \delta(t - kT)$, defines the amplitude of the reconstruction filter.

The reconstruction filter, $H(j\omega)$ is just an ideal low-pass filter, scaled by the fundamental period of the impulse train, T, to recover the amplitude of the continuous time signal $X(j\omega)$, in the frequency domain.

Note: Selection of the cutoff frequency ω_c is crucial in designing the reconstruction filter, $H(j\omega)$. The cutoff frequency ω_c should fully cover the bandwidth $2\omega_M$ of the signal $X(j\omega)$, for a correct reconstruction.

Once, we design the reconstruction filter $H(j\omega)$, with the parameters T and ω_c, all we need to do is to multiply the sampled signal with the reconstruction filter, in the frequency domain, as follows:

$$X_r(j\omega) = X_p(j\omega)H(j\omega) = X(j\omega). \tag{11.20}$$

Note that in the aforementioned theoretical derivation, the reconstructed signal $X_r(j\omega)$ is exactly equal to the original signal, $X(j\omega)$, provided that the reconstruction filter has the cutoff frequency, $\omega_M < |\omega_c| < (\omega_M - \omega_s)$ (see Figure 11.10).

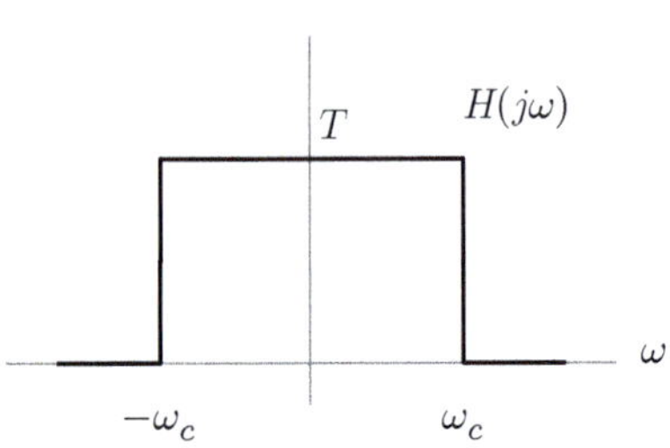

Figure 11.9 Reconstruction filter $H(j\omega)$, in frequency domain.

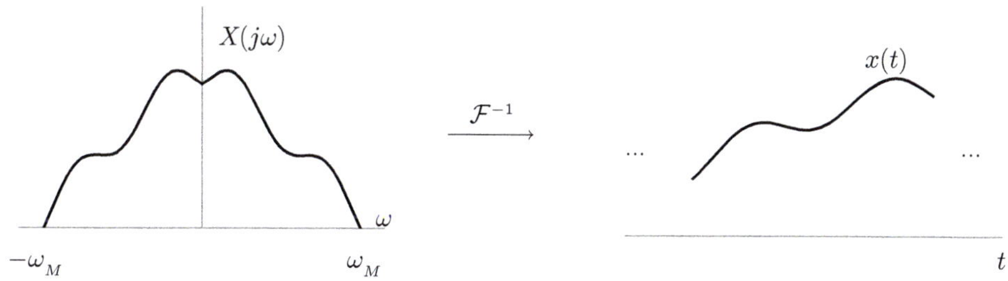

Figure 11.10 Reconstructed signal $X_r(j\omega) = X(j\omega)$ and its inverse Fourier transform.

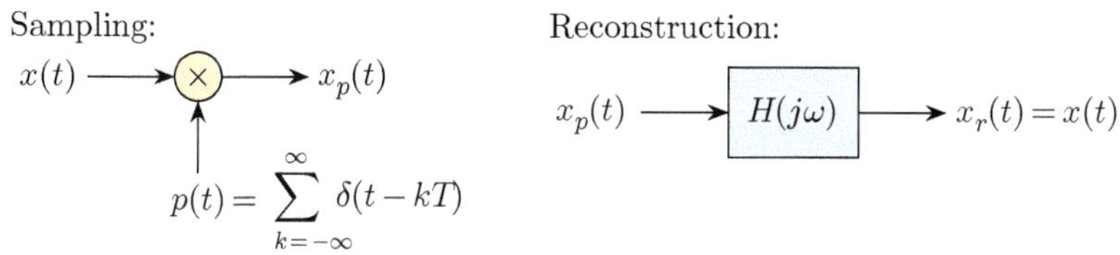

Figure 11.11 Sampling a continuous time signal $x(t)$ and reconstruction of the sampled signal $x_p(t)$ to obtain the original continuous time signal, $x(t)$.

Note also that, while the sampling process involves multiplying a continuous time signal by an impulse train in time domain, this corresponds to the convolution of them in the frequency domain. In addition, the reconstruction process involves multiplication of the frequency domain signal and the ideal low-pass filter for filtering the sampled signal, in frequency domain. This process corresponds to the convolution of the sampled signal with the time domain low-pass filter to recover the continuous time signal. Block diagram representation of the sampling and reconstruction system is shown in Figure 11.11.

Exercise 11.3 Consider the sampling and reconstruction system, shown in Figure 11.12, where the signal $x(t) = \cos(\omega_0 t + \theta)$ is sampled by the sampling period $T = 10^{-3}$ second and $H(j\omega)$ is a low-pass filter as follows:

$$H(j\omega) = \begin{cases} 10^{-3}, & \text{for } |\omega| < 1000\pi, \\ 0, & \text{otherwise.} \end{cases} \tag{11.21}$$

Find the reconstructed signal, $x_r(t)$, for the following angular frequencies and phases of the signal $x(t)$:

a) $\omega_0 = 500\pi$ and $\theta = \pi/4$.
b) $\omega_0 = 1000\pi$ and $\theta = \pi/2$.

Figure 11.12 Reconstructed signal $x_r(t)$ is obtained by filtering the sampled signal $x_p(t)$, in the frequency domain by an ideal low-pass filter, $H(j\omega)$ with height T.

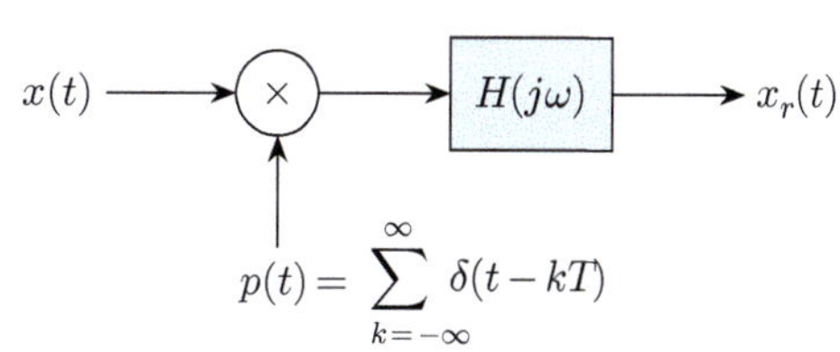

Solution

Fourier transform pair of the input signal is

$$x(t) = \cos(\omega_0 t + \theta) \longleftrightarrow X(j\omega) = \pi e^{j\theta}\delta(\omega + \omega_0) + \pi e^{-j\theta}\delta(\omega - \omega_0). \tag{11.22}$$

Fourier transform of the sampled signal is

$$X_p(j\omega) = \frac{1}{T}\sum_{k=-\infty}^{\infty} X(j(\omega - k\omega_s)), \tag{11.23}$$

where the sampling frequency of the input signal is $\omega_s = 2000\pi$.

a) For $\omega_0 = 500\pi$ and $\theta = \pi/4$: The Fourier transform of the sampled signal is

$$X_p(j\omega) = 1000\pi \sum_{k=-\infty}^{\infty} e^{j\theta}\delta(\omega - 500\pi - 2000k\pi) + e^{-j\theta}\delta(\omega + 500\pi - 2000k\pi). \tag{11.24}$$

The Fourier transform of the reconstructed signal is

$$X_r(j\omega) = X_p(j\omega)H(j\omega). \tag{11.25}$$

The low-pass filter $H(j\omega)$ suppresses the signal for $k \neq 0$ in $X_p(j\omega)$ (Figure 11.13). Hence,

$$X_r(j\omega) = \pi[e^{j\theta}\delta(\omega - 500\pi) + e^{-j\theta}\delta(\omega + 500\pi)]. \tag{11.26}$$

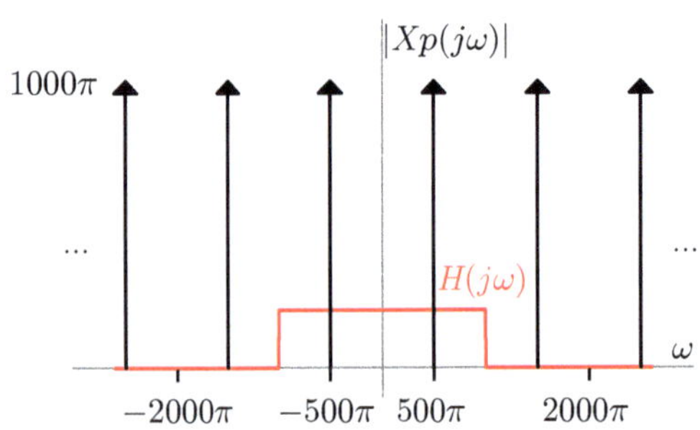

Figure 11.13 Fourier transform of the sampled signal $X_p(j\omega)$ and the ideal low-pass filter $H(j\omega)$ in Exercise 11.3. The amplitude of $H(j\omega)$ is 10^{-3}.

Taking the inverse Fourier transform of $X_r(j\omega)$ and inserting $\theta = \pi/4$, we obtain, $x_r(t) = \cos(500\pi t + \pi/4)$, which is equal to the input signal.

b) For $\omega_0 = 1000\pi$ and $\theta = \pi/2$: The Fourier transform of the sampled signal is

$$X_p(j\omega) = 1000\pi \sum_{k=-\infty}^{\infty} e^{j\pi/2}\delta(\omega - 1000\pi - 2000k\pi) + e^{-j\pi/2}\delta(\omega + 1000\pi - 2000k\pi). \tag{11.27}$$

The Fourier transform of the reconstructed signal is

$$X_r(j\omega) = X_p(j\omega)H(j\omega). \tag{11.28}$$

The cutoff frequency of the low-pass filter $H(j\omega)$ is $\omega_c < 1000\pi$. Thus, it does not cover the sampled signal. Hence,

$$X_r(j\omega) = 0. \tag{11.29}$$

Note that selecting the cutoff frequency of the low-pass filter is an important design issue. In order to be able to reconstruct the original signal from the sampled signal, the cutoff frequency of the low-pass filter should be selected to cover the bandwidth of the original signal.

11.4 Aliasing

In the aforementioned analysis and derivations of sampling and reconstruction, we made a very major assumption: We assumed that the sampling period $T = 2\pi/\omega_s$ is small enough, so that the sampling frequency ω_s becomes large enough to generate the sampled signal $X_p(j\omega)$ with nonoverlapping original signal, $X(j\omega)$, as it repeats itself at every sampling frequency.

Mathematically speaking, we assumed that $\omega_M < 2\omega_s$. This assumption assures that the sampled signal is made of **nonoverlapping** original signals, $X(j\omega)$, scaled by $1/T$ and is repeated every ω_s, in the frequency domain. Therefore, we can design a reconstruction filter with a cutoff frequency, which can cover the entire bandwidth of the continuous time signal, $X(j\omega)$ by a reconstruction filter.

Motivating Question: What if $\omega_s < 2\omega_M$?

When we enlarge the sampling period $T = 2\pi/\omega_s$, the sampling frequency ω_s gets smaller. If we keep enlarging the sampling period, at a certain point, the sampling frequency gets so small that $\omega_s < 2\omega_M$. This process is called **undersampling**.

In this case, the original signal $X(j\omega)$ starts to overlap as it repeats at each sampling frequency and the sampled signal cannot capture all the information embedded in the original signal, as indicated in Figure 11.14. When $\omega_s < 2\omega_M$, some of the information about the signal is shaded under the overlaps. Even if we design a low-pass filter, which covers the entire bandwidth of the original signal, the output of the filter does not provide the original signal. This phenomenon is called as **aliasing**.

In summary, **aliasing** is an effect that causes an information loss of the original signal, $x(t)$, during the sampling process due to **undersampling** of the original signal. It causes distortions or artifacts, when a signal is reconstructed from its samples using an ideal low-pass filter with any bandwidth. The reconstructed signal, $x_r(t)$, is no longer equal to the original continuous time signal, $x(t)$. In the following examples, we study the effect of aliasing on the reconstructed signal.

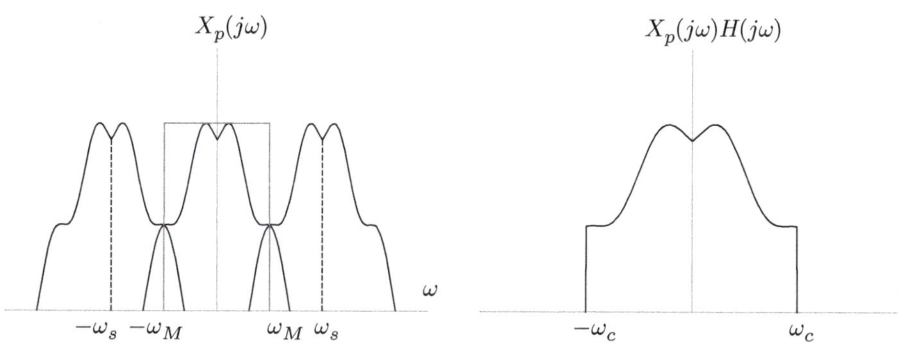

Figure 11.14 **Aliasing:** If $\omega_s < 2\omega_M$, then, $\frac{1}{T}X(j\omega)$'s in the sampled signal $X_p(j\omega)$ overlap with each other. The analytical form of the signal, in overlapped frequencies is distorted and it becomes impossible to recover the original signal from its sampled version by low-pass filtering.

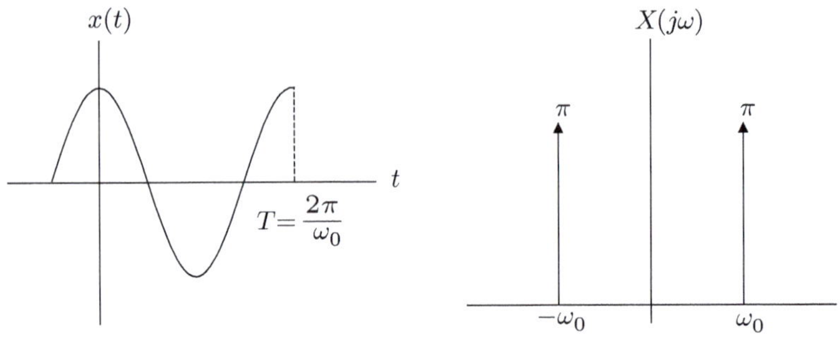

Figure 11.15 Cosine function with period T in time domain and its Fourier transform of two impulses at $|\omega_0| = 2\pi/T$.

Exercise 11.4 Suppose that we need to sample the periodic signal, represented in time and frequency domain, as shown in Figure 11.15:

$$x(t) = \cos(\omega_0 t) \longleftrightarrow X(j\omega) = \pi(\delta(\omega - \omega_0) + \delta(\omega + \omega_0)). \tag{11.30}$$

a) What is the maximum allowable sampling period T_s and the corresponding sampling frequency ω_s, so that we can uniquely reconstruct the signal $x(t)$ from its sampled version $x_p(t)$?
b) Suppose that the sampling frequency is $\omega_s = \frac{3}{2}\omega_0$. Can we reconstruct the original signal from its sampled version? Find a reconstructed signal, $x_r(t)$, which approximates the original signal $x(t)$, as much as possible.
c) Suppose that the signal $x(t)$ represents the motion of a turning wheel. What is the difference between the representations of signal $x(t)$ and its reconstructed version $x_r(t)$ when the signal is sampled and reconstructed with the sampling rate, $\omega_s = \frac{3}{2}\omega_0$?

Solution

a) The bandwidth of $X(j\omega)$ is $2\omega_0$. In order to avoid aliasing, we need to obtain nonoverlapping $X(j\omega)$'s in the sampled signal $X_p(j\omega)$. This requires that ω_s should be slightly larger than $2\omega_0$. Therefore, the sampling period should be $T_s < \pi/\omega_0$, Figure 11.16 shows the sampled and reconstructed signal, which satisfies this inequality.

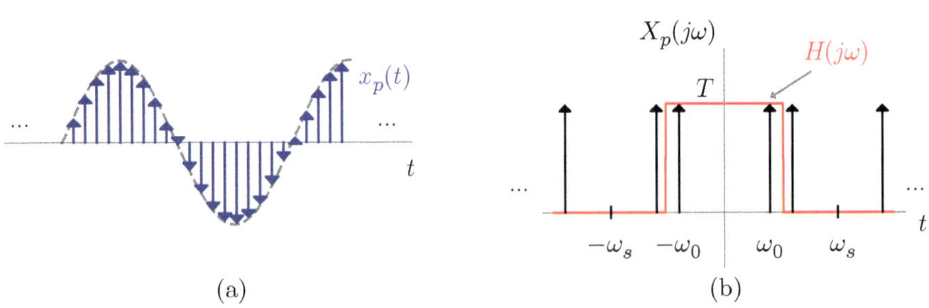

Figure 11.16 Impulse train sampling with period T_s. (a) Original signal $x(t) = \cos\omega_0 t$ and its sampled version $x_p(t) = \sum_n \cos(\omega_0 n T_s)\delta(t - nT_s)$. (b) Sampled signal $X_p(j\omega)$, in the frequency domain and the low-pass reconstruction filter, $H(j\omega)$, indicated by blue.

Note: As the sampling period, T_s gets smaller, in time domain; the sampling frequency, ω_s gets larger, in frequency domain. In other words, getting more samples in the time domain makes the original signal $X(j\omega)$ fall far apart from each other, in the sampled signal $X_p(j\omega)$ of the frequency domain.

b) When the sampling frequency is $\omega_s = \frac{3}{2}\omega_0$, the sampling period becomes $T_s = \frac{4\pi}{3\omega_0}$. The sampled signal in time domain has the following form:

$$x_p(t) = x(t)p(t) = \sum_{k=-\infty}^{\infty} x(kT_s)\delta(t - kT_s). \tag{11.31}$$

Fourier transform of the sampled signal has the following form:

$$X_p(j\omega) = \frac{1}{T_s}\sum_{k=-\infty}^{\infty} X(j(\omega - k\omega_s)) = \frac{3\omega_0}{4\pi}\sum_{k=-\infty}^{\infty} X\left(j\left(\omega - \frac{3}{2}k\omega_0\right)\right). \tag{11.32}$$

Note that, since $\omega_s = \frac{3}{2}\omega_0 < 2\omega_0$, the original signal has overlaps in the sampled signal (See: Figure: 11.17). Hence, there is aliasing. The original signal cannot be reconstructed from the sampled signal.

Let us now try to reconstruct the original signal by low-pass filtering the sampled signal in the frequency domain, using the following equation:

$$X_r(j\omega) = X_p(j\omega)H(j\omega),$$

when $\omega_s = \frac{3}{2}\omega_0$.

In order to reconstruct the signal from its sampled version, we design an ideal low-pass filter,

$$H(j\omega) = \begin{cases} T_s, & \text{for } |\omega| < \omega_c, \\ 0, & \text{otherwise,} \end{cases} \tag{11.33}$$

where the cutoff frequency is selected in $|\omega_s - \omega_0| < |\omega_c| < |\omega_0|$.

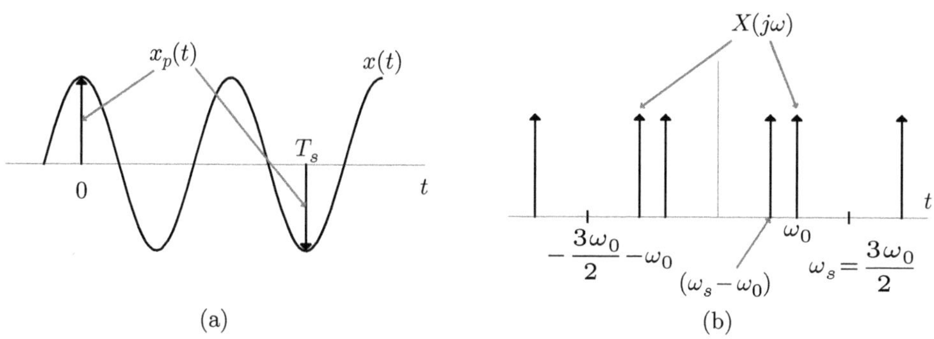

Figure 11.17 Impulse train sampling with sampling frequency, $\omega_s = \frac{3}{2}\omega_0$. (a) Original signal $x(t) = \cos\omega_0 t$ and its sampled version $x_p(t) = \sum_n \cos(\omega_0 nT_s)\delta(t - nT_s)$, where $T_s = \frac{4\pi}{3\omega_0}$ seconds. (b) Sampled signal $X_p(j\omega)$, in the frequency domain.

In this case, the reconstructed signal in the frequency domain will be

$$X_r(j\omega) = X_p(j\omega)H(j\omega) = \pi(\delta(\omega - (\omega_s - \omega_0)) + \delta(\omega + (\omega_s - \omega_0))). \tag{11.34}$$

Hence, the reconstructed signal in time domain is

$$x_r(t) = \cos((\omega_s - \omega_0)t). \tag{11.35}$$

Note: Although the reconstructed signal is still a cosine function, its angular frequency is not the same as that of the original signal,

$$x_r(t) = \cos((\omega_s - \omega_0)t) = \cos\left(\frac{1}{2}\omega_0 t\right) \neq x(t) = \cos(\omega_0 t).$$

The angular frequency of the reconstructed signal is, $\omega_r = \frac{1}{2}\omega_0$, and the period of the reconstructed signal is $T_r = \frac{2\pi}{\omega_r} = \frac{4\pi}{\omega_0} = 2T$.

The period of the reconstructed signal is two times more than that of the original signal.

c) Suppose that the original signal $x(t)$ represents a wheel, turning with a speed of angular frequency ω_0. The reconstructed signal $x_r(t)$ will represent a turning wheel with speed, two times slower than that of the original speed of the wheel.

The aforementioned analysis and example bring one of the most influential theorems of the modern age: sampling theorem.

WATCH Learn more about aliasing @ https://384book.net/v1101

11.5 Sampling Theorem

Let $x(t)$ be a band-limited signal with $X(j\omega) = 0$ for $|\omega| > \omega_M$. Then, $x(t)$ is uniquely recovered from its samples $x(nT)$, $n = 0, \pm1, \pm2, \ldots$, if $\omega_s > 2\omega_M$.

$\omega_N = 2\omega_M$ is called the Nyquist rate (Harry Nyquist, 1894–1976). Nyquist rate is the smallest possible sampling frequency to avoid aliasing.

The formal proof of this theorem is given in the original paper by Shannon.

Sampling theorem states that there are two important factors to sample a continuous time signal $x(t)$ and reconstruct this function from its sampled version, $x_p(t)$ without losing any information:

1) The continuous time signal $x(t)$ is to be band-limited with a finite bandwidth $2\omega_M$.
2) The sampling frequency ω_s should be greater than the Nyquist rate, $\omega_N = 2\omega_M$.

If these two conditions are satisfied, then it is theoretically possible to reconstruct the original signal exactly from its sampled version.

INTERACTIVE Sampling and reconstruction of a continuous time signal @ https://384book.net/i1101

11.6 Sampling with Zero-Order Hold

Sampling theorem is based on sampling a continuous time signal with an impulse train, which consists of finitely many impulses of zero width and infinite height. Although this theorem is quite elegant to prove that we can uniquely reconstruct a continuous time signal from its sampled version, in practice, it is not possible to realize the impulse train to sample a continuous time function. Therefore, we need some type of approximations to sample a continuous time signal.

One way of realizing the sampling theorem is to use zero-order hold. Loosely speaking, zero-order hold approximates a function with a set of piecewise constant functions using a sequence of sampled points from the signal. This approximated function can be used to reconstruct the original signal. Let us see how.

Consider the piecewise constant signal $x_0(t)$, shown in Figure 11.18. The signal $x_0(t)$, is quite easy to generate from $x(t)$, by a simple switch, which measures the value of $x(t)$ at every period T and keeps the value constant in between the measured values. **This process is called sampling with zero-order hold.**

Motivation Question: If we sample a continuous time signal $x(t)$ by zero-order hold and obtain a piecewise constant function $x_0(t)$, can we uniquely reconstruct $x(t)$ from the sampled signal $x_0(t)$?

Let us answer this question by formalizing the sampling and reconstruction processes with zero-order hold and analyzing the behavior of $x_0(t)$, theoretically, in time and frequency domains.

In time domain: Let us start by defining the **zero-order hold filter**, which is an LTI system represented by the impulse response of Figure 11.19, as follows:

$$h_0(t) = \begin{cases} 1, & \text{for } 0 < t < T, \\ 0, & \text{otherwise.} \end{cases} \tag{11.36}$$

Suppose that we feed the impulse train sampled signal,

$$x_p(t) = x(t)p(t) = \sum_{k=-\infty}^{\infty} x(kT)\delta(t - kT),$$

at the input of the zero-order hold filter $h_0(t)$. Then, the corresponding output becomes

$$y_0(t) = x_p(t) * h_0(t).$$

Motivation Question: What does $y_0(t)$ look like?

Let us evaluate the convolution of the input $x_p(t)$ and the impulse response $h_0(t)$:

$$y_0(t) = \left[\sum_{k=-\infty}^{\infty} x(kT)\delta(t - kT) \right] * h_0(t). \tag{11.37}$$

Recall that $\delta(t - kT) * h_0(t) = h_0(t - kT)$. Therefore,

$$y_0(t) = x(kt), \quad \text{for all} \quad k, \quad kT < t < k(T + 1).$$

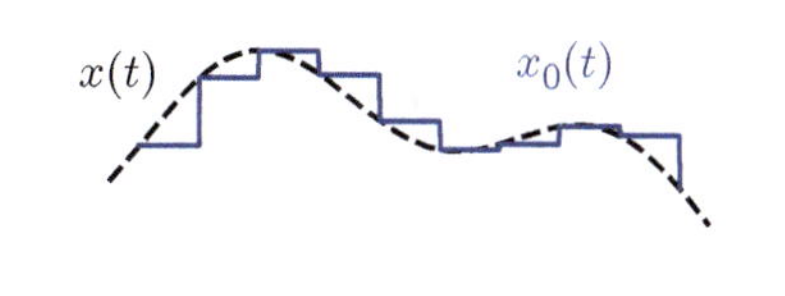

Figure 11.18 Sampling with zero-order hold: Generation of a piecewise constant function $x_0(t)$, from a continuous time function $x(t)$.

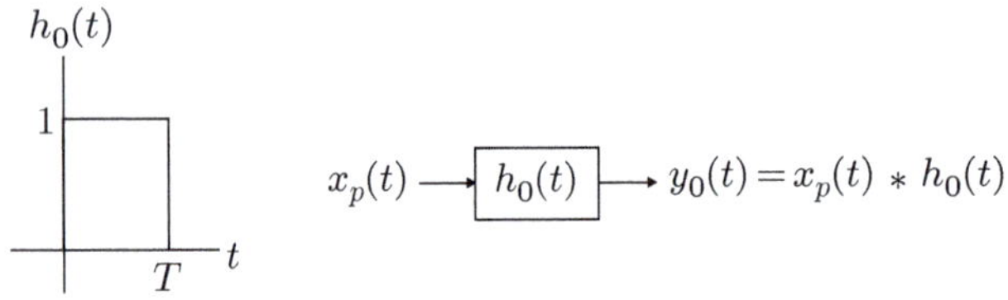

Figure 11.19 An LTI system represented by the impulse response $h_0(t)$.

Thus, the output, $y_0(t)$ of the zero-order hold filter $h_0(t)$ is just the piecewise constant function, $x_0(t)$:

$$y_0(t) = x_0(t). \tag{11.38}$$

The aforementioned derivations establish the relationship between the impulse train sampling, which outputs $x_p(t)$ and zero-order hold sampling, which outputs $x_0(t)$, in the time domain. Formally speaking, zero-order hold sampled signal $x_0(t)$ is the output of an LTI system represented by the zero-order hold filter $h_0(t)$, when it is fed by the impulse train sampled signal, $x_p(t)$:

$$x_0(t) = x_p(t) * h_0(t). \tag{11.39}$$

In frequency domain: Let us start by finding the Fourier transform of the zero-order hold filter,

$$H_0(j\omega) = \mathcal{F}\{h_0(t)\}.$$

Notice that zero-order hold filter is the shifted version of the following impulse response, indicated in Figure 11.20:

$$h(t) = \begin{cases} 1, & \text{for} - T/2 < t < T/2, \\ 0, & \text{otherwise.} \end{cases} \tag{11.40}$$

Recall that Fourier transform of $h(t)$ is the following Sinc function:

$$H(j\omega) = \frac{2\sin(\omega T/2)}{\omega}. \tag{11.41}$$

We can use the time shift property to compute the Fourier transform of $h_0(t)$ directly from $H(j\omega)$, as follows:

$$h_0(t) = h(t - T/2) \longleftrightarrow H_0(j\omega) = 2e^{(-j\omega T/2)}\frac{\sin(\omega T/2)}{\omega}. \tag{11.42}$$

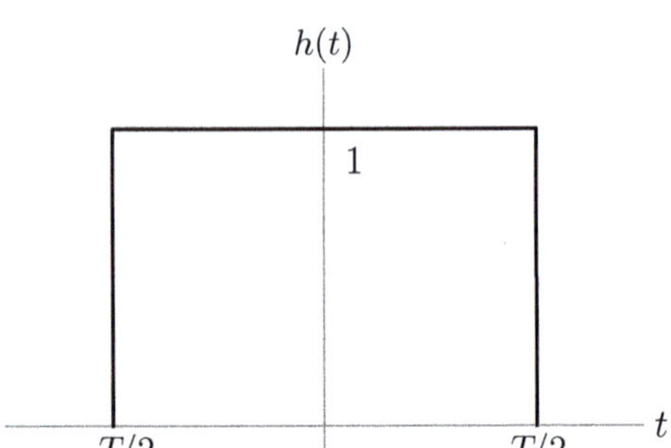

Figure 11.20 The impulse response function $h(t)$ is the shifted version of $h_0(t)$. In other words, $h_0(t) = h(t - T/2)$.

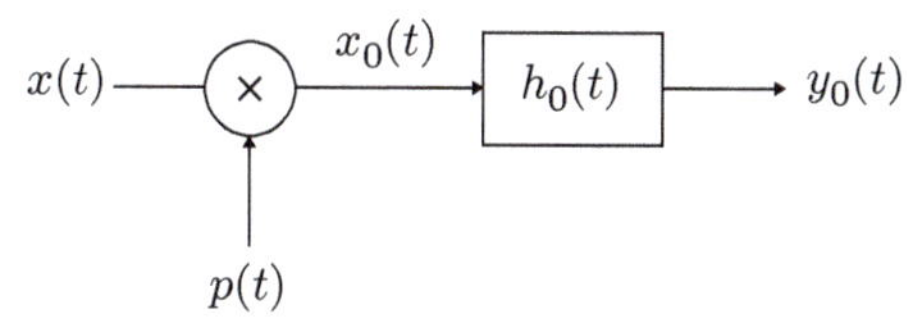

Figure 11.21 Block diagram representation of sampling with zero-order hold.

This equation shows that sampling with zero-order hold corresponds to filtering the impulse train sampled signal $X_p(j\omega)$ with a Sinc function,

$$H_0(j\omega) = \frac{2\sin(\omega T/2)}{\omega} e^{-j\frac{\omega T}{2}}. \tag{11.43}$$

There is a very elegant **duality** between the impulse train sampling and zero-order hold sampling:

Impulse train sampling involves multiplication in the time domain and convolution in the frequency domain:

$$x_p(t) = x(t) \cdot p(t) \longleftrightarrow X_p(j\omega) = X(j\omega) * P(j\omega).$$

Zero-order hold sampling involves convolution in time domain and multiplication in the frequency domain:

$$x_0(t) = x_p(t) * h_0(t) \longleftrightarrow X_0(j\omega) = X_p(j\omega) \cdot H_0(j\omega).$$

as depicted in Figure 11.21.

11.7 Reconstruction with Zero-Order Hold

In this section, our goal is to design an LTI system, represented by the impulse response $h_r(t)$, which reconstructs the original signal $x(t)$ from its sampled version $x_0(t)$, in time domain. Equivalently, in the frequency domain, we need to find a filter $H_r(j\omega)$, which outputs $X(j\omega)$ for the input $X_0(j\omega)$.

Motivating Question: How to define the LTI filter,

$$h_r(t) \leftrightarrow H_r(j\omega),$$

so that the output of this filter is

$$x_r(t) = x(t) \leftrightarrow X_r(j\omega) = X(j\omega),$$

when the input is the zero-order hold sampled signal?

The reconstructed signal can be obtained by convolution of the sampled signal $x_0(t)$ with the reconstruction filter $h_r(t)$, in the time domain.

$$x_r(t) = x_0(t) * h_r(t) = x_p(t) * h_0(t) * h_r(t)$$

Let us use the convolution property in time domain, which corresponds to the multiplication property in frequency domain:

$$x_r(t) = x_p(t) * h_0(t) * h_r(t) \longleftrightarrow X_r(j\omega) = X_p(j\omega)H_0(j\omega)H_r(j\omega).$$

Recall that reconstruction of $x(t) \leftrightarrow X(j\omega)$ from $x_p(t) \leftrightarrow X_p(j\omega)$ in impulse train sampling is accomplished by an ideal low-pass filter:

$$H(j\omega) = \begin{cases} T, & \text{if } -\omega_c < \omega < \omega_c, \\ 0, & \text{otherwise,} \end{cases} \tag{11.44}$$

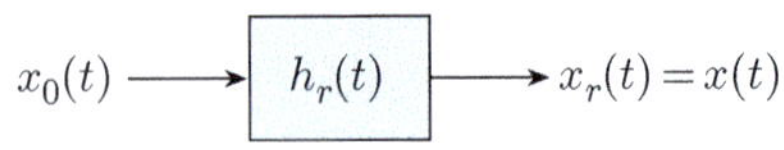

Figure 11.22 The reconstruction filter $h_r(t)$ receives the zero-order hold sampled signal $x_0(t) = x_p(t) * h_0(t)$ and outputs $x_r(t) = x(t)$.

where

$$X(j\omega) = X_p(j\omega)H(j\omega). \tag{11.45}$$

Therefore, if we set,

$$H(j\omega) = H_0(j\omega)H_r(j\omega), \tag{11.46}$$

then,

$$X_r(j\omega) = X(j\omega). \tag{11.47}$$

Hence, the reconstruction filter of zero-order hold, shown in Figure 11.22, we are looking for is, then,

$$H_r(j\omega) = \frac{H(j\omega)}{H_0(j\omega)} = \frac{e^{j\omega(T/2)}}{2\sin(\omega T/2)}\omega H(j\omega). \tag{11.48}$$

generates the original input $X(j\omega)$.

Motivating Question: What type of a filter is $H_r(j\omega)$?

In order to investigate the effect of $H_r(j\omega)$ on the sampled signal $X_0(j\omega)$, let us plot the magnitude and phase of $H_r(j\omega)$.

Let us set the cutoff frequency of $H_r(j\omega)$ as, $\omega_c = \frac{\omega_s}{2}$. Then, the magnitude and the phase of the reconstruction filter becomes

$$|H(j\omega)| = \frac{\omega T}{2\sin(\omega(T/2))} \quad \text{for} \quad -\frac{\omega_s}{2} \leq \omega \leq \frac{\omega_s}{2} \tag{11.49}$$

and

$$\measuredangle H_r(j\omega) = \frac{\omega T}{2} \quad \text{for} \quad -\frac{\omega_s}{2} \leq \omega \leq \frac{\omega_s}{2}, \tag{11.50}$$

respectively.

Figure 11.23 shows that the reconstruction filter $H_r(j\omega)$, for zero-order hold sampling is a low-pass filter. This filter slightly suppresses the lower frequencies around the origin.

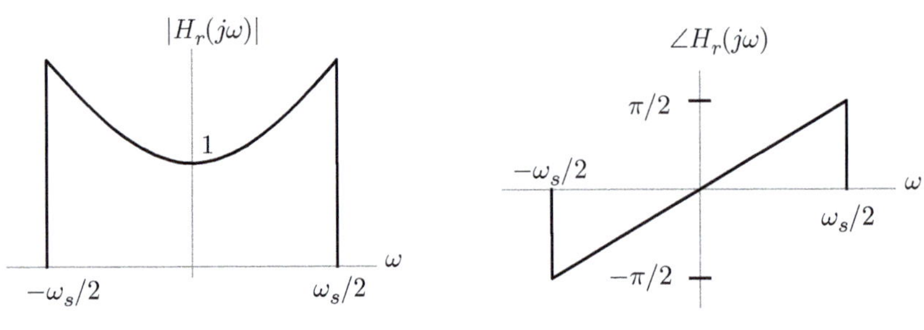

Figure 11.23 Magnitude and phase plots of the reconstruction filter, $H_r(j\omega)$, for zero-order hold sampled signal. $H_r(j\omega)$ is called the ideal compensation filter.

Let us compare the reconstruction filters $H(j\omega)$ for impulse train sampling and $H_r(j\omega)$ for zero-order hold sampling. Both of them are low-pass filters. However, $H(j\omega)$ is an ideal low-pass filter, whereas, $H_p(j\omega)$ somehow compensates for the process of zero-order hold.

11.8 Sampling and Reconstruction with First-Order Hold

So far, we have seen two types of sampling: **impulse train sampling** and **zero-order hold sampling**. While impulse train sampling provides us theoretically exact reconstruction of the sampled signal, it was not easy to realize it in practice. On the other hand, zero-order hold sampling simply converts the signal into a piecewise constant signal, which is more practical to realize sampling and reconstruction. However, zero-order hold filters still carry discontinuities at the cutoff frequency, which results in generation of high-frequency noise due to the Gibbs phenomenon, while we take the Fourier transform.

Motivating Question: Is it possible to define a more practical sampling method, which avoids the problems of impulse train and zero-order sampling? For example, can we somehow replace impulse train sampled signal

$$x_p(t) \longleftrightarrow X_p(j\omega)$$

by a realizable signal, so that instead of zero-order hold we can use the approximated form of impulse train sampling?

Let us start by analyzing the structure of the reconstruction filter $h(t) \leftrightarrow H(j\omega)$, for impulse train sampling, in the time domain.

Formally speaking,

$$x_r(t) = x(t) = x_p(t) * h(t) = \sum_{n=-\infty}^{\infty} x(nT)h(t-nT), \tag{11.51}$$

where T is the sampling period.

If we take the inverse Fourier transform of the ideal low-pass filter, $H(j\omega)$, we get the following sinc function as the impulse response of an LTI system, in time domain:

$$h(t) = \frac{\omega_c T \sin(\omega_c t)}{\pi \omega_c t}. \tag{11.52}$$

Inserting the impulse response into the convolution equation, $x_r(t) = x_p(t) * h(t)$, we get,

$$x_r(t) = \sum_{n=-\infty}^{\infty} x(nT) \frac{\omega_c T}{\pi} \frac{\sin(\omega_c(t-nT))}{\omega_c(t-nT)}. \tag{11.53}$$

Note: The reconstructed signal $x_r(t) = x(t)$ is just the superposition of shifted sinc functions, each of which is weighted by the value of $x(t)$ at nT, namely, $x(nT)$.

Rather than taking the superposition of the shifted sinc functions, shown in Figure 11.24a, we simply connect the peak values of the reconstructed signal to obtain a linear interpolation, shown in Figure 11.24b. This method of sampling is called **first-order hold**.

First-order hold sampling offers a practical method for sampling a continuous time signal in time domain. At the first step, we find the bandwidth, $2\omega_M$, of the signal $x(t)$. Then, set the sampling rate as the Nyquist rate, which is

$$\omega_s = \omega_N = 2\omega_M.$$

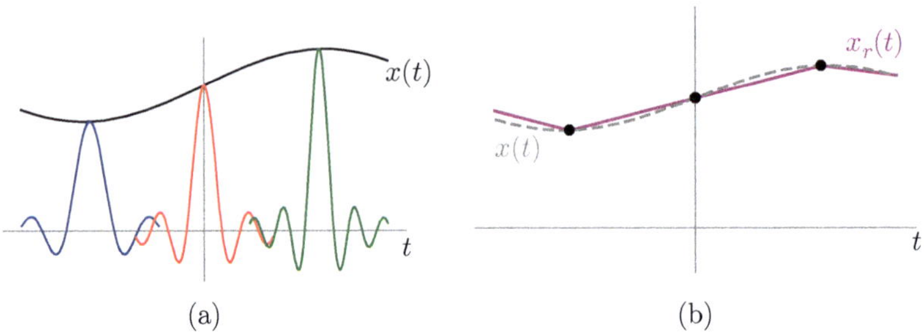

Figure 11.24 (a) Reconstructed signal in time domain from impulse train sampled signal, $x_p(t)$ which is obtained by the superposition of the shifted sinc functions. (b) Approximating the reconstructed signal by linear interpolation to obtain first-order hold sampled signal.

Then, we set corresponding sampling period to,

$$T = \frac{2\pi}{\omega_N}.$$

Finally, we simply create the sampled signal by connecting the selected points of $x(nT)$ for all n by a straight line, in time domain.

INTERACTIVE

Sampling and reconstruction with first-order hold @ `https://384book.net/i1102`

11.9 Chapter Summary

Can we select a set of time points $x(nT)$ from a continuous time function $x(t)$, which represents the function $x(t)$ without losing any information? As we all know, a continuous time function is represented by uncountably many points $(t, x(t))$, where $t \in \mathcal{R}$ is a real number. Thus, intuitively, neglecting infinitely many points at every interval between $x(nT)$ and $x((n \pm 1)T)$ reveals that we lose a great amount of information about the function. Fortunately, this is not a valid statement, provided that the signal in the frequency domain is bandlimited, in other words, the signal has nonzero values only in a finite interval of frequencies, in the transform domain.

In this chapter, we studied the famous sampling theorem proved by Claude Shannon, which states that a continuous time band limited signal can be sampled without losing any information. The original signal can be reconstructed from its sampled version uniquely, provided that the sampling rate $\omega_s = 2\pi/T$ is at least twice the bandwidth, ω_M, of the original signal, called the Nyquist rate, ω_N. Mathematically, for a unique sampling and reconstruction, the following inequality should be satisfied:

$$\omega_s = 2\pi/T > \omega_N = 2\omega_W.$$

We investigate sampling a continuous time signal to generate a more compact signal using three types of sampling methods.

First, we study the theoretical foundation of the sampling theorem by defining the impulse train sampling. In this method, we multiply the signal by an impulse train of period T, which is to be small enough to assure large enough sampling frequency, so that the Nyquist rate is satisfied. We reconstruct the original signal by simply using an ideal low-pass filter with a cutoff frequency, ω_C, which covers the entire bandwidth of the original signal.

Second, we define zero-order sampling, where we approximate the signal $x(t)$ by a piecewise constant function, between every sampling period T. The reconstruction filter in the frequency domain is another low-pass filter with the cutoff frequency ω_C, which covers the entire bandwidth of the original signal. However, this time, the low frequencies are slightly attenuated.

Both impulse train sampling and zero-order hold sampling require filtering in the frequency domain with a filter, which has discontinuities at the cutoff frequencies. Then, we take the inverse Fourier transform of the filtered signal to reconstruct the original signal from its sampled version. Taking the inverse Fourier transform of a signal with discontinuities results in high-frequency noise due to the Gibbs phenomenon, described in Chapter 6. Thus, a more practical method is required for sampling.

Finally, we introduce first-order hold sampling. This method avoids filtering in the transform domain and handles the sampling in time domain using the fact that impulse train sampling is nothing but the superposition of the Sinc function generated at every sampling point T. This superposition can be simply approximated by connecting the sampled signal with a straight line.

Problems

11.1 Consider a signal $x(t)$ whose Nyquist rate is ω_N. Find the Nyquist rate for each of the following signals in terms of ω_n.

a) $x(t-2)+x(t+2)$

b) $x(2t)+x(t-2)$

c) $\dfrac{dx(t)}{dt}$

d) $x^2(t)$

e) $x(t)\cos\omega_0 t$

11.2 Find the Nyquist rate for each of the following signals.

a) $x(t) = 1 + \cos(4000\pi t) + \sin(8000\pi t)$

b) $x(t) = \dfrac{\sin(8000\pi t)}{\pi t}$

c) $x(t) = \left(\dfrac{\sin(16,000\pi t)}{\pi t}\right)^2$

11.3 A continuous time signal,

$$x(t) = e^{-0.5t}u(t)$$

is fed to an ideal low-pass filter, $h(t) \longleftrightarrow H(j\omega)$ with cutoff frequency $\omega_c = 2000\pi$ to obtain the output signal $y(t) = h(t) * x(t)$. Suppose that impulse-train sampling is performed as $y_p(t) = y(t)p(t)$, where

$$p(t) = \sum_{n=-\infty}^{\infty} \delta(t-nT).$$

a) Find and plot the magnitude and phase of the Fourier transform of the output, $Y(j\omega)$.

b) Find the largest possible period T, which avoids aliasing.

c) Find and plot the reconstructed signal $y_r(t)$, when the sampling period is $T = 2\times10^{-3}$ and for $T = 2\times10^{-4}$.

d) Suppose that the sampling period is $T = 2 \times 10^{-4}$. What is the valid interval of cutoff frequency of the low-pass filter for reconstruction,

$$Y_r(j\omega) = Y_p(j\omega)H_r(j\omega),$$

which avoids aliasing.

11.4 A continuous time signal,

$$x(t) = \frac{\sin(3000\pi t)}{\pi t}$$

is fed to an ideal low-pass filter, $h(t) \longleftrightarrow H(j\omega)$ with cutoff frequency $\omega_c = 2000\pi$ to obtain the output signal $y(t) = h(t) * x(t)$. Suppose that impulse-train sampling is performed as $y_p(t) = y(t)p(t)$, where

$$p(t) = \sum_{n=-\infty}^{\infty} \delta(t - nT).$$

a) Find the largest possible period T, which avoids aliasing of $y_p(t)$.
b) What is the cutoff frequency of the low-pass filter $H_r(j\omega)$ for reconstruction,

$$Y_r(j\omega) = Y_p(j\omega)H_r(j\omega).$$

c) Find and plot the sampled signal $y_p(t)$, when the sampling period is $T = 2 \times 10^{-2}$. Is there aliasing?

11.5 The input to a continuous time system has the following Fourier transform:

$$X(j\omega) = \frac{1}{2}\left(\delta(\omega - \omega_0) + \delta(\omega + \omega_0)\right).$$

The Fourier transform of the corresponding output is

$$Y(j\omega) = 2X(j(\omega - \omega_0)).$$

a) Find the output $y(t)$.
b) Find an equivalent multiplicative system $z(t)$, which satisfies,

$$y(t) = x(t)z(t).$$

11.6 A continuous time band-limited input signal $x(t)$ is fed to a system, which generates the following output:

$$y(t) = x(t)\cos\omega_0 t.$$

a) Find and plot the magnitude and phase $Y(j\omega)$, when

$$X(j\omega) = \begin{cases} \frac{1}{j\omega+1}, & \text{for} \quad |\omega| < 100\pi, \\ 0, & \text{o.w.} \end{cases}$$

b) Find the interval of ω_0, which guarantees the reconstruction of $x(t)$ from $y(t)$.
c) Define a system, which reconstructs the input $x(t)$ from the output $y(t)$. Be specific about the type and the cutoff frequency of the reconstruction filter.
d) Find and plot $Y(j\omega)$, for $\omega_0 = 75\pi$. Can you reconstruct $x(t)$ from $y(t)$?

11.7 A continuous time system is given in Figure P11.7a.

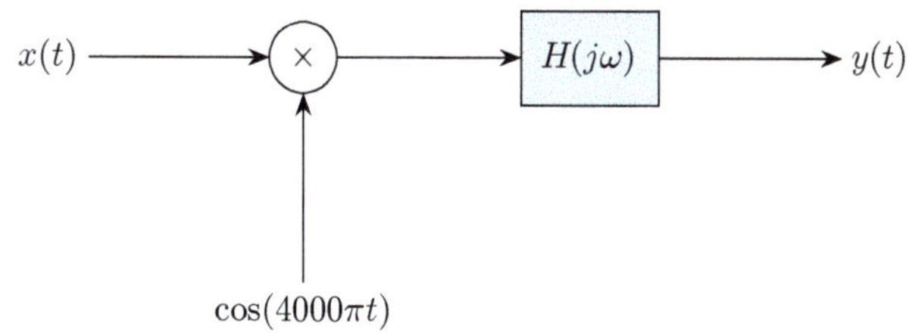

Figure P11.7a

The frequency response of the ideal low-pass filter is shown in Figure P11.7b.

Figure P11.7b

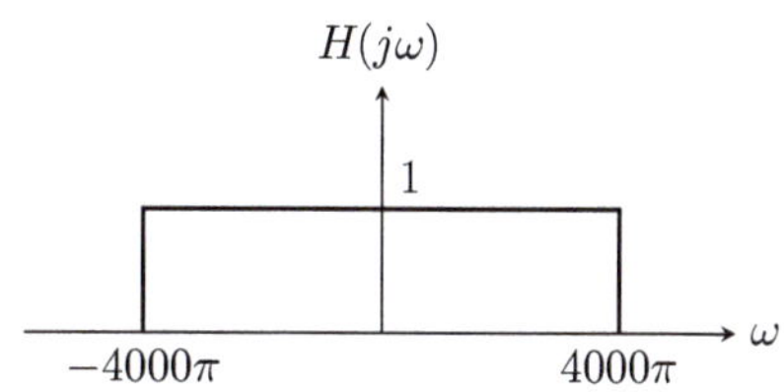

a) Find the Fourier transform of the signal obtained at the output of the multiplier, in terms of a band-limited input, $X(j\omega)$:

$$g(t) = x(t)\cos(4000\pi t).$$

b) Find the maximum bandwidth of $X(j\omega)$, which assures the reconstruction of it from $G(j\omega)$.

c) Find and plot the output signal $y(t) \longleftrightarrow Y(j\omega)$, when the input signal is

$$X(j\omega) = \begin{cases} 1, & \text{for} \quad |\omega| \leq 2000\pi, \\ 0, & \text{o.w.} \end{cases}$$

11.8 A continuous time system is given in Figure P11.8a, where the frequency response of the ideal low-pass filter is given in Figure P11.8b. The input to this system is

$$x(t) = \sin(400\pi t) + 2\sin(800\pi t).$$

a) Find and plot the Fourier transform of the signal obtained at the output of the multiplier:

$$g(t) = x(t)\sin(100\pi t).$$

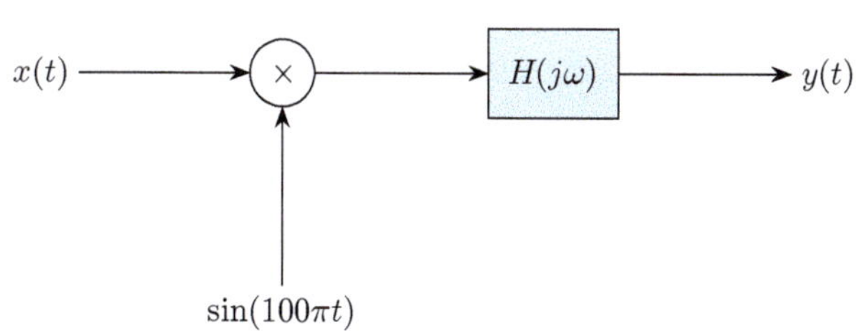

Figure P11.8a

b) Find the output $y(t)$ of the low-pass filter given in Figure P11.8a.

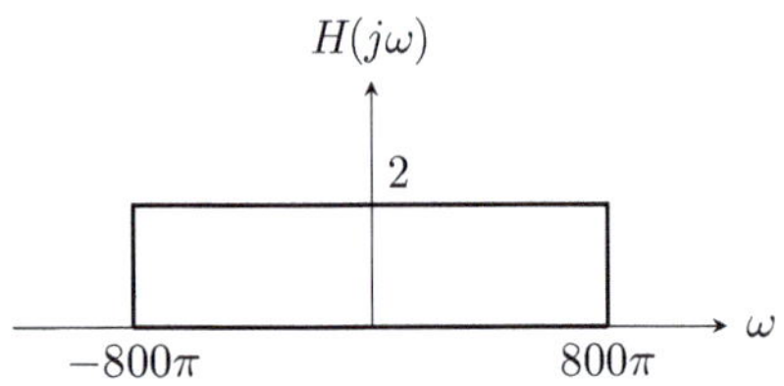

Figure P11.8b

11.9 Consider the following input signal,

$$x(t) = \frac{\sin(100\pi t)}{\pi t},$$

which is fed to a system to create the following output:

$$y(t) = x(t)\cos(500\pi t).$$

a) Find and plot the Fourier transform of the input, $x(t)$.
b) Find and plot the Fourier transform of the output, $y(t)$. What is the Bandwidth of $Y(j\omega)$?
c) Find the Nyquist rate for the impulse train sampled output $y_p(t)$.
d) Find and plot the sampled output $y_p(t)$.

11.10 Consider the impulse train sampling and reconstruction system of Figure P11.10, which is fed by the following input:

$$x(t) = \cos(1000\pi t) + \cos(5000\pi t).$$

$x(t) = cos(1000\pi t) + cos(5000\pi t).$

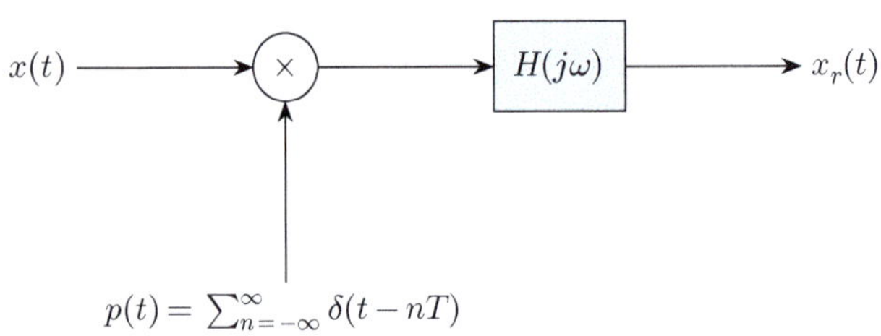

Figure P11.10

a) Roughly plot the input signal, $x(t)$.
b) Find and plot the sampled signal $x_p(t) \leftrightarrow X_p(j\omega)$, for the sampling frequency $\omega_s = 10,000\pi$.
c) Find the Nyquist rate of the sampled signal $x_p(t) = x(t)p(t)$.

d) Find the bandwidth of the low-pass filter, $H(j\omega)$, to reconstruct the original signal from its sampled version without losing any information.

11.11 Consider the system given in Figure P11.11a, which is fed by a band-limited signal,

$$X(j\omega) = \begin{cases} 1, & \text{for } \omega \leq \omega_m, \\ 0, & \text{o.w.} \end{cases}$$

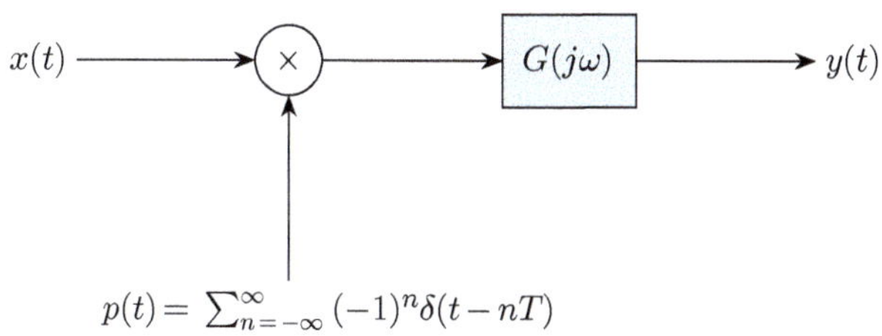

Figure P11.11a

The band-pass filter $G(j\omega)$ has the form, given in Figure P11.11b

Figure P11.11b

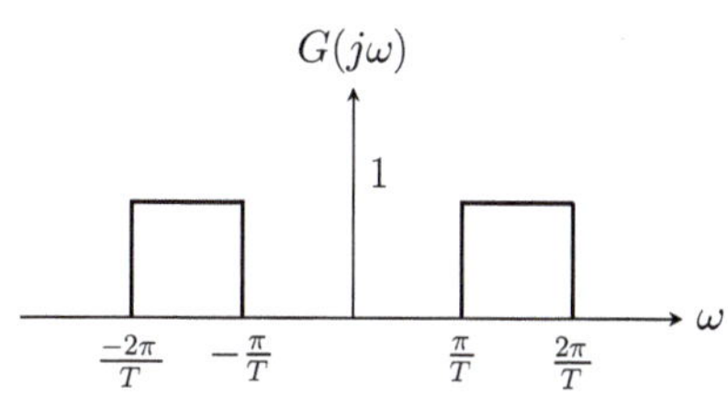

a) Find and plot the Fourier transform of $x_p(t) = x(t)p(t)$, when the sampling period is $T = \frac{\pi}{3\omega_m}$.
b) Find and plot the Fourier transform of $y(t)$ for $T = \frac{\pi}{3\omega_m}$.
c) Define a system, which reconstructs the input signal $x(t)$ from the output signal $y(t)$.

11.12 Consider the system shown in Figure P11.12, where the frequency response of the filter is given as follows:

$$H(j\omega) = \begin{cases} j, & \text{if, } \omega > 0, \\ -j, & \text{if, } \omega < 0. \end{cases}$$

The input signal $x(t)$, to this system, is band-limited with the Fourier transform $X(j\omega) = 0$, for $|\omega| > 1000\pi$.

a) Find the Fourier transform of the output signal, $y(t)$ in terms of the Fourier transform of the input signal $x(t)$.
b) Can we reconstruct $x(t)$ from $y(t)$, for $\omega_c = 500\pi$? Verify your answer.

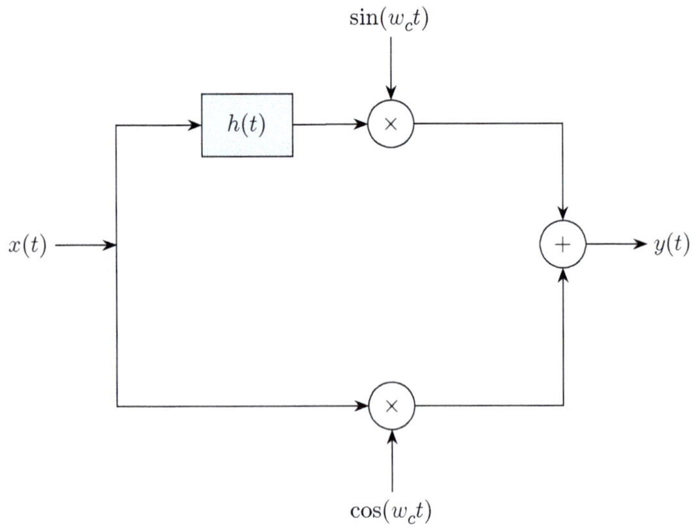

Figure P11.12

11.13 A sampling system is illustrated in Figure P11.13a,

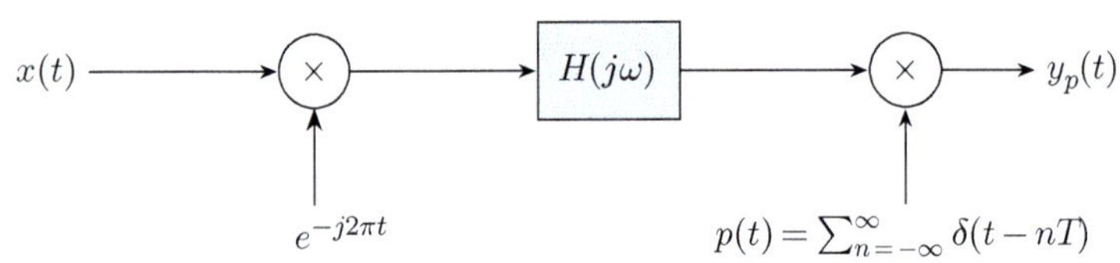

Figure P11.13a

where the low-pass filter $H(j\omega)$ has the cutoff frequency, $|\omega_c| = \pi$, as shown in Figure P11.13b.

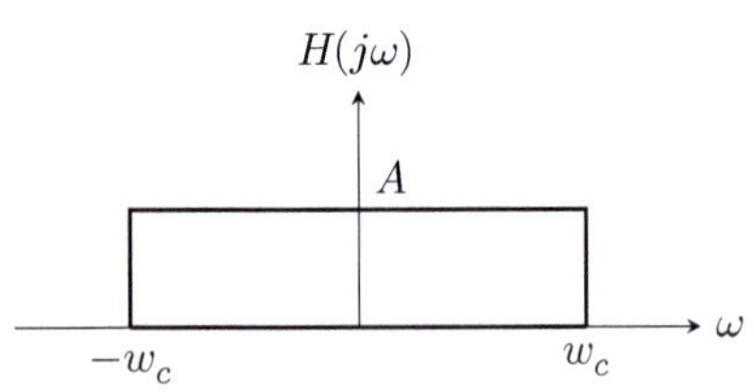

Figure P11.13b

The input to this system is shown in Figure P11.13c.

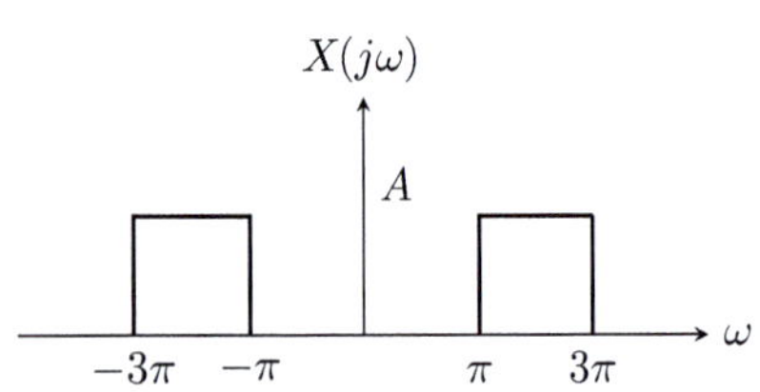

Figure P11.13c

a) Find the Fourier transform of the output of the multiplier, $g(t) = x(t)e^{-j2\pi t}$.
b) Find the Fourier transform of the output of the low-pass filter, $y(t) = g(t) * h(t)$.
c) Find the maximum sampling period T, which recovers the input signal $x(t)$ from the sampled signal $y_p(t)$.
d) Suggest a system, which recovers $x(t)$ from $y_p(t)$.

11.14 An LTI system, represented by a band-limited frequency response

$$H(j\omega) = \begin{cases} 1, & \text{for} \quad \omega \geq 1000\pi, \\ 0, & \text{o.w.} \end{cases}$$

a) Find the output $y(t)$ of this system for $x(t) = \sin 2000\pi$.
b) Find the maximum sampling period T_{max} to recover the signal $y(t)$ from the sampled signal $y_p(t) = y(t)p(t)$, where

$$p(t) = \sum \delta(t - nT).$$

c) Find and plot the sampled signal $y_p(t) = y(t)p(t)$ for $T = T_{max}/2$.

12

Discrete Time Sampling and Processing

"Information is the resolution of uncertainty."

Claude Shannon

In Chapter 11, we introduced the classical sampling methods based on the sampling theorem of Claude Shannon. This pioneering theorem allowed us to sample a continuous time signal into the sampled signal without losing any information.

WATCH Learn more about Claude Shannon, a hero of the digital revolution @ `https://384book.net/v1201`

We have seen three methods for sampling:

1) Impulse train sampling, where the sampled signal is represented by the superposition of the shifted impulse functions, weighted by the amplitude of the function.
2) Zero order hold sampling, where the sampled function is represented by piece-wise constant functions, weighted by the amplitude of the functions at each sampling interval.
3) First order hold sampling, where the sampled function is represented by piecewise linear functions fitted between two sampling points.

Note: In all of the sampling methods, we convert a continuous time signal into another continuous time signal, where we omit the points of the input function between the sampling points. We showed that we can reconstruct the original signal from its sampled version without losing any information, provided that the signal is band limited and the sampling frequency is at least twice as big as the bandwidth of the signal.

Suppose that we need to process a speech signal to reduce the noise or to decompose orchestral music into its instruments by a digital computer. The classical sampling theorem does not allow us to process a continuous time sampled signal in a digital machine. In addition, we cannot design a computer vision system, e.g., to extract an object from a given image dataset by classical sampling methods.

Motivating Question 1: Considering the fact that the sampled signal is still in continuous time, how do we process a continuous time signal by a digital computer?

The answer to this question requires conversion of a sampled signal into a discrete time signal, where we define the function only at integer values of time. This process is called **C/D conversion** (continuous to discrete time conversion).

Signals and Systems: Theory and Practical Explorations with Python,
First Edition. Fatoş Tunay Yarman Vural and Emre Akbaş.

Companion website: www.wiley.com/go/signalsandsystems

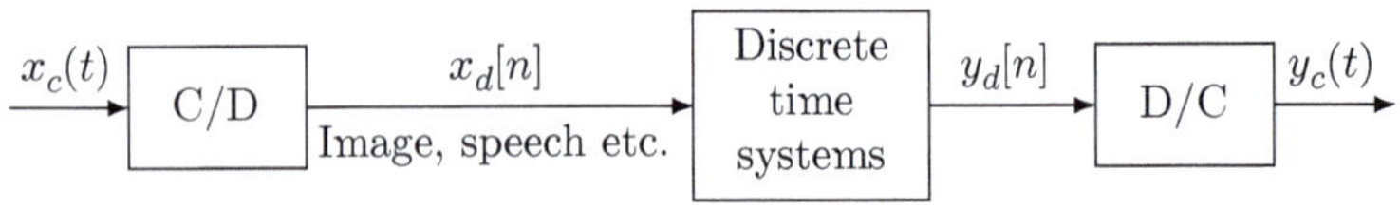

Figure 12.1 Block diagram representation of a general digital signal processing system.

Motivating Question 2: After we process the discrete time signal by a digital computer, how do we reconstruct the continuous time counterpart of the discrete signal?

The answer to these questions requires a conversion of a discrete time signal into a continuous time signal. This process is called **D/C conversion** (discrete to continuous time conversion).

The continuous time signal $x_c(t)$ is converted to a discrete signal, $x_d[n]$ by a C/D converter. The discrete time signal $x_d[n]$ is then processed by a digital computer. Finally, the discrete time signal, $y_d[n]$, obtained at the output of a digital computer is reconstructed by a D/C converter. Figure 12.1 illustrates this pipeline.

In this chapter, we study the methods for C/D and D/C conversions. Then, we will introduce simple methods to design discrete time filters from their continuous counterparts using the sampling theorem. Finally, we introduce the discrete version of the sampling theorem, where we sample a discrete time signal for further compressing the data without losing information.

12.1 Time Normalization

Recall that impulse train sampling replaces a continuous time signal by a set of weighted and shifted impulses, each of which is placed at every sampling period, T, of the signal.

A discrete time signal exists only at integer values of time. Therefore, one needs to convert a continuous time output of a sampled signal into a discrete time signal. The conversion is simple: we replace the continuous time impulse functions placed at every sampling period T with discrete time impulses placed at every integer value, as shown in Figure 12.2. Mathematically, given a continuous time signal $x(t)$, let us define the value of this function at every point nT as:

$$x_c(t) = x(nT), \quad \forall n. \tag{12.1}$$

Then, we define the discrete time counterpart of this continuous time signal as:

$$x_d[n] = x_c(nT), \quad \forall n. \tag{12.2}$$

This process is known as **time normalization.** The discrete time function $x[n]$ is called the time-normalized counterpart of the continuous time function $x(t)$.

In the time normalization process, the value of $x(t)$ is kept the same at every $t = nT$ value of time to generate the discrete time signal $x[n]$. However, the time axis of $x[n]$ is normalized by $1/T$. The upper row of Figure 12.2 shows that the continuous time impulses are placed at every sampling period $T = T_1$ (left). The corresponding discrete time impulses are placed at every integer value of time. The time axis is replaced by integer values at every period, T (right). In the bottom row of this figure, we double the sampling period as $T = 2T_1$. The corresponding discrete time function is presumably smoother (left). However, the time axis is replaced by integer values at every period $T = 2T_1$.

Note: As can be observed in Figure 12.2, when we make time normalization, the analytic form of the discrete time signal changes as we change the sampling period, T. Furthermore, the information about the sampling period T disappears after time normalization. No matter what the sampling period of the continuous time signal is, the time axis of the corresponding discrete time function consists of integer values of n.

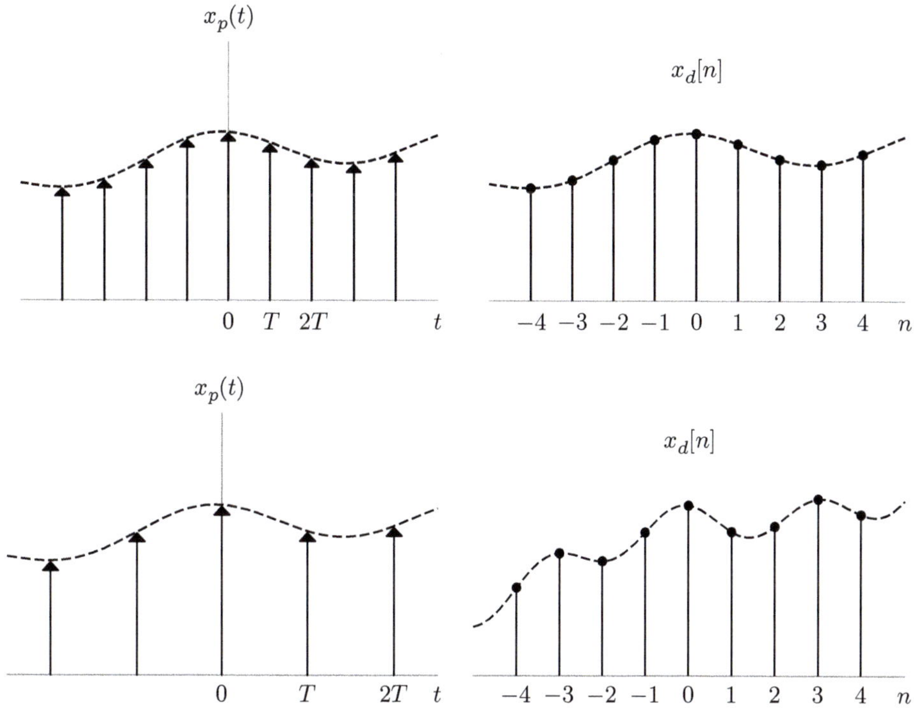

Figure 12.2 Converting a continuous time weighted impulse function of the sampled signal into discrete time impulse train by time normalization. In the upper row sampling period is $T = T_1$, whereas in the lower row $T = 2T_1$.

12.2 C/D Conversion: $x(t) \to x[n]$

Our goal is to find a discrete counterpart $x[n]$ of a continuous time signal $x(t)$, such that from this discrete function, we can uniquely recover the original continuous time function.

Motivating Question: How do we convert a continuous time signal into a discrete time signal without losing any information?

Let us realize time normalization to convert a continuous time-sampled signal into a discrete time signal, as shown in Figure 12.3.

Recall the continuous time sampling in the frequency domain,

$$x_p(t) = x(t)p(t). \tag{12.3}$$

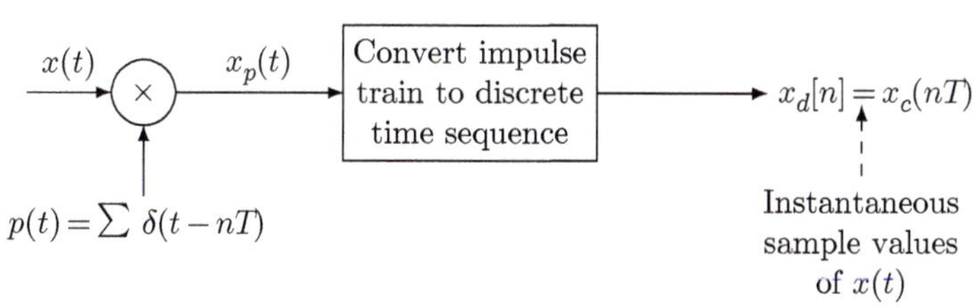

Figure 12.3 C/D conversion: Conversion of a continuous time signal into a discrete time signal. First, we apply continuous time impulse train sampling to a continuous time signal $x(t)$ to obtain $x_p(t)$. Then, we apply time normalization to obtain the discrete version of $x_p(t)$, namely $x_d[n]$.

Equivalently, we can express the sampled signal $x_p(t)$ as follows:

$$x_p(t) = \sum_{n=-\infty}^{\infty} x_c(nT)\delta(t - nT), \tag{12.4}$$

where $x_p(nT)$ is the value of the continuous time signal $x(t)$ at nT for all values of n. Recall also that $\mathcal{F}\{\delta(t - nT)\} = e^{-j\omega nT}$, then the Fourier transform of $x_c(t)$ is

$$X_p(j\omega) = \sum_{n=-\infty}^{\infty} x_c(nT)e^{-j\omega nT}. \tag{12.5}$$

Fourier transform of the discrete time signal is

$$X_d(e^{j\omega}) = \sum_{n=-\infty}^{\infty} x_d[n]e^{-j\omega n}. \tag{12.6}$$

Considering the fact that $x_c(nT) = x_d[n]$, for all n and comparing the Fourier transforms of $X_p(j\omega)$ and $X_d(e^{j\omega})$, we can obtain the relationship between the $X_d(e^{j\omega})$ and $X_p(j\omega)$ as follows:

$$\boxed{X_d(e^{j\omega}) = X_p\left(\frac{j\omega}{T}\right).} \tag{12.7}$$

Therefore, the discrete time counterpart $X_d(e^{j\omega})$ of a continuous time signal generated by time normalization is simply the frequency-scaled version of the continuous time sampled signal $X_p(j\omega)$. In other words, the original signal $X(j\omega)$ is repeated at every sampling frequency, $\omega_s = \frac{2\pi}{T}$, in $X_p(j\omega)$, whereas its discrete counterpart, $X_d(e^{j\omega})$, is repeated in every 2π.

Replacing,

$$X_p(j\omega) = \frac{1}{T}\sum_{n=-\infty}^{\infty} X(j(\omega - n\omega_s))$$

in Equation (12.7), we obtain a relationship between the Fourier transform of continuous and discrete time functions as follows:

$$\boxed{X_d(e^{j\omega}) = \frac{1}{T}\sum_{n=-\infty}^{\infty} X\left(j\left(\frac{\omega - 2\pi n}{T}\right)\right)} \tag{12.8}$$

Comparison of $X_d(e^{j\omega})$ and $X_p(j\omega)$ in Figure 12.4 shows that the only difference between these two functions is the scale T in the frequency axis. While the continuous sampled signal, $X_p(j\omega)$ is periodic with $2\pi/T$, the discrete counterpart, $X_d(e^{j\omega})$ is periodic with 2π.

Although $x(t)$ is a continuous time signal and $x_d[n]$ is a discrete time signal, their Fourier transforms are both continuous. Furthermore, the analytical form of $X(j\omega)$ is preserved in both $X_d(e^{j\omega})$ and $X_p(j\omega)$ in the frequency domain. However, due to time normalization, the sampling period and/or frequency disappears in $X_d(e^{j\omega})$. This reveals that we can recover the original continuous time signal from its discrete version, provided that the sampling period or sampling frequency is given.

Exercise 12.1 Consider a continuous time band-limited signal $x_c(t)$ and its Fourier transform, where $X_c(j\omega) = 0$ for $|\omega| \geq 6000\pi$. The discrete time counterpart of this signal is obtained by the following C/D conversion:

$$x_d[n] = x_c\left(\frac{10^{-3}}{3}n\right)$$

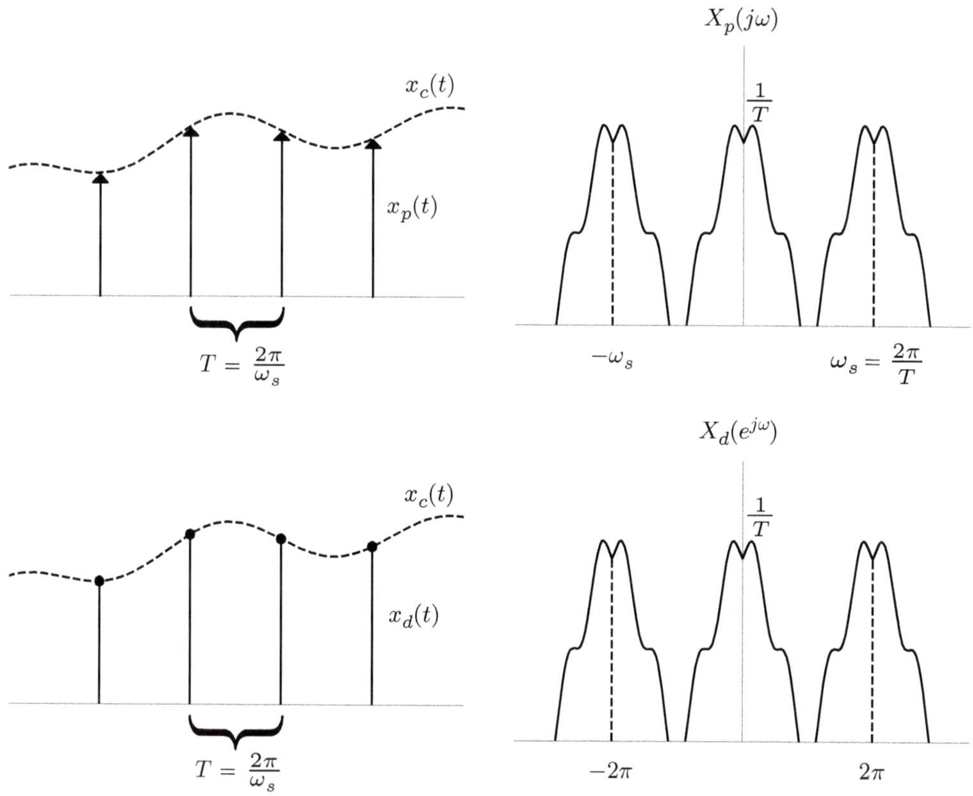

Figure 12.4 Comparison of the continuous time sampled signal $X_p(j\omega)$ and its discrete time counterpart $X_d(e^{j\omega})$, in the frequency domain.

a) What is the sampling frequency in Hertz and the angular frequency in rad/s?
b) Can we reconstruct $x_c(t)$ from its discrete time counterpart $x_d[n]$ without losing information?

Solution

a) The sampling period is $T = \frac{10^{-3}}{3}$ seconds. Therefore, the sampling frequency f_s is

$$f_s = \frac{1}{T} = \frac{1}{\frac{10^{-3}}{3}} = 3000 \text{ Hz}.$$

The angular frequency corresponding to the sampling frequency is

$$\omega_s = 2\pi f_s = 2\pi \cdot 3000 = 6000\pi \text{ rad/s}.$$

b) No, because there is aliasing, where the sampling frequency is smaller then the bandwidth of the signal:

$$\omega_s = 6000\pi < 2 \times 6000\pi.$$

12.3 D/C Conversion

Suppose that after converting the continuous time signal $x(t) = x_c(t)$ into a discrete time signal $x[n] = x_d[n]$ by the method described earlier, we perform some operations on it in a digital computer, such that reducing the embedded noise or filtering the signal. After all these digital signal processing applications, we generate a discrete time output signal, $y[n] = y_d[n]$.

Motivating Question: How do we reconstruct a continuous time version $y(t)$ from the discrete time signal $y_d[n]$?

In D/C conversion, the continuous counterpart of the discrete time signal is obtained in three steps.

In the first step, we convert the discrete time signal $y_d[n]$ into the continuous time sampled signal $y_p(t)$ by theoretically replacing the shifted impulses to their counterpart at each integer value n, as follows:

$$y_p(t) = \sum_{n=-\infty}^{\infty} y_d[n]\delta(t - nT). \tag{12.9}$$

Recall that impulse train sampled signal $y_p(t)$ is obtained by multiplying the signal $y(t)$ with the impulse train function:

$$y_p(t) = y(t)p(t), \tag{12.10}$$

where

$$p(t) = \sum_{n=-\infty}^{\infty} \delta(t - nT), \tag{12.11}$$

and T is the sampling period,

$$T \leq \frac{2\pi}{\omega_N}.$$

The Nyquist rate, ω_N is the bandwidth of the signal, which is the minimum allowable sampling rate to uniquely reconstruct the original signal from its sampled counterpart.

Recall also that the discrete time Fourier transform, $Y_d(e^{j\omega})$ is the frequency-scaled and repeated version of the continuous time Fourier transform, $Y(j\omega)$ with period 2π,

$$Y_d(e^{j\omega}) = \frac{1}{T}\sum_{n=-\infty}^{\infty} Y\left(j\left(\frac{\omega - 2\pi n}{T}\right)\right). \tag{12.12}$$

In the second step, we obtain the Fourier transform of $y_p(t)$,

$$Y_p(j\omega) = \frac{1}{T}\sum_{n=-\infty}^{\infty} Y(j(\omega - n\omega_s)),$$

where $\omega_s = 2\pi/T$ is the sampling frequency of the continuous time input signal $x(t)$. Then, we design a low pass filter, $H(j\omega)$ with the cutoff frequency ω_c, to cover the bandwidth of the discrete time signal $Y_d(e^{j\omega})$ and low pass filter the impulse train sampled signal $y_p(t) \leftrightarrow Y_p(j\omega)$, in the frequency domain to reconstruct the continuous counterpart of the discrete time signal $Y_d(e^{j\omega})$, as follows:

$$Y_r(j\omega) = H(j\omega)Y_p(j\omega). \tag{12.13}$$

The reconstructed signal, $Y_r(j\omega) = Y(j\omega)$, provided that the cutoff frequency of the low pass filter, $H(j\omega)$ covers the bandwidth of the signal at the center of $Y_p(j\omega)$.

Finally, we take the inverse Fourier transform of $Y_r(j\omega)$ to obtain the continuous time reconstructed signal, $y_r(t)$. The steps of D/C conversion are illustrated in Figure 12.5.

Note: In order to recover the continuous time signal $y_r(t) \leftrightarrow Y_r(j\omega)$, from its discrete time counterpart $y_d[n] \leftrightarrow Y_d(e^{j\omega})$ without losing any information, the output signal $Y_d(e^{j\omega})$ should also be

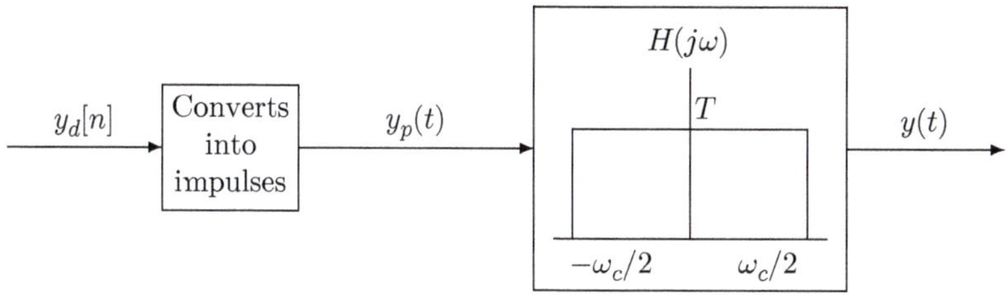

Figure 12.5 D/C conversion: Recovering a continuous time signal from its discrete counterpart.

band-limited and the sampled signal $Y_p(j\omega)$ should be periodic with nonoverlapping functions to avoid aliasing.

Interestingly, low pass filtering of the converted signal $Y_p(j\omega)$ yields an **interpolation** of the discrete time signal $Y_d(e^{j\omega})$, where we fill up the discrete time function $y_d[n]$ in between the sampled time instances to obtain the continuous time function $y(t)$. For this reason, D/C conversion is sometimes called **interpolation**.

Now, we know how to perform C/D and D/C conversion. Thus, we can design the discrete time counterpart of a given continuous time linear timeinvariant (LTI) system and vice versa. Let us give some examples as follows.

12.3.1 Band-Limited Digital Differentiator

Suppose that we are given a block diagram representation of a continuous time system, which consists of differentiators and adders. We can find the discrete counterpart of this LTI system by replacing the differentiators with their discrete counterpart using C/D conversion methods.

Recall the differentiation property of Fourier transforms in continuous time signals:

$$\frac{dx(t)}{dt} \longleftrightarrow j\omega X(j\omega). \tag{12.14}$$

Suppose that we have a subsystem in a block diagram representation, where the input and output are related by a differentiation operator, as seen in Figure 12.6,

$$y(t) = \frac{dx(t)}{dt} \longleftrightarrow Y(j\omega) = j\omega X(j\omega). \tag{12.15}$$

The frequency response of the differentiation subsystem is

$$H(j\omega) = \frac{Y(j\omega)}{X(j\omega)} = j\omega, \text{ for } -\infty < \omega < \infty. \tag{12.16}$$

Motivating Question: How can we find the discrete time counterpart $H_d(e^{j\omega})$ of a continuous time differentiator $H(j\omega)$?

The continuous time differentiator is not band limited. According to the sampling theorem, it cannot be sampled without losing information about the analytical shape of the frequency response function. One way of converting the continuous time differentiator into its discrete time counterpart is to chop the frequency spectrum after a predefined value to create a band-limited frequency response. Then, we can apply sampling to this band-limited function.

Figure 12.6 A continuous LTI system with a differentiation operator.

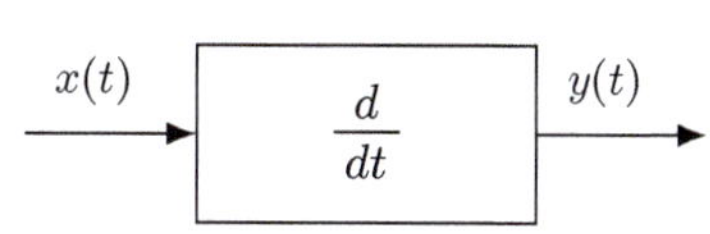

Let us define the band-limited differentiator as follows:

$$H_c(j\omega) = \begin{cases} j\omega, & \text{if } |\omega| < \omega_c, \\ 0, & \text{otherwise,} \end{cases} \tag{12.17}$$

where ω_c is the cutoff frequency and is determined by considering the design issues about the underlying physical phenomenon. Then, the magnitude of $H_c(j\omega)$ is

$$|H_c(j\omega)| = \begin{cases} \omega, & \text{if } |\omega| < \omega_c, \\ 0, & \text{otherwise} \end{cases} \tag{12.18}$$

and the phase is

$$\sphericalangle H_c(j\omega) = \begin{cases} \pi/2, & \text{if } 0 < \omega < \omega_c, \\ -\pi/2, & \text{if } -\omega_c < \omega < 0, \\ 0, & \text{otherwise.} \end{cases} \tag{12.19}$$

The plot of $H_c(j\omega)$ can be found in Figure 12.7. Since the function $H_c(j\omega)$ is now band-limited, we can sample it and reconstruct the original signal from its sampled version.

Now, we can apply the C/D conversion methods to find the discrete time version of a band-limited differentiator $H_d(e^{j\omega})$ from the continuous time band-limited frequency response $H_c(j\omega)$?

At the very first step, we need to find the Nyquist rate of the band-limited function $H_c(j\omega)$, which is

$$\omega_N = 2\omega_c.$$

Then, we select a sampling frequency $\omega_s \geq \omega_N$.

Let us set the sampling frequency to the Nyquist rate, $\omega_N = \omega_s = 2\omega_c$. The corresponding sampling period is

$$T_s = \frac{2\pi}{\omega_s} = \frac{\pi}{\omega_c}. \tag{12.20}$$

Finally, we apply Equation (12.8) for discrete version of band-limited differentiator as follows:

$$H_d(e^{j\omega}) = \frac{1}{T_s}\sum_{n=-\infty}^{\infty} H_c\left(j\left(\frac{\omega - 2\pi n}{T_s}\right)\right). \tag{12.21}$$

Notice that the discrete version of the band-limited differentiator is periodic with 2π. For one full period, the magnitude and phase of $H_d(e^{j\omega})$ are,

$$|H_d(e^{j\omega})| = \begin{cases} \frac{\omega}{T_s}, & \text{if } |\omega| < \omega_c, \\ 0, & \text{otherwise,} \end{cases} \tag{12.22}$$

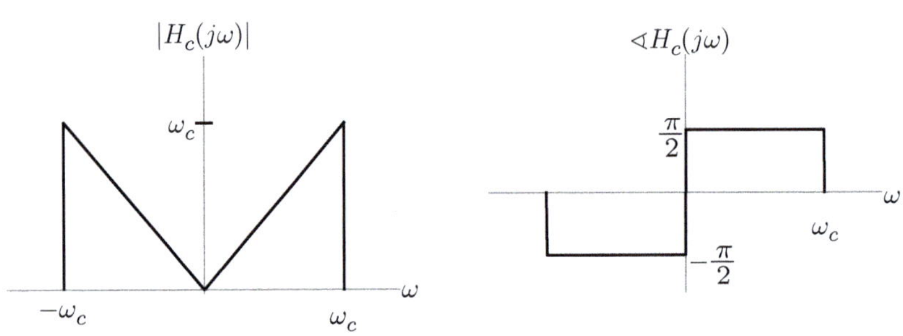

Figure 12.7 Magnitude and phase plots of the continuous time band-limited differentiator, $H_c(j\omega)$.

and

$$\measuredangle H_d(e^{j\omega}) = \begin{cases} \pi/2, & \text{if } 0 < \omega < \omega_c, \\ -\pi/2, & \text{if } -\omega_c < \omega < 0, \\ 0, & \text{otherwise.} \end{cases} \tag{12.23}$$

respectively, and they repeat at every 2π. The magnitude and phase spectrum of $H_d(e^{j\omega})$ are plotted in Figure 12.8.

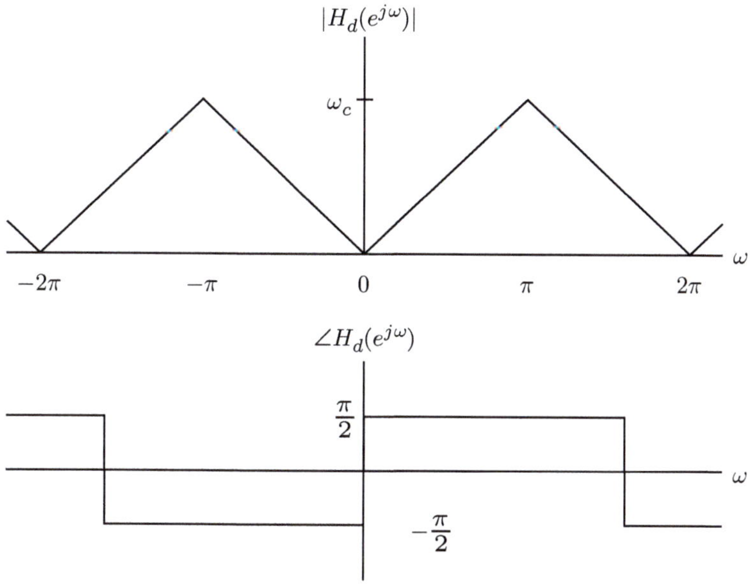

Figure 12.8 Discrete time counterpart of a band limited continuous time differentiator: Magnitude and phase spectrum.

Exercise 12.2 Consider the continuous time differentiator system and its discrete time counterpart given in Figure 12.9. Suppose the input signal $x(t)$ is

$$x(t) = \frac{\sin(\pi t/T)}{\pi t}. \tag{12.24}$$

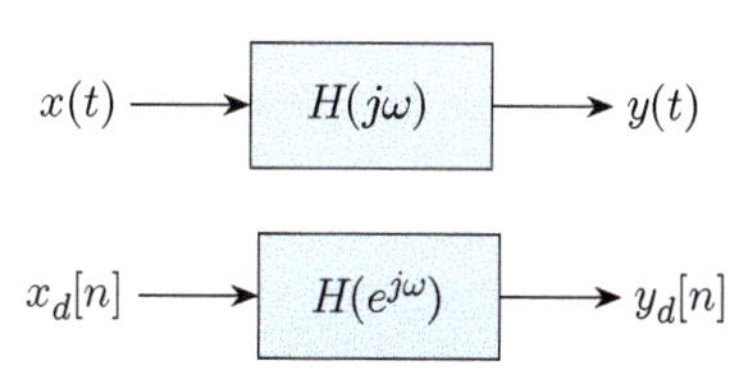

Figure 12.9 An LTI system represented by $H(j\omega) = j\omega$ and its discrete counterpart $H(e^{j\omega}) = j(\frac{\omega}{T})$, for $|\omega| < \pi$.

a) Find the discrete time counterpart, $x_d[n]$ of the input $x[n]$.
b) Find the output, $y(t)$ of the continuous time differentiator.
c) Find the discrete time output, $y_d[n]$ of the digital differentiator.
d) Find the discrete time impulse response of the digital differentiator.

Solution

a) The discrete time counterpart of $x(t)$ is simply obtained by time normalization of its continuous counterpart as follows:

$$x_d[n] = x_c(nT) = \frac{\sin(\pi n)}{\pi n T}. \tag{12.25}$$

Note: The discrete time function $x_d[n]$ is indefinite at $n = 0$, i.e., it approaches to ∞, for $n = 0$. In order to find the value of $x_d[n]$, as $n \to 0$, we use L'Hopital's rule: We take the derivative of the numerator and the denominator, which is

$$x_d[0] = \lim_{n\to\infty} \frac{\cancel{\pi}\cos(\pi n)}{\cancel{\pi}T} = \frac{1}{T}. \tag{12.26}$$

For the rest of the values $n \neq 0$, $x_d[n] = 0$. Therefore, the discrete time counterpart of the input is

$$x_d[n] = \frac{1}{T}\delta[n], \tag{12.27}$$

which is plotted in Figure 12.10.

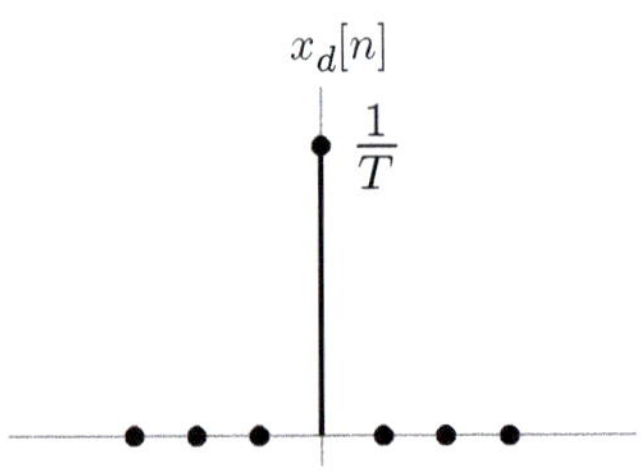

Figure 12.10 Discrete time counterpart $x_d[n]$ is time normalized version of the continuous time input $x(t)$, in Exercise 12.2.

b) The output of the continuous time differentiator is obtained by simply taking the derivative of the input:

$$y(t) = \frac{dx(t)}{dt} = \frac{\cos(\pi t/T)}{Tt} - \frac{\sin(\pi t/T)}{\pi t^2}. \tag{12.28}$$

c) In order to find the discrete time counterpart of the continuous time output all we need to do is time normalization:

$$y_d[n] = y_c(nT) = \frac{\cos(\pi n)}{\pi T n} - \frac{\sin(\pi n)}{\pi T^2 n^2}. \tag{12.29}$$

Note: There is a problem in the representation of the discrete time function $y_d[n]$, in Equation (12.29). It is indefinite, i.e., approaches to $0/\infty$ for $n = 0$.
In order to find the limiting value of $y_d[n]$ at $n = 0$, we apply L'Hopital's rule:

$$y_d[n] = -\frac{\pi\sin(\pi n)}{\pi T} - \frac{\pi\cos(\pi n)}{2\pi T^2 n} = 0, \tag{12.30}$$

and for $n \neq 0$, the discrete time counterpart of the output becomes

$$y_d[n] = y_c(nT_s) = \frac{(-1)^n}{nT^2}. \tag{12.31}$$

d) Finally, we need to find the discrete time counterpart of the impulse response, $h_d[n]$.
The discrete time output of a digital differentiator can be written in the following compact form:

$$y_d[n] = y_c(nT) = \begin{cases} \frac{(-1)^n}{nT^2}, & \text{if } n \neq 0, \\ 0, & \text{otherwise.} \end{cases} \tag{12.32}$$

Recall the discrete time convolution, which relates the input to the output through the impulse response as follows:

$$y_d[n] = x_d[n] * h_d[n] = \frac{1}{T}\delta[n] * h_d[n]. \tag{12.33}$$

Then, the discrete time impulse response is

$$h_d[n] = \begin{cases} \frac{(-1)^n}{nT}, & \text{if } n \neq 0, \\ 0, & \text{otherwise.} \end{cases} \tag{12.34}$$

12.3.2 Digital Time Shift

Recall the elementary signal operations on the time variable, such as time shift, time scale, and time reverse. These operations are very important for manipulating and generating a wide range of signals. We studied the operations on time variables in discrete time and continuous time signals separately. We stated that the values of the signals in between the integer time variable are undefined for discrete time signals. Is it really so?

Motivating Question: Can we switch the elementary time operations between the continuous time and discrete time signals?

Sampling theorem provides us a rigorous methodology to answer this question: Yes! It is possible to switch between the elementary time operations for the continuous time and discrete time signals, provided that we satisfy the following assumptions:

Assumption 1: The Input signal $X(j\omega)$ is band limited with the cutoff frequency ω_c.
Assumption 2: The sampling frequency is set at least to the Nyquist rate, in other words,

$$\omega_s \geq \omega_N,$$

where the Nyquist rate is

$$\omega_N = 2\omega_c.$$

Let us study a specific time operation, namely the time shift, in continuous time and discrete time cases:

Given a continuous time signal, which is the shifted version of a signal $x(t)$,

$$y(t) = x(t - t_0), \tag{12.35}$$

the discrete time counterpart, $y_d[n]$ can be directly obtained by time normalization,

$$y_d[n] = y(nT) = x(nT - t_0), \tag{12.36}$$

where $T = 2\pi/\omega_s$ is the sampling period, which is slightly bigger than the Nyquist rate.

Exercise 12.3 Consider an LTI system represented by the following equation:

$$y(t) = x(t - t_0), \tag{12.37}$$

where the input signal $x(t)$ is always band limited with the bandwidth, $2\omega_c$.

a) Is $y(t)$ also band limited?
b) Find the frequency response, $H(j\omega)$, of this system. Is the frequency response band limited?
c) Find the discrete counterpart, $H_d(e^{j\omega})$, of the frequency response.

Solution

a) Let us take the Fourier transform of $y(t)$, which is the direct application of time shift property,

$$y(t) = x(t - t_0) \longleftrightarrow Y(j\omega) = e^{-j\omega t_0}X(j\omega). \tag{12.38}$$

Since the Fourier transform of the shifted signal brings just a multiplicative factor of the complex exponential, the bandwidth of $Y(j\omega)$ is the same as that of $X(j\omega)$. Thus, the signal $y(t)$ is band limited.

b) Since it is the eigenvalue of the system, the multiplicative factor corresponds to the frequency response,

$$H(j\omega) = e^{j\omega t_0}. \tag{12.39}$$

This system is not band limited. However, since we assume that the input signal is always band limited, we can chop the frequency response at the cutoff frequency of $X(j\omega)$. Thus, the band-limited frequency response becomes

$$H_c(j\omega) = \begin{cases} e^{-j\omega t_0}, & \text{for } |\omega| < \omega_c, \\ 0, & \text{otherwise,} \end{cases} \tag{12.40}$$

which is plotted in Figure 12.11.

c) The discrete time counterpart of the continuous time frequency response can be obtained by frequency normalization of the band-limited frequency response, for one full period, as follows:

$$H_d(e^{j\omega}) = \begin{cases} H_c(j\omega), & \text{for } \ |\omega| < \omega_c, \\ 0 & \text{otherwise} \end{cases} \tag{12.41}$$

In Equation (12.41), the sampling period is $T = \frac{2\pi}{\omega_c}$. Considering the fact that counterpart of the band-limited frequency response is always periodic with $\omega = 2\pi$, we can extend the cut-off frequency to $\omega_c = 2\pi$. Since $Y(e^{j\omega}) = H(e^{j\omega})X(e^{j\omega})$, this extension will not change the bandwidth of the output signal, which is the same as that of the input signal. Hence,

$$H_d(e^{j\omega}) = \begin{cases} e^{-j\omega \frac{t_0}{T}}, & \text{for } \ |\omega| < \pi, \\ 0 & \text{otherwise,} \end{cases} \tag{12.42}$$

and repeats at every 2π period. This filter is plotted in Figure 12.12.

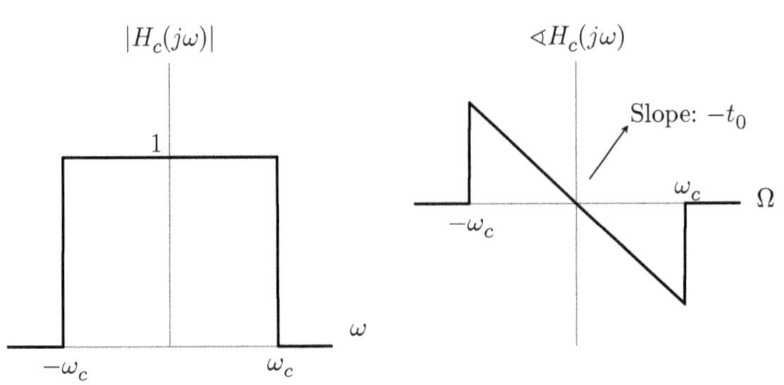

Figure 12.11 The magnitude and phase plots of the frequency response of the band-limited system, represented by $y(t) = x(t - t_0)$.

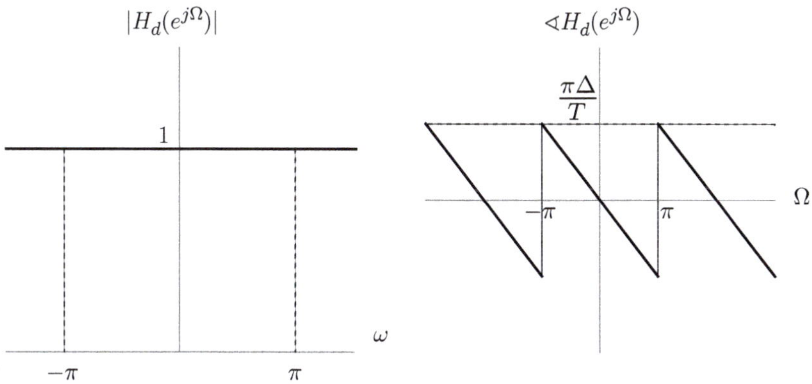

Figure 12.12 The magnitude and phase plots of frequency responses of the discrete time counterpart of $H_c(j\omega)$ in Exercise 12.3.

12.4 Sampling the Discrete Time Signals

Until now, we considered sampling the continuous time signals and systems to find their discrete time counterpart. This operation enabled us to process a continuous time signal in a digital computer and recover the processed continuous time signal from its discrete time counterpart.

Motivating Question: What if we need to sample a discrete time signal?

If we can manage to extend the sampling theorem to discrete time, we can further reduce the data used to represent signals and systems without losing their information content. Sampling in discrete time signals is sometimes called **down-sampling** because it results in a type of compression, when the signal is represented by a sequence of numbers, in the form of a time series.

12.4.1 Discrete Time Impulse Train Sampling

Let us now extend the continuous time sampling theorem to the discrete world. Impulse train sampling of a discrete time signal $x[n]$ is defined as:

$$x_p[n] = x[n]p[n] = \begin{cases} x[n], & \text{if } n \text{ is integer multiple of } N, \\ 0, & \text{otherwise,} \end{cases} \tag{12.43}$$

where N is the integer sampling period, and the impulse train is defined as:

$$p[n] = \sum_{k=-\infty}^{\infty} \delta[n - kN]. \tag{12.44}$$

This operation is schematically illustrated in Figure 12.13.

Figure 12.13 Impulse train sampling of a discrete signal $x[n]$.

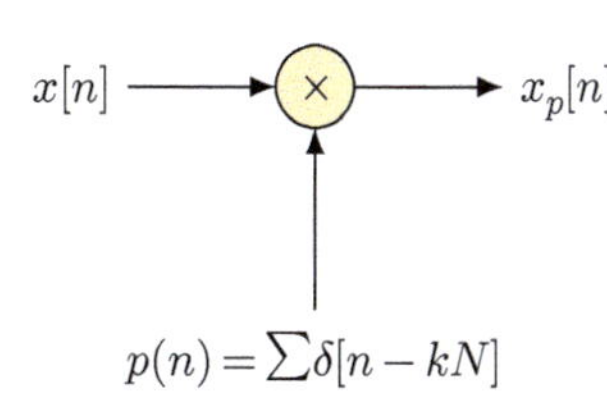

Note: The major difference between the continuous time and discrete time sampling is that sampled signal, $x[n]$ is set to 0 values between each sampling period N.

Motivating Question: What is the output of the discrete time impulse train sampling $x_p[n] \leftrightarrow X_p(e^{j\omega})$?

Although the discrete time impulse train sampling is an extension of the continuous time counterpart, there are three major restrictions, which shape the analytical form of $x_p[n] \leftrightarrow X_p(e^{j\omega})$:

1) Sampling period N should be integer valued and sampling is applied at integer multiples of N.
2) Since the Fourier transform of a discrete time function is periodic with 2π, the cutoff frequency of the band-limited signal cannot exceed 2π. Hence, the bandwidth of the discrete time signal should be small enough, i.e.,

$$\omega_M < 2\pi. \tag{12.45}$$

   This brings an upper limit to the bandwidth of the discrete time signal to be sampled.
3) Given the band-limited signal with cut-off frequency, ω_M, the sampling frequency $\omega_s = \frac{2\pi}{N}$ should be at least as big as the Nyquist frequency ω_N,

$$\omega_s \geq \omega_N = 2\omega_M. \tag{12.46}$$

   Otherwise, if $\omega_s < 2\omega_M$, then there are overlaps in $X_p(e^{j\omega})$, which results in aliasing.

The aforementioned constraints on the sampling frequency ω_s, Nyquist frequency ω_N and cutoff frequency ω_M of the signal shape up the analytical form of the sampled signal $x_p[n] \leftrightarrow X_p(e^{j\omega})$.

Let us now investigate the effect of the aforementioned constraints on the sampling of the discrete time signal $x[n] \longleftrightarrow X(e^{j\omega})$ on structure of the sampled signal, in both time and frequency domain.

Recall the Fourier transform pair of discrete time impulse train was

$$p[n] = \sum_{k=-\infty}^{\infty} \delta[n - kN] \longleftrightarrow P(e^{j\omega}) = \frac{2\pi}{N} \sum_{k=-\infty}^{\infty} \delta(\omega - k\omega_s). \tag{12.47}$$

Fourier transform of the sampled signal $x_p[n] = x[n]p[n]$ is

$$X_p(e^{j\omega}) = \frac{1}{2\pi} X(e^{j\omega}) * P(e^{j\omega}) = \frac{1}{N} \sum_{k=-\infty}^{\infty} X(e^{j(\omega - k\omega_s)}), \tag{12.48}$$

$P(e^{j\omega})$ and $X_p(e^{j\omega})$ are plotted in Figure 12.14.

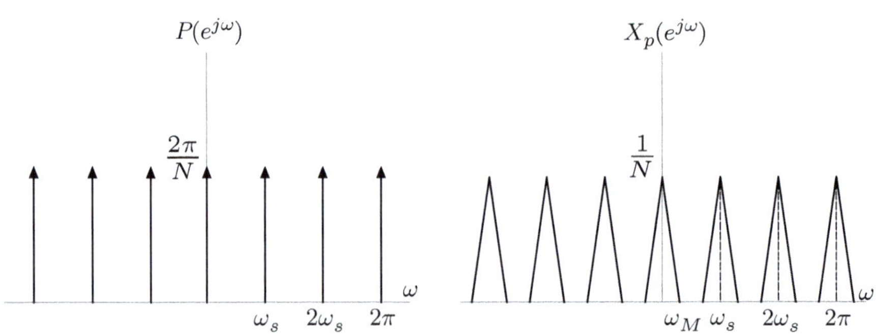

Figure 12.14 Fourier transform $P(e^{j\omega})$ of discrete time impulse train $p[n] = \sum \delta[n - kN]$ and Fourier transform of the sampled signal $X_p(e^{j\omega})$ of a discrete time signal $x_p[n]$.

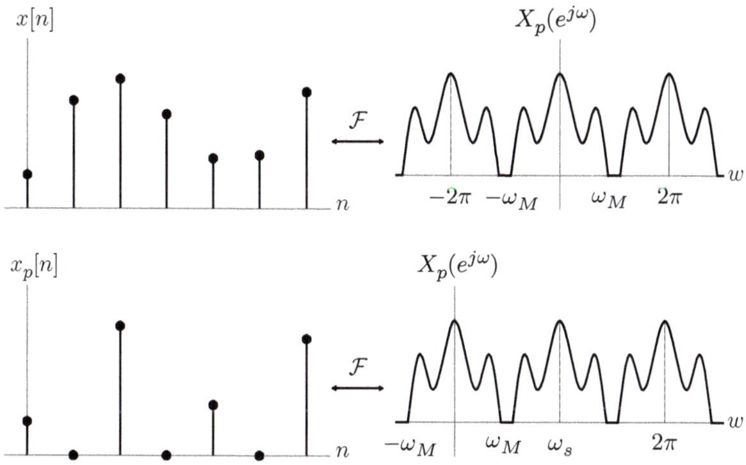

Figure 12.15 Time and frequency domain representations of the discrete time signal and its sampled version. We delete the values of $x[n]$ in between each sampling period N ($N = 2$ in this example). Hence, the sampled signal in time domain becomes $x_p[n] = 0$ in between each sampling period kN and $(k + 1)N$.

Note: There are two types of periodicities in both $P(e^{j\omega})$ and $X_p(e^{j\omega})$. The first periodicity comes from the nature of discrete time Fourier transform, which is 2π. The second periodicity comes from the repetition of the discrete time signal at each sampling frequency, $k\omega_s$ for all $k = 0, \pm 1, \pm 2, \ldots$. The relationship between these two periodicities affect the result of discrete time sampling. An example is given in Figure 12.15.

Note: We assume that

$$x[n] \longleftrightarrow X(e^{j\omega})$$

is a band-limited signal and the bandwidth $2\omega_M$ is small enough so that when we sample the signal with a sampling frequency ω_s, we can squish the repeated signals in the 2π period without any overlaps.

Exercise 12.4 A discrete time signal $y[n]$ is obtained by impulse train sampling of a signal $x[n]$ as follows:

$$y[n] = \sum_{k=-\infty}^{\infty} x[n]\delta[n - kN],$$

where the band limited signal $x[n]$ has the Fourier transform,

$$X(e^{j\omega}) = 0 \quad \text{for} \quad \frac{\pi}{4} \leq |\omega| \leq \pi.$$

Find the largest value for the sampling period N, satisfying the Nyquist rate.

Solution

Fourier transform of the output signal, $Y(e^{j\omega})$ is the periodic repetitions of $X(e^{j\omega})$ at every $\omega_s = \frac{2\pi}{N}$. The Nyquist rate is achieved, when

$$\omega_s = 2\omega_c,$$

where ω_c is the cutoff frequency of $X(e^{j\omega})$. Since the Fourier transform of the input is given as:

$$X(e^{j\omega}) = 0 \;\; \text{for} \;\; \frac{\pi}{4} \leq |\omega| \leq \pi,$$

which is periodic with 2π, the nonzero part of the input is

$$X(e^{j\omega}) \neq 0 \text{ for } -\pi/4 < \omega < \pi/4,$$

with the cutoff frequency, $|\omega_c| = \pi/4$. According to the sampling theorem, the sampling frequency $\omega_s = \frac{2\pi}{N}$, should satisfy

$$\frac{2\pi}{N} \geq 2\frac{\pi}{4}.$$

Hence, the Nyquist rate for the largest sampling interval is achieved at $N = 4$.

12.5 Reconstruction of Discrete Time Signal from Its Sampled Counterpart

Practically speaking, **down-sampling** deletes the values of the discrete time function between each sampling period kN and $(k+1)N$. Instead of the deleted values we insert 0s in the sampled signal $x_p[n]$. In order to recover the original signal $x[n]$ from the sampled signal $x_p[n]$, we would like to fill up the zero values with the original values of $x[n]$.

Mathematically speaking, given the sampled signal,

$$x_p[n] = x[n]p[n],$$

where $p[n]$ is an impulse train with sampling period N, our goal is to reconstruct the original discrete time signal $x[n]$, without losing any information. This reconstruction process is sometimes called **up-sampling**.

We assume that the signal is band limited with the cutoff frequency ω_M and it is properly sampled with the sampling frequency $\omega_s = \frac{2\pi}{N}$ to satisfy the Nyquist rate,

$$\omega_s = 2\omega_M. \tag{12.49}$$

If these constraints are satisfied, then we can design a reconstruction filter $H(e^{j\omega})$, which recovers the original signal $x[n] = x_r[n]$ from its sampled version $x_p[n]$ without losing any information (Figure 12.16).

When there is no aliasing, $\omega_M \leq \omega_c \leq \omega_s - \omega_M$, then

$$x_r[kN] = x[kN]. \tag{12.50}$$

Recall that the impulse response of the reconstruction filter can be obtained by taking the inverse Fourier transform of the frequency response:

$$h[n] = \frac{N\omega_c}{\pi}\frac{\sin(\omega_c\pi)}{\omega_c\pi} \longleftrightarrow H(e^{j\omega}). \tag{12.51}$$

The reconstruction filter is plotted in Figure 12.17.

Exercise 12.5 Consider the following discrete time signal:

$$X(e^{j\omega}) = \begin{cases} e^{-j\omega}, & \text{for } |\omega| \leq 2\pi/9, \\ 0, & \text{otherwise.} \end{cases} \tag{12.52}$$

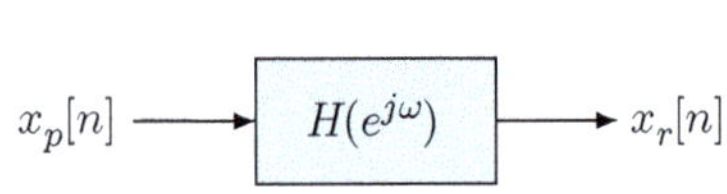

Figure 12.16 Reconstruction filter for discrete time sampling.

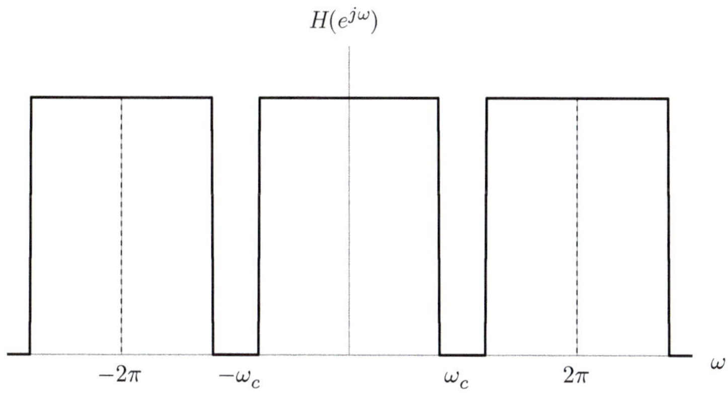

Figure 12.17 Frequency response, $H(e^{j\omega})$, of the reconstruction filter for discrete time sampling. This is a low pass filter which recovers the original signal by $X_r(e^{j\omega}) = X(e^{j\omega}) = X_p(e^{j\omega}) \cdot H(e^{j\omega})$.

a) What is the bandwidth of this signal?
b) Find the largest sampling period N to down-sample the signal without losing any information.

Solution

a) The bandwidth of the signal is $4\pi/9$.
b) The largest sampling frequency is to be the Nyquist rate:

$$\omega_s = \frac{4\pi}{9}.$$

Hence, the sampling period should satisfy

$$N \leq \frac{2\pi}{\omega_s} \leq 9/2.$$

The maximum integer, which satisfies this inequality is $N_{\max} = 4$.

12.6 Discrete Time Decimation and Interpolation

When we represent a down-sampled signal in a sequence of numbers in the form of time series, the array of numbers consists of zero values in between the integer multiples of the sampling period N. This representation carries a lot of redundancies during signal processing. One way to further compress the signal is to skip these repeated 0 values and just keep the signal at kN for all k. The operation of discarding the repeated zero values of the samples signal is called **decimation**.

Let us investigate the analytical structure of the decimation operation in time and frequency domains:

Recall that in discrete time sampling, we have $x_p[n] = x[n]p[n]$ where $p[n] = \sum \delta[n - kN]$ (Figure 12.13).

Fourier transform of the down-sampled signal consists of the repeated form of the Fourier transform of the original signal as follows:

$$X_p(e^{j\omega}) = \frac{1}{N} \sum_{k=-\infty}^{\infty} X(e^{j(\omega - k\omega_s)}). \tag{12.53}$$

If the signal is sampled with the Nyquist frequency, we obtain a plot similar to Figure 12.18.

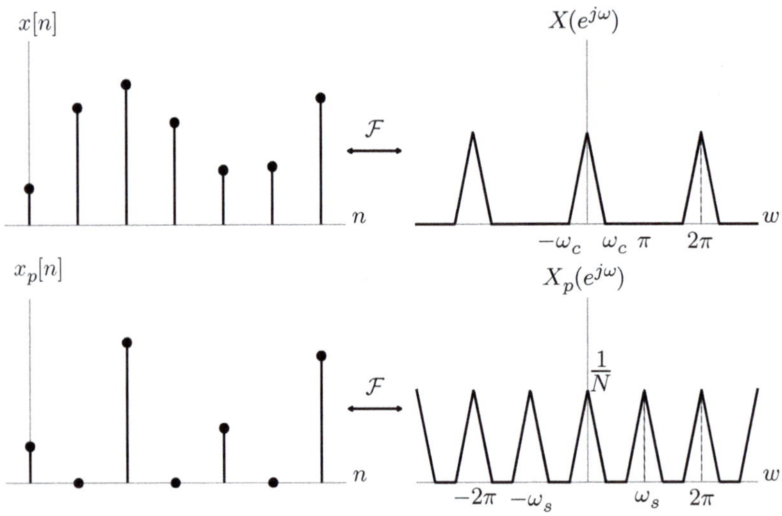

Figure 12.18 When a discrete time signal $x[n]$ is down-sampled, we only keep the values of this signal at integer multiples of N. In the frequency domain, this process corresponds to inserting the analytical form of the original function at every ω_s.

Let us define a decimated discrete time signal $x_{de}[n]$, as the sequence of the selected values of the sampled signal. Formally speaking, the decimated signal and its Fourier transform is

$$x_{de}[n] = x_p[nN] \longleftrightarrow X_{de}(e^{j\omega}) = \sum_{n=-\infty}^{\infty} x_{de}[n]e^{-j\omega n} = \sum_{n=-\infty}^{\infty} x_p[nN]e^{-j\omega n}. \tag{12.54}$$

Therefore, the relationship between the decimated signal $X_{de}(e^{j\omega})$ and its sampled version $X_p(e^{j\omega})$, in the frequency domain can be written as follows:

$$\boxed{X_{de}(e^{j\omega}) = X_p(e^{j\omega/N}).} \tag{12.55}$$

The Fourier transform of the decimated signal is plotted in Figure 12.19.

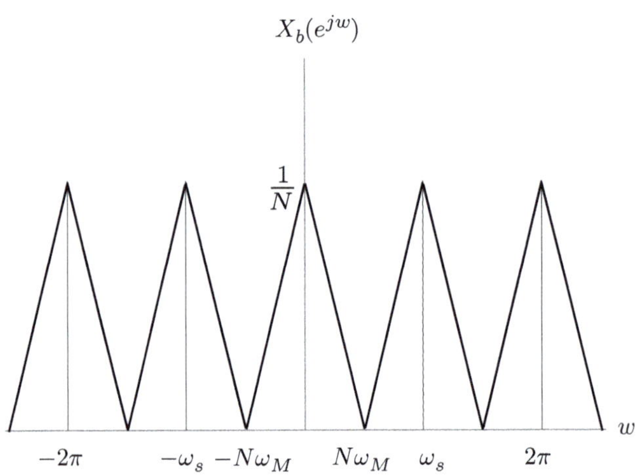

Figure 12.19 Fourier transform of the decimated signal $X_b(e^{j\omega})$.

Note: The only difference between the decimated signal and the down sampled signal is the scale in the frequency axis. When we decimate the sampled signal, all we need to do is to scale the frequency axis by $1/N$, in frequency domain. In order to recover the original signal, $x[n] \leftrightarrow X(e^{j\omega})$ from its decimated version $x_{de}[n] \leftrightarrow X_{de}(e^{j\omega})$, we need to first rescale the frequency axis by N, then, apply a low pass filter to the rescaled signal by $H(e^{j\omega})$, with the cut-off frequency of the original signal $x[n]$.

INTERACTIVE

Sampling and reconstruction of a discrete time signal @ `https://384book.net/i1201`

12.7 Chapter Summary

In this era, where almost the entire technology relies on digital computing, how can we ensure preserving the information content of signals and systems? How can we efficiently store the signals in the minimum space? How can we retrieve and transmit them without losing significant information? How can we process them for a predefined goal so that we can extract the desired information from the signals and design digital systems?

Discrete time sampling and decimation open a door to answer the aforementioned questions. In this chapter, we extended the continuous time sampling theorem of C. Shannon into discrete time signals and systems. We studied the methods for down-sampling the discrete time signals. We studied the constraints imposed by the discrete nature of signals in time and frequency domains: First of all, the sampling period is restricted to be integer valued. Secondly, as we sample the signal with period $N \geq 1$, we define a new signal, where we insert 0 values in between the integer multiples of the sampling period. Finally, we restricted the signal to be band limited with the cutoff frequency $\omega_c < 2\pi$.

We showed that a discrete time signal can be down-sampled with the sampling frequency ω_s, which is at least equal to the Nyquist rate $\omega_N = 2\omega_c$, without losing any information.

Reconstruction of a sampled signal is very similar to that of the continuous time signals. All we need to do is to design a low pass filter $H(e^{j\omega})$ with the cutoff frequency ω_c of the band limited signal $x[n] \longleftrightarrow X(e^{j\omega})$.

Problems

12.1 Consider an A/D converter system, shown in Figure P1.1, where the Fourier transform of the continuous time input is

$$X(j\omega) = \begin{cases} |\omega| & \text{for} \quad |\omega| < 10\pi \\ 0 & o.\omega. \end{cases}$$

a) Find and plot the Fourier transform of the sampled signal $X_p(j\omega)$, when the sampling periods are $T_1 = 0.05$ seconds and $T_2 = 0.1$ seconds.
b) Find and plot the Fourier transform of the discrete time signal $X_d(e^{j\omega})$ obtained at the output of the converter, for $T = 0.05$ seconds and $T = 0.1$ seconds.

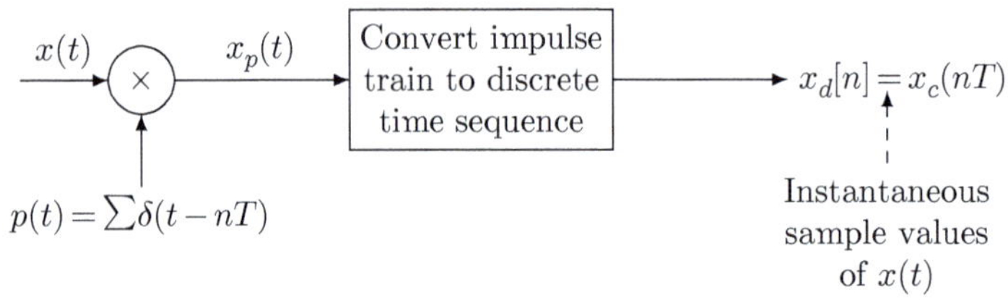

Figure P1

12.2 Consider a D/C converter shown in Figure P2, where the signal:

$$y_d[n] = x_d[n] * h_d[n]$$

is obtained at the output of an LTI system, when the input is

$$x_d[n] = \frac{\sin 10^4 \pi}{\pi n}$$

and the impulse response is

$$h_d[n] = \frac{\sin \frac{\pi}{4} n}{\pi n}$$

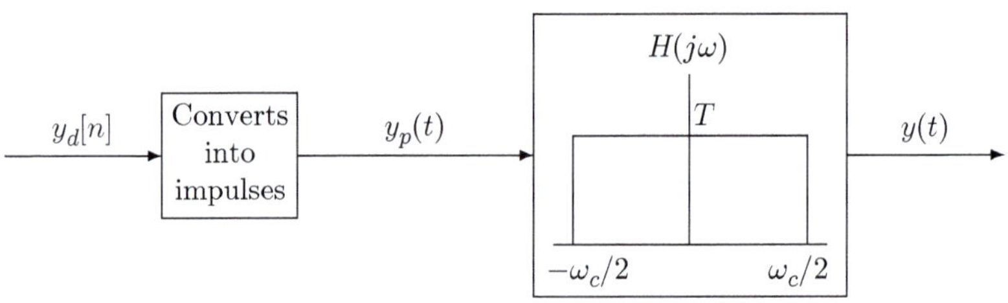

Figure P2

a) Find the Fourier transform of the output $Y_d(e^{j\omega})$.
b) Find the continuous counterpart $y_p(t) \leftrightarrow Y_p(j\omega)$, obtained at the output of the D/C converter for $T = 0.05$ milliseconds.
c) Find the Fourier transform of the continuous time output signal

$$Y(j\omega) = Y_p(j\omega)H(j\omega),$$

when the cutoff frequency of the low pass filter $H(j\omega)$ is $\omega_c = 2\pi/T$.

12.3 A discrete time band-limited signal $x[n]$, defined by the following Fourier transform:

$$X(e^{j\omega}) = \begin{cases} A, & \text{for } |\omega| < \pi/4, \\ 0, & \text{otherwise,} \end{cases}$$

and repeats at every 2π period. We down-sample the signal $x[n]$ to generate a signal $g[n]$, as follows:

$$g[n] = \begin{cases} x[n/2], & \text{for } n = 0, \pm2, \pm4, \pm6 \ldots, \\ 0, & \text{otherwise.} \end{cases}$$

a) Find and plot $x[n]$ and $g[n]$. Comment on the differences and similarities between the two signals.
b) Find and plot $G(e^{j\omega})$. What is the effect of down sampling on the signal in terms of the bandwidth?

12.4 A discrete time LTI system, which gives the following output, in $-\pi < \omega \leq \pi$ interval:

$$Y(e^{j\omega}) = \begin{cases} X(e^{j\left(\omega - \frac{\pi}{3}\right)}), & \frac{\pi}{3} < \omega \leq \frac{\pi}{2}, \\ X(e^{j\left(\omega + \frac{\pi}{3}\right)}), & -\frac{\pi}{2} < \omega \leq -\frac{\pi}{3}, \\ 0, & \text{o.w.}, \end{cases}$$

to the input given as follows:

$$X(e^{j\omega}) = \begin{cases} e^{-j\omega}, & \text{for} \quad |\omega| \leq \pi/6, \\ 0, & \text{otherwise}, \end{cases}$$

and repeats at every 2π.

a) Find and plot the output signal $Y(e^{j\omega})$. What are the cutoff frequencies and the bandwidth of the output $Y(e^{j\omega})$.
b) Find and plot the Frequency response of this system. What is the bandwidth of the frequency response.
c) Find the continuous time counterpart $H_c(j\omega)$ of the discrete time frequency response, $H_d(e^{j\omega})$, for the sampling periods $T = 0.1$ and $T = 1$ seconds.

12.5 A continuous time signal $x_c(t)$ is band limited with the Fourier transform $X_c(j\omega) = 0$ for $|\omega| \geq 3000\pi$ rad/s. The discrete time counterpart of this signal is,

$$x_d[n] = x_c\left(n\left(\frac{10^{-3}}{2}\right)\right).$$

Determine which of the following constraints is satisfied by $X_c(j\omega)$.

a) $X_d(e^{j\omega})$ is imaginary
b) $X_d(e^{j\omega}) = 0$ for $|\omega| \geq \frac{3\pi}{2}$
c) $X_d(e^{j\omega}) = X_d(e^{j(\omega - \frac{\pi}{2})})$

12.6 A discrete time band limited signal $x_d[n]$ has the Fourier transform $X_d(e^{j\omega}) = 0$ for $\frac{2\pi}{5} \leq |\omega| < \pi$. Its continuous time counterpart is

$$x_c(t) = 2 \times 10^{-4} \sum_{k=-\infty}^{\infty} \frac{\sin\left(10^4\pi(t - k \cdot 10^{-4})\right)}{\pi t - 2\pi \cdot 10^{-4}k}.$$

a) Find the discrete time counterpart, $x_d[n]$ of this signal.
b) Find the bandwidth of $X_c(j\omega)$.
c) Find the bandwidth of $X_d(e^{j\omega})$.

12.7 A continuous time LTI system is represented by the following equation:

$$y(t) = \frac{d}{dt}x\left(t - \frac{1}{3}\right).$$

a) Find the frequency response of this system.
b) Find the discrete time counterpart of the frequency response by defining a band-limited differentiator, where the cutoff frequency is $\omega_c = 2\pi/3$.
c) Find the discrete time impulse response $h[n]$ of this system.

12.8 A discrete time band stop filter is represented by the following frequency response:

$$H(e^{j\omega}) = \begin{cases} 1, & |\omega| \leq \frac{\pi}{6} \text{ and } |\omega| \geq \frac{5\pi}{6}, \\ 0, & \text{o.w.}, \end{cases} \qquad |\omega| \leq \pi.$$

a) Find the impulse response, $h[n]$ of this system
b) Find the frequency response of a decimated system with impulse response $h[3n]$.

12.9 Consider a system, which decimates an input $x[n]$ to generate the output $y[n]$, given as follows:

$$y[n] = x[7n].$$

a) Find and plot the output $y[n]$ of this system, when the input is

$$x[n] = \frac{\sin(\frac{\pi}{7}n)}{\pi n}.$$

b) Find a continuous counterpart of $y[n]$.
c) Find the output $g[n] = h[n] * y[n]$, where $h[n] \leftrightarrow H(e^{j\omega})$ is band stop filter with the following frequency response:

$$H(e^{j\omega}) = \begin{cases} 1, & |\omega| \leq \frac{\pi}{2} \text{ and } |\omega| \geq \frac{3\pi}{2}, \\ 0, & \text{o.w.,} \end{cases} \qquad |\omega| \leq \pi,$$

12.10 A band limited continuous time signal $x(t)$ with the Fourier transform $X(j\omega) = 0$ for $|\omega| > 2500\pi$ is sampled by an impulse train as follows:

$$x_p(t) = \sum_{n=-\infty}^{\infty} x(nT)\delta(t - nT),$$

where the sampling period is $T = 10^{-3}$.
a) Find the Nyquist rate of this signal.
b) Can we find a discrete counterpart $x[n]$ of this signal by C/D conversion method such that we can recover the original continuous time signal $x(t)$? Explain your answer.

12.11 A signal $x(t)$ with Fourier transform $X(j\omega)$ undergoes impulse-train sampling to generate the following:

$$x_p(t) = \sum_{n=-\infty}^{\infty} x(nT)\delta(t - nT),$$

where $T = 10^{-4}$.
If it is known that $X(j\omega) = 0$ for $|\omega| > 7500\pi$, does the sampling theorem guarantee that $x(t)$ can be recovered exactly from $x_p(t)$?

12.12 Consider the following signal obtained by downsampling of an input signal $x[n]$:

$$g[n] = \begin{cases} x[n/3], & \text{for } \; n = 0, \pm 3, \pm 6, \pm 9 \dots, \\ 0, & \text{otherwise,} \end{cases}$$

where the input is

$$x[n] = \frac{\sin \frac{4\pi}{5} n}{\pi n}.$$

a) Find and plot $G(e^{j\omega})$.
b) Find the low-pass filtered output $Y(e^{j\omega}) = G(e^{j\omega})H(e^{j\omega})$, where

$$H(e^{j\omega}) = \begin{cases} 1, & \text{for } \; |\omega| \leq \pi/5, \\ 0, & \text{o.w.,} \end{cases}$$

c) Find the decimated output $z[n] = y[5n]$.

Bibliography

J K Aggarwal. *Notes on Nonlinear Systems*, volume 1. Van Nostrand Reinhold, New York, NY, 1972.

Shaila D Apte. *Signals and Systems: Principles and Applications*, volume 1. Cambridge University Press, 2016.

S P Bhattacharyya, L H Keel, and D N Mohsenizadeh. *Linear Systems: A Measurement Based Approach*, volume 1. Springer, 2013.

Richards Bronson and Gabriel B Costa. *Schaum's Outline of Differential Equations*, volume 5. McGraw-Hill Education, 2021.

Andrew Burton. *Signals and Systems: An Engineering Perspective*, volume 1. Larsen and Keller Education, 2020.

Gordon E Carlson. *Signal and Linear System Analysis*, volume 1. Houghton Mifflin, Boston, MA, 1992.

Chi-tsong Chen. *System and Signal Analysis*, volume 2. Saunders College Publishing, 1994.

J S Chitode. *Signals and Systems: Signals, Systems and Transforms Analysis*, volume 1. Technical Publications, 2020.

C A Desoer. *Notes for A Second Course on Linear Systems*, volume 1. D. van Nostrand, 1970.

Bradley W Dickinson. *Systems Analysis, Design and Computation*, volume 1. Prentice Hall, Englewood Cliffs, NJ, 1991.

Milos Ercegovac, Tomas Lang, and Jaime H Moreno. *Introduction to Digital Systems*, volume 1. John Wiley & Sons, Inc., 1999.

Richard W Hamming. *Coding and Information Theory*, volume 2. Prentice Hall, Engelwood Cliffs, NJ, 1986.

Francis B Hildebrand. *Methods of Applied Mathematics*, volume 2. Prentice Hall, Engelwood Cliffs, NJ, 1969.

Franz E Hohn. *Elementary Matrix Algebra*, volume 3. The Macmillan Company, New York, NY, 1973.

Hwei P Hsu. *Schaum's Outline of Signals and Systems*, volume 3. McGraw-Hill Education, 2013.

Thomas Kailath. *Linear Systems*, volume 1. Prentice Hall, 1980.

V Kamaraju and R L Narasimham. *Linear Systems: Analysis and Applications*, volume 2. I. K. International, 2019.

A Anand Kumar. *Signals and Systems*, volume 3. PHI Learning, 2013.

B P Lathi. *Signal Processing and Linear Systems*, volume 1. Oxford University Press, 2000.

B P Lathi and Roger Green. *Linear Systems and Signals*, volume 3. Oxford University Press, 2017.

Jae S Lim. *Two-Dimensional Signal and Image Processing*, volume 1. Prentice Hall, London, UK, 1990.

Douglas K Lindner. *Introduction to Signals and Systems*, volume 1. McGraw-Hill, 1999.

Richard G Lyons. *Understanding Digital Signal Processing*, volume 3. Prentice Hall, 2010.

Anthony N Michel and Panos J Antsaklis. *Linear Systems*, volume 1. McGraw-Hill, Singapore, 1998.

Signals and Systems: Theory and Practical Explorations with Python,
First Edition. Fatoş Tunay Yarman Vural and Emre Akbaş.

Companion website: www.wiley.com/go/signalsandsystems

Jerry R Muir Jr,. *Complex Analysis: A Modern First Course in Function Theory*. Wiley, 2015. ISBN: 978-1-118-70522-2015.

Alan V Oppenheim, Alan S Willsky, Syed Hamid Nawab, and Jian-Jiun Ding. *Signals and Systems*, volume 2. Prentice Hall, Upper Saddle River, NJ, 1997.

Alan V Oppenheim, Ronald W Schafer, and John R Buck. *Discrete-Time Signal Processing*, volume 2. Pearson Education, 2013.

Abraham Peled and Bede Liu. *Digital Signal Processing*, volume 1. John Wiley & Sons, Inc., 1976.

John A Pierce. *An Introduction to Information Theory: Symbols, Signals and Noise*, volume 2. Dover Publications, 1980.

K Deergha Rao. *Signals and Systems*, volume 1. Birkher, 2018.

Donald G Schultz and James L Melsa. *State Functions and Linear Control Systems*, volume 1. Prentice Hall, Upper Saddle River, NJ, 1997.

C Senaras, M Ozay, and F T Yarman Vural. Building detection with decision fusion. *IEEE Journal of Selected Topics in Applied Earth Observations and Remote Sensing*, vol. 6, issue 3, pp. 1295–1304, 2013.

Pushpendra Singh, Amit Singhal, Binish Fatimah, Anubha Gupta, and Shiv Dutt Joshi. Proper definitions of Dirichlet conditions and convergence of Fourier representations [lecture notes]. *IEEE Signal Processing Magazine*, vol. 39, issue 5, pp. 77–84, 2022. https://ieeexplore.ieee.org/document/9869591.

Ken Steiglitz. *A Digital Signal Processing Primer with Applications to Digital Audio and Computer Music*, volume 1. Addison-Wesley Publishing Company, Menlo Park, CA, 1996.

Fred J Taylor and Anthony N Michel. *Principles of Signals and Systems*, volume 1. McGraw-Hill, Singapore, 1994.

David L Verbyla and Kang-tsung (Karl) Chang. *Processing Digital Images in Geographic Information Systems*, volume 1. OnWord Press, Santa Fe, NM, 1997.

F T Yarman Vural and G G Değirmendereli. *25th International Conference on Pattern Recognition (ICPR)*, 2020.

James S Walker. *Signals and Systems*, volume 2. John Wiley & Sons, Inc., 2002.

Wikipedia. The Free Encyclopedia. https://en.wikipedia.org/. (accessed November, 2024).

Index

Signals and Systems: Theory and Practical Explorations with Python,
First Edition. Fatoş Tunay Yarman Vural and Emre Akbaş.

Companion website: www.wiley.com/go/signalsandsystems